Mineral Exploration

Mineral Exploration
Principles and Applications

S. K. Haldar
Emeritus Scientist, Dept. of Applied Geology &
Environmental System Management,
Presidency University, Kolkata-700 073,
and IMX Resources Limited, Australia.
Formerly, Hindustan Zinc Limited,
Hindustan Copper Limited, ESSO INC
and BIL Infratech Ltd, India.

AMSTERDAM • BOSTON • HEIDELBERG • LONDON • NEW YORK • OXFORD
PARIS • SAN DIEGO • SAN FRANCISCO • SINGAPORE • SYDNEY • TOKYO

Elsevier
225 Wyman Street, Waltham, MA 02451, USA
The Boulevard, Langford Lane, Kidlington, Oxford, OX5 1GB, UK
Radarweg 29, PO Box 211, 1000 AE Amsterdam, The Netherlands

Library of Congress Cataloging-in-Publication Data
Mineral explorations : principles and applications / edited by Swapan Haldar.
pages cm
Includes bibliographical references and index.
ISBN 978-0-12-416005-7
1. Prospecting. 2. Mines and mineral resources. I. Haldar, S. K.
TN270.M6554 2013
622′.1–dc23
2012039837

British Library Cataloguing in Publication Data
A catalogue record for this book is available from the British Library

ISBN: 978-0-12-416005-7

Printed and bound in Republic of China

13 14 15 16 17 10 9 8 7 6 5 4 3 2 1

Dedicated to all my students in mineral exploration
— Past, Present and Future —
Who taught me the art of teaching, motivated me to keep learning
and galvanized me to write this book
And
Little Srishti and Srishta
Who inspired me to love Mother Nature!

Contents

Preface

I was often fascinated and enjoyed delivering occasional technical lectures to the students and research scholars of academic institutions throughout my 37 years of professional period with ESSO Standard Eastern Inc (1966–1967), Hindustan Copper Limited (1968–1975) and Hindustan Zinc Limited (1975–2001). I also took pleasure in addressing group of geoscientists, mining engineers and teaching faculties from industry and national institutions during various specialized training programs. But my dream of teaching students came true when I got the opportunity, after relinquishing my job, to join Presidency College, Kolkata as Emeritus Scientist. Department of Science and Technology (DST), Government of India, sponsored two of my research projects successively during 2003–2006 and 2008–2010. As a spin off activity and ethical commitments I started regular teaching, both theory and practical, on various aspects of mineral exploration to the postgraduate students of Applied Geology and Environmental System Management of the department. My teaching spree continued to M. Sc and M. Tech students of Indian School of Mines, Dhanbad, too.

Since 2004, I am retained by M/s Goldstream Mining NL/IMX Resources Limited, Perth, Australia, as exploration consultant looking for their interests in platinum, chromium, nickel, copper, gold and associated minerals in India with extensive field visits. Other recent assignments are consultancy with M/s McNally Bharat Ltd (2010) and BIL Infratech Ltd (2011–2012), India, for technical and economic evaluation of zinc–lead–copper projects in Saudi Arabia, Australia, South Africa, Tunisia, Mexico, Eritrea, Montenegro and Chile including couple of field visits.

During my formal class room teaching assignment, my endeavor was to mix my long practical multipurpose industrial experiences, technical consultancy to multinational companies and DST research projects with the essence of academic aspect of the subject. In this process, I also tried to read the students' attitude, aptitude, mind set, body language and their expectations. It helps me to improve my clarity of thoughts and expression day by day. I could visualize their level of understanding of the subject while evaluating their answer papers. It seemed to me that one of the expectations of the students from both the institutes is to gift them a comprehensive text book, with holistic approach, covering entire scope of mineral exploration and beyond during their academic study and as reference book while in job. The book **"Mineral Exploration—Principles and Applications"** is the outgrowth of my class lectures. The text is framed keeping the students pursuing both undergraduate and masters level courses in Pure and Applied Geology and allied subjects, teachers and professionals in mineral exploration, mining, beneficiation, environmental system management, mineral policies and acts behind the scene through out. This book will also endure the necessities of the candidates appearing for federal, state, regional and provincial geological services, forest and environment jobs, federal and state administrative and professional services, academic entrance tests and other competitive written examinations and interviews. My efforts in this attempt will be judged by the liking of the subject and success of my students—past, present and future.

"Mineral Exploration—Principles and Applications" divided into 15 chapters, discusses the basic elements of exploration on modern thoughts and information adopting the most desirable order and methods of presentation. Each chapter defines the theory (Principles) first followed by actual practices (Applications) in the field and industries. The chapters have been designed to set up with need of the subject, initial definitions, types and stages of mineral discovery, exploration methods, namely, geological, geochemical, geophysical, photo geology and remote sensing, sampling, mineral resource and ore reserve estimation, classification system and code, statistical and geo-statistical applications, mining practices, mineral beneficiation, environmental aspects and sustainable development, economic evaluation and feasibility study of mineral property. The chapters start with mineral policy, mining acts and end with 11 exploration case histories for the international readers. The book provides 350 illustrations and field-industrial images for systematic representation of the various concepts and actual procedures.

I have consulted many technical books during the preparation of the manuscript. The concepts of those authors have clarified many of my questions. I have specifically acknowledged those books and authors and recommended as additional study materials to my readers. Prof P. R. Mohanty and Prof Shalivan from ISM critically reviewed Chapter 5 on "Exploration Geophysics". I often shared with Prof A. B. Roy, FNA and Emeritus Scientist, Presidency University, on various aspects of the book. I acknowledge each one of peer reviewers, namely, Prof Martin Hale, University of Twente, The Netherlands, Dr D. B. Sikka, Barfanisai Enterprise Inc. Canada, Prof E. C. Nehru, Brooklyn College, NY, Prof B. C. Sarkar, HOD, ISM, Dhanbad, Dr William Petuk, Canada, Finn Barrett, IMX Resource Ltd, Perth, L. Moharana, BHP Billiton, Perth, Prof G. N. Jadhav, IIT Mumbai and Prof Asis Basu, Rochester University, NY, for their keen interest to evaluate the book proposal, decisive analysis in the context of global scenario and very best encouraging reflections articulated about the book. My inquisitive students, with whose able aid I applied the concept, supported me through and through, and I would like to thank each one of them. Ms Nandini Kar, one of my PG students, has gone through the entire manuscript for its acceptability among them. Ms Joyasree Sinha drafted most of the line drawings.

The book could be conceived and finished at the Presidency University with the opportunity for class room teaching and evaluation provided by Prof H. N. Bhattacharya, HOD, Department of Applied Geology and Environment System Management. Prof Prabir Das Gupta and Ms Ausmita Kaviraj took sincere interest for image conversion and software exposure, respectively.

Swami Buddhadevananda, President, Rama Krishna Math, Barisa, Kolkata, Prof Daya Kishore and Nandita Hazra, S. N. Medical College, Agra, Prof R. V. Karanth, M. S. University, Vadodara and Mr C. R. Chakraborty, Kolkata inspired me throughout the manuscript preparation.

All the chapters are edited by "Soumi", my daughter, while she came to relax with us on Durga Puja or summer vacations (2006, 2008, 2010 and 2012) and our visit to them at USA (2004, 2005, 2011) followed by online communication from Chicago, Florida and Studio City, CA. Suratwant, my son-in-law, who advocated the title of the book too, is my source of inspiration for all times in this endeavor.

Most of the examples and images, I have shared, are the life time academic collection during my association with M/s Hindustan Zinc Limited, Goldstream Mining NL and BIL Infratech Ltd. I sincerely acknowledge and state my indebtedness to each Institution. It was my learning experience with Elsevier and I enjoyed my association with Siddhartha Ghosh, Editorial Coordinator, Asia-Pacific, Mrs Linda Versteeg, Associate Acquisition Editor, The Netherlands, John Fedor, Senior Acquisitions Editor, NY, Ms Morrissey Kathryn (Katy), Editorial Project Manager, MA and Poulouse Joseph, Project Manager- Book Publication Division and team in India for their great friendliness throughout development of this book. I appreciate each member of the design, production and marketing team for sincere collaboration.

Acknowledgement is the privilege of the author. Many colleagues from industry, teachers and students, in particular, encouraged and enriched me with valuable suggestions for improvement. I do not wish to record specific names with a view not to miss any one. However, I am unable to resist mentioning my two little grand children, **"Srishti and Srishta",** who did not allow me to work during their vacations with us, but infused tremendous energy in me when they left for USA. And finally, I extend my whole hearted thanks to my wife, "Swapna", whose enthusiasm and criticism kept me on toes and sustained me to improve on every aspect of academic character.

"I do not know what I may appear to the world; but to myself I seem to have been only like a boy playing on the seashore, and diverting myself in now and then finding a smoother pebble or a prettier shell than ordinary, while the great ocean of truth lay all undiscovered before me.... If I am anything, which I highly doubt, I have made myself so by hard work"

– Sir Isaac Newton.

As always, I acknowledge the supreme entity that makes sense of it all, and urges us to seek for ourselves the ultimate truth. In that pursuit, this treatise is dedicated— "To strive, to seek, to find, and not to yield".

S.K. Haldar
Presidency University, Kolkata
29th June, 2012

List of Acronyms used in this book

GENERAL

Anglo American	Global leader in mining
BM	Bureau of Mines
BHP Billiton	A leading global resources company
BIL	Binani Infratech Ltd, India
CAPEX	Capital Expenditure
Goldstream	Goldstream Mining NL, Perth, Australia
GDP	Gross Domestic Product
GSI	Geological Survey of India
HCL	Hindustan Copper Limited, Kolkata, India
HZL	Hindustan Zinc Limited, Udaipur, India
IMX	IMX Resources Limited, Perth, Australia
IS	Indian standards
LBMA	London Bullion Market Association
LME	London Metal Exchange
JORC	(Australasian) Joint Ore Reserves Committee
MCDR	Mineral Conservation and Development Rules
MCR	Mineral Concession Rules
MECL	Mineral Exploration Corporation Limited, Nagpur, India
MMRD ACT	Mines and Minerals (Regulation & Development) Act
MPRDA	Minerals & Petroleum Resources Development Act
ML	Mining Lease/ Licence
MOE	Ministry of Environment
MOF	Ministry of Forest
MOM	Ministry of Mines
MSS	Multi-Spectral Scanner
Mt Isa	Mount Isa mine, Australia
MVT	Mississippi Valley Type
NMP	National Mineral Policy
OB	Over Burden
OMS	Ore Man Shift
OPEX	Operating Expenditure
PL	Prospecting Licence
RA	Rampura Agucha mine, India
RD	Rajpura Dariba mine, India
ROM	Run-of-Mine
RP	Reconnaissance Permit
SEDEX	Sedimentary Exhalative
SK	Sindesar Khurd Zn-Pb-Ag deposit, India
Sp. Gr	Specific Gravity
TM	Thematic Mapper
tpa / tpy	tones per annum / year
tpd	tones per day
UNFC	United Nations Framework Classification
USGS	United State Geological survey/
USBM	United State Bureau of Mines
UTM System	Universal Transverse Mercator System
VCR	Vertical Crater Retreat
VHMS	Volcanic Hosted Massive Sulfide
VMS	Volcanogenic Massive Sulfide
VRM	Vertical Retreat Mining
WHO	World Health Organization
ZM	Zawar Group of Mines, India

MINERALS

Cp	Chalcopyrite
Ga	Galena
Po	Pyrrhotite
Py	Pyrite
Sp	Sphalerite

METALS

Ag	Silver
Al	Aluminium
As	Arsenic
Au	Gold
Bi	Bismuth
Ca	Calcium
Cd	Cadmium
Ce	Cerium
Co	Cobalt
Cr	Chromium
Cu	Copper
F	Fluorine
Fe	Iron
He	Helium
Hg	Mercury
K	Potassium
La	Lanthanum
Li	Lithium
Mn	Manganese
Mo	Molybdenum
N	Nitrogen

Na	Sodium
Nd	Neodymium
Ni	Nickel
Pb	Lead
Pd	Palladium
Pm	Promethium
Pt	Platinum
Te	Tellurium
Rb	Rubidium
Rn	Radon
Sb	Antimony
Se	Selenium'
Si	Silicon
Sm	Samarium
Sr	Strontium
U	Uranium
Zn	Zinc

TIME SCALE

AD	Anno Domoni (year after Jesus Christ's birth)
BC	Before Christ (year)
Ga	Gigga (10^9) or Billion age (years)
Ma	Million (10^6) age (years)

CURRENCY

Country/Currency	Currency Code	Currency Image
Australia Dollar	AUD	$
Canada Dollar	CAD	$
Chile Peso	CLP	$
Euro Members	EUR	€
India Rupee	INR	₹
Saudi Arabia Riyal	SAR	ريال
South Africa	ZAR	R
Tunisia Dinar	TND	رانيد
United Kingdom Pound	GBP	£
United States Dollar	USD	$
Zimbabwe Dollar	ZWD	Z$

WEIGHTS

Bt	Billion tonnes
lb	pound
Kg	Kilogram
Mt	Million tonnes
t	tonne

LENGTH

m	metre
cm	centimetre

About the Author

S. K. Haldar (Swapan Kumar Haldar) has been a practicing veteran in the field of Mineral Exploration and metal mining for the past four and half decades. He received his B. Sc (Hons) and M. Sc degree from Calcutta University and Doctorate from Indian Institute of Technology, Kharagpur. The major part of his career from 1966 has been focused on base and noble metals exploration/mining with short stopover at ESSO Petroleum, Hindustan Copper Limited and finally Hindustan Zinc Limited where he undertook a varied set of technical roles and managerial responsibilities. Since 2003 he is associated as Emeritus Scientist with Department of Applied Geology, Presidency University, Kolkata and teaching mineral exploration to postgraduate students of the department and often at Indian School of Mines, Dhanbad. He is consultant with international exploration entities, namely, Goldstream Mining NL/IMX Resources Ltd, Australia and BIL Infratech Ltd, India. His profession has often required visits and interaction with experts of zinc–lead–gold–tin–chromium–nickel and platinum mines and exploration camps of Australia—Tasmania, Canada, Germany, Portugal, Saudi Arabia, France, Italy, Nepal, The Netherlands, Switzerland, Egypt and USA. He is a life-fellow of The Mining Geological and Metallurgical Institutes of India and Indian Geological Congress. Dr Haldar is recipient of "Dr J. Coggin Brown Memorial (Gold Medal) for Geological Sciences" by MGMI. He authored **"Exploration Modeling of Base Metal Deposits", 2007, Elsevier**. Dr Haldar has a unique professional blend of mineral exploration, evaluation and mineral economics with an essence of classroom teaching of postgraduate students of two celebrity Universities over the last one decade.

S.K. Haldar
Emeritus Scientist, Department of Applied Geology
Presidency University, Kolkata
Former Chief Manager—
Geology Hindustan Zinc Limited, India

Chapter 1

Mineral Exploration

Chapter Outline

Good exploration planning and decision making – measure risk and reward; persist where the geology is encouraging and where the rewards will be large; recognize when you have failed.

—Cameron R. Allen, Cominco Ltd.

1.1. INTRODUCTION

Minerals and metals are one of the essential components for the growth of human society. Needs of survival taught the prehistoric Paleolithic men the uses of stones as tools even before 20,000 years ago. The discovery of minerals, its exploitation and uses became many folds with the advent of civilization and is continuing till date.

A mineral deposit, more meaningfully concentration of specific mineral, is too small a size in comparison to the Earth's crust. Deposits near the surface had been discovered over the centuries, mined out and metals extracted. Future searches will be aimed at naturally occurring concealed types. It may rarely show surface signatures like weathered outcrop and are covered under transported soil. The new discovery will not be easy. It will require state-of-the-art exploration techniques, trained man power, scientific knowledge, ample experience, high-end data processing system and interpretation skill. The total procedure would be achieved step by step in a dynamic and logical sequence.

1.1.1. Why Mineral Exploration?

The mineral reserves and resources, annual production vs. consumption and index of per capita spending of any commodity are the measures that rank the status of a country as developed, developing or underdeveloped. The per capita consumption of zinc in India during 2008 was very low at 0.43 kg against a world average of 4.3 kg. The higher consumption during the same period was shared between Australia (12.7 kg), South Korea (11.3 kg), Canada (5.6 kg), Japan (5 kg), USA (4.1 kg) and China (2.7 kg). The policy makers in the Government and Private Sectors allocate funds for long- and short-term exploration plan programs guided by the demand-supply trend of all commodities as a whole. The fund allocation has special significance for strategic and deficient minerals. The annual percent satisfaction between consumption and indigenous production of zinc metal between 1992-1993 and

Mineral Exploration. http://dx.doi.org/10.1016/B978-0-12-416005-7.00001-5

2009-2010 at an annual growth rate of 8-10% has been depicted in Fig. 1.1.

The existing demand-supply disparity can be reduced by expanding the mining and smelting capacity with the on hand ore reserves as short-term ad hoc measure. The ultimate way out for long-term standpoint would be continuous efforts to enhance reserve and resource base. This is possible by new search, discovery and adequate exploration of mineral deposits, economic mining and smelting. The rate of reserve augmentation must commensurate with annual growth of the particular commodity (Fig. 1.2). The working group of sustainable mineral resource development program plans on the same analogy for future exploration investment in the country.

The process of mineral discovery and its development to a target production center takes a long gestation period of about 5-20 years. In terms of business requirements, this translates to a very high-risk tolerance at all levels, extensive period of time and rich pockets for a sustained cash flow. A small business unit in this field may often end its brief tenure with a total loss, in case of failure to make an economic return. Indeed, many of the discoveries are not viable at current market prices. Prima facie, the facts might indicate that investment in these ventures is a waste. However, one discovery out of 100 or even 1000 attempts may pay back the entire efforts. The task of policy maker is to plan timely allocation of funds for exploration and technology research of various mineral types, predicated on long-term demand and supply scenarios. Therefore, an investment-friendly environment, transparency and will of the Federal and the State Government and exploration companies and the political commitment of the regime are essential for mineral development in any country.

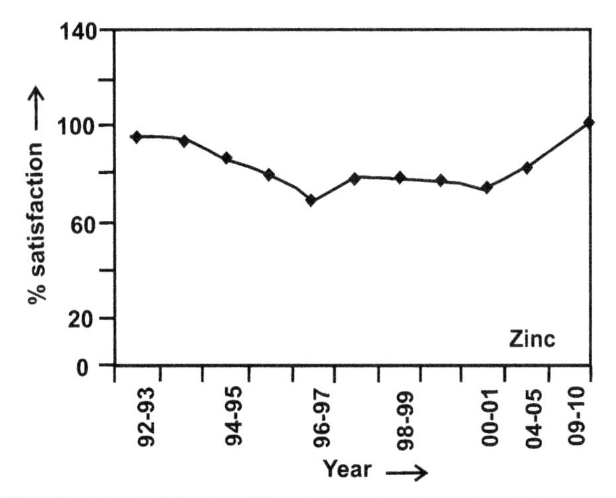

FIGURE 1.1 Satisfaction (%) of demand and supply of zinc metal between 1992-1993 and 2009-2010 in India.

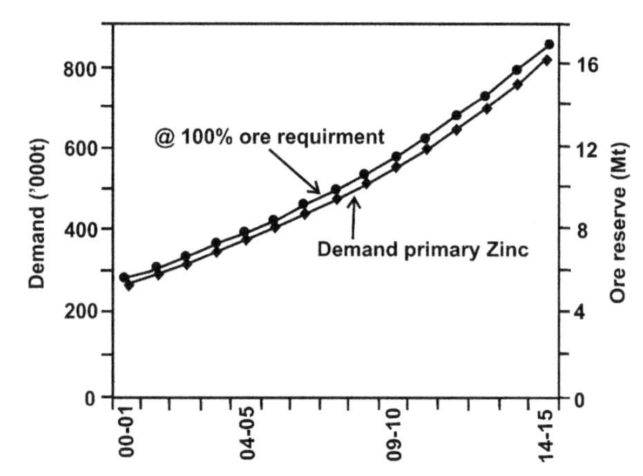

FIGURE 1.2 Projected reserve requirement of zinc at an annual growth rate of 8-10% up to 2015 at 90% mining-smelting capacity in India.

Some of the common basic terminologies that would be referred frequently in the subsequent chapters are defined here after.

1.2. DEFINITION

1.2.1. Mineral and Rock

"Mineral" is a homogeneous inorganic substance that occurs naturally, usually in crystalline form with a definite chemical composition. It is generally in solid form, the exceptions being mercury, natural water and fossil fuel. The common rock-forming minerals (RFM) are quartz (SiO_2), orthoclase feldspar ($KAlSi_3O_8$), plagioclase feldspar ($CaNaAlSi_3O_8$), albite ($NaAlSi_3O_8$), mica group such as muscovite (H_2KAL_3 $(SiO_4)_3$) and biotite ($H_2K(MgFe)_3Al$ $(SiO_4)_3$). The common ore-forming minerals (OFM) are hematite (Fe_2O_3), cassiterite (SnO_2), chalcopyrite ($CuFeS_2$), sphalerite (ZnS), galena (PbS), baryte ($BaSO_4$ $2H_2O$), gypsum ($CaSO_4$), apatite ($Ca_5(PO_4)_3$ (F,Cl,OH)), etc.

"Rock" is an assemblage of mineral(s) formed under natural process of igneous, sedimentary and metamorphic origin. The common rocks are basalt, granite, quartzite, sandstone, limestone, marble and mica-schist.

1.2.2. Ore

In the past, the word **"ore"** was restricted exclusively to naturally occurring material from which one or more types of metal could be mined and extracted at a profit. The economic deposits comprising of industrial minerals, rocks, bulk materials, gemstones and fossil fuel were excluded from ore. The concept has undergone radical changes over the years. The Institution of Mining and Metallurgy, UK, currently defines ***"Ore as a solid naturally occurring mineral aggregate of economic interest***

from which one or more valuable constituents may be recovered by treatment". Therefore, ore and orebody include metallic deposits, noble metals, industrial minerals, rocks, bulk or aggregate materials, gravel, sand, gemstones, natural water, poly-metallic nodules and mineral fuel from land and ocean bed (Fig. 1.3A-X). All ores are minerals or its aggregate, but the reverse is not true. The ore can be broadly classified as:

Metallic: Native-Pt, -Au, -Ag, -Cu, chalcopyrite, sphalerite, galena, hematite, magnetite, pyrite, pyrrhotite, bauxite.
Noble: Gold, silver, platinum, palladium.
Industrial: Quartz, garnet, phosphate, asbestos, barite.
Gemstones: Amethyst, aquamarine, diamond, emerald, garnet, opal, ruby, sapphire, topaz, zircon.
Rock: Granite, marble, limestone, rock salt.

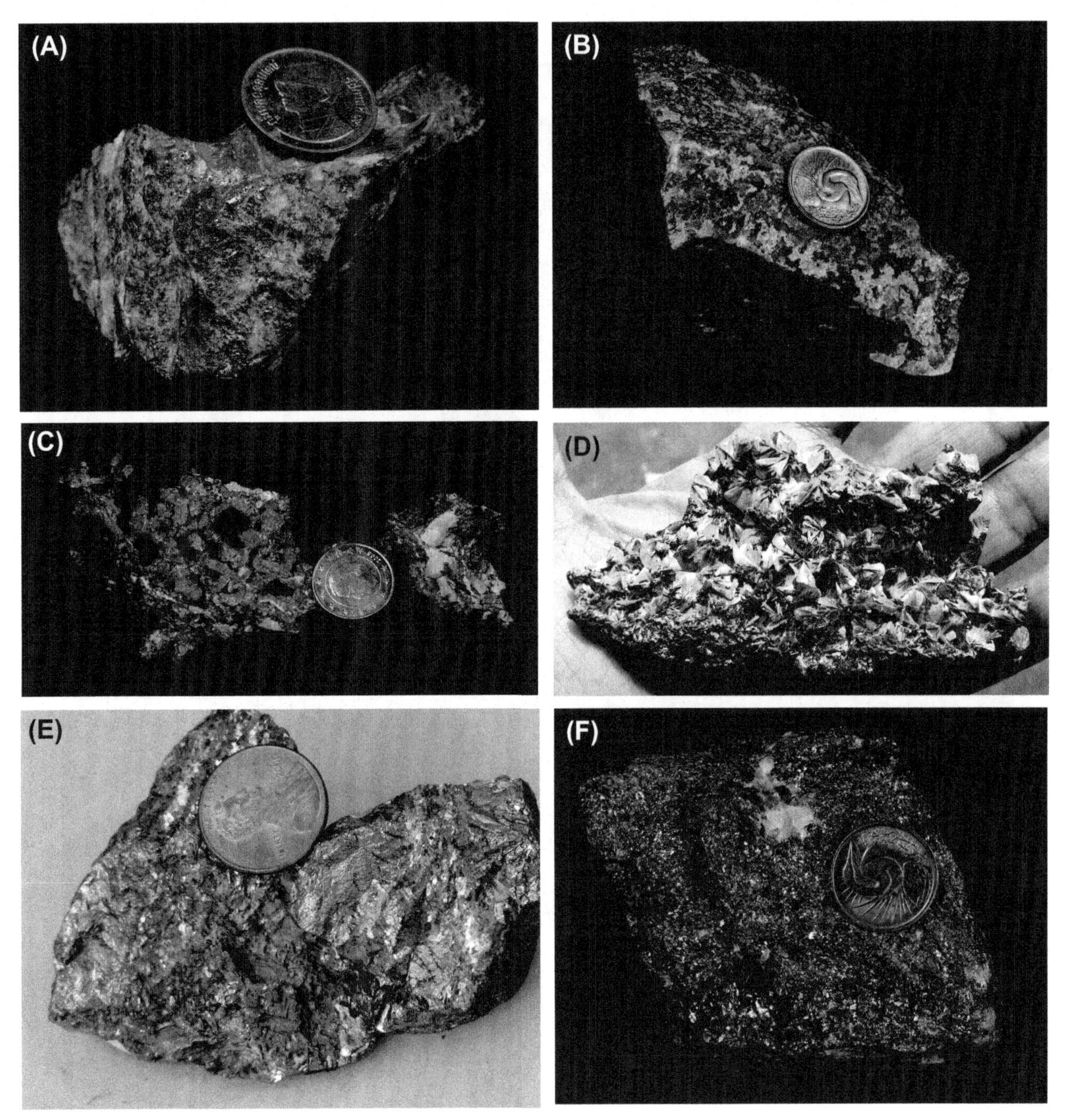

FIGURE 1.3 A-F Common ore minerals: (A) naive gold (Saudi Arabia), (B) native silver (Zawar mine), (C) native copper (Neves Corvo, Portugal), (D) malachite (credit: Prof. R. V. Karanth), (E) chalcopyrite (golden) and galena (steel gray) (Rajpura-Dariba), (F) brown sphalerite (Zawar mine), India.

FIGURE 1.3 G-L Common ore minerals: (G) honey yellow sphalerite, (Rajpura-Dariba, credit: V. K. Jhanjhri), (H) stratiform pyrite in graphite schist (Rajpura-Dariba), (I) massive pyrrhotite (Sindesar Khurd), (J) wolframite (Degana mine, India), (K) crystalline chromite (Sukinda, India), (L) pisolitic structure in bauxite (Bagru Hill, Jharkhond, India).

FIGURE 1.3 M-R Common ore minerals: (M) hematite (steel gray) and Jasper (red) (credit: Prof. A. B. Roy), (N) fluorite (Amba-Dungri, India), (O) barite embedded with pyrite crystals (Preislar mine, Germany), (P) amethyst, (Q) aquamarine, (R) ruby.

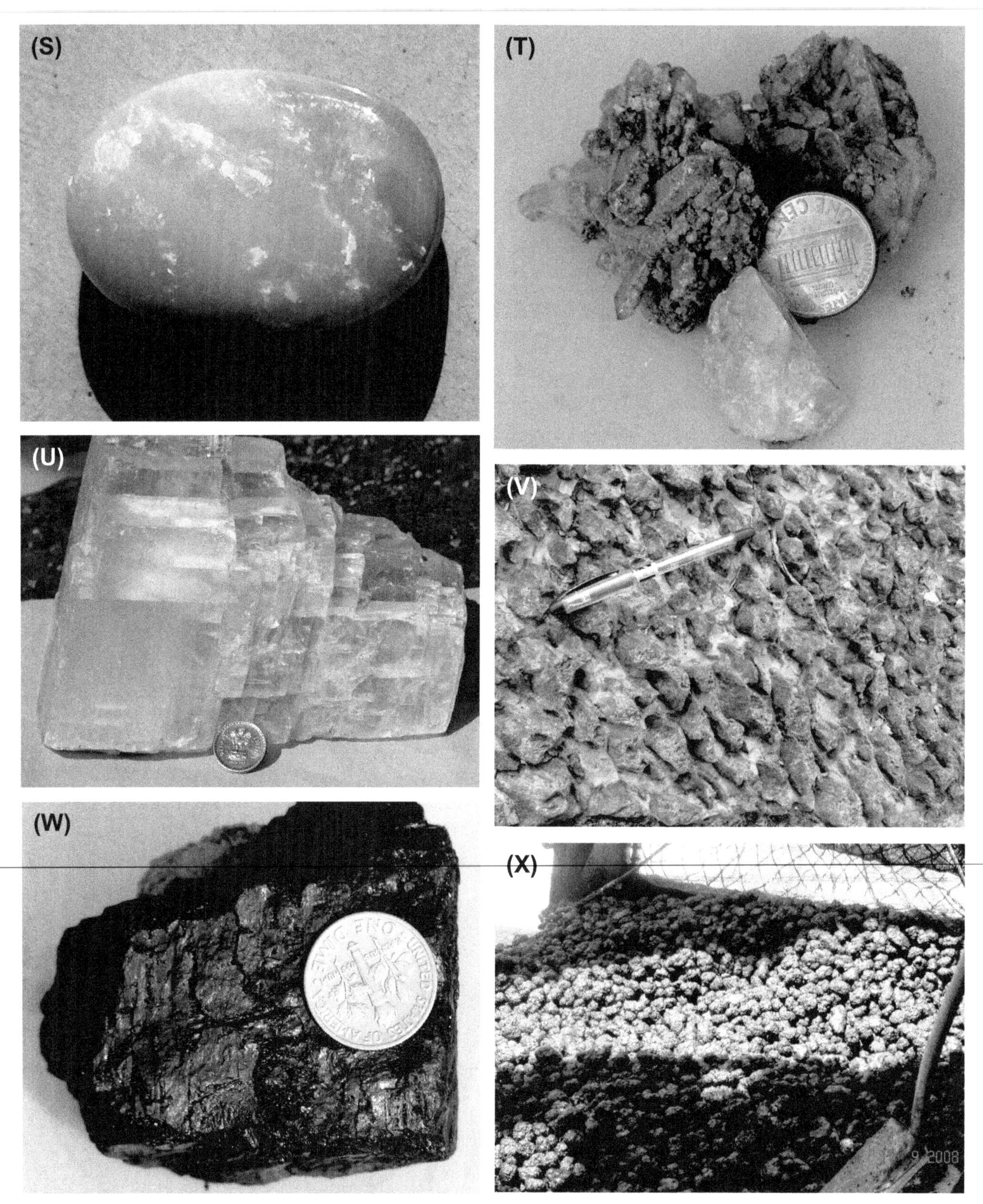

FIGURE 1.3 S-X Common ore minerals: (S) opal (image credit QP to S: Prof. R. V. Karanth), (T) quartz (Kolihan copper mine, India), (U) calcite, (V) rock phosphate (Jhamarkotra, India), (W) coal (Belatan mine, India), (X) poly-metallic nodules. (Indian Ocean, Credit: Ms Arpita De)

Bulk/aggregate: Sand, gravel.
Mineral fuel: Coal, crude oil, gas.
Strategic: Uraninite, pitchblende, thorianite, wolframite.
Life essential: Natural water.
Rare earth: Lanthanum (La), cerium (Ce), neodyminum (Nd), promethium (Pm).
Ocean: Poly-metallic nodules, coral, common salt, potassium.

1.2.3. Ore Deposit

An ore deposit is a natural concentration of one or more minerals within the host rock. It has a definite shape on economic criteria with finite quantity (tonnes) and average quality (grade). The shape varies according to the complex nature of the deposit such as layered, disseminated, veins, folded and deformed. It may be exposed to the surface or hidden below stony barren hills, agricultural soil, sand, river and forest (Fig. 1.4).

Some of the important ore deposits are Broken Hill, Mount Isa, McArthur, HYC, Century, Lady Loretta, Lenard Shelf zinc-lead, Munni Munni platinum and Olympic Dam copper-uranium-gold deposits, Australia; Neves Corvo copper-zinc-lead-tin deposit, Portugal; Sullivan zinc-lead deposit, British Columbia; Sudbury nickel-copper-platinum, Lac Des Iles palladium deposits, Canada; Pering zinc, Bushveld chromite-platinum deposits, South Africa; The Great Dyke platinum-nickel-copper deposit, Zimbabwe; Red Dog zinc-lead deposit, Alaska; Paguanta zinc-copper-silver deposit, South America; Stillwater platinum deposit, America; Bou Jabeur zinc-lead-fluorite-barite deposit, Tunisia; Hambok copper-zinc and Bisha copper-gold deposits, Eritrea; Noril'SK and Kola platinum deposits, Russia; Rampura-Agucha zinc-lead, Singhbhum copper, Bailadila iron ore, Sukinda chromium, Nausahi chromium-platinum, Kolar gold, Jaisalmer limestone, Jhamarkotra rock phosphate, Makrana marble and Salem magnesite deposits, India. There is no choice of preferential geographical location of orebody—it can be at a remote place or below a thickly populated city. It has to be accepted as it is and where it is. Moreover, ore deposits, being an exploitable nonrenewable asset, have to be used judicially at present and leaving sensibly for the future.

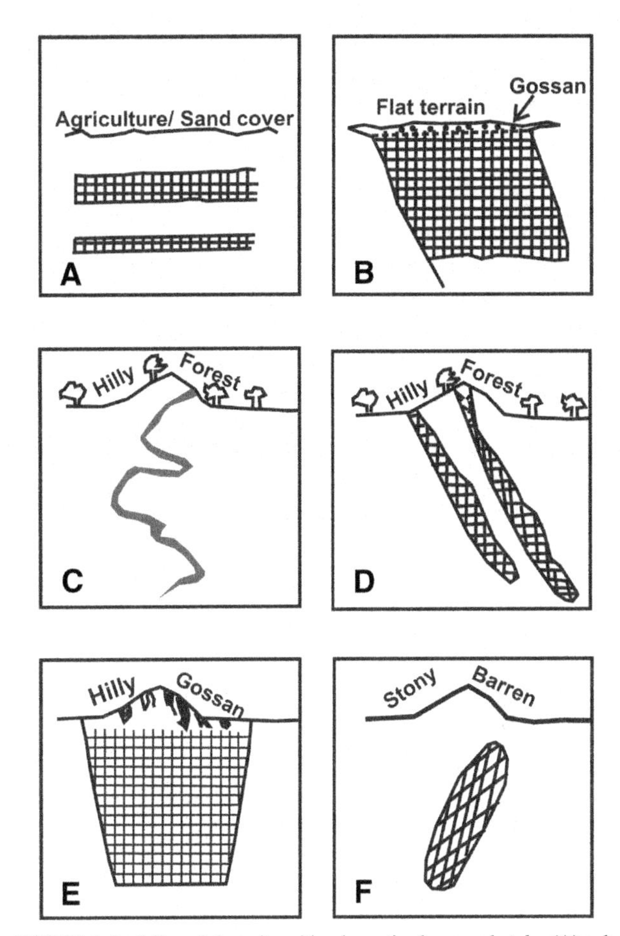

FIGURE 1.4 Mineral deposits with schematic shape and style: (A) sub-horizontal lignite body under agriculture/sand cover at Barsingsar, (B) Massive Zn-Pb-Ag orebody exposed to surface at Rampura-Agucha, (C) Intricately folded Pb-Cu deposit at Agnigundala, (D) En echelon Zn-Pb lenses under hilly terrain at Zawar, (E) Unique gossans signature of sulfide deposit at Rajpura-Dariba, (F) Concealed sulfide deposit under stony barren quartzite at Sindesar Khurd, India.

1.2.4. Prime Commodity, Associated Commodity and Trace Element

"Prime commodity" is the principal ore mineral recovered from the mines. "Associated commodities" are the associated minerals recovered as by-products along with the main mineral. In general all ore deposits contain number of valuable "trace elements" that can be recovered during processing of ore. The prime commodity of a zinc-lead-copper-silver mine is zinc, and the associated commodities are lead and copper. The expected value-added trace elements are cadmium, silver, cobalt and gold. The value of all prime commodity, by-products and trace elements are considered collectively for valuation of the ore/mine.

1.2.5. Protore

"Protore" is an altered rock mass or primary mineral deposit having uneconomic concentration of minerals. It may be further enriched by natural processes to form ore. These are low-grade residual deposits formed by weathering, oxidation, leaching and similar alteration. The protore can turn into an economic deposit with advance technology and/or increase of price. The examples are weathering of feldspathic granite to form kaolin deposits (Cornish kaolin deposit, England) or ultramafic rocks like peridotite to form laterite (Sukinda nickel deposit, India). It can be exploited for kaolin, iron and nickel due to sufficient enrichment of the respective metals.

1.2.6. Gangue Minerals and Tailing

Ore deposits are rarely comprised of 100% ore-bearing minerals, but usually associated with RFM during mineralization process. These associated minerals or rocks, having no significant or least commercial value, are called "gangue" minerals. Pure chalcopyrite having 34.5% Cu metal in copper deposit and sphalerite with 67% Zn metal in zinc deposit are hosted by quartzite/mica-schist and dolomite respectively. The constituent minerals of quartzite, mica-schist and dolomite are called the gangue minerals. A list of common gangue minerals are:

Quartz	SiO_2
Barite	$BaSO_4$
Calcite	$CaCO_3$
Clay minerals	All types
Dolomite	$CaMg(CO_3)_2$
Feldspar	All types
Garnet	All types
Gypsum	$CaSO_42H_2O$
Mica	All types
Pyrite	FeS
Pyrrhotite	$Fe_nS_{(n+1)}$

Processing plants are installed in every mine to upgrade run-of-mine ore having ~1% Cu and ~8% Zn to produce ideal concentrate of +20% Cu and +52% Zn. The raw ore is milled before the separation of the ore minerals from the gangue by various beneficiation processes (Chapter 12). The concentrate is fed to the smelter and refinery to produce 99.99% metal. The rejects of the process plant is called "tailing" which are composed of the gangue minerals. Tailings are used as support system by backfilling of void space in the underground mines. Alternatively, it is stored in a tailing pond and is treated as waste. High-value metals can be recovered by leaching from tailing in future. Tailing of Kolar gold mine, India, historically stored at tailing dam, is being considered to recover gold by leaching without any mining cost.

1.2.7. Deleterious Substances

Metallic ore minerals are occasionally associated with undesired substances that pose extra processing cost and penalty on the finished product. Arsenic in nickel and copper concentrate, mercury in zinc concentrate, phosphorus in iron concentrate and calcite in uranium concentrate impose financial penalties to a custom smelter for damaging the plant. Similarly extra acid leaching cost is required for processing limonite-coated quartz sand used in glass making industry.

1.3. EXPLORATION

1.3.1. Discovery Type

The discovery of a mineral occurrence or a deposit is characterized by a measurable quantity and grade, which indicate an estimated amount of contained minerals or metals. The discovery may be immediately useful and the exploitation profitable at that point of time, in which case it is classified as a reserve or more specifically ore reserves. On the other side, if the potential is known but not immediately extractable or profitable, it is classified as a resource. Explorationists must first find and then engineers convert theoretical resources into minable reserves. Often, the technology exists, but it toggles between one type and other with the changes in market price. An uneconomic discovery may become economic tomorrow and vice versa.

1.3.1.1. Greenfield Discovery

"Greenfield discoveries" are the findings from a broad base grassroots exploration program well away from known orebodies or known mineralized belts—in essence, pioneering discoveries in new locales. The term comes from the building industry, where undeveloped land is described as Greenfield and previously developed land is described as Brownfield. The knowledge base basin model discovery of Kanpur-Maton-Jhamarkotra rock phosphate deposit in Lower Aravalli formation during 1968 in India was Greenfield type (Chapter 15.10). Similarly the discovery of world's largest and richest zinc-lead deposit at Broken Hill, Australia, during 1883 was Greenfield type (Chapter 15.3).

1.3.1.2. Brownfield Discovery

"Brownfield discoveries" are assigned where discovery is made by enhancing the reserve in strike and dip continuity of known orebody or in the vicinity of an existing mine. In such cases the economics of development are improved by existing infrastructures. This is an important distinction in the analysis of discovery trends though both types contribute to the rate of development and depletion. The rediscovery of world-class largest and richest zinc-lead orebody in India at Rampura-Agucha during 1977 was a Brownfield type (Chapter 15.9).

1.3.2. Stages of Exploration

Any exploration program can be classified by successive stages—each stage is designed with a combined specific objective to achieve within the time schedule and allocated fund. The outputs of each stage provide inputs to the next successive stage. The stages for major and with limited application for minor minerals are placed in ascending order with respect to increasing geological confidence in the context of any exploratory procedure.

1.3.2.1. Reconnaissance

"Reconnaissance" is the "grassroots" exploration for identifying enhanced mineral potential or initial targets on a regional scale. The preparations at this stage include literature survey, acquisition of geophysical data, if any, synthesis of all available data and concepts, and obtaining permission (Reconnaissance License/Permit—RP) from the State/Provincial/Territorial Government. The activities encompass remote sensing, airborne and ground geophysical survey, regional geological overview, map checking/mapping on 1:250,000, 1:50,000 scales, geochemical survey by chip/grab sampling of rocks and weathered profiles, broad geomorphology and drainage, pitting and trenching to expose mineralized zone at ideal locations, and limited scout/reverse circulation/diamond drilling to know the possible extent of mineralization. Petrographic and mineragraphic study will help to determine principal host and country rocks and mineral assemblages. The prime objective is to scan the entire area under leasehold within stipulated time frame and to identify probable mineralized area (targets) worth for further investigation. The targets are ranked on the basis of its geological evidences worth for further investigation toward deposit identification. The initial leasehold area is thus substantially reduced to smaller units at the end. Estimates are preliminary resource status. This enables to focus concentration of maximum exploration efforts to the target area in the next stage. The total area and duration permissible for RP vary between states and countries.

A comprehensive work program by year during RP can be envisaged for execution in a sequential manner (Table 1.1). The subsequent activities are planned and suitably modified based on the results achieved. The definite physical targets of various exploration activities would depend on end result.

1.3.2.2. Large Area Prospecting

"Large Area Prospecting", a blend of Reconnaissance and Prospecting License (PL), is initiated in some countries. It combines Reconnaissance and Prospecting activities including general and detailed exploration. It is systematic exploration of potential target anomalies after obtaining Large Area Prospecting License (LAPL) from the State/Provincial/Territorial Government. The activities encompass detailed geological mapping, rock chip and soil samplings, close-spaced ground geophysics, Reverse Circulation (RC) and diamond core drilling on wide-spaced section lines, and resource estimation of Inferred or Possible category. Other information like rainfall, climate, availability of infrastructures, and logistic facilities including health care and environmental implications are collected. The main objective is to identify a suitable deposit that will be the target for further definitive exploration. The permissible area and duration will be between Reconnaissance and Prospecting.

1.3.2.3. Prospecting

"Prospecting" is the systematic process of searching promising mineral targets identified during Reconnaissance. The objective is more definitive exploration for increasing geological confidence leading to further

TABLE 1.1 Work Program during Reconnaissance License by Years

Year	Proposed work program
Year-1	(1) Regional geological check, mapping and rock chip sampling. (2) Acquisition and interpretation of available airborne geophysical data from previous surveys. (3) Identification of prospective geological packages/structures. (4) Regional geochemical surveys—soil/stream sediment sampling as required. (5) Regional airborne geophysics, ground magnetic and electromagnetic traverses as required.
Year-2	(1) Integration and interpretation of geological, geophysical and geochemical data to identify anomalies/targets (could be geological, geochemical and geophysical). (2) First pass follow-up of anomalies/targets by detailed mapping, infill soil/rock chip sampling and ground geophysics. (3) Prioritization of anomalies/targets for drill testing. (4) Scout drilling of interesting targets.
Year-3	(1) Second pass follow-up and target definition. (2) Reverse circulation/diamond core drilling. (3) Down-hole geophysics and drilling, if required. (4) Reports/recommendations. (5) PL application—if encouraging results obtained.

exploration. The program starts on obtaining PL from State/Provincial/Territorial Government within the framework of area and duration. PL is granted to conduct prospecting, general exploration and detail exploration. PL shall be deemed to include "LAPL", unless the context otherwise requires. The activities include mapping on 1:50,000-1:25,000 scale, linking maps with Universal Transversal Mercator (UTM), lithology, structure, surface signature, analysis of history of mining, if exists, ground geophysics, geochemical orientation survey, sampling of rock/soil/debris of background and anomaly area, pitting/trenching, Reverse Circulation (RC) and diamond drilling at 100-1000 m section at one level depending on mineral type, core sampling, petrographic and mineragraphic studies, borehole geophysical logging and baseline environment. Estimates of quantities are inferred, based on interpretation of geological, geophysical and geochemical results.

The year-wise exploration activities are programmed similar to Table 1.1 with emphasis on more objective-oriented results for broad delineation of orebody.

1.3.2.4. General Exploration

"General exploration" is the initial delineation of an identified deposit. Methods include mapping on 1:25,000, 1:5000 or larger scale, for narrowing down the drill interval along strike (100-400 m) and depth (50-100 m), detail sampling and analysis for primary and secondary Commodities, value-added trace and deleterious penalty elements, ~10% check sampling and analysis for Quality Assurance/Quality Control (QA/QC), borehole geophysical survey, bulk sampling for laboratory and bench scale beneficiation tests and recoveries and collection of geo-environmental baseline parameters The objective is to establish the major geological features of a deposit, giving a reasonable indication of continuity and providing an estimate of size with high precision, shape, structure and grade. Estimates are in the Indicated and Inferred category. The activity ends with preparation of broad order of economic or "Pre-Feasibility" or "scoping" study.

1.3.2.5. Detail Exploration

"Detail exploration" is conducted before the start of mining phase or mine development. It involves three-dimensional (3D) delineation to outline firm contacts of the orebody, rock quality designation (RQD) for mine stability, planning and preparation of samples for pilot plant metallurgical test work. The works envisaged are mapping at 1:5000, 1:1000 scale, close space diamond drilling (100 × 50, 50 × 50 m), borehole geophysics, trial pit in case of surface mining and sub-surface entry with mine development at one or more levels in case of underground mining. The sample data are adequate for conducting 3D geostatistical orebody modeling employing in-house or commercial software for making Due Diligence reports. The reserves are categorized as Developed, Measured, Indicated and Inferred with high degree of accuracy. The sum total of Developed, Measured and Indicated reserves amounts to 60% of total estimated resources for investment decision and preparation of Bankable Feasibility Study report.

The Mining Lease (ML) is obtained at this stage for the purpose of undertaking mining operations in accordance under the Act for major minerals. It shall also include quarrying concessions permitting the mining of minor minerals. ML is granted by competent authority, i.e. the State/Provincial/Territorial Government with the clearance from Federal Ministry of Mines (MOM), Ministry of Forest and Environment (MOFE) and Bureau of Mines (BM). The permissible area under ML will be negligible and may be 1/100th of the Reconnaissance area.

A total span of ~15-50 or more years, from beginning to closure of the mining, is conceived for project schedule. Conditions change in this time, and it is a combination of founding foresight, steady perseverance and agility in adaptation, which, along with a good measure of statistical providence provide for project success.

1.3.2.6. Ongoing Exploration

Diamond drilling is a continuous process throughout the entire life of the mine to supplement reserve for the depleted ore. Exploration continues during the mine development and production. This is primarily conducted by underground diamond drilling to enhance reserve down dip and in the strike direction. The aim of the mine geologist is to replace depleted ore of 1 tonne by 2 tonnes new reserve at the end of each year. This will increase the mine life and continue mining operation. It also upgrades the category of reserve from Inferred to Indicated, Measured and Developed. The drill information helps to precisely delineate ore boundary, weak rock formation and shear zones for mine planning. This also provides additional data on RQD. Geochemical sampling of soil, debris and ground water in wells, streams, and rivers within and around the mining license area is carried out at regular interval for monitoring environment hazards. The process continues beyond the closure of the mines.

1.3.2.7. Exploration Scheme

Diamond drilling acts as a prime exploration tool. It is expensive and is carried out in a planned logical sequence. The scheme can be divided into a number of phases depending upon the extension, size and complexity of the mineral deposit. The activity in each phase is defined in clear vision with respect to interval of drilling, number of drill holes, meters to drill, time and cost required to achieve the specified objectives (Table 1.2 and Fig. 1.5). At the close of each phase of activity the results are reviewed with

TABLE 1.2 Design of an Exploration Scheme with Identification of Work, Time and Cost at Million US$ to Achieve the Anticipated Results

Phase	Drilling interval (m × m)	Total meter	No. of holes	Time (year)	Cost (m US$)	Objectives
1	400 × one hole	Mtr_1	N_1	Y_1	$\$_1$	To establish the existence of target mineralization in space
2	200 × 50	Mtr_2	N_2	Y_2	$\$_2$	To establish broad potential over the strike length and laboratory scale metallurgy
3	100 × 50	Mtr_3	N_3	Y_3	$\$_3$	To establish firm reserves, grade, bench and pilot scale metallurgy, Detail Project report (DPR) for conceptual mine planning, mine investment decision
4	50 × 50	Mtr_4	N_4	Y_4	$\$_4$	To create database for detail production planning and grade control
5	Close space near surface and wide space at lower levels	Mtr_5	N_5	Y_5	$\$_5$	To delineate near surface features like weathering limit, extension of orebody down depth
Total		Mtr	N	1-5	mUS$ (m is million)	

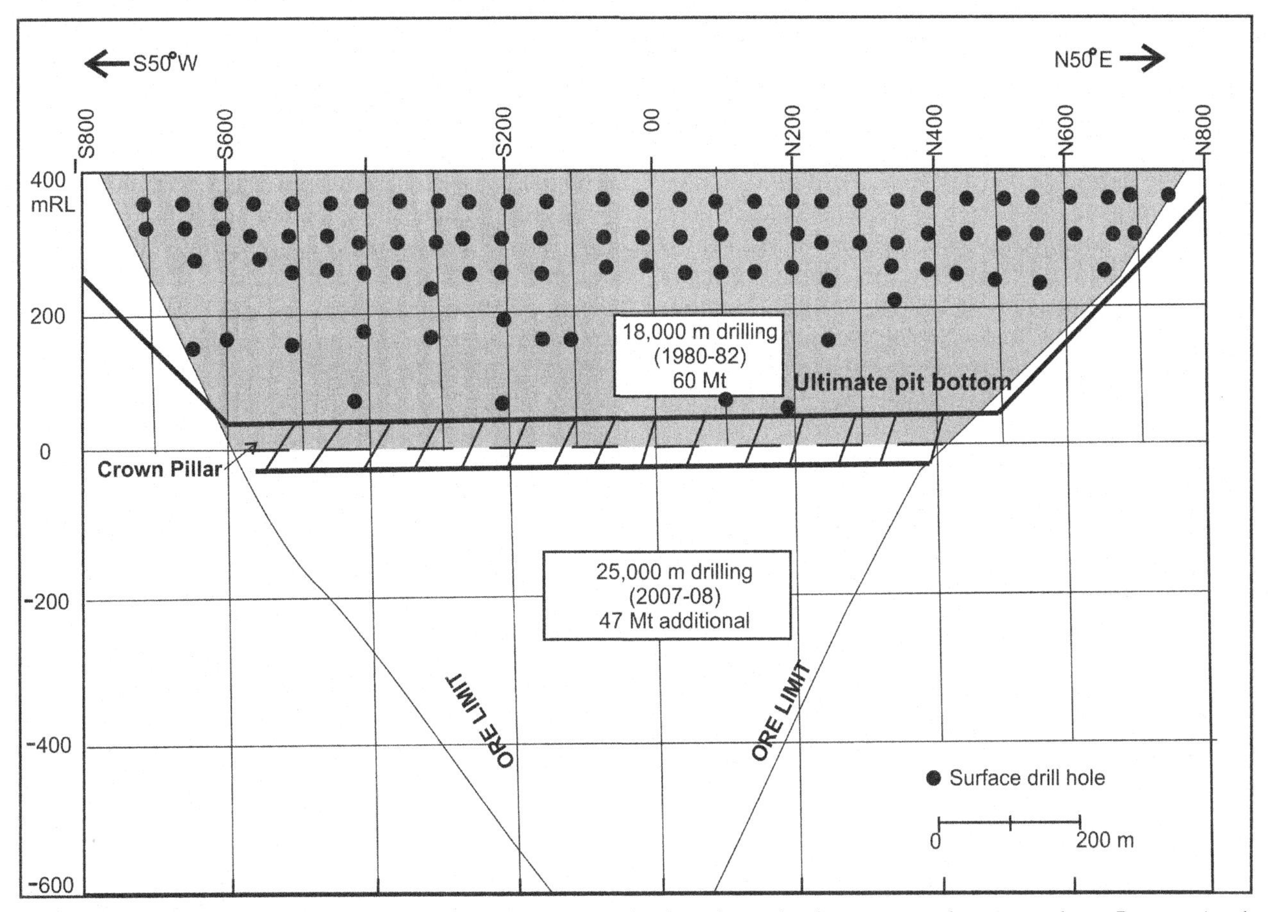

FIGURE 1.5 Sketch of Longitudinal Vertical Projection showing phased and ongoing exploration program and resource update at Rampura-Agucha zinc-lead-silver mine, India.

economic benchmark. If necessary, the activities of next phase are modified or withdrawn from the project for the time being.

1.4. MINERAL POLICY AND ACT

Mother Earth is endowed with variety of valuable mineral resources. These natural resources provide vital raw material for economic and overall growth of a country through development of infrastructure, basic industries and capital goods. Governments, particularly under Federal and State structure, are committed to maintain Gross Domestic Product (GDP) at around 10-15% share from mineral sector. The efficient management and scientific development of mineral sector through integration of exploration, exploitation, beneficiation, extraction and sustainable uses are to be guided within the framework of short- and long-term national goals, priority and prosperity. There is sufficient natural mineral resources in the mother Earth for human need but not enough for man's greed.

Mineral Policy or more precisely National Mineral Policy (NMP) is a reflection in that trend taking care of public interest. The Federal Government necessitates development of minerals in the interest of the nation as a whole and creates mechanism to benefit the local population in particular. The policy ensures linkages between forward and backward states of Union/Federal Government for setting of mining projects. Individual State or Region can formulate own policy under national outline for State's interest.

The national policy is the principle, philosophy, vision and mission for short- and long-term sustainable development of the mineral sector. The Rule, Regulation and Act are the framework formulation of actual enforceable laws authorized by major legislation for legislative applications of pronouncing the efficient policy. The common examples are Mineral Concession Rule (MCR), Mineral Conservation and Development Rules (MCDR), Mines (Health and Safety) Rules, Mines and Mineral (Development and Regulation) Act (MMDR), Mines Act, Environment (Protection) Act, Forest Act, etc.

The national goals and perspectives, the policy and Acts, are dynamic and responsive to the changing need of the industry in the context of the domestic and global economic scenario. All Policies and Acts are amended at interval as and when require. A healthy and transparent Policy and Acts encourage greater focused and sustained investment climate in exploration and mining by both national and foreign players. A favorable working environment by removing the bottlenecks and red tapeism which hinder the long-term productivity and efficiency must be reflected in the Act. The increasing competition on account of globalization, highest level of technology employed and initiative for the growth in mining sector have assumed critical significance. The Mineral Policy and Acts visualize the future national need and acceptability within the country. It has to ensure the will, stability, sustainability and internationality irrespective of the ruling Government.

The Policy and Acts as defined in developed and developing countries are discussed in brief.

1.4.1. Australia

The commonwealth of Australia is a highly developed country with world's seventh highest per capita income. The country is comprised of mainland, Australian continent (7.618 million km^2 surface area), surrounded by more than 8000 islands in Indian and Pacific Ocean that includes Tasmania (68,000 km^2). The population is estimated to be 22.70 million as on 2011. Australia follows constitutional monarchy with federal parliamentary system of government with Queen Elizabeth II at its apex. Australia has six states and two major mainland territories. The Australian dollar is the currency for the nation (1 AU$ = 1.05 US$ or 1 US$ = 0.95 AU$) as on March 2012. The mineral resource base industries are the key pillar of Australian economy. Australia is the world's leading producer of bauxite (65 Mt) and iron ore (393.9 Mt), the second largest producer of alumina (19.6 Mt), lead (0.57 Mt) and manganese (4.45 Mt), the third of brown coal (66 Mt), gold (~250 t), nickel (185 kt), zinc (1.29 Mt) and uranium (1.224 Mt U), the fourth of aluminum (2 Mt), black coal (445 Mt) and silver (1.63 kt), and finally the fifth largest producer of tin during 2009.

Australia possesses large landmass with sparse population. It has significant mineral endowment supported by strong advance mineral exploration technology along with scientific and mechanized mining tradition. The country is equipped with technical institution and vocational training centers, well-trained engineers and workforces. The exploration and mining companies are spread over the globe at different capacities. Australian mineral policy instruments and regulations are aimed at rational development of the country's mineral resources. The wide goals of the Government are to achieve community development, employment generation, lowering inflation rate, overall economic growth maintaining high environmental standards and sustainability. The policy has been balanced to satisfy the company's perspective, i.e. interests of its shareholders.

The mineral policy and mining legislation are largely provincial. Mines and minerals are a state subject in Australia and hence each of the six states and two major territories has their own mining legislation. Although there are many similarities, differences in legislation from state to state are also very significant. The policy framework and Acts are powerful with clarity, efficiency and competitive in the process to make the fortune. But the system is quite complex on certain issues.

The legislation as framed in Western Australia (Table 1.3) can be considered as model for discussion, with some deviation from other states and territories.

The Act may be cited as Mining Act 1978 and updated 2009. The mining tenement or concession includes PL, exploration license, retention license, ML, general purpose lease and miscellaneous license granted or acquired under this Act or by virtue of the repealed Act. The Act includes the specified piece of land in respect of which the tenement is so granted or acquired. The application for all types of license in prescribed format is to be submitted to the office of the mining registrar or warden of the mineral field or district in which the largest portion of the land to which the application relates is situated. The application must be accompanied by the following documents.

(a) Written description of the area.
(b) A map with clearly delineated tenement boundaries and coordinates.
(c) Detail program of work proposal.
(d) Mining proposal or mineralization report prepared by a qualified person.
(e) Estimated amount of money to be expended.
(f) Stipulated fee and the amount of prescribed rent for the first year or portion thereof.

The mining registrar may grant the license if satisfied that the applicant has complied in all respects with the provisions of this Act or refuse the license if not so satisfied. The holder of prospecting and exploration license will have priority for grant of one or more mining or general purpose leases or both in respect of any part or parts of the land while the license in force. The license and lease are transferable.

1.4.2. Canada

Canada is the world's second largest country by total area of 9.985 million km^2 bordered by Atlantic, Pacific and Arctic Ocean in the east, west and north respectively. The country consists of 10 provinces and 3 territories with a total population of 31.613 million (2006). Canada is a federal state, governed by parliamentary democracy and a constitutional monarchy with Queen Elizabeth II as its Head of State. It follows English and French as official language at the federal level. The currency is Canadian Dollar (1 CAD = 1.00492 US$) as on March 2012. Canada

TABLE 1.3 Western Australia—Tenement Type Summary (2009)

Tenement	Area	Term	Objective	Condition
PL	Not to exceed 2 km^2 for one license, eligible for more than one license	4 years from date of grant, discretionary renewal for one period of 4 years and further 4 years on license retention status	Prospecting for all minerals of economic interest	No significant ground disturbance other than drilling, pitting and trenching
Exploration License	1-70 blocks, each block of 2.8 km^2	5 years, discretionary renewals for one period of 5 years in prescribed circumstances, and by further 2 years in exceptional circumstances	All minerals of economic interest—exploratory tenure—may extract up to 1000 tonnes of material.	Applicant is technically competent and financial resources available, exploration reports to Minister/Geological Survey
Retention License	Whole or part of primary tenement, eligible for more than one license	5 years, discretionary renewals for one period of 5 years in prescribed circumstances	Further exploration for all mineral of economic interest	Statutory declaration of identified resource, uneconomic and impracticable mining at present
ML	Not exceeding 10 km^2, eligible for more than one ML	21 years, renewable as of right for one period of 21 years, discretionary thereafter	All minerals—may work and mine the land and remove, take and dispose minerals.	Feasibility report, mineralization report prepared by a qualified person, approved mine plan, Environment Mine Plan (EMP), pay rent and royalties
General purpose lease	$\leq$0.10 km^2, eligible $\geq$1 lease	21 years, discretionary renewals for 21 years.	Mining and related operations	Discretions of Minister, Registrar/Warden
Miscellaneous Licenses	—	5 years, renewals for 5 years and more	Rights for water	—

remains among the foremost top five producers of iron ore (31.70 Mt), nickel (181,000 t), zinc (698,901 t), copper (606,999 t), silver (608 t), gold (97,367 kg), lead (68,761 t), molybdenum (8836 t), Platinum Group of Elements (PGE) (6500 kg), coal, petroleum, natural gas and the leading exporter during 2009.

The Minerals and Metal Policy was formulated in 1996 after an extensive participation and deliberation between Federal Government and Provincial and Territorial mines ministries, industries, environmental groups, labor, and Aboriginal Communities. The policy aimed at:

(a) Promoting economic growth and job creation,
(b) Affirming provincial jurisdiction over mining,
(c) Delineating a new role for the Federal Government in minerals and metals that is tied to core federal responsibilities,
(d) Meeting the challenges of sustainable development of minerals and metals at the international level,
(e) Committing the Government to pursue partnerships with industry, the stakeholders, the provinces and territories, and others in addressing issues within its jurisdiction,
(f) Green environment, and
(g) Globalization of Canadian mineral sector through active, effective and influential partner on the international stage.

It plays an essential role in country's economy. It enjoys the leadership of exploration and mine production within the country and in export of mining technology and equipments all over the world. Canada has become one of the world's principal venture capital markets for mineral exploration and development with the globalization and liberalization of investment regimes. The mineral and mining Act and Regulation can be grouped as:

(a) Mineral rights in Canada are primarily owned by the Provincial or the Territorial Government.
(b) An individual or a Company is entitled to apply for an exploration, development and mining rights in prescribed format, fees and documents comprising of maps, descriptions, annual work plan and anticipated expenditures. The lease is granted by the state department authority under its Statutes and Regulations.
(c) The mining legislations for each of the 13 Canadian jurisdictions are separate, except for Nunavut. Nunavut is newly formed territory covering east and north portions of Northwest Territories. Nunavut exploration and mining activities continue to be regulated by the Department of Indian Affairs and Northern Development's office based in Northwest Territories.
(d) The surface and mineral rights on the same property can be held by different owners for different minerals/end products.
(e) There is no competitive bidding for mineral exploration rights in Canada.
(f) Locations for mineral rights are selected by companies or individuals according to their choice in freehold land.
(g) Individuals and companies must obtain a prospector's license before engaging in exploration for minerals. This is applicable in Northwest Territories, British Columbia, Manitoba, Ontario, Quebec, New Brunswick and Nova Scotia.
(h) However, one can conduct prospecting or exploration without a license in other states. A license will be required later to acquire mineral rights to protect his claim of discovery. A special permit is required to obtain the right to fly an airborne geophysical survey over an area not covered by a mineral claim.
(i) PL and recording fees are imposed at variable rates across jurisdictions.
(j) It is expected that the leaseholder carryout certain amount of assessment work each year to keep the claims in good standing. The technical activities include geological mapping, geochemical sampling, diamond drilling, assaying and related work of certain value. Copies of geological maps, reports, drill-logs and assay values must be submitted to the mining recorder. The reports are kept for future access by any interested party, after the end of a confidentiality period.
(k) Mining claim units are normally 0.16-0.25 km^2, with a maximum individual claim dimension varying between 2.56 and 5 km^2. It can be even larger, especially where claims are registered by way of "map staking". Map staking is the process of recording the claims on a surveyed map directly at the mining recorders' office without visiting the location.
(l) Holders of claims in good standing must obtain a ML to develop the property into a mine. MLs require that claim boundaries be surveyed by a Registered Land Surveyor. Mining rights are valid in most provinces/territories for 21 years and renewable till the ore persists.

1.4.3. Chile

The republic of Chile, stable and prosperous nation in South America, occupies a long narrow coastal strip between the Andes Mountain in the east and Pacific Ocean in the west. The land area covers 0.753 million km^2. The country is a multiethnic society with 15 million Spanish spoken populations. The currency is Chilean peso (CLP) (I US$ = 489.68 CLP or 1 CLP = 0.002 US$, as on March 2012). The climate varies widely, ranging from the world's driest Atacama Desert in the north, through subtropical Mediterranean climate in the center, to a rainy temperate climate in the south. Chile is divided into 15 regions from north to south. The northern desert contains great mineral wealth, primarily of copper. Chile is the largest producer (3.357 Mt) and exporter of copper in the world and also produces a large quantity of molybdenum (0.035 Mt), iron ore (8.24 Mt), gold (40,834 kg) and nitrates in 2009.

The destination for worldwide investment in exploration and mining became trade and industry friendly for Latin America (Chile, Peru, Argentina and Bolivia). These countries formulated massive legal reform to safeguard the private investment players in mining sector. It also emphasizes the protection of environment, human rights, community development and sustainability in mining. The exploration and mining Acts are framed under:

(a) The provisions of the legal framework are derived from the Constitution of Chile and by Basic Constitutional Laws, Codes and Regulations which apply specifically to the mining industry. Legal framework guaranties assuring ownership of mineral holdings for both exploration and mining concessions.

(b) The Mining Code provides State ownership and Mineral Rights. The State has absolute, exclusive, inalienable and regulatory ownership of all mines, including coal and hydrocarbon, except surface clays.

(c) Surface clay, sand, rocks and similar materials which can be used as such directly for construction are not considered mineral substances and not governed by the provisions of this Code.

(d) It is regardless of property rights of natural or legal individuals over lands wherein the minerals may be found. All individuals and companies may file exploration and mining concessions and acquire the right after granted.

(e) However, applications are debarred for Justices, Judges, Secretaries, Registrars, officials and employees of Appellate and Civil Courts, State agencies or companies who, because of their positions, office or duties participate in granting mining concessions or have access to geological discoveries of minerals and related information or mining data. This restriction is limited up to 1 year after they have ceased to hold their position. The spouse and legitimate children of any of the above-mentioned individuals are not entitled because of reasons based on considerations of national interest.

(f) A mining concession may be granted for exploring and/or mining mineral deposit. Any person is entitled to drill test holes and take samples during exploration irrespective of surface rights over the tenement. However, any damage of crop or building shall be compensated to the private landowner and obtain prior permission for government land.

(g) Mining concessions may be granted on claimable mineral substances including discarded mine production, tailings and slag of earlier mining and smelting within the lease boundary.

(h) In the event of any new mineral is found in the leasehold in significant quantity and that can be technically and economically mined and beneficiated, the State shall be informed. The new product shall be sold on behalf of the State. The cost incurred by the producer shall be reimbursed by the State.

(i) A concession shall be comprised of a north-south oriented parallelogram with coordinates of intersection in most précised UTM system. The sides of the exploration tenement shall be measured at a minimum of 1000 m and multiples thereof. The area of exploration concession shall not exceed 50 km^2. The legislation contemplates two periods for the expiration. The first period is for 2 years and discretionary renewal for an additional 2 years reducing 50% of the initial area. The annual taxes must be paid to maintain the property rights in good standing.

(j) The right to mine mineral substances is the legal continuity of exploration concessions. The sides for mining concession shall be at a minimum of 100 m or multiples thereof. The area of mining concessions shall not exceed 0.10 km^2. Once the filing process for mining concession has been completed, granted and as long as the annual property tax payments have been made, the system provides the legal tools to maintain the concession till the mine closure.

1.4.4. India

The Federal Constitutional Republic of India is the largest democracy in the world with 1200 million people. The total land area of 3.287 million km^2 consists of 28 states and 7 union territories. It is the fourth largest economy by purchasing power parity (PPP). The first official language is Hindi and English is equally recognized. The currency is Indian Rupee (₹) (1 US$ = 50 [₹] or 1 [₹] = 0.02 US$ as on March 2012). India is the second largest producer of chromium (3.90 Mt ore) and one of the leading producers of iron ore (245 Mt), manganese, bauxite (16 Mt), coal, zinc (0.695 Mt) and lead (0.092 Mt) metals during 2009. Nonmetallic minerals include limestone, rock phosphate, Kaolin and dolomite.

The Indian mining history dates back to more than 5000 years and naturally gone through the various processes of changes over the periods. In the ancient time king was the supreme authority. The mine owners had limited freedom. Safety and environment were not matter of concern. Subsequently, British Ruled states formulated Indian Mines Act (IMA) to take care of safety and welfare of miners. However, Rulers of Princely states were unconcerned with the Act and the Government was satisfied with indirect return. The nation witnessed major developments after the postindependence era (1947) by abolition of Princely states. The first Industrial Policy Resolution and Industrial Development Act were formed in 1948 and1951 respectively with active role of State for development of industries in socialistic pattern. The participating agencies were Federal and State Sector Public Enterprises for major and strategic minerals. The Private Sectors were entrusted with minor minerals. NMP was formulated in 1990 with following salient features:

(a) Explore and mineral identification on land and offshore,
(b) Mineral resource development at national and strategic considerations, ensure adequate supply for present and future need,
(c) Linkage for smooth and uninterrupted development of mineral industry to meet need of the country,
(d) Promote research and development in minerals,
(e) Ensure establishment of appropriate education and training facilities for Human Resource Development (HRD) to meet man power requirement in mineral industry,
(f) Minimize adverse effect of mineral development on forest, environment and ecology through protective measures,
(g) Ensure mining operations with due safety and health care of all concerned,
(h) State Government to grant and renew of License and lease,
(i) Federal and State Enterprises primarily responsible for mining and processing of thirteen minerals of basic and strategic importance:
   (i) Iron
   (ii) Chromium
   (iii) Gold
   (iv) Copper
   (v) Zinc
   (vi) Tungsten
   (vii) Platinum group of metals
   (viii) Manganese
   (ix) Sulfur
   (x) Diamond
   (xi) Lead
   (xii) Molybdenum
   (xiii) Nickel
(j) Mining and processing of minor minerals by Private Sectors,
(k) Geological Survey of India responsible for regional mineral resource assessment,
(l) Detailed exploration by State Directorate of Mines and Geology (DMG), State and Central Undertakings,
(m) Seabed exploration, mining and processing by Department of Ocean Development (DOD),
(n) National Mineral Inventory—Creation, Update and Dissemination by Indian Bureau of Mines (IBM).

Globalization of Indian economy (1991) persuaded to revise NMP in 1993 to de-reserve the above 13 minerals and open to domestic and foreign private Mineral Exploration Groups. The NMP, 1993, aimed at encouraging the flow of private investment and introduction of state-of-the-art technology in exploration and mining. The eligibility for grant of concessions is open to such person of Indian National or a Company incorporated in India under Company Act, 1956, and has registered himself with IBM or State DMG. Such person or Company shall undertake any Reconnaissance, Prospecting, general exploration, detailed exploration or mining in respect of any major and minor minerals under Reconnaissance License, LAPL, PL or ML. Finally, the NMP (2008) is revised with finer tuning and transparency. The summary of the mineral rights (Table 1.4) and additional salient features are:

(a) Mineral Policy (2008) [56] aspires to develop a sustainable framework for optimum utilization of the country's natural mineral resources for the industrial growth in the country. At the same time it emphasizes to improve the life of people living in the mining areas which are generally located in the backward and tribal regions.
(b) The policy aims at strengthening the framework/ institutions supporting the Indian Mining sector, which include the IBM, Geological Survey of India and State DMG.
(c) The Government agencies will continue nationwide survey, exploration and mineral resources assessment.
(d) IBM will maintain database in digitized form comprising of Resource Inventory in United Nations Framework Classification (UNFC) system and Tenement Registry that will be available at cost.
(e) Major thrust will be development of infrastructural facilities in mineral-bearing remote areas.
(f) Removal of restriction on foreign equity holding in the mining sector enabling any Company registered in India irrespective of foreign equity holding to apply for Reconnaissance and PL or ML. The Foreign Direct Investment (FDI) or Direct Foreign Equity participation is raised to 100%.
(g) It is aimed to attract large FDI in Reconnaissance and Large Area Prospecting and right of entry to improved technology in scientific and mechanized mining.
(h) It emphasizes security of tenure along with transparency in allocation of concessions, seamless transition from prospecting to mining and transferability.
(i) It promotes competitive auction of orebodies fully prospected at public expenses.
(j) Cluster approach will be adopted by grouping small deposits within single lease.

TABLE 1.4 Summaries of Mineral Rights Applicable in India (2012)

Tenement	Objective	Area	Term	Requirements	Conditions
Reconnaissance License/RP	Grassroot exploration to identify mineral targets in regional scale for further prospecting, preliminary resource	Not to exceed 10,000 km^2 within a state	Between 1 and 3 years, nonrenewable	Indian National or Company incorporated in India under Company Act, prescribed application	Well-defined work program, technically competent and financially strong, 100% FDI permissible
LAPL	Exploration and identify deposit, Inferred resource.	Between 500 and 5000 km^2	Between ≥3 and 6 years, discretionary renewal for 2 years	-Do-	-Do-
PL	Systematic searching of deposits, outline Inferred resource	Between 1 and 500 km^2	Between 2 and 3 years, discretionary renewal for 2 years	-Do-	-Do-
General exploration	Indicated	Under PL	—	-Do-	-Do-
	Inferred				
	Pre-Feasibility				
Detailed exploration	Measured	Under PL	—	-Do-	-Do-
	Indicated				
	Inferred				
	Feasibility				
ML	Feasibility, mine development, production	Between 100 and 0.10 km^2	Between 20 and 30 years, discretionary renewal for 20 years at a time, ensure full ore utilization	-Do-; mine plan, Environment Mine Plan (EMP) complete, forest clearance, land detail	-Do-
	—Developed				
	—Measured				
	—Indicated				
	—Inferred				

(k) Relief and rehabilitation of displaced and affected persons will be extended with human face.

1.4.5. Portugal

Portugal or Republic of Portuguese extends north-south in a rectangular area of 92,391 km^2 that includes continental Portugal, the Azores (2333 km^2) and Madeira Islands (828 km^2). The country is located in southwest corner of Europe and bordered by Spain in the north and east and the Atlantic Ocean in the west and south. The country has a population of 10.76 million (2011 est.) with 95% literacy and 92% belong to Roman Catholic. The official language is Portuguese. The currency in Portugal is Euro (EUR) with currency image € at exchange rate of 1 EUR = 1.3206 US$ as on March 2012.

The 85-km long Iberian Pyrite Belt (IPB) exists between Portugal in the west and Spain in the east. IPB is one of the rich mineralized geological provinces of Western Europe and hosts number of deep-seated volcanogenic massive sulfide and gold deposits in Portugal part. The production of metallic minerals includes copper (Neves Corvo, refer Chapter 15, Section 15.6), tungsten (Panasqueira mine, sixth largest in the world), lead, zinc, lithium (sixth largest), tin (10th largest), silver and gold. Portugal is one of the leading producers of copper metal and tungsten concentrate in Europe. Mineral industry represents ~1% of GDP. The country also possesses important uranium deposits, with ~4200 tonnes of U_3O_8 produced between 1950 and 1990.

The new mineral policy (amended in 2003) on spatial planning was created at the national level for a sectorial planning for minerals which must be produced by the national government. The legal nature of this plan is considered as policy guidance to assist in the decision making process, both for the preparation and approval of the low-level plans (all minerals) and total operation.

The prime target for exploration activities is focused around the IPB because it appears to have a good potential for success on the basis of the large sulfide deposits discovered so far. The IPB is a focal point of interest for the mining companies. Several gold and base metal projects are under feasibility studies. Portugal is also known for lower cost of production in Europe and Asia. These factors attract significant opportunities for FDIs.

1.4.6. South Africa

The Republic of South Africa (SA) with Parliamentary democracy is divided into nine provinces. The land area covers 1.20 million km^2 with 50 million people. The official languages are Afrikaans and English. The South African currency is Rand (R) with code as ZAR (1 US$= 7.65488 ZAR or 1 ZAR = 0.130636 US$ as on March 2012). South Africa is the world's largest producer of platinum (75 tonnes) and chromium (9.683 Mt) and one of the leading producers of gold (197.698 tonnes), diamond, copper (0.109 Mt), zinc (0.028 Mt), lead (0.049 Mt) and coal during 2009.

Prior to May 2004 mineral rights in South Africa were owned by individuals or legal entities. Since then the government enacted the Minerals and Petroleum Resources Development Act (MPRDA) and all minerals now vest in the State as long-term objective. The Act defines the State's legislation on mineral rights and mineral transactions. The Act also emphasizes that the government did not accept the dual State and private ownership of mineral rights. The government has entrenched a "use it or lose it" principle that is applied to leaseholders who own mineral and prospecting rights but are not making use of it. Privately held mineral rights had to be transferred under the provisions of the Act into licenses to prospect and mine that are granted by the State. Ultimately, all minerals in South Africa vest in the State.

The Act persuades that the government policy is to furthering Black Economic Empowerment (BEE) for mineral industries within the country. It encourages the mineral exploration and mining players to enter into equity partnerships with BEE companies. The Act made provision for implementation of social responsibility procedures and programs. The details of these criteria are required to be provided in applications for permits and licenses under Schedule II (Transitional Arrangements). Existing prospecting and mining rights continued in force for a period of 2 and 5 years respectively from the 1 May 2004. Holders of any unused old-order rights had the exclusive right to apply for a prospecting or a mining right within 1 year of the Act coming into effect, failing which the right ceased to exist.

The State has confirmed its commitment to guaranteeing security of tenure in respect of prospecting and mining operations. The Act states that a company has 5 years (due date 30 April 2009) to apply for a new-order mining license and the new-order mining license will be granted for a maximum of 30 years. Summary of mineral rights applicable in South Africa are given in Table 1.5.

1.4.7. Tunisia

Tunisia is a small North African country (0.155 million km^2 land cover) located on the Mediterranean coast of Africa, between Algeria and Libya. The Capital, Government, Head of State and languages are Tunis, Republic, President and Arabic-French respectively. Tunisia with a population of ~10 millions is considered the most liberated in the Arab world and has peaceful relations with both Israel and Palestine. State investment in infrastructure for the last 10 years has ensured that, despite inhospitable terrain, there are 20,000 km of good-quality roads linking all parts of the country, 18,226 km of which are paved. Fuel in the country is cheap and readily available. The mineral resources are phosphates, iron ore, zinc-lead and salt. The rail system is good, regular and reliable services. There are six international airports in Tunisia and the national airline, Tunis-air, flies to European and the Middle East. There are eight commercial seaports and 22 smaller ports within Tunisia. Power is generated by the state-owned Société Tunisienne de l'Electricité et du Gaz (STEG). The telephone network and Internet facility are above the standard of most African countries. The Tunisian Dinar (TND) is soft currency of Tunisia (1 US$ = 1.510 TND or 1 TND = 0.664055 US$ as on March 2012).

All exploration and mining activities in Tunisia are governed by the Directorate General of Mines (DGM) and regulated under the Mining Code (Code Minier) promulgated under Law 2003-30, 28 April 2003. All mineral resources belong to the public domain of the State of Tunisia with full rights. All exploration and exploitation rights are vested in the State as represented by the Minister of Mines.

All prospecting, exploration or exploitation rights are granted to natural Tunisians or foreign persons on the basis that they can demonstrate adequate financial resources and technical capacity to undertake the proposed activities in an

TABLE 1.5 Summaries of Mineral Rights Applicable in South Africa

Tenement	Objective	Duration	Requirements	Condition
RP	Exploration at grassroot level	2 years (nonrenewable)	Financial ability, technical competency and well-defined work program	Holder does not have the exclusive right to apply for a Prospecting Right
Prospecting Right	Exploration at target definition stage.	Initially 3 years. Renewable once for 3 years.	-Do- plus economic program and environmental plan.	Payment of Prospecting fees.
Retention Permit	Hold on to legal rights between Prospecting and mining stages.	3 years initially. Renewable once for 2 years.	Prospecting stage complete; Feasibility study complete; project currently not feasible; and EMP complete.	May not result in exclusion of competition, unfair competition or hoarding of rights. May not be transferred, ceded, leased, sold, mortgaged or encumbered in any way.
Mining Right	Mine development	30 years initially. Renewable for further periods of 30 years. Effective for Last Ore Mining (LOM).	• Financial ability, • Technical ability, • Prospecting complete, • Economic program, • Work program, social, labor and environmental plan	Payment of royalties. Compliance with Mining Charter and Codes of Good Practice on Black Economic Empowerment (BEE).
Mining Permit	Small-scale mining.	Initially 2 years, renewable for three further periods of 1 year at a time.	Life of project must be <2 years, area not to exceed <0.01 km^2 and environmental plan completed.	Payment of royalties. May not be leased or sold, but is mortgageable.

optimal manner. All rights are granted only provided that written permission has been obtained from the landowner (Article 79 Code Minier) and in the case of dispute, final authority is vested in the State. The holder of a mining right is responsible to mark out the perimeter.

The following rights are granted by the DGM:

(a) PL: granted in order to permit the applicants to conduct the investigation necessary to prepare application documents for Exploration Permits but precludes drilling or mining activities. A PL grants access to areas covered by an Exploitation Concession or Exploration Permit for substances covered in the existing rights;

(b) Exploration Permit: initially granted for 3 years (Article 30 Code Minier), renewable twice thereafter, with each renewal period for a maximum of 3 years. A work program and detail of financial commitment by the permit holder is required on application and renewals are contingent upon successful completion of the commitment and fulfillment of conditions set out in the Code Minier.

(c) An Exploration Permit can be renewed for a further 2-year period after the expiry of the second renewal period and such a renewal is referred to as a Special Renewal and is granted to the permit holder to enable it to complete Pre-Feasibility Studies. At the end of the Special Renewal period, the permit holder has to apply to convert the permit into an Exploitation Concession (mining license) or relinquish the property. Exploration activities must commence within 12 months of the approval of the right. The permit holder is obliged to submit progress reports as a number continuing obligation;

(d) Exploitation Concession: granted by the Minister of Mines after approval of the Mining Consultative Committee. The concession is granted for a period consistent with the quantity of mineral reserves and the holder is required to commence development work no later than 2 years after authorization of the

TABLE 1.6 Summary of Comparative Royalty Rates in Various Countries

Country	Commodity		
	Copper	Zinc	Gold
India	4.2% of London Metal Exchange Cu metal price chargeable on the contained Cu metal in ore produced.	8 or 8.4% of London Metal Exchange Zn metal price on ad valorem basis chargeable on contained Zn metal in ore or concentrate produced respectively.	2% of London Bullion Metal Association price chargeable on the contained Au metal in ore produced.
Australia	Royalty rates are fixed at 30 cents/tonne (aggregate, clays, dolomite, gravel, gypsum, construction limestone, rock, salt, sand and shale); 50 cents/tonne (building stone, metallurgical limestone, pyrophyllite, silica and talc). All other minerals are rated as a percentage of the realized value at 2.5% (Co, Hg, PGE, Ag), 7.5% (bauxite, calcite, diamond, gems, precious and semiprecious stones, iron ore, manganese and quartz crystal) and 5% for all other minerals with some minimum value per tonne for garnet, ilmenite, leucoxene, rutile, nickel and zircon.		
Canada	Mining tax/royalty varies between Provincial/Territorial regime at 12-20% of net profits after full cost recovery.		
Chile	Progressive royalty rates which range between 5 and 14% levied on the margin of profits obtained on sales of nonrenewable mining products.		
South Africa	Royalty tax rates = 0.5% + X/9.0 (max 7%) for metals in concentrate, or, 0.5 + X/12.5 (max 5%) for refined metals. X = EBIT/gross sales × 100. EBIT = Earnings Before Interest and Taxes. Royalties will be paid biannually in accordance with Minerals and Petroleum Resources Development Act (MPRDA), May 2004 and Mineral and Petroleum Resources and Royalty Bill (MPRRB).		
Tunisia	Mining royalty equal to 1% of the gross revenue from extracted ore, paid biannually during the 2 months following the previous quarter.		

concession. The holder is responsible for rehabilitation liabilities and has social obligations in terms of the training and employment of Tunisian employees.

(e) The holder of the above rights is required to furnish the DGM with annual progress reports, as well as monthly statistical reports of employees, production, revenue and equipment.

1.4.8. Royalties and Taxation

The fiscal policy related to application and registration fees, royalty, income tax, compensation to landowners, rehabilitation cost, annual and dead rent applicable for holders of mineral exploration and exploitation rights vary between countries and even states. Mineral royalties are exclusively State/Provincial/Territorial earnings and constitute a significant revenue source for the State/Provincial/Territorial Government. Mineral-rich states like Queensland, Western Australia, California, Ontario, Northern Province of South Africa, Rajasthan receive more than 10% revenue from mineral sector. Neither the rates nor the methods of calculating royalty are uniform. The rate of royalty in respect of copper, zinc and gold ore, removed or consumed by the leaseholder or his agent, has been compared in Table 1.6. The rates are variable so as to enhance or reduce and amended by the Federal Government from time to time.

1.4.9. Lease Application

Mineral concessions, in general, are State/Provincial/Territorial responsibility with the approval of the Federal Government. The model format of lease application pertaining to Reconnaissance, Large Area Prospecting, Prospecting, Mining and all other lease-related matters can be obtained from State/Provincial/Territorial Department of Mines and Geology. The applications are submitted to the State/Provincial/Territorial Department of Mines and Geology along with supporting documents (reports, work plan, proposed expenditure, map marked with tenement borders and coordinates, etc.) and stipulated fees. It must be within the framework of Mineral Concession Rule [39] and Mines and Minerals (Regulation and Development) MMRD Act of the home country. The lease is granted or denied giving reasons within specified time by the State Department with concurrence from the Federal Government.

1.5. MINERAL TO METAL—A FULL CIRCLE

The significant and salient steps from discovery of mineral deposits to delivery of finish products to the end users have been articulated in a circle of 1-12.

(1) Mineral exploration to discover a deposit—Reconnaissance, Prospecting and modeling—Chapters 1, 2 and 10.
(2) Exploration geology, geochemistry, geophysics, photo-geology-Remote Sensing (RS) and Geographic Information System (GIS)—Chapters 3, 4, 5 and 6.
(3) Sampling and reserve—resource estimation with precision—Chapters 7, 8 and 9.
(4) Feasibility study to prove its commercial viability—Chapter 13.
(5) Mine development, infrastructure and extraction of ore from ground—Chapter 11.
(6) Mineral processing—milling and separation of ore from gangue to produce concentrates, refinement of industrial mineral product—Chapter 12.
(7) Caring for ecosystem and sustainable future for now and then—Chapter 14.
(8) Smelting—recovery of metals from mineral concentrate—Reference only.
(9) Refining—purifying the metals—Reference only.
(10) Marketing—Shipping the product (concentrates, metals and minerals) to the buyer: in-house or custom smelter and manufacturer—Chapter 13.
(11) Experience from exploration and mine case studies—Chapter 15.
(12) New search for minerals continues—Chapters 1-15.

FURTHER READING

Roy (2010) [61] can be the basic study material to understand the fundamentals of any branch of Geo-science, and in this case mineral exploration. The technical aspects of mineral exploration are covered in greater detail by Evans (1999) [24]. The book on exploration modeling of base metal deposits by Haldar (2007) [33] will be meaningful reading for understanding and development of exploration concepts. Readers are suggested to consult Websites of Internet services for latest Mineral and Metal Policy and Act of various countries.

Chapter 2

Economic Mineral Deposits and Host Rocks

Chapter Outline

Economic strength of a country is known by its mineral wealth.

—Author.

2.1. DEFINITION

A mineral deposit becomes economic when it has a profitable commercial value attached to it. The concentration of minerals or metals in deposits vary widely and range from few parts per million (1-100 g/t or ppm) in noble metals like platinum, palladium, gold, silver to low percentage (1-10%) for copper, zinc, lead, and higher grade (40-60%) for aluminum, chromium, iron and aggregates. A mineral can be termed economic or uneconomic depending on its industrial use. The mineral quartz is economic as silica sand used in glass or optical industry. The same mineral is uneconomic when it hosts gold as auriferous quartz vein or occurs as a constituent of

Mineral Exploration. http://dx.doi.org/10.1016/B978-0-12-416005-7.00002-7

TABLE 2.1 Forms of Occurrences and Common Minerals

Forms	Minerals
Native element	Antimony (Sb), silver (Ag), gold (Au), sulfur (S), copper (Cu), Platinum (Pt), Palladium (Pd).
Oxide	Quartz and amethyst (SiO_2), hematite (Fe_2O_3), cassiterite (SnO_2).
Carbonate	Calcite ($CaCO_3$), magnesite ($MgCO_3$), dolomite ($CaMg(CO_3)_2$), ankerite $Ca(Fe, Mg, Mn)(CO_3)_2$, smithsonite ($ZnCO_3$), cerussite ($PbCO_3$), rhodochrosite ($MnCO_3$).
Silicate	Andalusite-kyanite-sillimanite-(Al_2SiO_5), beryl ($Be_3Al_2Si_6O_{18}$), amazonite ($KAlSi_3O_8$), garnet group-pyrope ($Mg_3Al_2(SiO_4)_3$), almandine ($Fe_3Al_2(SiO_4)_3$).
Sulfide	Chalcopyrite ($CuFeS_2$), sphalerite (ZnS), galena (PbS), Pyrite (FeS_2).
Sulfate	Barites ($BaSO_4$ $2H_2O$), gypsum ($CaSO_4$), anglesite ($PbSO_4$).
Sulfosalts	Bournonite ($PbCuSbS_3$), tetrahedrite ($Cu_{12}Sb_4S_{13}$), tennantite ($Cu_{12}As_4S_{13}$).
Phosphate	Apatite ($Ca_5(PO_4)_3(F, Cl, OH)$), rock phosphate.

rocks hosting copper, zinc and iron ore. It is then processed and discarded as gangue, tailing or waste. The ore deposits are generally composed of a main product, one or more by-products and trace elements such as zinc-lead-silver, copper-gold, chromium-nickel-platinum-palladium. Sometimes a single mineral forms the valuable deposit such as calcite in marble. The same mineral can be designated as metallic or industrial depending on its use. Bauxite ore is "metallic" when aluminum is produced and "industrial" when used directly for refractory bricks and abrasives. An ore deposit can be composed of metallic and nonmetallic minerals, mined together and processed to produce separate products. An example can be Bou Jabeur deposit, Tunisia, containing galena and sphalerite along with fluorite and barite.

The economic minerals occur in various forms such as native elements to compounds of oxide, carbonate, silicate, sulfide, sulfate, sulfosalts, phosphate etc. (Table 2.1).

2.2. COMMON ECONOMIC MINERALS

A list of common economic Ore Forming Minerals (OFM), formulas, percent content and major uses are listed in Table 2.2.

2.3. CLASSIFICATION OF MINERAL DEPOSITS

The understanding of various types of mineral deposit can facilitate to formulate appropriate and successful exploration program from grassroot to detail stage. In order to assess a deposit type more effectively, it is subdivided into classes. The classification can be based on single or multiple criteria e.g. as geographic localization, depth of occurrence, relation to host rocks, structural control, nature of mineralization, morphology, genetic features and contained metal. It is unlikely for two mineral deposits to be exactly identical, but in a broad sense, it will fall into one or another group or class, perceivable and comparable. Therefore, a largely acceptable physical description is attempted that can serve to design an exploration scheme.

2.3.1. Geographical Localization

Mineral deposits can broadly be described based on geographic location and dimension.

2.3.1.1. Province

"Province" or "Metallogenic Province" is a large specific area having essentially notable concentration of certain characteristic metal or several metal assemblages or a distinctive style of mineralization to be delineated and developed as economic deposits. The metallogenic province can be formed on various processes such as plate tectonic activity, subduction, igneous intrusive, metal-rich epigenetic hydrothermal solution and expulsion of pore water enriched in metals from sedimentary basin. The examples of metallogenic provinces are Zn-Pb-Ag-bearing McArthur-Mount Isa inlier in Northern Territory, Australia, gold province in Canadian shield, Pt-Pd-Ni-Cu-Au deposits in Sudbury basin, Canada, Bushveld Igneous Complex with Pt-Pd-Cr deposits, South Africa, Katanga and Zambian copper province, tungsten province of China, Zn-Pb-Ag deposits of Aravalli Province and diamond-bearing Kimberlite province of Wajrakarur-Narayanpet, India.

2.3.1.2. Region

"Region" is similar to province but relatively smaller in size, controlled by stratigraphy and/or structure, for occurrence of specific mineral(s) at commercial quantity. The examples are Kalgoorlie Goldfield, Esperance region of Western Australia, Zn-Pb region of Mississippi Valley, United States, copper region of Chile and Peru, diamond-bearing region of northern Minas Geraes, Brazil, diamond-bearing region of Kimberley, South Africa, Pacific and Central coal-bearing region of US and rubies in high-grade metamorphic rocks of Kashmir region of India.

TABLE 2.2 List of Common Metallic and Nonmetallic OFM and Uses

Principal ore mineral	Mineral formula	% Content	Major uses
Andalusite	Al_2SiO_5	63.2 Al_2O_3; 36.8 SiO_2	Gemstone, porcelain spark plug.
Apatite	$Ca_5(PO_4)_3(FClOH)$	41-42 P_2O_5	Fertilizer, occasionally gemstone.
Argentite	Ag_2S	87.0 Ag; 13.0 S	Jewelry, photo processing, currency and investment.
Arsenopyrite	FeAsS	46.0 As; 34.3 Fe; 19.7 S	Herbicide, alloys, wood preservative, medicine, insecticide, rat poison.
Asbestos Gr.	$CaMg_3Si_4O_{12}(OH)_2$	—	Building and pipe material.
Barite	$BaSO_4$	65.7 $Ba(OH)_2$; 34.3 SO_3	Drilling mud, filler, paper, rubber industry, radiology x-ray.
Bauxite	$Al_2O_3.2H_2O$	73.9 Al_2O_3; 26.1 H_2O	Construction, transport, consumer durables, packaging, electrical, machinery equipment, refractory bricks, abrasives.
Bentonite	$Al_2O_34SiO_2H_2O$	66.7 SiO_2; 28.3 Al_2O_3	Drilling mud, geotechnical, pellets, metal casting and medical.
Beryl	$Be_3Al_2Si_6O_{18}$	67 SiO_2; 19 Al_2O_3; 14 BeO	Gems, alloys, electronics, ceramics.
Bismuthinite	Bi_2S_3	81.2 Bi; 18.8 S	Alloy and pharmaceutical.
Bornite	Cu_5FeS_4	63.3 Cu; 11.1 Fe; 25.6 S	Source of copper metal.
Braggite	(Pt, Pd, Ni)S	Variable	Source for Pt and Pd.
Cassiterite	SnO_2	78.6 Sn; 21.4 O	Tin plate, solder, alloys.
Cerussite	$PbCO_3$	83.5 PbO; 16.5 CO_2	Lead ore
Chalcocite	Cu_2S	79.8 Cu; 20.2 S	Electricity, alloys, currency, medicine.
Chalcopyrite	$CuFeS_2$	34.5 Cu; 30.5 Fe; 35.0 S	Electricity, alloys, currency, medicine.
Chromite	$FeCr_2O_4$	68.0 Cr_2O_3; 32.0 FeO	Hard rustles steel, chrome plating, refractory bricks, pigments and dyes.
Cinnabar	HgS	86.2 Hg; 13.8 S	Primary source of mercury, pigment.
Coal	C, O and H	60-91; 2-34	Fuel and energy.
Cobaltite	CoAsS	35.5 Co; 45.2 As, 19.3 S	Strategically and industrially useful metal, high-temperature alloy, steel tools.
Copper native	Cu	100.0 Cu	Electricity, alloys, currency, medicine.
Corundum	Al_2O_3	52.9 Al; 47.1 O	Gemstones, abrasive, grinding media.
Covellite	CuS	66.4 Cu; 33.6 S	Insecticide, computer chips.
Cuprite	Cu_2O	88.8 Cu; 11.2 O	Source of copper metal.
Diamond	C	Pure carbon	Gem, abrasive, cutting tool, drill bit.
Feldspar	$NaAlSi_3O_8$ $KAlSi_3O_8$	18.4 Al_2O_3; 16.9 K 64.7 SiO_2	Ceramics, glass manufacture, fillers, paints, plastics and rubber.
Fluorite	CaF_2	51.1 Ca; 48.9 F	Flux in steel manufacture, opalescent glass, enamels for cooking utensils, hydrofluoric acid, high-performance telescopes and camera lens.
Galena	PbS	86.6 Pb; 13.4 S	Battery, electrodes, ceramic glazes, stained glass, shielding radiation.
Gold native	Au	100.0 Au	Jewelry, dentistry, electronics, coin and investment.

(Continued)

TABLE 2.2 List of Common Metallic and Nonmetallic OFM and Uses—cont'd

Principal ore mineral	Mineral formula	% Content	Major uses
Graphite	C	70-85 C	Steelmaking, crucibles, refractory, foundry, pencil, electrodes.
Gypsum	$CaSO_4.2H_2O$	32.5 CaO; 46.6 SO_3; 20.9 H_2O	Plasterboard, cement, insulation.
Halite	NaCl	39.4 Na; 60.6 Cl	Salt and preservative, soda ash for glass, soap and bleaching industry.
Hematite	Fe_2O_3	70.0 Fe; 30.0 O	Iron and steel industry.
Ilmenite	$FeTiO_3$	31.6 Ti; 36.8 Fe; 31.6 O	Alloy for high-tech in space and medical application, pigments.
Kaolin	$Al_4Si_4O_{10}(OH)_8$	46.5 SiO_2; 39.7 Al_2O_3	Paper, rubber manufacture, coating clay, linoleum, paints, inks, leather, refractory, pottery, insecticide, plastics and fertilizers.
Kyanite	$3Al_2O_3$, $2SiO_2$	63.2 Al_2O_3; 36.8 SiO_2	Heating element, electrical insulation, ceramic industry, gemstone.
Laurite	$(Ru, Ir, Os)S_2$	—	Source of ruthenium and osmium.
Lepidolite	$KLi_2Al(Al,Si)_3O_{10}(F,OH)_2$	~5 Li_2O	Battery, coloring of glass and ceramics, rocket propellant and nuclear fusion.
Magnesite	$MgCO_3$	47.6 MgO; 52.4 CO_2	Refractory bricks, cement industry, slag former in steelmaking furnaces.
Magnetite	$FeO.Fe_2O_3$	72.4 Fe; 27.6 O	Iron and steel industry.
Marcasite	FeS_2	46.6 Fe; 53.4 S	Iron and steel industry.
Marmarite	(ZnFe)S	46-56 Zn; <20 Fe	Source of zinc metal.
Millerite	NiS	64.7 Ni; 35.3 S	High-grade source for Ni.
Mineral oil	H and C	—	Most important energy source.
Molybdenite	MoS_2	60.0 Mo; 40.0 S	Corrosion resistance ferroalloy, alloy in stainless steels, electrodes.
Monazite	$(CeLaTh)PO_4$	48 Ce; 24 La; 17 Nd	Source for rare earth elements Ce, La, Th, Nd.
Niccolite	NiAs	43.9 Ni; 56.1 As	Minor source of nickel, blended with "clean" ore.
Palladium native	Pd	100.0 Pd	Emission control catalyst in automobiles, chemical-petroleum refining, electronic industries, coins.
Pentlandite	$(Fe, Ni)_9S_8$	22 Ni; 42 Fe; 36 S	Stainless tarnish resistant steel plating.
Platinum native	Pt	100.0 Pt	Jewelry, electrical and electronic industries, dental-medical fields.
Psilomelane	MnO_2	50.0 Mn	Drier in paints, steelmaking.
Pyrite	FeS_2	46.6 Fe; 53.4 S	Sulfur for sulfuric acid.
Pyrolusite	MnO_2	63.0 Mn	Batteries, coloring in bricks, decoloring in glass, pottery.
Pyrrhotite	Fe_nS_{n+1}	60.4 Fe; 39.6 S	Sulfur source, often nickel bearing.
Quartz	SiO_2	46.7 Si; 53.3 O	Gemstone and building material, porcelain, glass, paint, mortar industry and acid flux in smelting furnaces.
Rutile	TiO_2	60.0 Ti; 40.0 O	Source of titanium.
Scheelite	$CaWO_4$	80.6 WO_3; 19.4 CaO	Electric bulb, alloy, cutting material, defense.

TABLE 2.2 List of Common Metallic and Nonmetallic OFM and Uses—cont'd

Principal ore mineral	Mineral formula	% Content	Major uses
Silver native	Ag	100.0 Ag	Jewelry, electrical-electronics, photography, dentistry.
Skutterudite	$(Co,Ni,Fe)As_3$		High-temperature alloy, tool steel.
Smithsonite	$ZnCO_3$	64.8 ZnO; 35.2 CO_2	Same as sphalerite.
Sperrylite	$PtAs_2$	57.0 Pt	Jewelry, catalyst, electrical.
Sphalerite	ZnS	67.0 Zn; 33.0 S	Galvanizing, alloys, cosmetics, pharmaceutical, micronutrient for human, animals and plants.
Stannite	$Cu_2S.FeS.SnS_2$	27.5 Sn; 29.5 Cu	Source of tin and copper.
Stibnite	Sb_2S_3	71.8 Sb; 28.2 S	Textile, fiber, alloy with lead.
Sulfur nat.	S	100.0 S	Sulfuric acid, fertilizer.
Sylvite	KCl	52.4 K; 47.6 Cl	Source of potash as fertilizers.
Sylvanite	$(AuAg)Te_2$	24.5 Au; 13.4 Ag; 62.1 Te	Source of gold, silver, tellurium.
Talc	$3MgO,4SiO_2H_2O$	31.7 MgO; 63.5 SiO_2	Cosmetics, paint, plastic, paper, rubber, ceramics, pharmaceutical, detergents.
Uraninite	UO_3	88.0 U	Nuclear fuel, military.
Wolframite	$(Fe.Mn)WO_4$	76.0 W	Electric bulb, alloy, cutting material, defense.
Wollastonite	$CaSiO_3$	48.3 CaO; 51.7 SiO_2	Principal ingredient in ceramics industry, paint, paper, polymers and metallurgical applications.
Zircon	ZrSiO4	67.2 ZrO_2; 32.8 SiO_2	Alloy in nuclear reactors, gems and radiometric dating.
Dolomite*	$CaMg(CO_3)_2$	21.7 Ca; 13.2 Mg	Building stones, refractory bricks, cement, ornamental stone, ore of metallic magnesium, glassmaking.
Limestone*	$CaCO_3$	—	Cement industry, flux in steel purifying blast furnace.
Marble*	$CaCO_3$	<56 CaO	Building and decorative stone.
Silica sand*	SiO_2	<100 SiO_2	Building, glass manufacture
Rock phosphate*	$3Ca_3(PO_4)_2.CaR_2$	15-35 P_2O_5	Fertilizer, phosphoric acid.

**Dolomite, limestone, marble and silica sand are rocks/aggregates with variable composition. These deposits are mined for specific richness of elements of economic value. Similarly, rock phosphate is a natural lithified rock and mined for phosphorus. It has no definite chemical composition and is formed due to bacterial accumulation of phosphorus and calcium along with other chemical sediments in the seafloor.*

2.3.1.3. District

"District" is comprised of one geographical area popularly known for occurrence of particular mineral e.g. Aeolian soils of Blayney District, NSW, Australia, Baguio mineral district in Philippines for copper deposits, New Mexico for uranium deposits, Singhbhum district for copper and Salem district for magnesite, India.

2.3.1.4. Belt

"Belt" is a narrow linear stretch of land having series of deposits of associated minerals, such as, Colorado gold-molybdenum belt, US, Grant uranium mineral Belt, New Mexico Khetri copper belt, Rajpura-Dariba-Bethumni zinc-lead-silver belt, Rajasthan, Sukinda chromite belt, Orissa, India.

2.3.1.5. Deposit

"Deposit" is comprised of a single or a group of mineral occurrences of sufficient size and grade separated by natural narrow barren parting e.g. Broken Hill group of zinc-lead deposits, Australia, Zawar group of zinc-lead deposits, India, Red Dog zinc-lead deposit, Alaska, OK Tedi copper deposit, Papua New Guinea, Olympic Dam copper-gold-uranium-silver deposit, South Australia, Neves Corvo poly-metallic deposit, Portugal and Stillwater group of platinum deposit, US.

2.3.1.6. Block

"Block" is a well-defined area having mineral concentration wholly or partly of economic value, such as Broken Hill main, Australia, Bailadila deposit-14, Central Mochia, India. The blocks in underground mining are subdivided to "Level" (say: upper level, lower level, 500-700 and 300-500 mRL). The levels are further split into "Stope" (say: West 301 stope, North 101 stope, Valley stope). These terms are locally convenient to use for attention and allocation of work activities in mineral exploration and sequencing mine production block.

2.3.2. Depth of Occurrence

2.3.2.1. Exposed to Surface

Mineral deposits like iron ore, bauxite, chromite, copper, limestone and magnesite are exposed to the surface and easy to explore. Although most of the significant exposed ore deposits, namely, zinc-lead-silver Rampura-Agucha, India, Red Dog, Alaska, OK Tedi copper-gold, Papua New Guinea and Olympic Dam copper-gold-uranium-silver, Australia, have been discovered and being exploited, there is ample possibility of finding new deposits under glacial or forest cover. The prospecting efforts should get emphasis on looking for fresh rock exposure and newly derived boulders. The examples are Adi Nefas Zn-Cu-Au-Ag deposit, Madagascar, El Abra Cu deposit, Chile, and chromite deposits in Orissa, Tamil Nadu, India (Fig. 2.1).

2.3.2.2. Shallow Depth

Deposits like base metals, coal and gypsum are covered by altered oxidized capping or exist at shallow depth or under thick overburden of bedrock. The deposits are Cerro de Maimon copper-gold deposit at Dominican Republic, Zawar zinc-lead-silver, Ranigange coalfield, West Bengal, and gypsum deposits, Rajasthan, India. Geochemical prospecting and ground geophysical survey will be helpful for discovery of deposits at shallow depth.

FIGURE 2.1 Massive chromite orebody exposed to surface near Karungalpatti village at Sitampundi belt, Namakkal District, Tamil Nadu, India (Finn Barrett, Goldstream Mining NL, Australia, during Reconnaissance).

2.3.2.3. Deep-Seated Hidden Deposit

The hidden poly-metallic deposits discovered in the past are Neves Corvo copper-zinc-tin, Portugal, at 330-1000 m depth (Fig. 15.9), Portugal and Sindesar Khurd zinc-lead-silver at 130 m depth (Fig. 15.8), India. Near surface deposits are mostly discovered. Deep-seated hidden deposits will be the future target of mineral exploration. The key exploration procedures suitable for discovery of an orebody at a depth range of 300-700 m require clear understanding of regional structure, applications of high penetrative geophysical methods and interpretation by simulation tools to identify, describe and delineate. Exploration for such deposits is expensive and associated with considerable economic risk. The high costs result from the necessity of expensive instrumentation and extensive drilling at depth.

2.3.3. Relation to Host Rocks

2.3.3.1. Host Rocks

There are three types of rocks that host the mineralization namely igneous, sedimentary and metamorphic. The examples of igneous rocks are porphyry copper deposits in granite, platinum-palladium-chromium-nickel deposits in dunite, peridotite, gabbro, norite and anorthosite, tantalite, columbite, cassiterite in pegmatite. Ore deposits can exclusively be formed under sedimentation process like iron ore as Banded Iron Formation (BIF), and Banded Hematite Quartzite (BHQ), zinc-lead deposits in dolomite, copper-gold in quartzite, diamond in conglomerate and limestones. The deposits show bedded, stratabound and often stratiform features having concordant relation with country rocks. Metamorphic rocks host important ore deposits generated as contact metamorphic aureoles. The ore deposits are garnet, wollastonite, andalusite, and graphite. The metamorphic equivalent of sedimentary and igneous rocks forms large deposits of marble, quartzite, and gneisses and commonly used as building stones and construction materials.

2.3.3.2. Identical with Host

Mineral deposits like granite, limestone, marble, quartzite, and slate are indistinguishable with the host rock. The mineral deposit represents the host rock. The examples are Keshariyaji green marble, Mekrana white marble, Jaisalmer limestone, Rajasthan, India.

2.3.3.3. *Different from Host*

Gold-bearing quartz veins act as an exclusive host for Au and different from the surrounding rocks such as Kolar gold deposit, Karnataka, India.

2.3.3.4. *Gradational Contact*

Gradational deposits are often formed around the vein systems with characteristics disseminated mineral distribution. Bulldog Mountain vein systems, Colorado, show abundance of fine-grained sphalerite and galena, with lesser tetrahedrite and minor chlorite and hematite. The mineralization becomes progressively richer of barite and silver with increasing elevation. Some mineral deposits, particularly those containing disseminated Cu, depict gradational contact to form an economic deposit. Sargipalli lead-copper deposit, Orissa, India, is an example.

2.3.3.5. *Metal Zoning*

The metal zoning occurs in a multiple series of hydrothermal depositional source. Mineralization zoning is characterized by Fe-Ba-Cu-Pb-Ag-Au. This is obviously a gradational transition of mineralization from vent-proximal mineralization to more distal mineralization. Metal zoning is an indication of metal deposition in relative order during primary crystallization or sedimentation. It may be modified by deformation and remobilization at later stage. The metal zoning can be within a single orebody and between orebodies occurring in a group. The common metal zoning is in massive sulfide deposits: Cu → Zn/Pb → Pb/Zn → Fe or alternate rich → poor → rich bands, e.g. El Guanaco gold-copper in Chile, Zn-Cu-Au-Ag deposits of Scuddles, Golden Grove, Gossan Hill, Western Australia and Rajpura-Dariba Zn-Pb-Cu-Ag and Zawar Zn-Pb-Ag deposits, India.

2.3.3.6. *Wall Rock Alteration*

The mineral deposits formed under epigenetic condition, magmatic intrusion and hydrothermal depositional environments cause changes in mineralogy including formation of new minerals, chemical composition, color and texture of the host rock at the contacts and some distance from the orebody. This alteration halo is known as "alteration zone" or "zone of wall rock alteration". The size of the alteration halo around the orebody varies from narrow to wider depending on the physical and chemical condition of the process of alteration. The most common form of wall rock alterations are silicification, chloritization, sericitization and serpentinization. Presence of pyrite, siderite, titanium, manganese, potassium, lithium, lead, silver, arsenic, rubidium, barium, calcium, epidote and carbonaceous material is common and characteristic features enveloping most of the sedimentary exhalative (SEDEX) type of copper-zinc-lead ± silver-gold deposits in the world. If these alteration halos are identified properly it adds considerable value to the mineral exploration in general and particularly for planning drilling targets. The good examples of ore deposits with alteration halo are Broken Hill, Mount Isa, Hilton, Century, HYC and Lady Loretta in Australia, Sullivan in Canada, and Rampura-Agucha, Rajpura-Dariba and Khetri in India

2.3.4. Structural Control

Structure, tectonics and surface weathering play a vital role over geological time as a passage for hydrothermal flow of mineralized fluids, accumulate and concentrate at suitable location, remobilize and reorientate as postgenetic activity. The features related to mineralization control are deformation, weathering, joints, fractures, folds, faults, breccias and plate tectonics.

2.3.4.1. *Un-Deformed*

Most of the residual and placer deposits are un-deformed type such as East Coast bauxite deposit, India.

2.3.4.2. *Joints and Fractures*

Many deposits show varied degree of deformation, contemporaneous to formation or aftereffect. "Joints" and "fractures", caused by regional stress, break in the rock along which little or no movement have occurred. Mineralization often concentrates along these regular and irregular planes. Magnesite accumulation can be seen along road cutting near Salem town, Tamil Nadu, India (Fig. 2.2). Lennard Shelf zinc-lead deposit, Western Australia, is an example of cavity filled along major fault zone.

FIGURE 2.2 Magnesite veins deposited along joints, faults and fractures in ultramafic host rocks, Salem Road, Tamil Nadu, India.

FIGURE 2.3 Stratiform pyrite-zinc-lead mineralization folded with mineral concentration at crests presenting saddle reef structure, Rajpura-Dariba deposit, India.

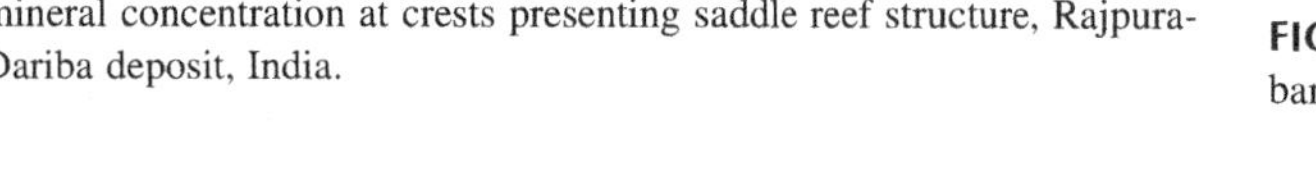

2.3.4.3. Fold

Directed compression of the crust, resulting in a semi-plastic deformation, creates "folding" of strata ("fold"). The fold closure, limb in-flex zone and axial planes are suitable for mineral localization. Mineral deposits are often folded during or after formation e.g. Rajpura-Dariba zinc-lead-copper deposit (Fig. 2.3), Agnigundala lead-copper deposit, Sukinda chromite belt, India.

2.3.4.4. Fault

Joints and fractures along which noticeable movements have occurred are called "fault" (Fig. 2.4). Deposits can be faulted with displacement from millimeters to kilometers thus creating challenges for exploration. Fault zones are favorable avenue and localization of mineralized solution for movement and concentration. Manto Verde Cu in Chile and many of the coal deposits are faulted.

2.3.4.5. Shear Zone

"Shear" is the outcome of rock deformation generating particular texture like intense foliation, deformation and micro folding due to compressive stress. A "shear zone" is a wide zone of distributed shearing in crushed rock mass with width varying between few centimeters and several kilometers. The interconnected openings of shear zone serve as an excellent channel ways for mineral-bearing solutions and subsequent formation of deposits. Many shear zones in orogenic belts host ore deposits. Shingbhum shear zone hosts copper-uranium mineralization. Fig. 2.5 shows chromite-magnesite veins developed in shear zone, Sinduvally, Karnataka, India.

FIGURE 2.4 Massive chromite lode depicting sharp faulted contact with barren ultramafic rock, Kathpal underground mine, Sukinda belt, India.

2.3.4.6. Breccia

"Breccia" is commonly used for clastic sedimentary rocks composed of large sharp-angled fragments embedded in fine-grained matrix of smaller particles or mineral cement. Breccia generated by folding, faulting, magmatic intrusions and similar forces are called "tectonic breccia". Tectonic breccia zones are represented by crush, rubble, crackle and shatter rock mass. Breccia and conglomerate are similar rocks with difference in shape of the larger particles due to transportation mechanism. "Igneous", "Flow" or "Pyroclastic" breccias are rocks composed of angular fragments of preexisting igneous rocks of pyroclastic debris ejected by volcanic blast or pyroclastic flow. An outstanding example would be intrusion of gabbroic magma within the preexisting ultramafic rocks hosting layered chromite at Nausahi, India. The sharp-angled fragments of Cr in host rock formed the angular fragments embedded in a matrix of fine-grained gabbro containing PGE (Fig. 2.6).

FIGURE 2.5 Layered chromite (black) and magnesite (white) veins developed in shear zones, Sinduvally, Karnataka, India.

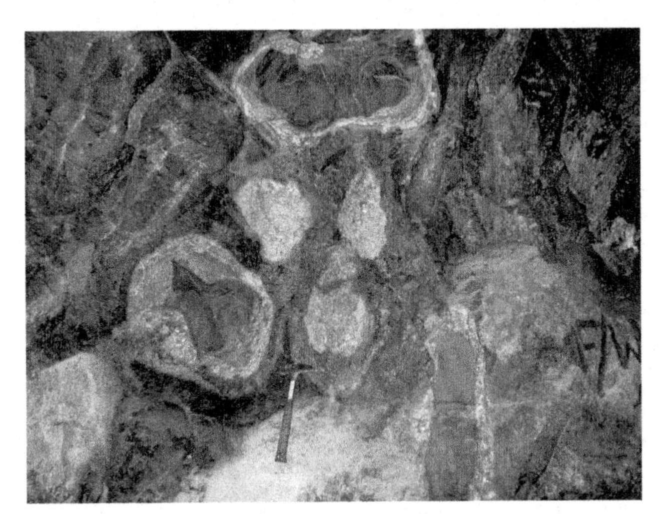

FIGURE 2.6 Irregular fragmented chromite (black with white rims) in matrix of Pt-Pd-bearing gabbro from the tectonic breccia zone, Boula-Nausahi underground mine, Orissa, India.

Zinc-copper-gold deposits of Saudi Arabia are hosted in volcanic-clastic breccia. Fossil Downs Zn-rich ore, Lennard shelf deposit, Western Australia, is closely related to the major N-S trending fault, brecciated cavity filled in limestone reefs.

2.3.4.7. Subduction

"Subduction" is the process of two converging tectonic plate movement. The plates of Continental Margin Arcs, Oceanic Lithosphere and Volcanic Island Arcs collide and one slides under the other. In the process the heavier Oceanic Crust stoops under the lighter Continental Crust or the Volcanic Island Arc forming a "Subduction Zone". The formation of Subduction Zone is closely associated with multidimensional tectonic activities like shallow and deep focuses Earthquakes, melting of mantle, volcanism, rising magma resulting volcanic arc, plutonic rocks of ophiolite suites, platinum-chromium-bearing peridotite-dunite-gabbro-norite, movement of metal-bearing hydrothermal solution and metamorphic dewatering of crust. The great belt of porphyry copper-gold that extends north from central Chile into Peru is a good example associated with the subduction of the Pacific Ocean floor beneath the South American plate. The main Chilean porphyry copper belt hosts some of the largest open-cut copper mines in the world.

2.3.5. Nature of Mineralization

2.3.5.1. Dissemination

"Disseminated" types of mineralization are formed by crystallization of deep-seated magma. The early-formed in situ valuable metallic and nonmetallic minerals are sparsely disseminated or scattered as fine grains throughout or part of the host rock. Good examples are diamonds in kimberlite pipes in South Africa, porphyry copper deposits at El Salvador, Chile, porphyry tungsten-molybdenum deposit at Yukon, Malanjkhond copper and Sargipalli lead-copper deposit, India.

2.3.5.2. Massive

"Massive" deposits with more than 60% sulfides [volcanogenic massive sulfide (VMS), volcanic-hosted massive sulfide (VHMS) or SEDEX] are formed due to accumulations on or near the seafloor in association with volcanic activity or hydrothermal emanations along with sedimentary deposition. Zinc-lead-silver deposit of Red Dog, Northwest Alaska, Neves Corvo, Portugal, and Gorubathan, India. Fig. 2.7 shows massive chromite deposits hosted by ultramafic complex with sharp contact.

2.3.5.3. Veins and Stringers

"Veins", "fissure-veins" and "lodes" are tabular deposit usually formed by deposition of ore and gangue minerals in open spaces within a fault, shear and fracture zones. Veins often have great lateral and/or depth extent but which are usually of narrow width that portray veins and stringers. Veins frequently pinch and swell out in all directions. The pinch and swell structure type of deposits pose problem both during exploration and mining. Proper delineation of orebody, dilution control and planning for large-scale mining are difficult. There are several examples such as poly-metallic deposits of Silvania, Silver Cup, Lucky Jim, Highland Lass Bell in British Columbia, Zawar zinc-lead-silver (Fig. 2.8), Kolihan copper deposit, chromite-magnesite deposit at Sinduvally, India.

FIGURE 2.7 Massive chromite ore, Tamil Nadu, India.

FIGURE 2.8 Sheeted veins and fine stringers of sphalerite in dolomite host rock at Zawar deposit, India.

FIGURE 2.10 Quartz-filled ladder vein structure in compact dolomite mass at the footwall of main lodes within graywacke rocks, Zawar group of mines, India.

"Stringers" are large numbers of thin, tiny and closely spaced mineralized veins originating from the main orebody and often described as "Stringer Zone" (Fig. 2. 9).

2.3.5.4. Ladder Veins

"Ladder veins" are regularly spaced, short and transverse fractures confined wall to wall within dikes or compact rock mass (Fig. 2.10). The fractures are nearly parallel to each other and occur for considerable distance along the host dike or rock. The fractures are generally formed by contraction joints and filled with auriferous quartz or valuable mineral matter to form an economic deposit. The examples of commercial ladder vein-type deposits are Morning Star gold mine in Victoria, molybdenite veins in New South Wales, Australia, and copper ladder veins in Norway.

FIGURE 2.9 Stratification and stringers of sphalerite, galena and pyrite hosted by carbonaceous calc-silicate rock at Sindesar Khurd orebody, India.

2.3.5.5. Stock Work

The "stock work" styles of metalliferous deposits are characterized by a large mass of rock impregnated by dense interlacing network of variously oriented irregular ore-bearing veins and grouped vein-lets. Stock works are formed by group of hydrothermal systems of metal-bearing substance from hot mineralized solutions circulating through the fissured rock or deposited in the basin. The veins contain metallic minerals. The stock work style of mineralization commonly occurs in porphyritic plutonic igneous intrusions. These kinds of deposits are especially common with platinum-bearing sulfides, copper, gold, molybdenum, tin, tungsten, beryllium, uranium, mercury and other metal ore. The stock work mineralization may occur as separate body or in association with other style. The ore is mined as chambers or stories when the stock work style of mineralization occurs exclusively as solid massive form outside the host strata or veins. The examples of stock work are disseminated gold-bearing Trinity Mine, Nevada, copper- and tin-rich stock work at Neves Corvo mine, Portugal, platinum-palladium-chromite mines at Boula-Nausali (Fig. 2.11), India.

2.3.6. Morphology

2.3.6.1. Stratiform

Hydrothermal, volcanogenic and "SEDEX"-type mineralization closely resembles stratification of sedimentary formation. Stratification is formed by upward moving metal-bearing hydrothermal solution through a porous

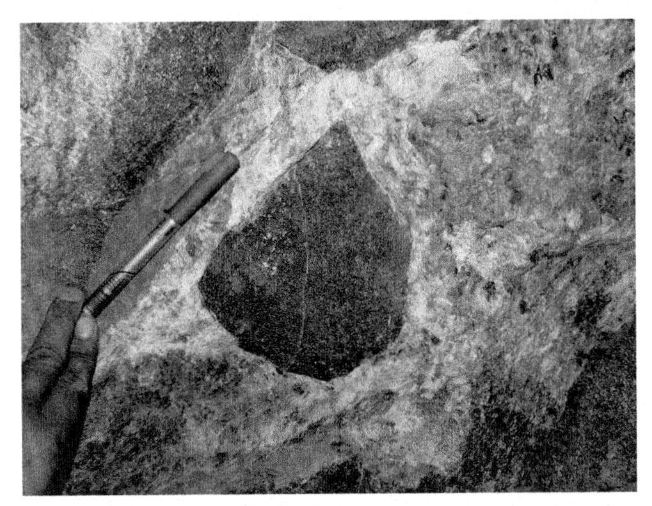

FIGURE 2.11 Stock work of Pt-Pd-bearing magma (greasy white color gabbro) around sharp-angle chromite (black) from the tectonic breccia zone, Boula-Nausahi underground mine, Orissa, India.

aquifer and deposits ore minerals in the overlying pile of sedimentary strata of shale and dolomite. These deposits may contain a significant amount of organic matter and fine pyrite. There are several worlds' largest and most famous stratiform base metal deposits: they are copper deposit at White Pine, Michigan, copper deposits of Zambia, lead-zinc-copper deposits at Sullivan in British Columbia and Rajpura-Dariba in India (Fig. 2.12), and lead-zinc deposits of Broken Hills in New South Wales, Mount Isa in Queensland and McArthur River in Northern Territory, Australia.

2.3.6.2. Stratabound

Ore minerals in "stratabound" deposits are exclusively confined within a single specific stratigraphic unit. Stratabound deposits will include various orientation of mineralization representing layers, rhythmic, stratiform, vein-lets, stringers, disseminated and alteration zones, strictly contained, within the stratigraphic unit, but that may or may not be conformable with bedding. There are several world-class stratabound zinc-lead-silver deposits: they are Proterozoic Mt Isa-McArthur Basin System of Northern Territory, Australia (Mt Isa, George Fisher, Hilton, Lady Loretta, Century, and McArthur River), and Proterozoic Middle Aravalli System in India (Zawar, Rajpura and Rampura-Agucha).

2.3.6.3. Layered, Rhythmic and Bedded

"Layered", "rhythmic" and "bedded" types of deposits are formed generally by deposition and consolidation of sediments that may or may not be metamorphosed. The type of ore deposit will depend on the composition of the transported sediments. The deposits showing these features are iron ore (BHQ/BIF), lignite, and coal (Fig. 2.13).

The layered and rhythmic features are also developed during the differential crystallization and segregation of the mafic and ultramafic magma in a huge chamber over a prolonged time. The early crystallization, settling and consolidation of heavy metal-rich layers are composed of Cr-Ni-Cu-Pt-Pd and disseminated sulfides $\pm$ Au and Ag forming economic mineral deposits. The late crystallization and solidification of residual magma form alternate layers of dunite, peridotite, gabbro and anorthosite. The process repeats with addition of fresh magmatic cycles. The examples are Bushveld platinum-chromite deposits, South Africa, Sittampundi Cr-Pt-Pd (Fig. 2.14), Sukinda Cr-Ni and Nausahi Cr-Pt-Pd, India.

2.3.6.4. Porphyry

"Porphyry" is a diversity of igneous rock consisting of large-grained crystal such as quartz and feldspar scattered in a fine-grained groundmass. The groundmass is

FIGURE 2.12 Stratiform sphalerite (honey brown) mineralization in calc-silicate (bluish gray) host rock at Rajpura-Dariba deposit, India.

FIGURE 2.13 Alternate bands of coal (shining black) and shale (brownish gray), Belatan mine, Jharia coalfield, India.

FIGURE 2.14 Layers of chromite (black) and Pt-Pd-bearing gabbro (white), Sittampundi Igneous Complex, Tamil Nadu, India.

composed of indistinguishable crystals (aphanites as in basalt) or easily distinguishable crystals (phanerites as in granite). Porphyritic refers to the texture of the rocks and suffix as granite-, rhyolite-, and basalt-porphyry. The porphyry deposits are formed by differentiation and cooling of a column of rising magma in stages. The different stages of cooling create porphyritic textures in intrusive as well as in sub-volcanic rocks. In the process it leads to a separation of dissolved metals into distinct zones and responsible for forming rich deposits of copper, molybdenum, gold, tin, zinc and lead in the intrusive rock itself. There are several large porphyry copper deposits in the world: Chuquicamata (690 Mt at 2.58% Cu), Dexing Cu-Au-Ag Distirct, South China, Escondida and El Salvador, Chile, Toquepala, Peru, Lavender pit, Arizona and Malanjkhand, India.

2.3.6.5. *Lenticular*

Magmatic segregation deposits are formed by fracture filling within the host rock or contained within them. They are generally irregular, roughly spherical, more often "tabular" or "lenticular" in shape. The width and thickness ranges between few centimeters and few meters. The length may exceed kilometers. Examples are Sukinda chromite deposits in dunite-peridotites and Balaria zinc-lead-silver deposit in dolomite, India

2.3.6.6. *Pipe*

"Pipe"-like deposits are relatively small in horizontal dimension and extensively large in vertical direction. These "pipes" and "chimneys" are orientated in vertical to sub-vertical position. Pipes may be formed by infillings of mineralized breccias in volcanic pipes e.g. copper-bearing breccia pipes of Messina, South Africa. Another common type of volcanic pipes is a deep narrow cone of solidified intrusive magma characteristically represented by Kimberlite or lamproite. Kimberlite is high in magnesium, carbon dioxide and water. Kimberlite is the primary source of diamond, precious gemstone and semiprecious stones. The best example is diamond pipe at Kimberly, South Africa, and Panna, India.

2.3.7. Genetic Model

2.3.7.1. *Magmatic*

"Magmatic" deposits are genetically linked with the evolution of magma that emplaced into the Continental or Ocean Crust. The mineralization is located within the rock types derived from the differential crystallization of the parent magma. The most significant magmatic deposits are related to mafic (gabbro, norite) and ultramafic (peridotite, dunite) rocks originated from the crystallization of basaltic and ultramafic magma. Ore minerals are formed by the separation of metal sulfides and oxides in molten form within an igneous melt before crystallization. The deposit types include chromite, nickel-copper and platinum group of elements. There are several largest magmatic deposits: they are Cr-PGE deposits at Bushveld Igneous Complex, South Africa, Ni-Cu-PGE deposits at The Great Dykes, Zimbabwe, Ni-PGE-Cr deposits at Sudbury, Canada, Ni-Cu-PGE deposits at Stillwater Igneous Complex, Montana, US and Cr-Ni $\pm$ PGE deposits at Sukinda and Nausahi, India.

2.3.7.2. *Sedimentary*

"Sedimentary" rocks are formed by the process of deposition and consolidation of loose materials under aqueous condition. The sedimentary deposits are concordant and may be integral part of stratigraphic sequence. It is formed due to seasonal concentration of heavy minerals like hematite on the seafloor. The structures consist of repeated thin layers of iron oxides, hematite or magnetite, alternating with bands of iron-poor shale and chert. The examples are Pilbara BIF, Northwestern Australia, Bailadila and Goa iron ore, India. Similarly economic limestone deposits are formed by chemical sedimentation of calcium-magnesium carbonate on the seafloor. Coal and lignite are formed under sedimentary depositional condition.

Evaporite deposits form through the evaporation of saline water in lakes and sea, in regions of low rainfall and high temperature. The common evaporite deposits are salts (halite and sylvite), gypsum, borax and nitrates. The original character of most evaporite deposits is destroyed by replacement through circulating fluids.

2.3.7.3. *Metamorphic*

"Metamorphic" rocks are transformed alteration product of preexisting igneous or sedimentary materials. The reconstruction is formed under increasing pressure and temperature caused by igneous intrusive body or by tectonic events. Metamorphic mineral deposits are formed due to regional prograde or retrograde metamorphic process and hosted by metamorphic rocks. Minerals like garnet, kyanite, sillimanite, wollastonite, graphite and andalusite are end product of metamorphic process. Copper deposits of Kennicott, Alaska and White Pine, Michigan are formed by low-grade metamorphism of organic-rich sediments resting over mafic or ultramafic rocks. The low copper values of underlying source rocks liberate during a leaching process caused by passing of low-temperature hydrothermal fluids. The fluids migrate upward along the fractures and faults and precipitate high-grade copper in the rocks containing organic matter.

2.3.7.4. Volcanogenic Massive Sulfide and Volcanic-Hosted Massive Sulfide

VMS and VHMS type of ore deposits contribute significant source of Cu-Zn-Pb sulfide ± Au and Ag, formed as a result of volcanic associated hydrothermal events under submarine environments at or near the seafloor. It forms in close time and space association between submarine volcanism, hydrothermal circulation and exhalation of sulfides, independent of sedimentary process. The deposits are predominantly stratabound (volcanic derived or volcano-sedimentary rocks) and often stratiform in nature. The ore formation system is synonymous to black smoker type of deposit. Kidd Creek, Timmins, Canada, is the largest VMS deposit (35 Mt @ 2.2% Cu, 5.3% Zn, 0.22% Pb and 60 g/t At at 2005) in the world. Kidd is also the deepest (+1000 m) base metal mine. The other notable VMS/ VHMS deposits are IPB of Spain and Portugal, Wolverine Zn-Cu-Pb-Ag-Au deposit, Canada and Khnaiguiyah Zn-Pb-Cu, Saudi Arabia.

2.3.7.5. Black Smokers Pipe Type

"Black smokers" pipe-type deposits are formed on the tectonically and volcanically active modern ocean floor by superheated hydrothermal water ejected from below the crust. The water with high concentrations of dissolved metal sulfides (Cu, Zn, Pb) from the crust precipitates to form black chimney-like massive sulfide ore deposits around each vent and fissure when it comes in contact with cold ocean water over time. The formation of black smokers by sulfurous plumes is synonymous with VMS or VHMS deposits of Kidd Creak, Canada, formed 2400 million years ago on ancient seafloor.

2.3.7.6. Mississippi Valley Type

The "Mississippi valley type" (MVT) deposits are epigenetic, stratabound, rhythmically banded ore with replacement of primary sedimentary features predominantly carbonate (limestone, marl, dolomite and rarely sandstone) host rocks. The mineralization is hosted in open space filling, collapse breccias, faults and hydrothermal cavities. The deposits formed by diagenetic recrystallization of carbonates creating low-temperature hydrothermal solution that migrates to suitable stratigraphic traps like fold hinge and faults at the continental margin and intra-cratonic basin setting. The OFM are predominantly sphalerite, galena and barite. Calcite is the most common gangue mineral. Low pyrite content supports clean concentrate with high metal recovery of +95%. Some deposits are surrounded by pyrite/ marcasite halo. Prospects can be defined by regional stream sediment, soil and gossan sample anomaly supported by aeromagnetic and gravity survey. There are numerous Zn-Pb-Ag sulfide deposits along the Mississippi river in US, Pine Point, Canada, San Vicente, Central Peru, Silesia, Southern Poland, Polaris, British Columbia, Lennard Shelf (Fig. 2.15) and Admiral Bay, Western Australia.

2.3.7.7. SEDEX

SEDEX ore deposits are formed due to concurrent release of ore-bearing hydrothermal fluids into aqueous reservoir mainly ocean, resulting in the precipitation of stratiform zinc-lead sulfide ore in a marine basin environment. The stratification may be obscured due to postdepositional deformation and remobilization. The source of metals and mineralizing solutions are deep-seated superheated formational brines migrated through intra-cratonic rift basin faults which come in contact with sedimentation process. In contrast the sulfide deposits are more intimately associated with intrusive or metamorphic process or trapped within a rock matrix and not exhalative. The formation occurred mainly during Mid-Proterozoic period. SEDEX deposits are the most important source of zinc, lead, barite and copper with associated by-products of silver, gold, bismuth and tungsten. The examples are zinc-lead-silver deposits of Red Dog, Northwest Alaska, McArthur River, Mt Isa, HYC, Australia, Sullivan, British Columbia, Rampura-Agucha, Rajpura-Dariba (Fig. 2.16), India, and Zambian copper belt.

2.3.7.8. Skarn

The "skarn"-type deposits are formed in the similar process of porphyry orebodies. Skarn deposits are developed due to

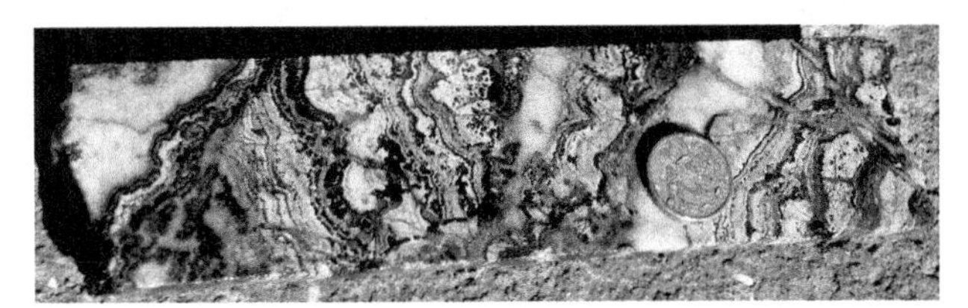

FIGURE 2.15 Sphalerite (yellow) and galena (black) mineralization in calcite (white) bands indicating different fluid phase events, Lennard Shelf MVT deposit, Western Australia.

FIGURE 2.16 Massive sphalerite (iron brown) and galena (shining) in carbonaceous calc-silicate host rock sedimentary exhalative deposition at Sindesar Khurd SEDEX type in India.

replacement, alteration and contact metasomatism of the surrounding country rocks by ore-bearing hydrothermal solution adjacent to a mafic, ultramafic, felsic or granitic intrusive body. It is most often developed at the contact of intrusive plutons and carbonate country rocks. The latter are converted to marbles, calc-silicate hornfels by contact metamorphic effects. The mineralization can occur in mafic volcanics and ultramafic flows or other intrusive rocks. There are many significant world-class economic skarn deposits: they are Pine Creek tungsten, California, Twin Buttes copper, Arizona and Bingham Canyon copper, Utah, US, OK Tedi gold-copper, Papua New Guinea, Avebury Nickel, Tasmania and Tosam tin-copper, India (Reconnaissance stage).

2.3.7.9. Residual

"Residual" deposits are formed by chemical weathering process like leaching which removes gangue minerals from protore and enrich valuable metals in situ or nearby location. The most important example is formation of bauxite under tropical climate where abundance of high temperature and high rainfall during chemical weathering of granitic rocks produces highly leached cover rich in aluminum. Examples are bauxite deposit of Weipa, Gove Peninsula, Darling Range and Mitchel Plateau in Australia, Awaso and Kibi, Ghana, East Coast, India, Eyre Peninsula Kaolin deposit, Australia. Basic and ultrabasic rocks tend to form laterites rich in iron and nickel respectively. Nickel-bearing laterites, may or may not be associated with platinum group of elements, are mined at New Caledonia, Norseman-Wiluna greenstone belt of Western Australia and Central Africa, Ni-bearing limonite overburden at Sukinda, India. The other residual-type deposits are auriferous laterites in greenstone belts (Western Australia), Ni-Co and Cr in laterites on top of peridotites (New Caledonia and Western Australia respectively), and Ti in soils on top of alkali igneous rocks (Parana Basin, Brazil).

2.3.7.10. Placer

"Placer" deposits are formed by surface weathering and ocean, river or wind action resulting in concentration of some valuable heavy resistant minerals of economic quantities. The placer can be an accumulation of valuable minerals formed by gravity separation during sedimentary processes. The type of placer deposits are namely, alluvial (transported by a river), colluvial (transported by gravity action), eluvial (material still at or near its point of formation), beach placers (coarse sand deposited along the edge of large water bodies) and paleo-placers (ancient buried and converted rock from an original loose mass of sediment). The most common placer deposits are those of gold, platinum group minerals, gemstones, pyrite, magnetite, cassiterite, wolframite, rutile, monazite and zircon. The California gold rush in 1849 began when someone discovered rich placer deposits of gold in streams draining the Sierra Nevada Mountains. Recently formed marine placer deposits of rutile, monazite, ilmenite and zircon are currently being exploited along the coast of eastern Australia, India and Indonesia.

2.3.8. Contained Metal

The deposits can be classified on the basis of concentration of mineral or metal grade.

2.3.8.1. High Grade

High-grade deposits are economically the most encouraging type for the mineral industry such as Red Dog (22% Zn + Pb), Alaska, Sullivan (12% Zn + Pb), Canada, Lady Loretta, (27% Zn + Pb), HYC (20% Zn + Pb), Broken Hill (15% Zn + Pb), Mt Isa (13% Zn + Pb), Australia, Rampura-Agucha (15% Zn + Pb), India.

2.3.8.2. Medium Grade

Medium-grade deposits are also equally important as source of metal, such as, Lennard Shelf (9.5% Zn + Pb), Australia, San Felipe (10% Zn + Pb), Mexico, Rajpura-Dariba (10% Zn + Pb), Sindesar Khurd (10% Zn + Pb), India.

2.3.8.3. Low Grade

Large low-grade deposits are exploited on account of available existing infrastructure e.g. Bou Jabeur (5.6% Zn + Pb), Tunisia, Scotia (5% Zn + Pb), Canada, Hambok (3% Zn + Cu), Eritrea, Zawar Group (5% Zn + Pb), India. These deposits are workable with high mechanization for large production, improved process recovery with existing infrastructures.

2.3.8.4. Very Low Grade

Very low-grade deposits like Suplja Stijena (2% Zn + Pb), Gradir, Pering (1.4% Zn + Pb), South Africa, Sindeswar-Kalan East (2.5% Zn + Pb), India, are explored and kept in abeyance as future resource with technology upgradation in low-cost mining and mineral beneficiation.

2.4. HOST ROCKS

Mineral deposits are integral part of the parent rock bodies formed under certain physicochemical processes at definite time and space. The mineral bodies, more specifically orebodies, are concentration of particular mineral or metal or group, which is techno-economically exploitable from host rock mass. Therefore, an exploration geologist must possess adequate knowledge of the favorable stratigraphy, structure and rock association of the region to design the exploration program. For example, if one looks for coal his focus of prospecting will be traversing rock formation of Gondwana age. Similarly if the search is for platinum group of elements the attention should be focused on layered mafic and ultramafic rocks with associated trace elements like nickel, copper, and chromium. If one is interested for SEDEX-type zinc-lead-silver ore, then rocks like dolomite, carbonaceous black schist of Proterozoic age should be considered as that host most of such deposits in the world. The OFM, their broad affinity to host rocks, associated elements and type of deposit are given in Table 2.3.

2.5. INDUSTRY SPECIFICATIONS

The mineral sectors, in general, work on standard industrial specifications. If required the raw material is processed for market-finish commodity. Some minerals can directly be sold involving negligible processing such as quartz, feldspar and limestone. The others may require processing through few steps with intermediate salable goods. Zinc (4-10%), lead (1-2%), and copper (0.5-2.0%) ore at run-of-mine grade can either be transferred to in-house beneficiation plant as a separate profit center or sold to a third party process plant to produce respective concentrate (refer Chapter 12 on Mineral Processing). The average concentrate grades are +50% for zinc and lead, and +20% for

TABLE 2.3 Host Rock of Common Economic Minerals and Type of Deposits

Element	Host rock	Associated elements	Deposit type
Ag	Dolomite, carbonaceous schist, bright white quartz, ultramafics	Pb, Cu, Py, Cd, Ba	Base/precious metal
Al	Khondalite, laterite, Carbonates	Fe, Mn, SiO_2	Bauxite
As	Schist, graywacke	Au, Fe	Base metal
Au	Quartz reefs and veins, Archaean auriferous Greenstone Belt	Ag, Cu, As, Sb	Gold
C	Schist, khondalite		Graphite
C	Kimberlite, conglomerate	Cr	Diamond pipes and placer
Coal	Shale and Sandstone	Pyrite, methane	Coal
Cr	Layered mafic-ultramafic intrusive	Cu, Ni, Au, PGE	Chromite
Cu	Granite, schist, quartzite	Au, Ag, Ni, Zn	Copper
Fe	BHQ, BIF	SiO_2	Iron ore
Mn	Veins and nodules	Fe	Manganese ore
Ni	Mafic and ultramafic	Cu, Cr, PGE	Nickel
Pb	Dolomite, carbonaceous schist	Zn, Cu, Ag, Cd	Base metal
P_2O_5	Dolomite		Phosphate
Sn	Pegmatite and granite	W	Tin
U	Black shale, sandstone, hematite breccias, quartz and pebble conglomerate	Cu, Mo, Fe, Au, Ag, RE	Uranium
Zn	Dolomite, carbonaceous schist	Pb, Cu, Ag, Cd	Base metal

copper. The bulk concentrate (copper + zinc + lead) is of lower grade produced from complex type of mineralization. The concentrate is further processed either in the in-house or by a third party smelter and subsequently refined to make 99.99% metal grade. This refined metal is the input for the manufacturing industry for making consumer serviceable goods. The specifications for some of the minerals are generalized and described.

(1) Bauxite
- **(a)** Metal grade: >50% Al_2O_3, <5% SiO_2
- **(b)** Refractory grade: >55% Al_2O_3, <3% SiO_2 and Fe_2O_3 each
- **(c)** Chemical grade: >58% Al_2O_3, <3% Fe_2O_3

(2) Chromite
- **(a)** Metallurgical grade: >48% Cr_2O_3, Cr:Fe, >2.8:1
- **(b)** Refractory grade: 38-48% Cr_2O_3, $Cr_2O_3 + Al_2O_3$, >60%
- **(c)** Chemical grade: 48-50% Cr_2O_3, Cr:Fe, 1.6:1
  Fe as FeO: <15%, MgO: <12-16%
  SiO_2: <5%, P as P_2O_5: <0.005-0.20%
  CaO: <5-12%, S as SO_3: 0.1%

(3) Copper ore
Run of Mine (ROM) Cu grade: 0.50-2.00%
Concentrate grade: >20.00%
Refined copper grade: 99.99%

(4) Fluorite
- **(a)** Metallurgical grade: >85% CaF_2, <5% SiO_2, <0.03% S
- **(b)** Ceramic grade: >95% CaF_2, <3% SiO_2, <1% $CaCO_3$, entirely free from Pb, Zn, Fe, S.
- **(c)** Acid grade: >97% CaF_2, <1% SiO_2 and $CaCO_3$, entirely free from Pb, Zn, Fe.

(5) Graphite
- **(a)** Lumpy:
  - **(i)** Lump—walnut to pea
  - **(ii)** Chip—pea to wheat grain
  - **(iii)** Dust—finer <60 mesh
- **(b)** Amorphous: >50% graphitic carbon
- **(c)** Crystalline flacks: >85% graphitic carbon (−8 to 60 mesh in size)

(6) Gypsum
- **(a)** Cement grade: $CaSO_4 \cdot 2H_2O$: >70%
- **(b)** Fertilizer grade: >85% $CaSO_4 \cdot 2H_2O$, <6% SiO_2, <0.01% NaCl, no clay
- **(c)** Plaster of Paris: 80-90% $CaSO_4 \cdot 2H_2O$

(7) Glass sand
- **(a)** Normal glass: >96% SiO_2
- **(b)** Optical glass: 99.8% SiO_2, <0.02% iron oxide, <0.1% CaO + MgO, <0.10% Cr, Co, Al_2O_3, TiO_2, <1 ppm Mn.

(8) Iron ore
- **(a)** Grade classification:
  Very high grade: >65% Fe
  High grade: 62-65% Fe
  Medium grade: <62% Fe
  Unclassified: Inadequate sampling
  Phosphorus: <0.18%
- **(b)** Size classification
  Lump ore: Particles >8 mm
  Sinter feed: Fines >100 mesh
  Fines (pallet feed): Fines <100 mesh

(9) Limestone
- **(a)** Cement grade: 45% CaO, <3% MgO
- **(b)** Blast furnace grade: 46-48% CaO, <11.3% total insoluble
- **(c)** Steel melting grade: >48% CaO, <4% total insoluble
- **(d)** Conventional open-hearth steelmaking: <4% SiO_2
- **(e)** Basic oxygen furnace: <1% SiO_2

(10) Dolomite
- **(a)** Blast furnace grade: >28-33% CaO, >18-20% MgO, <7% total insoluble
- **(b)** Steel melting grade: >29% CaO, >20% MgO, <4% total insoluble
- **(c)** Glass grade: Consistent chemical composition, <0.2% Fe_2O_3

(11) Manganese ore
Manganese ore grade: >35% Mn
Ferruginous Mn ore: 10-35% Mn
Manganiferous iron ore: 5-10% Mn
Metallurgical grade: >44% Mn
Battery grade: >78% MnO_2, <4% HCl soluble Fe
Chemical grade: >80% MnO_2

(12) Rock phosphate
P_2O_5: >24% (preferably + 30%)
Si_2O_2: <20%
Fe: <3%
Al_2O_3: <7%

(13) Sillimanite and kyanite
Al_2O_3: >59%
Si_2O_2: <39%
Fe_2O_3: 0.75%
TiO_3: <1.25%
CaO + MgO: <0.20%

(14) Talc
Talc is classified according to its color and softness.
Grade-I: Pure white appearance with smooth feel and free from grit.
Grade-II: Tinted variety with smooth feel and without grit.
Grade-III; Off-color variety with smooth feel and without grit.
Grade-IV: White or colored with grit.

(15) Zinc-lead ore
ROM grade: >8% Pb + Zn
Fe content as Py, Po: Lesser the better
Graphite content: Lesser the better
Zinc concentrate: >52% Zn
Lead concentrate: 56-60% Pb
Refined metal: >99.99% Zn, Pb

(16) Coal

(a) Noncoking coal
Grade-A: Useful heat value >6200 kcal/kg
Grade-B: Useful heat value >5600 and <6200 kcal/kg
Grade-C: Useful heat value >4940 and <3600 kcal/kg
Grade-D: Useful heat value >4200 and <4940 kcal/kg
Grade-E: Useful heat value >3360 and <4200 kcal/kg
Grade-F: Useful heat value >2400 and <3360 kcal/kg
Grade-G: Useful heat value >1300 and <2400 kcal/kg

(b) Coking coal
Steel grade-I: Ash content <15%
Steel grade-I: Ash content >15 and <18%
Washery grade-I: Ash content <18 and <21%
Washery grade-II: Ash content <21 and <24%
Washery grade-III: Ash content <24 and <28%
Washery grade-IV: Ash content <28 and <35%

(c) Semi-coking coal
Semi-coking-I: Ash + moisture content <19%
Semi-coking-II: Ash + moisture content between 19 and 24%

(d) Hard coke
Premium: Ash content <25%
Ordinary: Ash content between 25 and 30%
Beehive Premium: Ash content <27%
Beehive Superior: Ash content between 27 and 31%
Beehive Superior: Ash content between 31 and 36%

FURTHER READING

Bateman (1950) [4] and Evans (1999) [24] gave a comprehensive account of economic mineral deposits, host rock, composition and uses of major and minor minerals. Sinha and Sharma (1993) [69] provide a detail specification of industrial minerals. Bureau of Mines (BM) United State Geological Survey (USGS), Canada, and India publish commodity-wise monograph of minerals every year with total mineral statistics. Dana's New Mineralogy by Genies et al. (1997) [25] described a new outlook to the system of mineralogy. The books of Chatterjee (2004 [10], 2007 [11] and 2008 [12]) on uses of minerals, metals, rocks and fresh water are suggested.

Chapter 3

Exploration Geology

Chapter Outline

One discovery out of 100 or even 1000 attempts will pay back the entire efforts.

—Author.

3.1. DEFINITION

Exploration is a complete sequence of activities. It ranges between searching for a new prospect (Reconnaissance) and evaluation of the property for economic mining (Feasibility study). It also includes augmentation of additional ore reserves in the mine and whole of the mining district. There are various exploration techniques being followed over the centuries. Exploration is conducted by one or a combination of many of the techniques. It all depends on the availability of infrastructures, funds with the state agencies and private players, size and complexity of the deposit, price of the minerals, government policy and good will. The programs are carried out by multidisciplinary data generation in a sequential manner. In addition to technical inputs, the activities encompass collection of information about the infrastructure around the area, such as accessibility (road, rail, nearest rail head, airport and sea port), average rainfall, and availability of potable and industrial water, power grid and supply system, local community, living condition, health care, security, forest and environment. The information about agencies from Federal and State/Regional/Provincial Government and Public Sector and Private Sector (PS) including multinational companies (MNCs) engaged with any mineral exploration program in the area will be useful.

3.2. REGIONAL PLANNING AND ORGANIZATION

The regional planning for all mineral resource development for the nation as a whole is carried out in the form of Five-Year Plan program. A multidisciplinary expert group selected from Federal Administrative Services, Planning Commission, Technocrats, Economists and Statisticians from Public and Private institutions collect information about past trend of consumption-supply, import-export and strategic importance of various mineral commodities. The information is projected for futuristic demand and optimum resource generation on the principle of scientific and sustainable use for each mineral (refer Figs 1.1 and 1.2). The activities are identified and responsibilities are earmarked to respective organizations/institutions along with allocation of funds and budgeting. The government continues routine monitoring, reconciliation and corrective

Mineral Exploration. http://dx.doi.org/10.1016/B978-0-12-416005-7.00003-9

measures at appropriate stage. The key implementing organizations are:

3.2.1. Bureau of Mines

The prime role of the Bureau of Mines (BM) is to create and maintain an up-to-date onshore and offshore mineral database, promote scientific development, encourage sustainable conservation and protect environment in mineral industries. BM is responsible for approval of mining plan and regulatory inspecting authority for mining and environment management plans to ensure maximum use of in-situ mineral resource at minimal adverse impact on environment. BM publishes "Mineral Year Book" every year highlighting the total mineral-related statistics and disseminates free online service as well as in printed form at cost. The various institutions are Department of Mines and Petroleum (DMP), Australia, Bureau of Mines, Canada, Bureau of Mines, Chile, Indian Bureau of Mines (IBM) and United States Bureau of Mines (USBM).

DMP, Australia, ensures a stronger focus on the resources sector, maintaining a mining and petroleum regulatory role incorporating the resources safety and responsibilities. DMP processes and provides an efficient and timely approval of mining title, which is essential for guaranteeing the sustainability of the resources sector and the future prosperity of the States/Region/Province.

The Mining sector in Chile is the major support of national economy. Copper export alone contributes ~40% to the government exchequer.

IBM, established in 1948, is a multidisciplinary national organization with headquarter at Nagpur and several regional/district offices. It is responsible for the mineral policy planning, conservation and mining research. IBM is the custodian of total mineral statistics for the Government of India. The functions of IBM include promoting conservation of mineral resources by way of inspection of mines, geological studies, scrutiny and approval of mining plans and schemes, conducting environmental audits, evolving technologies for mineral beneficiation, preparation of feasibility reports for mining and beneficiation projects, preparation of mineral maps and National Mineral Inventory of minerals resources, providing technical consultancy services to mineral industry, and maintains data bank for mines and minerals, and preparing of technical and statistical publications. IBM compiles and publishes "Mineral Year Book" every year covering statistics of total exploration, geological reports, mine production, export and import, price trend and related matters of all minerals. IBM maintains database of entire RP, PL and MLs in the country and available on price. IBM has six technical divisions with its headquarters at Nagpur supported by a Modern Mineral Processing Laboratory and Pilot Plant. IBM has 3 Zonal Offices, 12 Regional Offices and 2 Subregional Offices, 2 Regional Ore Dressing Laboratories and Pilot Plants spread over the country located at Ajmer, Bangalore, Bhubaneswar, Kolkata, Chennai, Dehradun, Goa, Guwahati, Hyderabad, Jabalpur, Nagpur, Nellore, Ranchi, and Udaipur.

USBM was established in 1910 as primary government agency with a mission to conduct scientific research to enhance the mineral resources, safety, health care and environment impact, and disseminate information on mining, processing, extraction, use and sustainable conservation of minerals. Since inception, USBM was viewed as the nodal point for the new and emerging science and technology in the mineral sector, both nationally and internationally. The government closed the BM during 1995–1996 and merged certain functions to other interrelated federal agencies. Mineral Year Book of USBM is the pioneering effort by collection, analysis, and dissemination of information about mining and processing of more than 100 mineral commodities across the Nation and in more than 185 countries around the world.

3.2.2. Geological Survey

The Geological Survey (GS) is the national geo-scientific and academic institution responsible for specialized multithematic mapping of the entire country to develop basic research in earth science and to target new mineral discoveries that includes water and energy resources to meet the demands of growing population. The other functions are assessment of earth science-related environmental impacts and geological hazards like landslide, earthquake, flood, coastal zone instability and desertification. It creates and updates geo-scientific database, reports and maps for dissemination to the government departments, user agencies and individuals free or at cost as the case may be. There are several national geological institutions, namely Geological Survey of Australia (GSA), Geological Survey of Canada (GSC), Geological Survey of India (GSI), United States Geological Survey (USGS), and National Service of Geology and Mining (NSG & M), Chile.

The Geological Survey of Western Australia (GSWA) is a division within the DMP. Since 1880, the function of GSWA is to collect and synthesize information on the State's geology, exploration, and mineral and petroleum resources. GSWA publishes reports, maps, and state-of-the-art database documenting all information to enable building blocks to design exploration programs. Reports, maps and Geographical Information System (GIS) information on the geology, geophysics, geochemistry, geochronology, mineral exploration, resources as well as reports of Exploration Company can be downloaded free of charge from WAMEX system. Mineral deposit and mine information is available from MINEDEX. The department is critical in government decision making, particularly on economic and land use issues. Every State or Region in Australia functions in a way suitable to that State or Region or Province.

The GSC established in 1842. It is a part of the Earth Science Sector of Natural Resources Canada. The GSC is the premier agency for geo-scientific information and research, with world-class expertise focusing on geosciences surveys, sustainable development of Canada's resources, environmental protection and technology innovation.

GSI, established in 1851, is an all India national organization with Central Headquarter at Kolkata and six Regional Offices, several operational wings in each region and six training institutes. The Regional Offices are spread over at Kolkata (Eastern Region and Coal Wing), Shillong (Northeastern Region), Hyderabad (Southern Region), Lucknow (Northern Region), Jaipur (Western Region) and Nagpur (Central Region). The headquarter of AMSE Wing is located at Hyderabad with five Zonal Offices at Hyderabad, Nagpur, Ranchi, Shillong and Jaipur. GSI is premier institution in the country imparting state-of-the-art training in different disciplines of earth science. The headquarter of training institute is at Hyderabad and conducts various training programs through its six centers located at Hyderabad (Andhra Pradesh), Lucknow (Uttar Pradesh), Zawar (Rajasthan), Ranchi (Jharkhand), Raipur (Chhattisgarh), and Chitradurga (Karnataka). The Institute has specialized divisions such as Photo Geology and Remote Sensing, Geophysics and the Center for Geo-information Management Training, located at Hyderabad.

The main functions and responsibilities of GSI is preparing and updating geological, geophysical and geochemical maps, exploration and assessment of mineral and energy resources of the country and its offshore areas, systematically developing and maintaining national drill core libraries, conducting research in earth sciences and promoting application of the new knowledge for effecting management of the earth system and its resources. It also works on promoting understanding of geological knowledge to reduce risk to life and property from geological hazards to enhance quality of life. The other function is creating and maintaining earth science databases and acting as the national repository of earth science data generated by various organizations and disseminating these in public domain for developmental, educational and societal needs. It maintains national geological monuments, museums and parks. GSI participates in international collaborative scientific projects and developing data sharing net works with other countries. The maps and reports are available from central and regional offices to everybody at cost.

Survey of India (SOI) was established in 1767 with its headquarter at Dehradun, Uttarakhand. The principal responsibilities of the SOI are geodetic control (horizontal and vertical) and geodetic and geophysical survey, topographical control, surveys and mapping within India, mapping and production of geographical maps and aeronautical charts, surveys for developmental projects. The organization prepares topo-sheets at 1:250,000, 1:50,000, 1:25,000 scales for the entire country. The survey sheets that are not of strategic importance (restricted) can be purchased by anybody from the SOI offices located at all the state capitals. The restricted topo-sheets can be obtained by the special permission of authorized Central and State Administrative Officials and/or the Ministry of Defense.

Mineral Exploration Corporation Limited (MECL) was established in 1972 as an autonomous Public Sector Company to bridge the gap between the initial discovery of a prospect and its eventual exploitation. MECL conducts geo-scientific services like survey, geology, geophysics, detailed exploration, mine development, metallurgical tests and feasibility reports for mine planning and allied activities for major minerals. The institution works on promotional as well as payment basis in India and abroad.

The USGS was established in 1879 as a scientific agency of the government. USGS headquarter is at Reston, Virginia, with major offices at Lakewood, Colorado (Denver Federal Center) and Menlo Park, California. The primary mission was to study of landscape, topographic mapping, natural resources, earthquake and volcanic hazards, geomagnetism program and related matter. USGS has prepared Geological Quadrangle (GQ) Maps, Geophysical Investigations (GP) Maps, Hydrologic Investigations Atlas (HIA), Land Use and Land Cover (L) Map, Mineral Investigations Resource (MR) Maps, etc. The organization has four major science disciplines comprising of biology, geography, geology and hydrology. The USGS is a fact-finding research organization with no regulatory responsibility.

3.2.3. State/Regional Department of Mines and Geology

State/Regional/Provincial Department of Mines and Geology (DMG) is responsible for limited mineral exploration, granting of RP, LAPL, PL and ML with the approval/consent of Federal Government, and the inspection and collection of royalty from various mining units.

3.2.4. Public Sectors Undertakings

In the mineral sector, there are many government-owned companies (GOCs), state-owned companies, state enterprise, publicly owned companies [Public Sector Undertakings (PSU)] with legal entity created by the government whose primary focus is to explore, operate and expand mineral properties in the country. GOCs can be fully or partially owned by the government. The major shares of these companies are held by the government with limited share by the individual public. The GOCs are Aerosonde Ltd (Australian Platinum Mines), Alkeno Exploration.

Since independence, the government of India focused on development of essential minerals through PSU such as: Coal India Limited (CIL, 1951), Orissa Mining Corporation Limited (OMCL, 1956), a State and Federal joint sector company, Oil and Natural Gas Commission (ONGC, 1960), a Fortune Global 500 company, Bharat Gold Mines Ltd (BGML, 1972), National Mineral Development Corporation (NMDC, 1958), Manganese Ore India Limited (MOIL, 1962), Bharat Aluminum Company Ltd (BALCO, 1965), now a part of Vedanta Resources, Hindustan Zinc Ltd (HZL, 1966), now a part Vedanta Resource plc, Hindustan Copper Ltd (HCL, 1967), National Aluminum Company Ltd (NALCO, 1981). These PSUs are formed for exploration mining and to promote the growth of specific minerals even in the remote areas and economic development of the tribal community. Most of these PSU have exploration wing to carry out own requirements. Exploration and sampling are continuous activities for mine planning, grade control and reconciliation.

3.2.5. Private Sectors

Mining industry site covering exploration through to mining, processing and transport of minerals are privately owned or PS activity in countries like: Osisko Mining, Argentina, Red Back Mining in Canada, Australia, Europe, United States, Latin America, Anglo American plc, Anglo Platinum, De Beers, Exxaro Resources Ltd, Gold Fields, Harmony Gold, Impala Platinum, JFPI Corporation, Kumba Resources, Rio Tinto and Wesizwe Platinum in South Africa and Equinox, Meteopex, Mopani Copper, Konkola Copper, Zambia Consolidated Copper Mines, First Quantum Minerals, Venturex Resources Ltd, and Vale, Zambia. The other international exploration companies are CSR Ltd, Forteseue Metals Group, Rio Tinto Group, Santos, Zinifex Ltd, Goldstream mining NL, IMX Resources Inc., Australia, Cameco Uranium, Goldcorp, Exeter resource Corporation (2002), Canada, Codelco Mining Corporation (1976), and Capstone Mining Corporation, Chile. The major Indian PS exploration and mining companies are: ACC Ltd, ADI Gold Mining Pty Ltd, Binani Zinc Ltd, Birla Corporation Ltd, BHP Billiton, De Beers India Pvt Ltd, ESSAR Mineral Resources Ltd, Ferro Alloys Corp Ltd, Rio Tinto India Pvt Ltd, Sesa Goa Ltd, Tata Steel Ltd and Vedanta Resources plc. The companies are permitted to explore and mine nonrestricted minerals. The PS are equipped with technical personnel and machineries to conduct exploration and evaluation of mineral deposits in on going mining as well as in virgin areas.

3.2.6. Multinational Limited Companies

The concept of open market policy gave way to radical opportunity to Multinational Limited Companies (MLC) or MNC worldwide. Any MNC registered in the country, individually or as Joint Venture partner, is allowed to explore and mine all nonstrategic minerals with 100% FDI. The participation of MNC in mineral sector has far-reaching effects and benefits to the receiving country by way of readily available specialized experienced skilled technocrats and most advance technology. The MNCs working all over the world are Anglo American plc, BHP Billiton, De Beers Private Ltd, Freeport-McMoRan Copper and Gold Inc., Rio Tinto Group, etc.

3.3. SURFACE GUIDE

Most of the mineral deposits portray surface signature like favorable stratigraphy and host rocks, weathering effects of metallic and nonmetallic mineralization, presence of earlier mining and smelting remnants, shear zone, lineaments, etc., that can be identified by experienced eyes. If the features are recorded properly during geological traverses in the field followed by exploration, a new deposit may be discovered.

3.3.1. Favorable Stratigraphy and Host Rocks

The existence and identification of favorable stratigraphy and complimentary host rocks are the essential prerequisites to initiate any exploration program for a specific mineral or group of minerals. Geological maps are available from GS or BM or Department of Mines. This can be downloaded free of cost or purchased at small price. The layered ultramafic-mafic assemblage of Archean-Proterozoic age is the most suitable target for chromium-nickel-platinum-copper and gold association (Bushveld chromium-platinum in South Africa and Sudbury nickel-platinum in Canada, Stillwater nickel-copper-platinum in United States, The Great Dyke platinum-chromium in Zimbabwe, and Sukinda-Nausahi chromium in India).

3.3.2. Weathering

Weathering and leaching of near-surface metallic deposits is an indicator of probable existence of mineral deposit down depth. This has been described in detail at Chapter 4.5.2.4 and (Figs 4.4–4.9). Presence of gossans above Broken Hill zinc-lead-silver deposit in Australia, Adi Nefas, zinc-copper-gold-silver deposit Madagascar, Rajpura-Dariba zinc-lead-silver deposit and Khetri copper deposit in India are good examples of base metal deposits.

3.3.3. Ancient Mining and Smelting

Ancient mining and smelting has been reported and radiocarbon dated of woods in the Indus valley (3000 BC for gold mining), along Lake Superior, North America

(3000 years Before Christ (BC) for copper mining), Egypt (2613 and 2494 BC for Cu mining), India (3000 BC for copper, 1400 BC for iron, 1000–100 BC for zinc-lead-silver mining) and Spain (25 AD for gold mining). One can find ancient mine debris around open pits with wooden wall supports, entry system to shallow and greater depth for underground mining, abandoned underground gallery with wooden ladder and platform, in situ potholes rock grinder at surface for ore dressing, smelting furnaces, enormous heaps of slag and retorts reused for wall making, ruined places of worship and deserted township in and around the ancient mine-smelting sites. These evidences easily suggest existence of rich ore deposits and guide for modern-day exploration for depth and strike continuity in the region.

The remnants of ancient mines play significant role in mineral exploration. Once the mineralization was located at the surface based on the presence of gossans or fresh mineralized veins, the ancient miners excavated the area as open pit mine with wall supports wherever required as observed in case of East Lode of Rajpura-Dariba mine. The miners follow the downward extension along dip and pitch of narrow ore shoot making vertical or inclined entry system (Fig. 3.1). The excavations are generally carried out by fire setting and sudden cooling by water.

The entry system can be of multiple in nature in case of large exposed orebody and particularly extends over great depth (Fig. 3.2).

The miners develop huge stopes and underground chambers within rich part of the mineralization. They have demonstrated competent skill of engineering by making arching of the pillars, use of wooden ladders and platforms, wooden launders for underground drainage, timber support to prevent roof collapse and clay lamps for mine illumination (Fig. 3.3).

The presence of small pits on the surface nearby the smelting sites and vast heaps of crushed debris near the mine opening indicate that the zinc ore was crushed, richer portion handpicked and later ground, before smelting. At Rajpura-Dariba, north of East Lode in the hard calc-silicate outcrops, potholes of 30 cm in diameter and 60–70 cm deep with rounded bottoms are observed (Fig. 3.4).

The ancient zinc smelting process was resolved through distillation and condensation technology (pyrometallurgy) of zinc ore using moderately refractory clay retorts. Archaeometallurgical excavation at Zawar discovered intact ancient zinc distillation furnaces containing their full spent charge of 36 retorts (Fig. 3.5). Each furnace is 60 cm

FIGURE 3.1 Ancient entry system to underground mine at shallow depth of orebody without any plunge during the 3rd and 2nd millennium BC at Khetri copper belt, Rajasthan, India.

FIGURE 3.2 Ancient entry system to underground mine at greater depth with orebody plunge to northeast at Rajpura-Dariba copper-zinc-lead-silver mine, India.

FIGURE 3.3 Ancient underground mining in rich zinc orebody at a depth of 172 m from surface. The arching of stope chamber evidenced high engineering skill. The wooden ladder (left) and platform (right) are still at abandon work site. Radiocarbon age dating indicates 3000 years from now at Rajpura-Dariba Mine, India.

FIGURE 3.4 Ancient potholes at surface used for crushing, grinding and concentration of rich zinc ore are still preserved at Rajpura-Dariba mine, India.

in height and divided in two parts, a lower condensing chamber, separated by perforated plate from the upper main furnace of distillation chamber. The smelting was carried out at 1100−1200 °C for 4−6 h. This site is recognized and preserved by the American Society of Metals as International Zinc Smelting Heritage in 1988.

A complete retort is cylindrical in shape, tappers at one end (Fig. 3.6) and fitted with an extended cylindrical hollow tube for channeling distilled hot zinc vapor from the retorts at upper chamber and condensed at the lower cooling chamber. The condensed zinc metal is dropped at collecting pot.

The exhausted retorts are either dumped or reused for making walls of hutments for the mining community (Fig. 3.7) in the township.

There are ample evidences of antique mining and smelting history as seen at the old deserted ruined Industrial Township in the valley area of Zawar Mine, India. The presence of abandoned mine entry system in the surrounding hills, clay and sand covered intact smelting furnaces, part of broken down residential walls, and Hindu

FIGURE 3.6 Ancient cylindrical distillation clay retorts from the smelting site at Zawar mine area, India.

temples (Fig. 3.8) of thirteenth century AD in the center of the ancient zinc smelting site. Some of the historical monuments are maintained by Indian Archeology Department.

3.3.4. Shear

Shear zones is the result of huge volume of rock deformation due to intense stress in the region, typically in the zones of subduction at depths down to few kilometers. It may occur at the edges of tectonic blocks, forming discontinuities that mark a distinct structure. Shear zones often host orebodies as a result of syngenetic or epigenetic hydrothermal flow through orogenic belts. The rocks are commonly metasomatized, and often display some retrograde metamorphism assemblage. An intense fractured or shear zone is a favorable structure to trap mineralization. The Hyde-Macraes shear zone, New Zealand, is a low-angle thrust system in which gold-bearing quartz veins have been deposited. Copper sulfide vein-type mineralization associated with migmatization in the southeastern part of the Singhbhum shear zone, Jharkhand, India, is an example of shear-controlled copper-uranium mineralization.

FIGURE 3.5 Ancient smelting furnace unearthed with 6 × 6 numbers of retorts placed in an inverted position for distillation of zinc by heating and condensation process at Zawar mine, during 2180 ± 35 years from today, India.

FIGURE 3.7 Ancient smelting retorts reused for walls for living shelter in the mining township at Zawar, India.

FIGURE 3.8 Ancient Hindu temple of thirteenth century AD in the center of the ancient deserted zinc smelting site at Zawar Mine area, India (Credit: Prof Martin Hale, The Netherlands).

3.3.5. Lineament

In general, mineral deposits occur in groups and follow a linear pattern along fold axis, shear zone and basement fracture traps. The linear alignment can be traced in the regional map of Aravalli Mountain, India, and McArthur-Mt Isa Province, Australia. Lineament mapping of different terrains using Remote Sensing Imageries is capable to guide ground water flow. Analysis of surface lineament with the help of Geoinformatics became significant in oil and gas exploration as at Sabatayn mature basin in Yemen. A satellite image Enhanced Thematic Mapper-based analysis was conducted for extracting surface and subsurface lineaments overlaying the seismic, magnetic and gravity data.

3.4. TOPOGRAPHIC SURVEY

An accurate topographic map is essential for long-term and short-term purposes of any type of projects. This is more relevant during all stages of mineral exploration, mine development, mining and related activities. The simplest way of topographic surveying is carried out by a tape and a compass with low level of accuracy. The accurate topographic surveys are carried out using Electronic Total Stations (ETS) to capture three-dimensional (3D) observation data (x, y, z) on site. The data is processed using commercial software to generate Digital Terrain Model (DTM). The DTM is capable to produce contours, volumes, sections and 3D wireframe view and plots.

An accurate topographic map (topo-sheets) in 1:250,000 or 1:50,000 scales, marked with reference survey station or triangulation points, contours and all other land-related features is easily available on cost from the National Survey departments. Topo-sheets can also be downloaded internationally from US Army Mapping Service in Universal Transverse Mercator (UTM) system at nominal price. The sheets are base maps and significantly useful for geological mapping, sampling and borehole locations. In the advance stages of exploration and detail mine planning, highest accuracy is maintained in topographic surveys (1:5000, 1:1000 scales) by using Leica total station and Leica global positioning system (GPS) equipments. The underground mine survey is routinely cross-checked by closing the survey from and to the known surface station.

3.5. GEOLOGICAL MAPPING

The first work in mineral exploration is the preparation of a high-quality geological map. The precision and scale of map depends on the stages of exploration, technical infrastructure and finance available for the program. A geological map is a record of geological facts such as occurrence of rocks in space and their contacts, weathering effects such as leaching or gossan, and structure in their correct space relations. There are sharp distinctions between observations and interpretations. The inferences are confirmed by opening through pits and trenches. It can also be authenticated by subsurface information as obtained from drill holes and mine workings. The map must have a scale, direction and index describing various features shown on it.

3.5.1. Surface Map

Surface maps are prepared by taking traverses on the surface at various intervals and plotting the records like rock types and all other observations including strike, dip, plunge, etc. The government agencies prepared maps of the entire country on a regional as well as district scale for future planning purposes. The maps are becoming précised and meaningful with the advent of facilities like Remote Sensing Technology.

3.5.1.1. Regional Scale

Regional scale mapping starts with the study and interpretation of geological features from satellite imageries, aerial photographs. These base maps along with topo-sheets at 1:250,000 or 1:50,000 scales are used for selection of boundaries for Reconnaissance License. The regional survey is based on widely spaced traverse and cross-verification of broad geological contacts, shear zones and weathering features. The map represents an overall regional

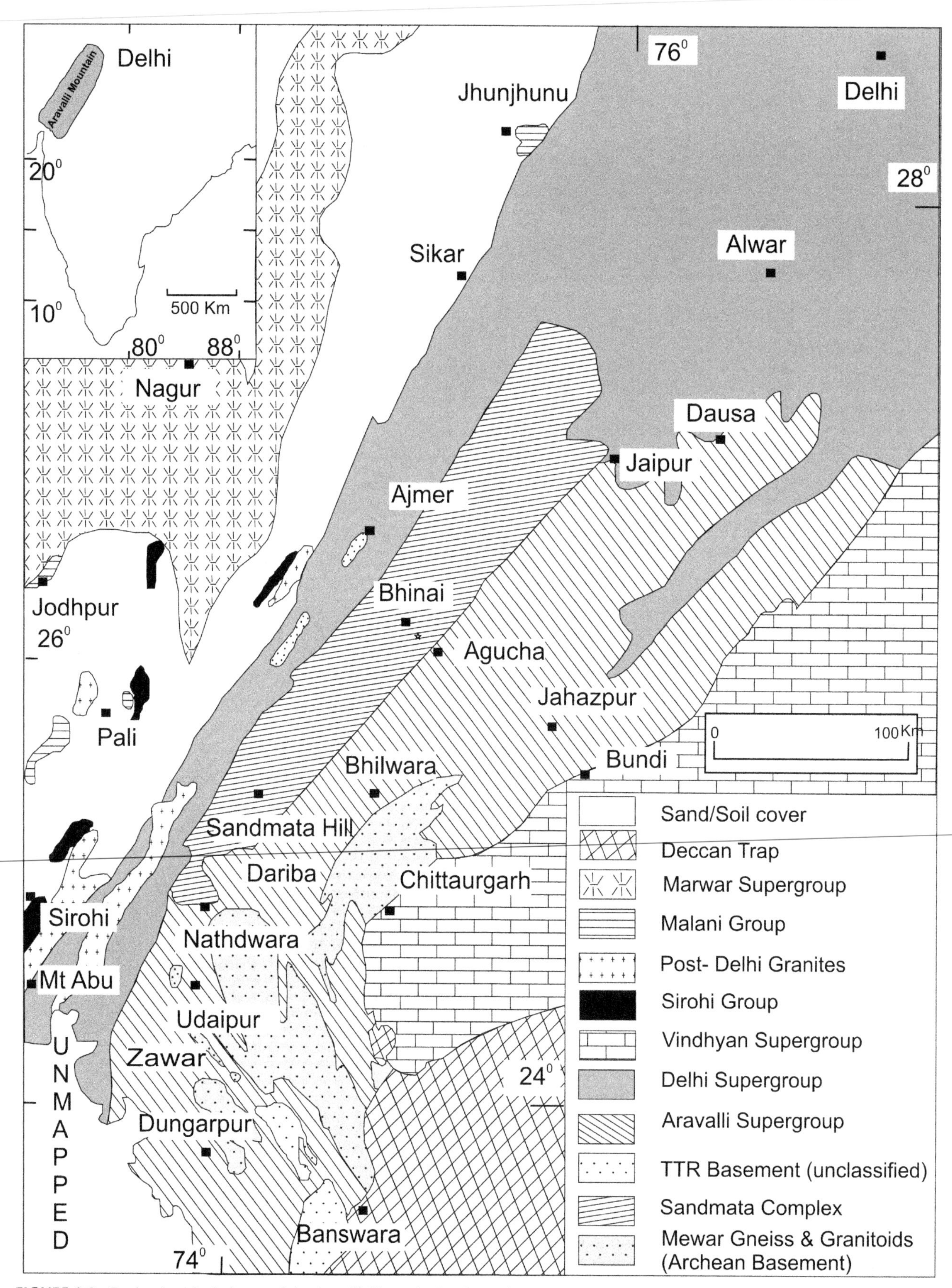

FIGURE 3.9 Regional geological map of the Aravalli Mountain showing major tectonostratigraphic units. BGC/MGC stands for Banded Gneissic Complex/Mewar Gneissic Complex, respectively showing location of mineral deposits (Credit: Prof. AB Roy).

picture including geomorphology, drainage pattern and major structures. Regional scale maps are not for representing precise specific features. Soil, grab and chip samples of rocks and weathered profiles are collected at this stage. The purpose is to provide a base map for further detail study from search of economic minerals, design of roads, dam and other infrastructures (Fig. 3.9).

3.5.1.2. District/Belt Scale

The district/belt map represents a part of the regional map having a cluster of mineral deposits. The features are recorded with close space traverses and rock sampling. The scale of map may be in 1:25,000 and 1:10,000 scale using theodolite and GPS survey instrument. The geological map is linked with topography. The map will be useful for selection of Prospecting Lease boundary and to formulate exploration program. The map makes an assessment of right stratigraphy, accurate lithology, detail structures, surface show of mineralization and analysis of ancient history of mining and smelting (Fig. 3.10).

3.5.1.3. Deposit Scale

The deposit map contains maximum detail information and is more focused at large scale like 1:5000 to 1:1000 with triangulation station benchmarks. The area is under Prospecting or Mining License. The map helps to design the drilling program to delineate the mineral or orebody (Fig. 3.11). The map stands for detail surface geological features, extent of deposit, location of trenches, pits and boreholes. The map helps to draw sections along and across the elongation of mineralization.

3.5.2. Underground Mapping

The surface geology related to lithology, structure and mineralization can be correlated for down-depth continuity with subsurface features as the mine progresses through service and stope development. The walls and back of mine entry system like adit, incline, raises, winzes and shafts, and level development in the form of drives and crosscut

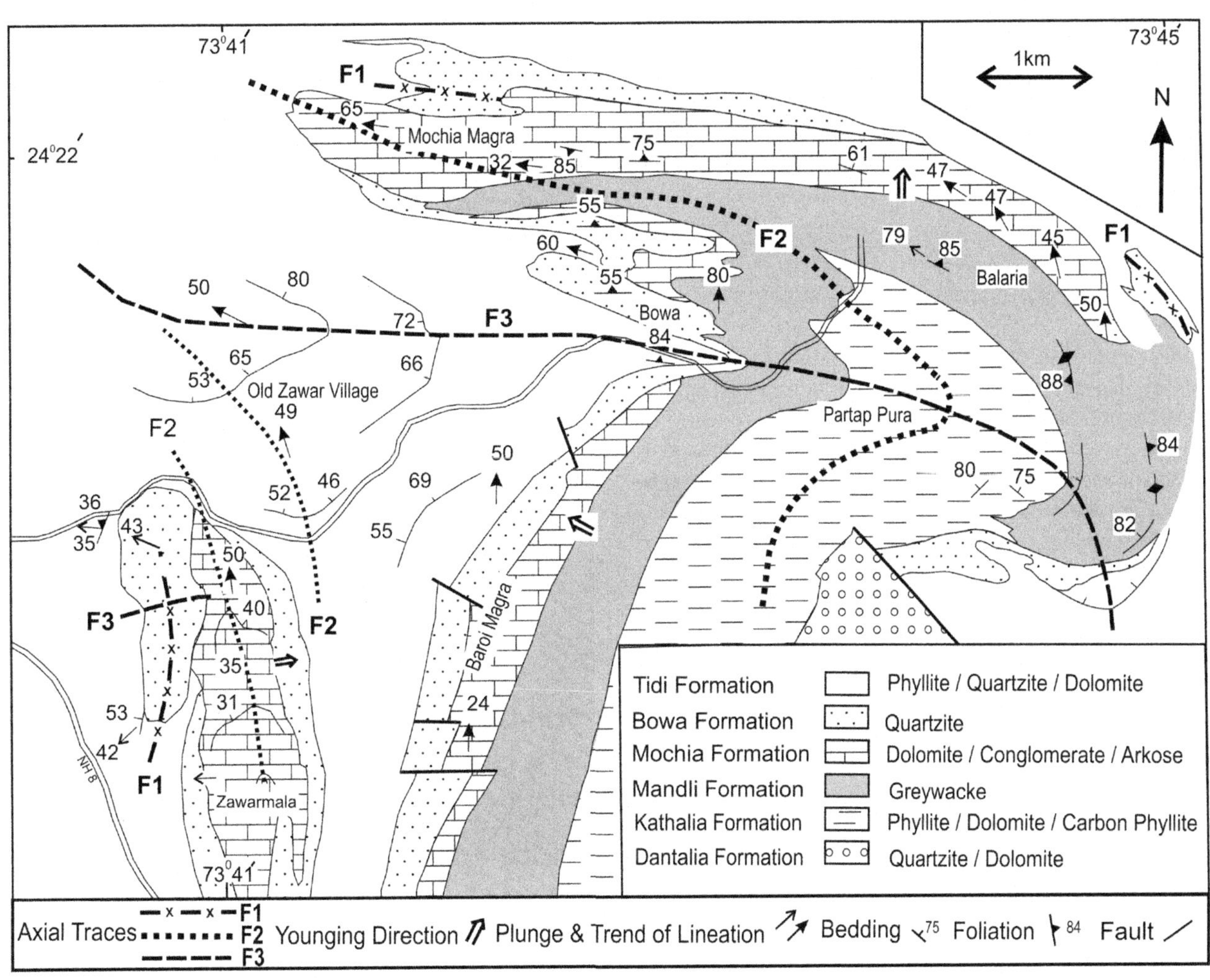

FIGURE 3.10 Surface map of Zawar Group of deposits showing all mining blocks, namely Balaria, Mochia Magra, Baroi, and Zawarmala, India (Credit: Prof. AB Roy).

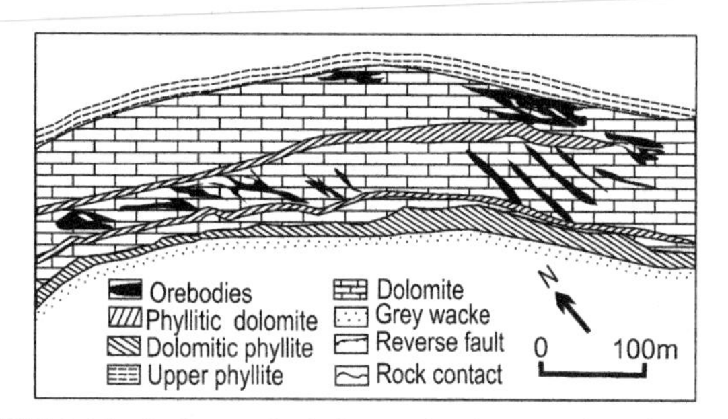

FIGURE 3.11 Surface geological maps showing strata-bound en echelon orebodies of Balaria zinc-lead-silver deposit, Rajasthan, India.

FIGURE 3.12 Scan line underground mapping of crosscut at Balaria mine, India.

are mapped (Fig. 3.12) carefully in very large scale such as 1:200 or 1:50.

In all underground workings, survey stations are marked in the roof of the working maintaining the line of sight to guide the successive development. The surface is cleaned by high-pressure water jet or compressed air. Cloth tape is stretched between two survey stations. Pocket metallic tape is used for taking off-set of working profile, rock contacts, foliation, bedding, joints, fractures, shears, void fills and mineralization (Fig. 3.13). The information is plotted on graph sheets in the underground and later transferred to master plan in the office. These maps help in drawing rocks and orebody contacts with confidence, planning of underground drill holes and finally stope design.

3.6. STRATIGRAPHIC CORRELATION

The geological maps of regional, basin and deposit scale can be analyzed and correlated to establish a comparative statement of the lithological succession with metallogeny of the region (Figs 3.14 and 3.15). This helps to generalize the ore-hosting horizon for future search in the in-between blocks and extension on either side.

3.7. EXPLORATION ACTIVITY

The geological exploration can be divided in to three broad groups, namely regional scale, district scale and deposit scale. The overall activities can be identified as:

3.7.1. Regional Scale

- Survey of existing literature, examination of aerial photographs, satellite imageries, acquisition of geophysical data, if any, and geological maps of prospective region, understanding the stratigraphic setting and structural architecture, synthesis of all available data and concepts and submission of RP. The resource at this stage is of preliminary nature.
- The work plan includes preparation of organization plan, exploration scheme with fund allocation and budgeting and time schedule to achieve specific objective (refer Table 1.1).
- The task encompasses aerial geophysical and broad geochemical survey, ground check, wide space soil and rock chip sample collection, pitting trenching, few scout drilling to establish existence of mineralization and demarcation of priority and ranking of targets.
- Investment decision for next phase.

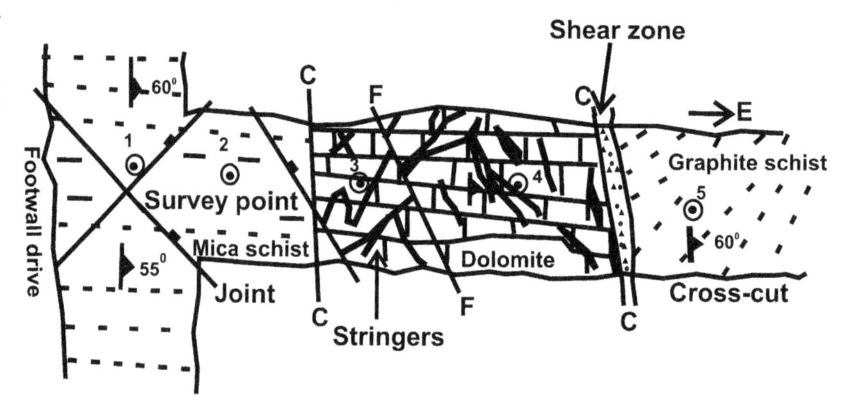

FIGURE 3.13 A typical underground map of development working showing survey stations (points and circle), rock contacts (C-C), structures and mineralized stringers (black) in dolomite host rock at Balaria mine, India.

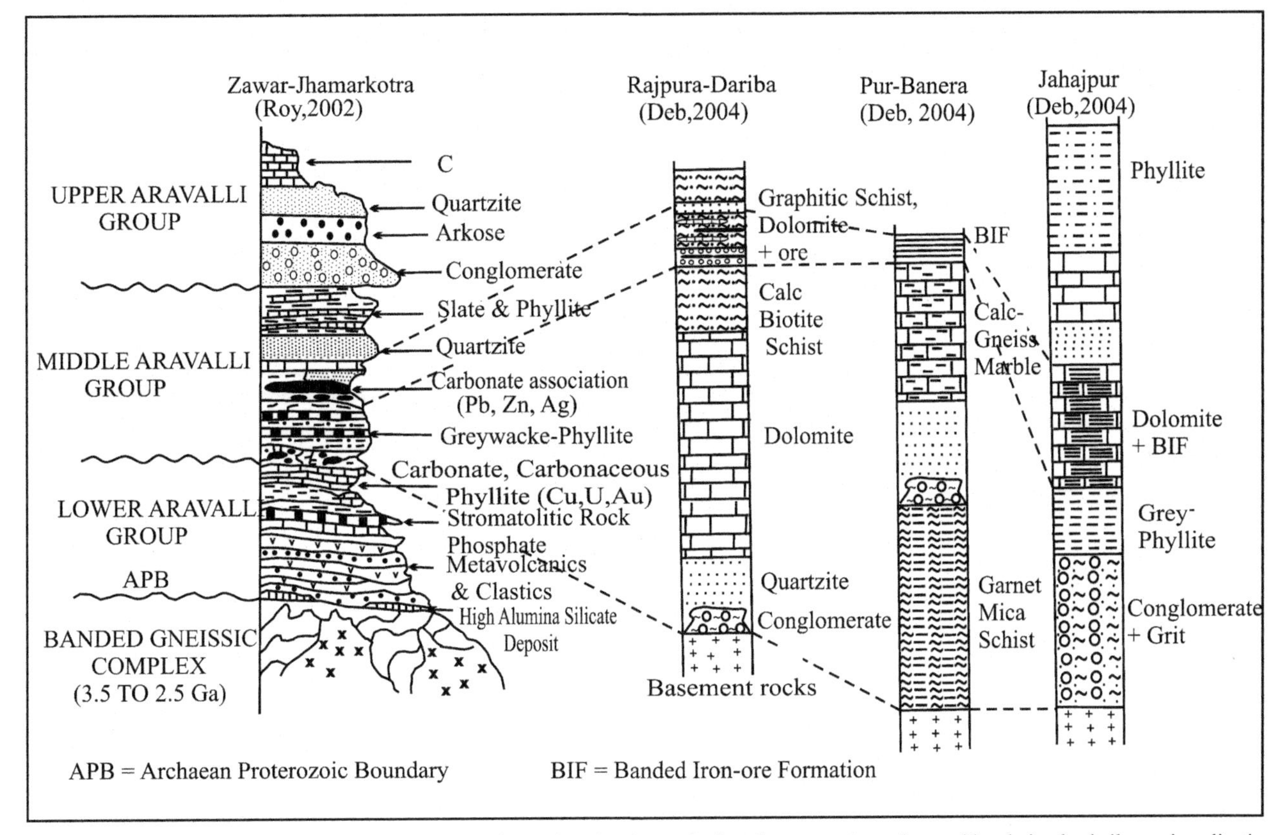

FIGURE 3.14 Lithostratigraphic correlations of Aravalli Province hosting rock phosphate, copper-uranium-gold and zinc-lead-silver mineralization from lower to higher horizons (Source: Haldar, 2007) [33].

3.7.2. District Scale

- Submission of Prospecting Lease, designing the exploration scheme with respect to work component, technology, type of exploration method, time and fund requirement (refer Table 1.2).
- Geological mapping of the target areas, recognition of surface signatures like presence of weathering and alterations, identification of host rock, structural settings and control.
- Ground geophysics and geochemistry, pitting, trenching, data synthesis and interpretation for reinforcing the drill targets (refer Fig. 4.2). Drilling continues to delineate proved, probable ore reserves and possible resources.
- Baseline environment plan.
- Investment decision and next phase.

3.7.3. Local Scale

- Detail geological mapping of host rock and structure controlling the mineralization, close-spaced surface directional drilling (refer Figs 15.2 and 15.7) to compute reserve with high confidence, pitting, trenching and entry to subsurface for level development, underground drilling for precise ore boundary, reserve of higher category, metallurgical test work and environmental baseline reports.
- Scoping study, Pre-feasibility study and Feasibility study.
- Submission of ML application along with Environment Management Plan.
- Ore production, mineral processing, extraction of metal or salable commodity.
- Cash inflow-out flow (Fig. 13.3).

3.7.4. Exploration Components

The various exploration components that can be summarized as:

- Sampling: soil, pitting, trenching, grab, chip, channel, directional drilling, sample reduction, check studies, tests of Quality Control and Quality Assurance that has been elaborated at Chapter 7.
- Optimization of drilling.
- Preparation of cross-section, longitudinal vertical section, level plan, 3D orebody modeling, estimation and categorization of reserves and resources (Chapter 8).
- Environment Management Plan.
- Sustainable development in mining (Chapter 14).

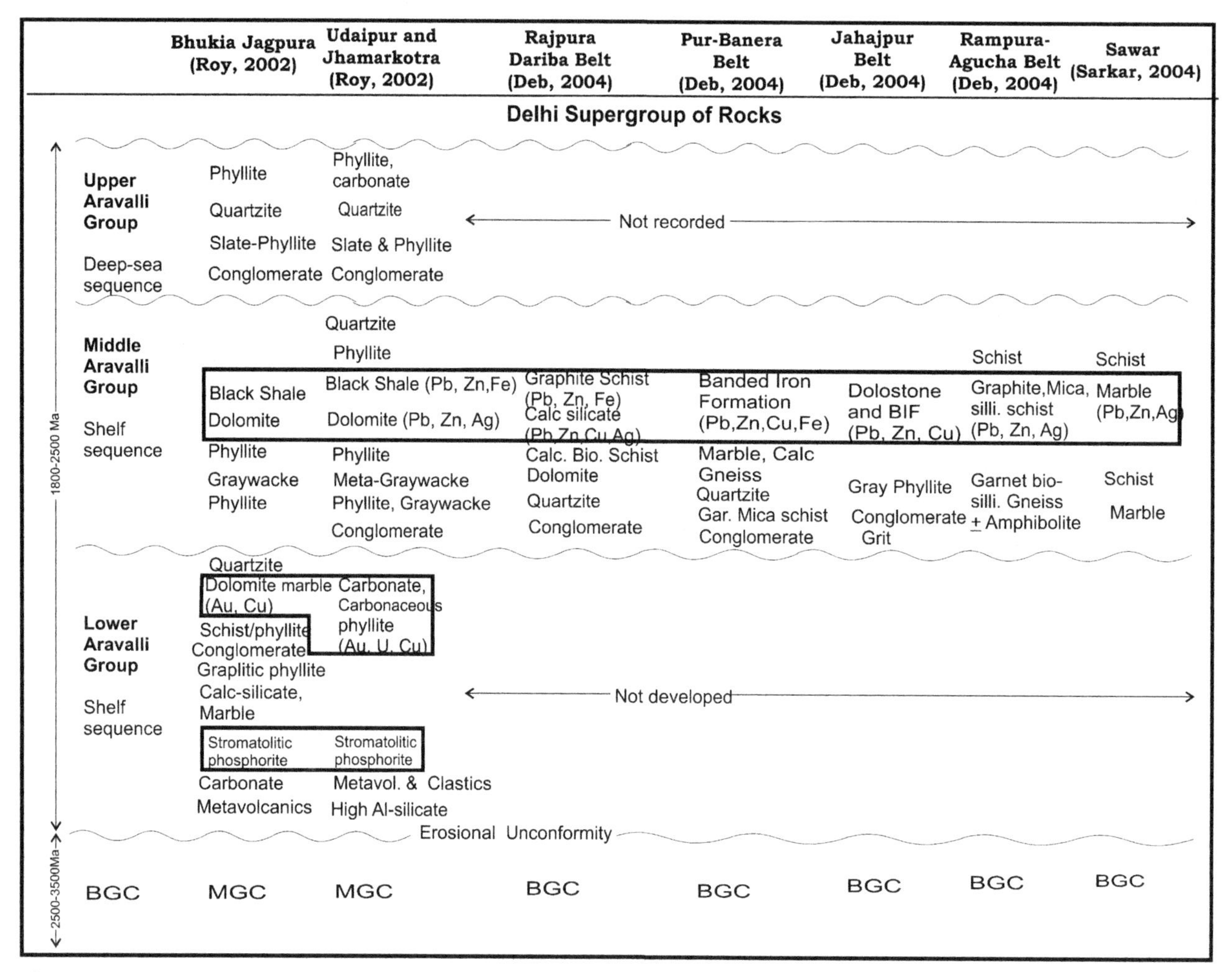

FIGURE 3.15 Lithostratigraphic correlations and comparative statement of the Aravalli rocks hosting rock phosphate, copper-uranium-gold and zinc-lead-silver mineralization (Source: Haldar, 2007) [33].

3.8. EXPLORATION OF COAL AND COAL BED METHANE

Coal is stratified carbonized remains of plant material transformed over millions of years. It is formed, first, by heavy growth of vegetable matter followed by accumulation and in situ burial under sediments. The next process is the transformation to coal by chemical and thermal alteration of organic debris. Coalification can also happen by drifting of the plant material to distance lakes or any water body and submersion under sediments. Plant materials tend to alter progressively through peat-lignite-subbituminous-bituminous and anthracite coal during the transformation process. Water, carbon dioxide, nitrogen and methane gas are produced along with coal in this process.

Coal bed methane (CBM) is a clean burning fuel for domestic and industrial uses. Its extraction reduces explosion hazards in underground coal mines. These gases are part of the coal seam at different percentages. It occurs as free gas in the fractures or absorbed into the micro pore surfaces in the matrix of the coal beds. The amount of methane held in coal seam depends on the age, moisture content and depth of the coal seam. The excess gas migrates into the surrounding rock strata and sand reservoirs that may overlie the deeply buried seams. The gas is being tapped and sold in commercial quantities using recent suitable technology. It is identified as a cleaner natural gas (CNG) form of energy than traditional coal and petroleum.

The exploration for coal and CBM includes geological mapping, study of geological setting of coal basins, remote sensing, surface geophysical inputs, core and noncore drilling technology, digital down-the-hole logging, use of GPS and microcomputers. Remote sensing data can identify the major lineaments, faults and other tectonic setup useful to explore coal-bearing area. High-resolution seismic survey can define the basin configuration, its tectonic style, thickness of coal-bearing formation, lateral continuity and approximate depth of different coal seams.

The design and procedure for core and noncore drilling program for coal seam must be performed on a sequential approach by successively narrowing the drilling interval along the elongation of expected coal seam. Down-hole geophysical logging will be helpful in proving the continuity of seam in strike and dip directions. The use of bentonite drill mud is substituted by high-density polymer foams to facilitate removal of cuttings and stabilize sidewall of drill hole, thereby allowing ready conversion of drill holes to monitoring wells. The core recovery in and around the coal seam should be achieved over 85% by using split tube core barrels. Samples of drill cuttings are taken at regular intervals for analysis. Formal core descriptions are made and the core is frequently photographed by digital cameras that are then appended to computerized drilling reports.

The operational aspects must achieve the aims of maintaining wells and prevent formational damage in case of CBM exploration. Excessive pressure of gas/water induces disproportionate disruption within the formation and imposes primary concern for drilling in the presence of high permeability. It often results in loss of circulation fluid, damage of coal formation and finally sloughing of holes. Another major problem during drilling could be the excessive water flow. Escape of large quantities of water from the coal seam generally obstructs drilling with pressure. The rigs, commonly employed, are portable, self-propelled and hydraulically-driven.

The total recoverable coal reserves in the world as on end 2006 stand at 9,09,000 million tonnes (Mt). This is shared by United States (2,47,000 Mt, 27%), Russia (1,57,000 Mt, 17%), China (1,15,000 Mt, 12.6%), India (92,000 Mt, 10%) and Australia (78,000 Mt, 8.6%). The world coal production during 2009 was 6940 Mt. The major coal producing countries during the same year were China (3050 Mt, 46%), United States (973 Mt, 16%), India (558 Mt, 8%), Australia (409 Mt, 6%) and European Union (537 Mt, 7.7%).

FURTHER READING

Haldar (2007) [33] and Banerjee et al. (1997) [2] on mineral exploration practices will be a good reading. A brief reference of coal in India is available from Deshmukh (1998) [21, 22].

Chapter 4

Exploration Geochemistry

Chapter Outline

Abnormality in chemical signature on surface is defined as geochemical anomaly which may show the way to discovery of large mineral deposits.

—Author.

4.1. DEFINITION

Exploration geochemistry fundamentally deals with the enrichment or depletion of certain chemical elements in the vicinity of mineral deposits other than barren regions. Geochemical prospecting is performed by systematic measurements of one or more chemical parameters, usually in traces, of naturally occurring material in the Earth's crust. The samples are collected from rocks, debris, soil, gossan, stream or lake sediments, groundwater, vapor, vegetation and living things. The results of samples, at times, may show abnormal chemical pattern over the regional properties. This abnormality is defined as geochemical anomaly which may end up with discovery of near-surface or deep-seated hidden mineral deposits.

4.1.1. Elemental Dispersion

Traces of metallic elements are usually found in soil, rocks, and groundwater in the proximity of an orebody. The fundamental dispersion pattern can be due to distribution of elements disseminated in rocks, which occurred during ore-forming process. It can also be formed by migration and redistribution of elements in solution during oxidation, weathering and erosion of mineral deposit. The process usually occurs around the surface environment. The degree of concentration of specific elements logarithmically diminishes away from the deposit till it reaches to the background value of the enclosing rock. The geochemical envelope, which is an expression of alteration and zoning conditions surrounding metalliferrous deposit, is called "primary dispersion halo." The formation of primary halo is synchronous to the mineralization. It is essentially identical with geochemistry of unweathered rocks and minerals, irrespective to how and where the orebody itself was formed. The halos are either enriched or depleted in several elements as a result of introduction or redistribution related

Mineral Exploration. http://dx.doi.org/10.1016/B978-0-12-416005-7.00004-0

to ore-forming phenomena. The shape and size of the halo is exceptionally variable due to diverse mobility characteristics of the elements in solution and microstructures in the rocks.

"Secondary dispersion halo" is the dispersed remnants of mineralization caused by surface processes of chemical and physical weathering and redistribution of primary patterns. The halo can be recognized in samples taken from the soil, rocks, sediments, groundwater and volatile matters at a distance of meters to tens of kilometers. Minerals are often unstable in secondary environment.

Chemical weathering involves elemental breakdown of rocks and minerals in abundance of water, oxygen and carbon dioxide. It can move considerable distances from the source. Multicolored gossan above the base metal deposits at Broken Hill in Australia, Rajpura-Dariba in Rajasthan and Malanjkhand in Madhya Pradesh are unique examples of chemical weathering.

Physical weathering implies disintegration of rocks and minerals with little or no chemical and mineralogical changes. It may often involve long distance transportation from the source. These minerals are chemically resistant. They are oxides (cassiterite, rutile, magnetite, chromite) and native form of gold, platinum, diamond, etc. The example of physical weathering in geochemical exploration is found in "till" and "glacial" deposits.

4.1.2. Pathfinder Elements

"Pathfinder" or "indicator" elements are characteristic parameter in geochemical prospecting. These are relatively mobile elements due to physicochemical conditions of the solutions in which they are found or in volatile state (gaseous). They occur invariably in close geochemical association of the primary minerals being searched. These elements can be more easily found either because they form a broader halo around the deposit or because they can be detected more easily by simpler and less expensive analytical methods. Pathfinder elements play great role in locating concealed deposits due to these special properties. The choice of pathfinder necessitates that the elements used must occur in the primary association with the element being sought, for example copper, nickel and chromium as pathfinders for deposits containing Platinum Group (Platinum, Palladium, Osmium, Iridium, Rhodium and Ruthenium) of Elements (PGE) (Bushveld chromium-PGE deposit, South Africa, Sudbury nickel-PGE deposit, Canada and Nausahi chromium PGE deposit, India), zinc for lead-silver-zinc deposits of Rajasthan and scheelite in Kolar gold field, Karnataka, India. The elements can also be derived from its parent by radioactive decay, such as the use of radon as a pathfinder for uranium deposits. Some common pathfinder elements are listed at Table 4.1.

TABLE 4.1 Common Pathfinder Elements in Geochemical Exploration

Type of deposits	Pathfinder elements
Gold, silver, gold-silver-copper-cobalt-zinc and complex sulfide ores	As
Copper-zinc-lead-silver and complex sulfide deposits	Hg, Zn
Wolframite-tin deposits	Mo
Porphyry copper, barium-silver deposits	Mn, Mo, Au, Te
Platinum-Palladium in mafic-ultramafic rocks	Cu-Ni-Cr-Pd
Uranium (all types)	Rn, Cu, Bi, As
Sulfide deposits	SO_4

4.1.3. Background and Threshold Value

"Background" values are characterized by the normal range of concentration of elements in regional perspective rather than localized mineral occurrences. It is significant to establish the background value of the area against which the anomalies due to economic mineral accumulations, if any, can be identified. Large number of samples comprised of rock, soil, sediments, groundwater and volatile matters are analyzed for multiple elements separately for each area before the exploration begins. Each type of sampled material should be treated separately. The values may vary significantly between samples. The frequency distribution is usually positively skewed. The arithmetic average (mean) is evidently biased by few scattered high value. The most frequently occurring value (mode) tends to lie within relatively restricted range and considered to be the normal abundance or background for that particular element of the area (Fig. 4.1A).

"Threshold" value is defined as the probable upper or lower limit of the background value (Fig. 4.1B) at some statistically precise confidence level. Any sample that exceeds this threshold is considered as possibly anomalous and belongs to a separate population. It may vary for each element, each rock type, different types of samples and in each area. The negative anomalous threshold defines the lower limit of background fluctuation. A geochemical section of a traverse-line is given in Fig. 4.1C.

4.1.4. Orientation Survey

The success of exploration geochemistry depends on the appropriate detection of the pathfinder elements in primary

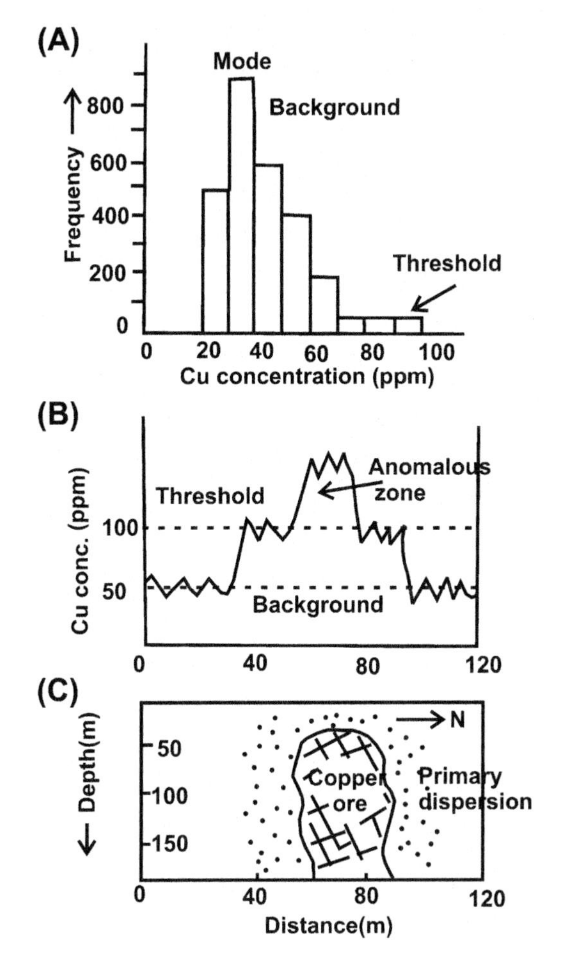

FIGURE 4.1 A typical illustration of (A) Histogram of Cu values showing the most frequently occurring samples, (B) geochemical profile of soil samples viewing background, threshold values and anomalous zone and (C) geological section of a concealed copper deposit.

and secondary environment. In practice, a first round "orientation survey" is conducted in every new area to draw detail work plan that adequately distinguish anomaly from background value. The important parameters to be considered in combination during orientation survey are:

(a) host rock environment,
(b) identify criteria that influence dispersion,
(c) possible local contamination if any,
(d) effect of topography,
(e) best sample medium,
(f) optimum sample interval,
(g) depth of soil sample,
(h) size fraction,
(i) analysis of group of elements,
(j) anomaly enhancement and
(k) analytical techniques for establishing the background and threshold value.

In the reconnaissance phase, regional geological knowledge, possible presence of economic mineral association and previous experience elsewhere will be a guiding factor to plan the orientation survey. The activities are focused around the probable targets with better knowledge of geochemical characteristics of the area for detailed exploration. The procedure should be simplified and finite. Orientation survey is always justified for any new area. It will optimize the sampling program and increase efficiency of interpretation with higher confidence. In turn, it saves considerable time and money in the long run.

"Anomaly enhancement" techniques are commonly practiced during orientation surveys with weak anomalies, particularly for deep-seated mineralization. Value enhancement can be obtained by physical, chemical and statistical means. Physical methods are enrichment of metallic, magnetic and heavy minerals by panning, magnetic and heavy media separation. Chemical methods include selective leaching of iron and manganese oxides in the host environment. Statistical means are anomalous to background and trace elements ratio, additive and multiplicative halo concept discussed at Section 4.5.3. This technique highlights insignificant values of elements of interest in locating concealed mineral deposits.

4.1.5. Regional, District and Local Scale Geochemistry

"Geochemical province" can be defined as a relatively large segment of area in which the chemical composition of bedrock is significantly different from the surrounding. It is manifested by certain suite of rocks relatively enriched in certain specific elements such as: Southern Australia and southeastern Tanzania favorable for locating copper, chromium, nickel and platinum group of elements, Aravalli Mountain province for base metals, Chhattisgarh and Goa for iron ore, India. Similarly, "metallogenic province" represents large area characterized by an unusual abundance of particular type of metal in the country rocks such as: copper-producing area of Peru, Chile, Singhbhum and Khetri (India), lead-zinc-silver-producing area of Mt Isa (Australia) and Sullivan province (Canada), Zawar, Dariba, Rampura Agucha (India), tin at northwestern Europe and Baster in India. There is no certain or unique sample interval. Traditionally low-density stream sediment surveys of one sample per 50–200 or 5–20 km^2 will be adequate for selection of a province depending on the regional geological knowledge and terrain. The analysis is done for 16–25 elements.

A "mineral district" is defined by the presence of known characteristic mineralization such as chromium mineralization in Jajpur-Keonjhar district, India. Stream sediment and limited soil and rock chip samples at 3–6 or even 1–2 km^2 grid interval are followed during prospecting stage depending on the geological heterogeneity and signatures.

The aim of "local-scale" geochemistry is to outline the location and broad extent of mineralization by detailed stream sediments sample at interval of 30−300 m. Rock and soil geochemistry can be exercised in the absence of inadequate drainage system. Once the target is identified more close spaced traverses at 100−300 m apart are sampled for soil and rocks at an interval 10−50 m across. The interpretation is further upgraded and precised by addition of pitting and deep trenching in close grid pattern. The target is now ready for drill testing.

The mission and extent of geochemical exploration is generalized by progressively diminishing the size of search area in which an economic deposit may exist. The activity continues from grass root reconnaissance to detailed sampling until a target is defined that can be tested by drilling. The sequential program demands increasingly more detail and expensive techniques with a sole objective of maximum probability of discovery at the lowest possible time and cost. The stages of exploration, i.e. Reconnaissance → Prospecting → Detail exploration → Feasibility → Mining, have been discussed at Chapter 1 on "Mineral Exploration," Chapter 13 on "Mineral Economics" and Chapter 11 on "Elements of Mining" in this book.

4.2. FIELD PROCEDURE

The field procedure covers the collection of samples. It includes best sample material and optimum sampling design that identifies the presence of mineralization. The major sources of samples are from stream sediments, soils, rocks, groundwater and volatile matter.

During field survey, 250−1000 g samples are collected by hand or with a plastic or aluminum scoop at specified interval from unconsolidated soil. Rock chips are collected by chisel and hammer. It is kept in plastic bags for soil and cloth bag for rocks with code numbers indicating project name, location, and sample type with description. The geological observations and sketches are recorded in field notebook. The grain size is preferred to be fines, say (−) 80 mesh in case of soil. The sample is reduced to 50 g by drying, screening, grinding, cone and quartering at camp site; 5 g will be sufficient for laboratory, and about 1 g for analysis. The remaining sample part is preserved for future references. The quantity of sample to be collected from surface and groundwater varies between 100 and 1000 ml depending upon the number of elements and type of analysis. The water is collected in an exceptionally clean "hard" polyethylene sampling bottle as the elements are in the variation of parts per billion.

4.3. ANALYTICAL METHODS

The methods for analyzing geochemical samples require utmost accuracy and precision, so as to detect background and related anomaly levels by rapid and inexpensive techniques. The analytical methods used in geochemical survey are visual colorimetry, emission spectrograph (ES), atomic absorption spectrometry (AAS), X-ray fluorescence (XRF), X-ray diffraction (XRD), radon atomic emission counter and radiometric analysis, mercury detector, inductively coupled plasma-atomic emission spectrometry (ICP-AES), instrumental neutron activation analysis (INAA), scanning/transmission electron microscope (SEM/TEM), electron microprobe and secondary ion mass spectrometers (SIMS). A brief description follows.

4.3.1. Visual Colorimeter

The "visual colorimeter (VC)" is a rapid and inexpensive method to reduce the bulk of samples collected during the field survey. It is primarily a semi-quantitative and hence less accurate technique to determine Pb, Zn, Cu, Ni, Co, Mo, As, and W, etc., from soil, bedrock and stream sediments. About 10 mg of sample at 80−100 mesh is decomposed by acid digestion and the digested mass is extracted with water or acid. The extract is masked by suitable masking agents for interfering radicals at suitably adjusting pH. An organic chelating agent (say, dithizone and furidioxime, etc.) is added to the solution to develop a colored compound with one of the expected elements. The color intensity of the solvent layer is visually compared with pre-calibrated solution. The lower level of detection (LLD) is 10−50 g/t. The reliability of the results can be accepted with practice.

4.3.2. Emission Spectrograph

The "emission spectrograph (ES)" is also a rapid and low-cost semi-quantitative analytical technique capable of analyzing broad spectrum of elements of widely varying composition with tolerable accuracy. About 10−15 mg of sample at (−) 80 mesh is thoroughly mixed with equal weight of pure graphite powder. The processed sample is completely burned between two carbon electrodes. The arc light passes through a dispersion medium with simultaneous recording of isolated radiation on a photographic plate. The sample plate is compared visually with standards to determine the qualitative presence of an expected element. The intensity of blackening of plate emulsion is indicative of concentration of element present. The LLD is between 100 and 150 g/t.

4.3.3. Atomic Absorption Spectrometry

The "atomic absorption spectrometry (AAS)" is the most suitable, rapid, precise and commonly used method for quantitative determination of large number of samples for multielemental analysis. About 50 mg of sample is treated with 5 ml of aqua regia and digested over hot plate. The solution is aspirated in to an air-acetylene or N_2O-acetylene flame for complete atomization of test metals. Absorbance of a characteristic radiation of the desired metal is measured for computation of elemental concentration. About 30 elements can be determined form one solution, one after another, by AAS using specific hollow cathode lamps with background correction.

4.3.4. X-ray Fluorescence

The "X-ray fluorescence (XRF)" technique uses high-energy X-ray photons from an X-ray generation analyzer to excite secondary fluorescence characteristic X-ray from the sample. The characteristic line spectra emitted by different elements of the sample are detected in the analyzer. The intensity of each line is proportional to the concentration of individual elements.

4.3.5. Inductively Coupled Plasma-Atomic Emission Spectrometry

The "inductively coupled plasma-atomic emission spectrometry (ICP-AES)" works on optical emission method excited by inductively coupled plasma. It is a promising emission technique which has been successfully used as a powerful tool for fast multielemental analysis since 1975. The flame for this technique consists of incandescent plasma of argon heated inductively by radio-frequency energy at 4–50 MHz and 2–5 kW. The energy is transferred to a stream of argon through an induction coil, whereby temperature up to 10,000 °K is attained.

The sample solutions are forced through capillary tube, nebulizer system and spray chamber to relatively cool central hole of the plasma torus. The spray chamber reduces the particle size of the aerosol to an ideal 10 μm. The sample atomizes and ionizes. The radiation from the plasma enters through a single slit and is then dispersed by a concave reflection grating. The light from each exit slit is directed to fall on the cathode of a photomultiplier tube, one for each spectral line isolated. The light falling on the photomultiplier gives an output which is integrated on a capacitor; the resulting voltages are proportional to the concentration of the elements in the sample. Multichannel instruments are capable of measuring the intensities of emission lines of up to 60 elements simultaneously.

4.3.6. Instrumental Neutron Activation Analysis

The "instrumental neutron activation analysis (INAA)" technique utilizes the high-energy neutrons for irradiation of a sample to produce gamma radiation that can be analyzed for detection of elements. The method is suitable for detection of trace elements and Rare Earth Elements (REE), namely, Lanthanides, Scandium and Yttrium with high level of accuracy.

4.3.7. Scanning/Transmission Electron Microscope

The detailed knowledge of physical nature of the surfaces of solids is of great importance in the field of geology, chemistry, and material science. Finer surface information at considerably higher resolution is obtained by scanning electron microprobe. In obtaining an image of "scanning electron microscope (SEM/TEM)" technique, the surface of a solid sample is swept in a raster pattern with a finely focused beam of electrons. The beam is swept across a surface in a straight line (the X-direction), then returned to its starting position and shifted downward (the Y-direction) by standard increment. The process is repeated until a desired area of the surface has been scanned. During the scanning process, a signal is received above the surface (the Z-direction) and stored in a computer system, where it is ultimately converted to an image.

In case of "transmission electron microscope" the instrument is used either in biological application for ultra high-resolution transmission electron photomicroscope studies of thin slice of cell and tissue material or in metallurgical studies, including, for example the investigation of defect structures in alloy material. Subject to some limitation, equivalent studies are carried out on geological samples. However, samples must be prepared as thin foils. As well as photomicrograph data, quantitative measurements may be derived in two ways, first crystal structure data may be derived from the pattern resulting from electron diffraction within the sample. Second, compositional data may be obtained from characteristic X-ray emissions excited by the electron beam as it is transmitted through the foil.

4.3.8. Electron Microprobe and Secondary Ion Mass Spectrometers

The "electron microprobe" provides a wealth of information about the physical and chemical nature of surfaces. With the electron microprobe method, X-ray emission is stimulated on the surface of the sample by a narrow focused beam of electrons, the resulting X-ray emission is detected and analyzed with either a wavelength or energy-dispersive spectrometer.

"Secondary ion mass spectrometers (SIMS)" has proven useful for determining both the atomic and molecular composition of solid surface. SIMS is based upon bombarding the surface of the sample with a beam of 5–20 KeV ions such as Ar^+, CS^+, N_2^+, and O_2^+. The ion beam is formed in an ion gun, in which gaseous atoms or molecules are ionized by an electron impact source. The positive ions are then accelerated by applying a high Direct Current (DC) potential. The impact of these primary ions causes the surface layer of atoms of the sample to be tripped off, largely as neutral atom. A small fraction, however, forms as positive (or negative) secondary ions that are drawn into a spectrometer for mass analysis.

4.3.9. Choice of Analysis

The choice of analytical technique for a particular element depends upon number of factors, not all related to analytical sensitivity. Such factors might include the stage of exploration, availability of specialized facilities, capital cost of instrumentation, speed and convenience to user for undertaking the analysis. The primary objective is to identify the techniques that are most widely used to determine individual elements and free of bias. The second aim is to examine the analytical potential of individual techniques.

VC and ES are used for determination of selected major elements, including, SiO_2, TiO_2, Al_2O_3, Fe_2O_3 and P_2O_5. These techniques are usually advantageous during reconnaissance survey for availability of results at shortest time in the field. It can identify the presence of certain trace elements for target selection.

A comparative analysis of interlaboratory standard rock/mineral analysis indicated that XRF and AAS analysis are the most widely used and acceptable for determination of major elements as a group (except P_2O_5, H_2O). Since 1984, ICP-AES has become much better established technique for the analysis of trace elements including rare earth elements. There is a considerable overlap in the element coverage of ICP-AES and AAS, where as the former is versatile and the later is specific.

XRF, AAS, and ICP-AES can determine a core of element including Ba, Co, Cr, Cu, Ni, Rb, Sr, Th, V and Zn. AAS is moderately expensive but precise and specific method for Co, Cr, Cu, Cd, Li, Ni, Rb, Sr, V and Zn at trace level. ICP-AES are costly and fast method. SEM, TEM and electron microprobes are excellent, efficient and accurate method for surface- and structure-related compositional analysis. However, the instruments are very costly and may not be within the reach of every one.

4.4. DATA INTERPRETATION

The multielemental data from the geochemical survey can be analyzed by standard statistical methods to express the characteristics of each element. Analysis of frequency distribution, median, mode and mean of the sample population can identify and distinguish the background, threshold and anomalous value. Trend surface analysis and moving average method can produce the regional distribution pattern of the elements being investigated. The isograde contour map will guide the detailed exploration program such as drill testing around the most anomalous zone (Fig. 4.2). Correlation coefficient matrix of multielement data will clearly show the maximum affinity between elements. Cluster analysis deals with mutually correlated elements. It exhibits the greatest within-group correlation relative to the between group, taking in to account all possible combinations of the given elements. Various statistical methods have been discussed in Chapter 9.

4.5. GEOCHEMICAL METHODS

Geochemical prospecting can broadly be classified into the following types depending on the stages of survey, nature of terrain, signal associated with the mineralization, type of analytical instrumentation available and finally time and cost permissible for the program:

1. Pedo-geochemical (soil survey),
2. Consolidated weathered cover prospecting,

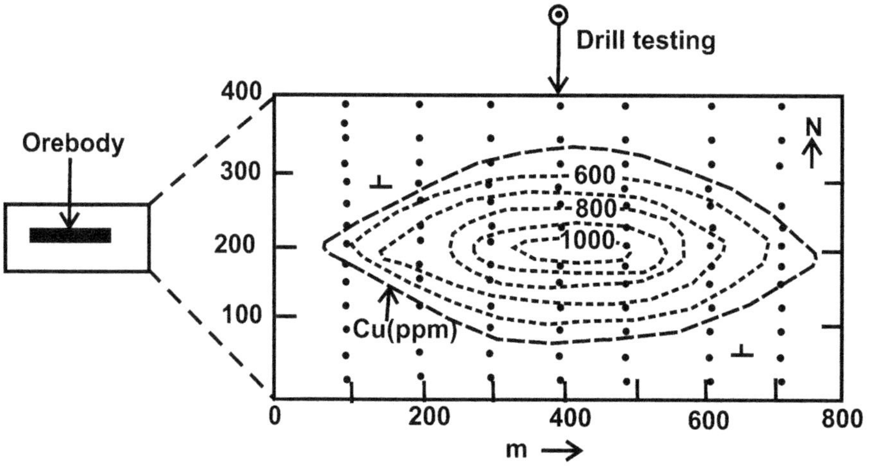

FIGURE 4.2 Data interpretation of soil and pit geochemical samples by contouring copper values to identify best target for drill testing.

3. Litho-geochemical (rock survey),
4. Drift or till geochemical survey,
5. Stream sediment survey,
6. Hydro-geochemical survey,
7. Vegetation survey
   (a) Geobotany,
   (b) Biogeochemical,
8. Geo-zoological/homo-geochemical survey,
9. Atmo-geochemical (vapor survey),
10. Electro-geochemical survey,
11. Radiogenic isotope geochemistry,
12. Heavy mineral survey,
13. Polymetallic-polynodule survey.

4.5.1. Pedo-geochemical

"Pedo-geochemical survey" is also known as "soil survey." Soil is the unconsolidated weathering product. It generally lies on or close to its source of formation such as residual soil. It may be transported over large distances forming alluvial soil. Soil survey is widely used in geochemical exploration and often yields successful results. Anomalous enrichment of elements from underlying mineralization may occur due to secondary dispersion in the overlying soil, weathered product and groundwater during the process of weathering and leaching. The dispersion of elements spreads outward forming a larger exploration target than the actual size of the orebody.

Soils display layering of individual characteristic horizons differing in mineralogy and trace element composition. Therefore, sampling of different horizons will present different results. The soil profile can be classified in three broad groups such as A, B and C horizons (Fig. 4.3). "A" horizon is composed of humus charged topsoil with minerals. "B" zone represents accumulation of clays enriched with trace elements. "C" horizon consists of bedrock fragments in various stages of degradation and gradually changes to hard parent rock. The metal content is generally higher in B horizon than in the A zone. The anomalous behavior of C zone is similar to the parent bedrock. Therefore, samples from B layer enriched with trace elements are most preferred during soil survey.

Soil samples, from residual or transported material, play significant role during reconnaissance survey (refer Fig. 7.20). It can provide a quick idea about presence or absence of target metals in the environment. Soil geochemistry as a successful exploration tool was demonstrated in discovery and development of the Kalkaroo Cu-Au-Mo deposit, South Australia, large PGE mineralization at Ural Mountain, Russia. Soil sampling had extensively been used for locating base metal deposits in Khetri, Pur-Banera, Zawar, Rajpura-Dariba, Agucha (Rajasthan), Malanjkhand (Madhya Pradesh), Singhbhum copper-uranium belt (Jharkhand) and Sukinda chromium-nickel deposit (Orissa), India.

4.5.2. Consolidated Weathered Cover

Weathering of cover sequence undergoes various types of chemical fractionation over millions of years due to paleoclimatic setup of the region. The consequential resistant residual weathering component of rocks and

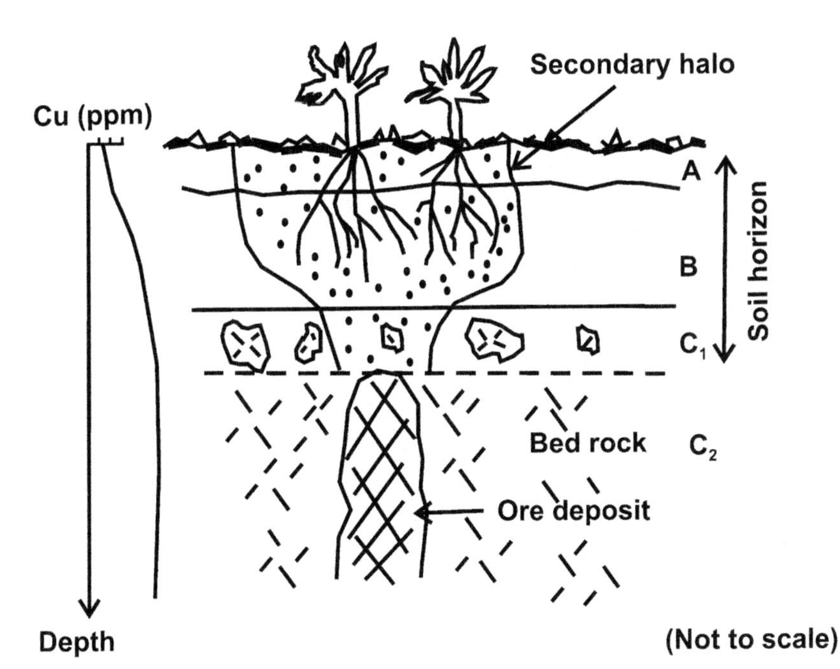

FIGURE 4.3 Diagrammatic presentation of soil horizons, relative vertical and lateral spread of dispersion and associated metal contents.

soils consolidates to form landscape geochemistry. The weathered cover can broadly be classified in to four types depending on their composition, type of weathering and can guide in mineral search.

4.5.2.1. Calcrete

The weathered crust in arid and semiarid region is represented by mixture of sand and silt cemented by calcite, dolomite, gypsum, halite, and ferric oxide simulated mainly by near-surface groundwater and vertical-lateral concentration of minerals like uranium, vanadium, potassium, calcium, magnesium, base and noble metals. Economic calcrete-type uranium deposits occur typically in Australia, Namibia, South Africa, Botswana, China, and recently at desert/semi-desert terrain of Jodhpur and Bikaner, western Rajasthan (vast drainage area of Luni and extinct Saraswati river basin). Calcrete-type weathering indicates presence of carbonates nearby that may be metalliferrous as in case of Rajpura-Dariba base metal deposit in Rajasthan, India. This is also formed near ultramafic intrusions where carbonate nodules are common in soil along river banks like Subarnarekha, east of Singhbhum copper belt, Jharkhand, indicating presence of Cu-Ni-Co-PGE.

4.5.2.2. Silcrete

This is a surface crust of residual weathering where sand and silt are cemented by silica. This is formed in the semiarid region in a stable groundwater condition. Silcrete is commonly found in association with gossan over the copper deposits of Khetri region, Rajasthan.

4.5.2.3. Laterite

Laterite is a consolidated product of humid tropical weathering predominantly composed of goethite, hematite, kaolin, quartz, ± bauxite and other clay minerals. It is red, brown to chocolate colored at the top showing hollow, vesicular, and botryoidal structure. It changes progressively from a nodular iron oxide-rich zone at the top to structureless clay-rich zone and ultimately merges with the partially altered to unaltered bedrocks. Laterite carry enriched grade of Fe, Al, Mn, Ni, Cu, Ti, and V. Lateritic cover can turn in to low-grade iron, aluminum, nickel-copper and gold deposits with the increase of metal content. Geochemical studies of laterite have been successfully used in exploration for Ni-Cu and gold deposits in Western Australia and Ni deposit of Kansa at Sukinda chromite belt, Orissa, India.

4.5.2.4. Gossans

"Gossans are signboards that point to what lies beneath the surface" – Bateman (1950) [4]. Gossans are highly ferruginous rock which is the product of the oxidation by weathering and leaching of a sulfide body. The color is significantly dependent on the mineralogical composition of the iron hydroxides and oxides phases and varies between reddish (hematite), yellow (jarosite), brown, black with stains of azure blue, malachite green and peacock color (copper). Their texture can be brecciated, cleavaged, banded, diamond mesh, triangular, cellular, contour, sponge and colloform with box work of primary sulfides. When grains of sulfide are oxidized and residual limonite remains in the cavities, the texture assumes a honeycomb pattern, called box work of various colors that exist in the capping. The characteristic relic textures and colors resulting from the weathering of certain primary sulfide minerals, like sphalerite, galena, and chalcopyrite, will be specifically diagnostic. The identification can be corroborated by microscopic observation and chemical analysis. The field observations make it easy to detect gossan in prospecting areas with good outcropping conditions. The study of color air photos and satellite images is of much use to focus on certain dark reddish bodies which have to be checked in the field. The depth of gossan may extend from few centimeters to hundreds of meter.

The weathering of sphalerite above massive primary sulfide deposits usually depicts yellow-brown color with coarse cellular box work and sponge structure (Fig. 4.4). The primary mineral sphalerite (ZnS) often changes to willemite (Zn_2SiO_4). The gossans of multisulfide deposits are often associated with typical contour box work from silver-rich tetrahedrite (copper-stibnite) and tennantite (copper-arsenic) minerals.

The dark chocolate color with cleavages, crust, radiate structures and cubic diamond mesh cellular box work (Fig. 4.5) in the gossan zone indicate the presence of galena and tetrahedrite as primary mineralization of the sulfide deposit at depth. The primary galena (PbS) often changes to anglesite ($PbSO_4$), cerussite ($PbCO_3$), pyromorphite ($Pb_5(PO_4)_3Cl$) and mimetite ($PbS(AsO_4)_3Cl$).

The chalcopyrite often changes into native copper, melaconite (CuO), azurite ($Cu_3(CO_3)_2(OH)_2$) and malachite ($CuCO_3Cu(OH)_3$). The dark shades of peacock blue, green, black color (Fig. 4.6) and triangular cellular

FIGURE 4.4 The yellow-brown color, cellular box work and sponge structure of unique gossans formation above Rajpura-Dariba zinc-lead-copper-silver deposit indicate presence of sphalerite underneath.

FIGURE 4.5 The dark chocolate color, crust radiate structure and diamond mesh cellular box work of unique gossans formation above Rajpura-Dariba deposit indicate presence of galena and tetrahedrite underneath.

structure is easily recognizable to expect primary copper sulfide at depth.

Massive sulfide deposits contain usually large quantities of iron sulfides and carbonates (pyrite, pyrrhotite, marcasite, siderite and ankerite) which oxidize and produce an exceptional acidic environment above groundwater table. The acidic environment induces formation of characteristic gossan. Disseminated sulfides with less amounts of pyrite will produce a less acidic environment and gossan with a different geochemical signature.

Geochemical studies of gossan have been successfully used in exploration of Ni-Cu at Malanjkhand Copper deposit, Madhya Pradesh, India. It shows complete alteration and enrichment profile (Fig. 4.7) to form a typical textbook gossan. The thin oxidized cap is represented by limonite with stains of malachite, azurite, and native copper. This is followed by zone of secondary sulfide enrichment in central and southern part of orebody with predominance of covellite, bornite, chalcocite and chalcopyrite. This is the most copper-enriched horizon of the deposit. The secondary enrichment grades into the primary orebody with gradational decrease in secondary minerals with predominance of chalcopyrite and pyrite.

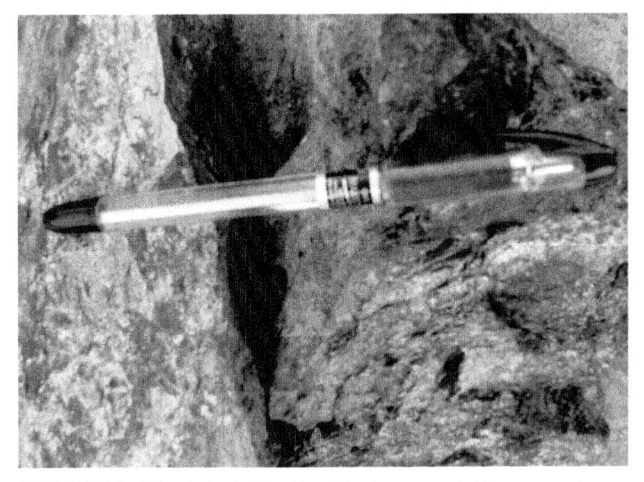

FIGURE 4.6 The dark shades of multicolor peacock blue, green, brown and black, tiny box work of unique gossans formation above Rajpura-Dariba deposit indicate presence of primary chalcopyrite, tetrahedrite and tennantite ore deposit.

Most significant examples can be cited from unique textbook type images as: World-class gossans formation at Rajpura-Dariba Pb-Zn-Cu-Ag deposit India, Zn-Pb-Ag deposit at Broken Hill (Australia), Ashanti Cu-Au deposit, Ghana, Zn mineralization in Togo (western Africa), Rouez gold deposit (France), Hassai Cu-Zn-Au deposit (Sudan), and Al Hazar Cu-Au deposit (Saudi Arabia), (refer Figs 4.8 and 4.9) and Khetri (refer Fig. 4.10).

All the facies of the gossan have to be sampled as they may correspond to different type of hosting mineralization. The weight of sample is about few kilograms scooped from the surface depending upon the homogeneity and chips collected in grid pattern for consolidated mass. Multi-elemental analysis in the reconnaissance stage is recommended. True gossan can be difficult to distinguish from ironstone and other iron oxide accumulation such as laterite.

4.5.3. Litho-geochemical Survey

"Litho-geochemical surveys" (rock survey) are useful during regional work to recognize favorable geochemical province and favorable host rocks. Most of the epigenetic and syngenetic mineral deposits show primary dispersion around the mineralization by presence of abnormal high value of target elements. Litho-geochemistry aims at identifying that primary dispersion, other diagnostic geochemical features and trace element association which are different from barren rocks. Some granite possesses above-average contents of Mn, Mo, Au, and Te, indicating the potential of hosting porphyry copper deposits as in case of Malanjkhand, Madhya Pradesh. Other granite may contain high uranium. Rocks associated with tin deposits in Tasman geo-syncline content 3–10 times Sn value than barren rocks. Most of the SEDEX-type lead-zinc-silver deposits show primary pyrite halo.

The survey is based on analysis of unweathered whole-rock samples or individual minerals. The sampling is carried out on a uniform grid across a geological terrain that includes several rock types from fresh outcrops, wall rocks and drill cores. Rock chip sample consists of five to six small broken fragments which individually or collectively represent a sample. Sample weighing around 1 kg will be adequate for uniform fine- to medium-grained rocks. The weight increases with larger grain size. Primary halos in rocks play a great role in discovering deep-seated hidden deposits particularly having favorable structure. This is possible due to advanced analytical methods for correct

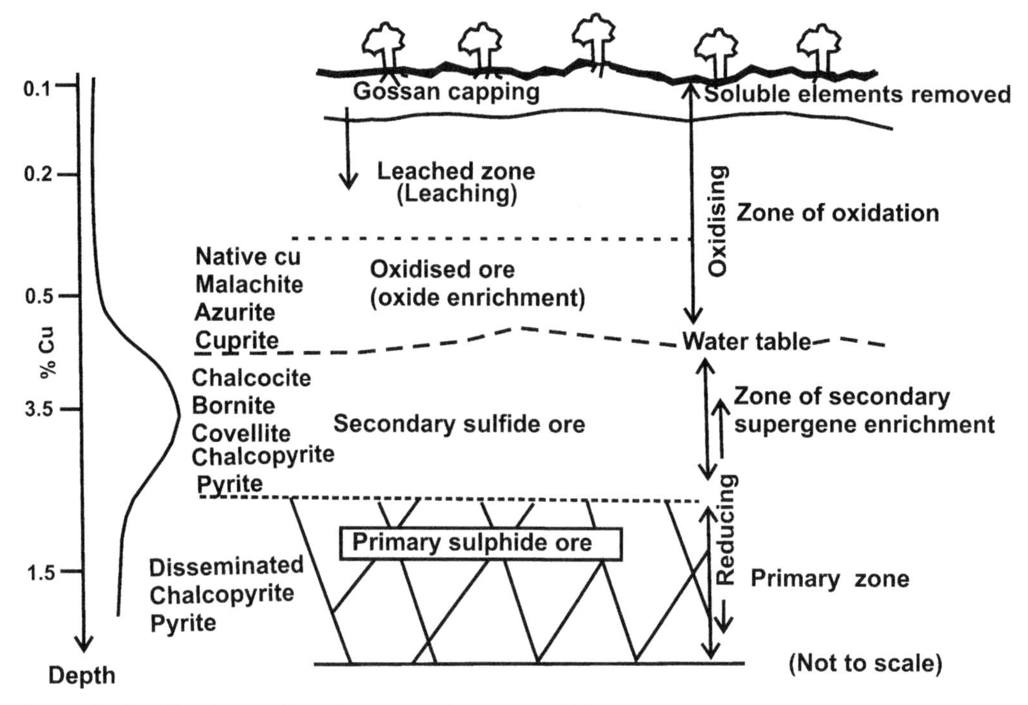

FIGURE 4.7 A generalized profile of gossan formation over massive copper sulfide deposits conceptualized after Malanjkhand copper deposit, India.

FIGURE 4.8 Unique World-class gossans hill above the massive silver-rich zinc-lead-copper deposit at Rajpura-Dariba. In 1977, the site dedicated to the Nation as "National Geological Monument" by Government of India.

FIGURE 4.9 Part of the unique gossans collapsed in the eastern part of the hill during active underground mining operation leaving fresh exposure (Photo: Author, January 2010).

assaying of very low vale and knowledge of multielemental statistical techniques.

The key guiding factors in litho-geochemical survey are:

1. Maximal concentration area with target elements around 10 times higher than background values.
2. Additive halo procedure by adding all the anomalous content of group of indicator elements.
3. Multiplicative halo concept by the ratio between product of all economic elements and product of impurity elements.
4. Vertical zoning of metal distribution.
5. Linear productivity is the product of width of anomaly with % content of the element.
6. Anomaly ratio is the ratio of anomalous to the background value.

4.5.4. Drift or Till Geochemical Survey

Drift prospecting is a broad term for sediments created, transported and deposited under the influence of moving ice

FIGURE 4.10 Gossans are very common features as surface exploration guide at Khetri copper belt, India. The dimension is small compared to Rajpura-Dariba belt.

of glaciers particularly in steep mountain terrain. The various sizes of rock fragments travel longer distances to form the drift sequences. The size and shape of the mineralized boulders along with stream sediments would reveal the extent of transportation and to trace back the source of the parent deposit a few kilometers away at higher elevation. The deposits are classified as glaciofluvial gravels and sands, silts and clays, and till or moraine. Till is a favored sample medium for locating mineral deposits in glaciated terrain. The basal till sequence is studied for the presence of mineralized clasts, heavy minerals and relative abundance of major, minor and trace elements to assess the potentiality. The sample density, depth and method should be selected according to the needs of the exploration program. Concentration of oxidized ore minerals and their product of decomposition can be detected from the fine fractions in surface till sampling. Portable reverse circulation drills are used for collection of till samples at depth and to determine the vertical and lateral variation in till geochemistry. The survey also traces the detrital dispersal of the bedrock mineralization to the primary source. The activity components of till survey are collection of basal till sample, determination of ice flow history and data interpretation.

Till geochemical exploration is extensively conducted in Canada (Thompson Ni-Cu-PGE and Bell copper deposit), North America (Eagle Bay Cu-Mo-Au-Ag deposit) and Glenlyon and Carmacks Cu-Zn-Pb-Ag-Au deposits of Yukon-Tanana terrain and Finland. Gorubathan massive multimetal deposit of Himalaya at Darjeeling district, India, was discovered by the presence of float mineralized boulders in the downstream. The parent body was located few kilometers uphills.

4.5.5. Stream Sediment Survey

Stream sediment sampling is most widely used in all reconnaissance and detailed survey of drainage basins. Many minerals, particularly the sulfide minerals, are unstable in weathering environment and will break down as a result of oxidation and other chemical reactions. The process will motivate secondary dispersion of both ore and indicator elements. The elements will move in solid and solution form to relative further distances within the drainage basin. The mobility of the different elements will vary significantly and eventually detrital fine-grained particles of rocks, minerals, clay, solutions, organic and inorganic colloids enriched in ore and indicator elements will be deposited downstream.

The samples represent the best possible composite of weathered and primary rocks of the upstream catchments area. It is comprised of unconsolidated material in a state of mechanical transportation by streams, springs and creeks. Sample density of 1 in 200 km^2 to close space following the course of the stream depends on the stages of exploration and results obtained.

The optimum size fraction varies in different environment and generally (−) 80-mesh size is recommended. The samples are generally collected in dry season from natural sediment traps along stream courses below confluences or in three-point set around the confluence (Fig. 4.11). The choice of samples from first-, second- and third-order stream will depend on the terrain, climate and nature of weathering of the region. Two sets of samples are collected at each location. The first set is panned for heavy mineral concentration and the second one is the wet-sieved (−) 200-mesh (−75 micron) or (−) 80-mesh fraction of stream sediments. The second set is allowed to settle for 10 min., decanted and transferred in high-quality plastic bag without any contamination, decanted again and air-dried.

The discovery of large porphyry copper deposit of Bougainvillea, Papua New Guinea, gold deposits in British Columbia, lateritic nickel deposit of Sukinda belt, Orissa, and diamond-bearing kimberlite deposit at Wajrakarur, Andhra Pradesh, India, by heavy mineral concentration are few examples of stream sediment survey.

4.5.6. Hydro-geochemical Survey

There are two types of water sources, i.e. groundwater and surface water, having distinctly different chemical and physical properties. Groundwater occurs in dug wells, springs, and boreholes. It has a better potential in exploration geochemistry particularly if it is acidic (low pH) to dissolve and transport metal elements like Cu, Pb, Zn, Mo, Sn, S, U, Ni, and Co than surface water caused by chemical weathering and oxidation followed by leaching. Sample collection will be better if it is shallow enough for easy approach. Surface water from streams, river and oceans has less dissolving power and fine-grained sediments adsorb much of the metal carried by the water. The samples from stream water and sediments are collected simultaneously for analysis.

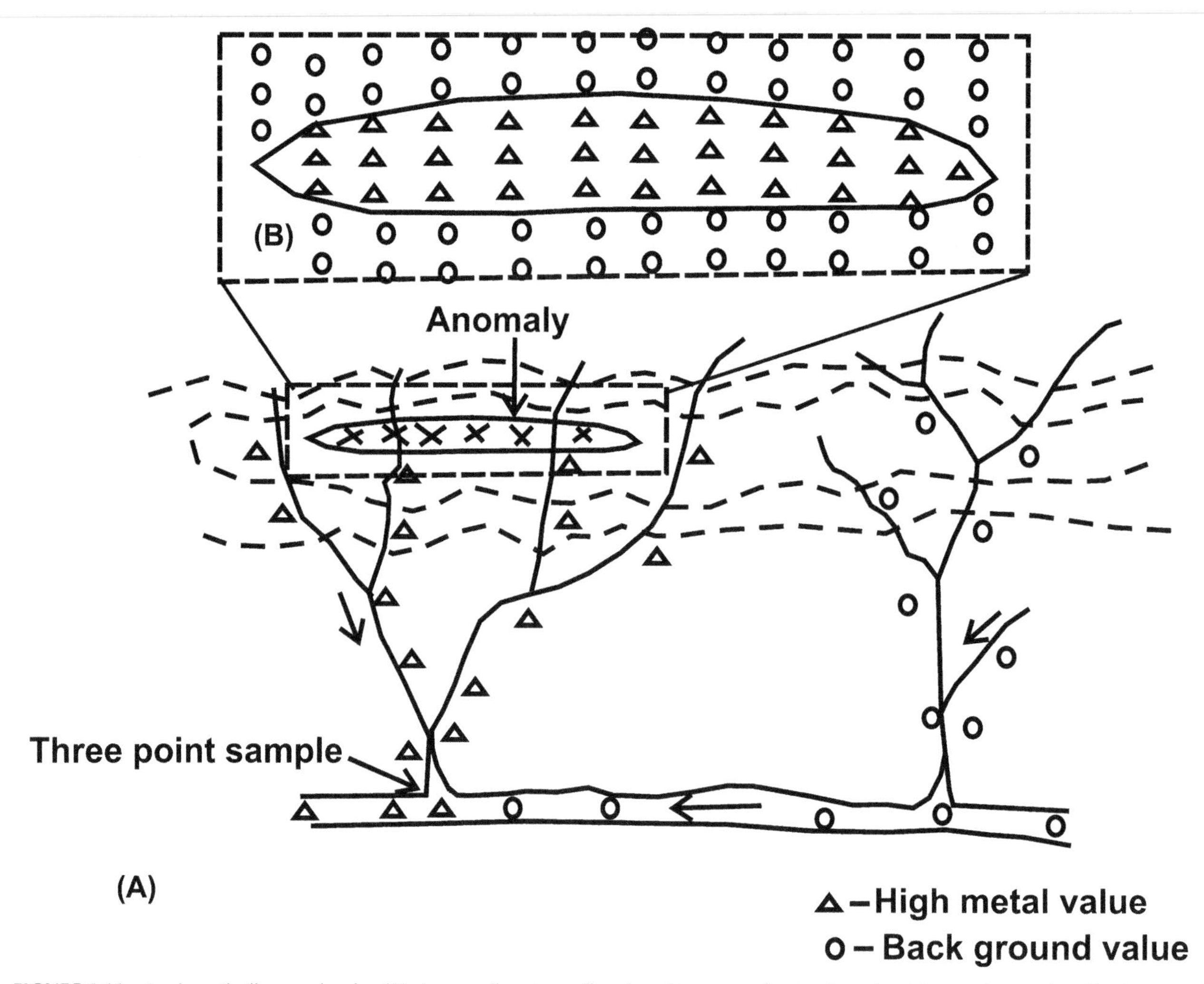

FIGURE 4.11 A schematic diagram showing (A) stream sediment sampling along the course of water channel and three-point sample collection around the confluence and (B) detailed pit or soil samples for drill testing (top inset).

Water samples are easy to collect. About a liter of water is collected and stored in a special quality container. Solubility of metals reduces with increase of pH from 4 to 7. Therefore, the pH is recorded at the time of sample collection. The suspended solids are filtered before analysis. The elemental value changes with time and season. It is desirable to analyze the sample before 48 h of collection. Sample cannot be preserved for future check studies.

Hot springs are probable location for B, Li and Hg mineralization. Geochemical methods are also applied for search of mineral deposits under the sea such as manganese and phosphate nodules on the ocean floor. Water sample is becoming benchmark information on natural dispersion of toxic elements and for identifying pollution.

4.5.7. Vegetation Survey

Vegetation survey can broadly be grouped in two types such as geobotany and biogeochemistry. Vegetation survey will receive prominence as an exploration guide in future as much of the world's mineral resources are hidden beneath the vegetation.

4.5.7.1. Geobotany

A plant will response to its geologic environment in which it grows and may show characteristic variations with respect to form, size, color, rate of growth and toxic effects. Geobotany uses these environmental variations. It involves visual survey of vegetation by recognition of a specific plant population and presence and absence of certain plant varieties associated with particular elements. "Alamine violet" thrives only on zinc-rich soils in the zinc district of Central and Western Europe. *Viola calaminaria* spp. acts as an indicator plant for base metal prospecting. Prolific growth of *Impatiens balsamina* and *Nyctanthes arbortristis* (Seuli) in the rainy seasons, exactly over the outcrops of lead-zinc deposits at Zawar (Fig. 4.12 and Fig. 4.13) and *Leucas aspera* in the ancient mine dump of Rajpura lead-zinc-silver deposit are location-specific indicator

FIGURE 4.12 "*Impatiens balsamina*" or "Garden Balsam" or "Rose Balsam" (Balsaminaceae family) often grow over the outcrops of zinc-lead deposits and act as a natural geobotanical guide for mineral exploration as evidenced at, Zawar belt, India.

plants. Some time normal growth of certain plant variety suffered from malformed or oddly colored caused by excess presence of certain deleterious toxic trace elements on or near the mineralization. Dwarfing of plants and total absence of Sal (*Shorea robusta*) over the Kansa nickel deposit, Orissa, is significant. In contrast, it is abundant in rest of the valley. Bryophyte moss has been a good indicator of U mineralization in the Siwalik sandstone of Himachal Pradesh, India.

4.5.7.2. Biogeochemical

Biogeochemistry involves the collection and chemical analysis of whole plants, selected parts and humus. During chemical weathering, mobilized elements dissolved and enrich in soils. As the plants and trees grow, these dissolved elements, including metals, from the soil are extracted by the roots of the trees which act as a sampling agent. The elements migrate to various parts of the tree such as roots, trunk, stem and finally to the leaves. When the leaves fall to the ground they enrich humus in the metal and the cycle becomes complete (Fig. 4.14). Anomalies indicating buried mineralization can be detected by judicial selection of appropriate parts of the plants (such as roots, bark, twigs, needles, leaves) and subject to analysis.

FIGURE 4.13 Growth of "*Nyctanthes arbortristis*" or "Night Jasmine" or "Seuli" (Oleaceae family) often play as geobotanical guide for exploration of sulfide deposits, Zawar belt, India.

Widely distributed plants of the same species, age and same parts should be sampled from point to point and compared for definition of anomaly. Samples should be washed thoroughly, cleaned and dried before analysis. The quantity should be large enough to generate adequate ash for trace element analysis. *Artemisia* (Sagebrush) accumulates high Cu in British Columbia and Pb, Zn, and Ba in Kazakhstan. *Curatella americana* L. is known to be potentially reliable indicator tree for epithermal gold-quartz veins in Costa Rica.

4.5.8. Geo-zoological Survey

Human and animals of certain territory suffer from specific diseases due to excess intake or deficiency of certain elements enriched in the surrounding rocks and soils. The common transfer routes are through drinking water, milk, and vegetables; cattle feed grass from the local area. People suffering from arsenosis, arthritis, fluorosis, sclerosis, and goiter indicate presence of anomalous trace elements of As, Cd, F, Hg and I deficiency, respectively in the surrounding area. The copper, zinc and lead contents of the trout livers have been used as pathfinder elements for mineralization.

4.5.9. Vapor Survey

"Vapor surveys" (atmo-geochemical) helps in the location of buried deposits through detection of halos of mercury, helium, nitrogen, sulfur dioxide, hydrogen sulfide, hydrocarbon, radon, methane and other gases and volatile elements, often at a considerable distance from the source of mineralization. Vapors can be detected from air and soils and in groundwater. Volatile elements are released through oxidation of ore deposits. Common types of anomalies are as follows.

Mercury vapor anomalies can be determined over structurally controlled mineralization in arid terrain. Hg anomalies are associated with concealed deep-seated high-temperature geothermal system, lead-zinc-bearing sulfide assemblages, and over hydrocarbon gas and oil fields. Hg gas from the soil can be sampled (Fig. 4.15) by precipitation of Hg as an amalgam on extra pure noble metal foils (Ag) in couple of hours and analyzed.

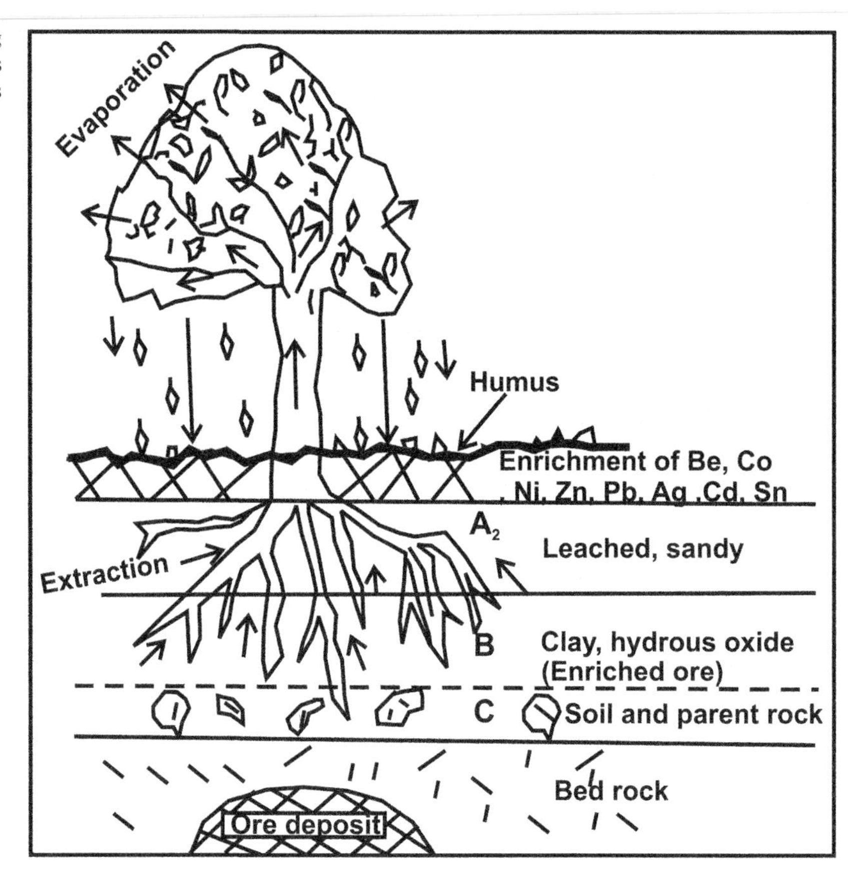

FIGURE 4.14 Schematic illustration showing growth of plants by extraction of metallic elements from soil and migration from root acting as a sampling agent to the leaves.

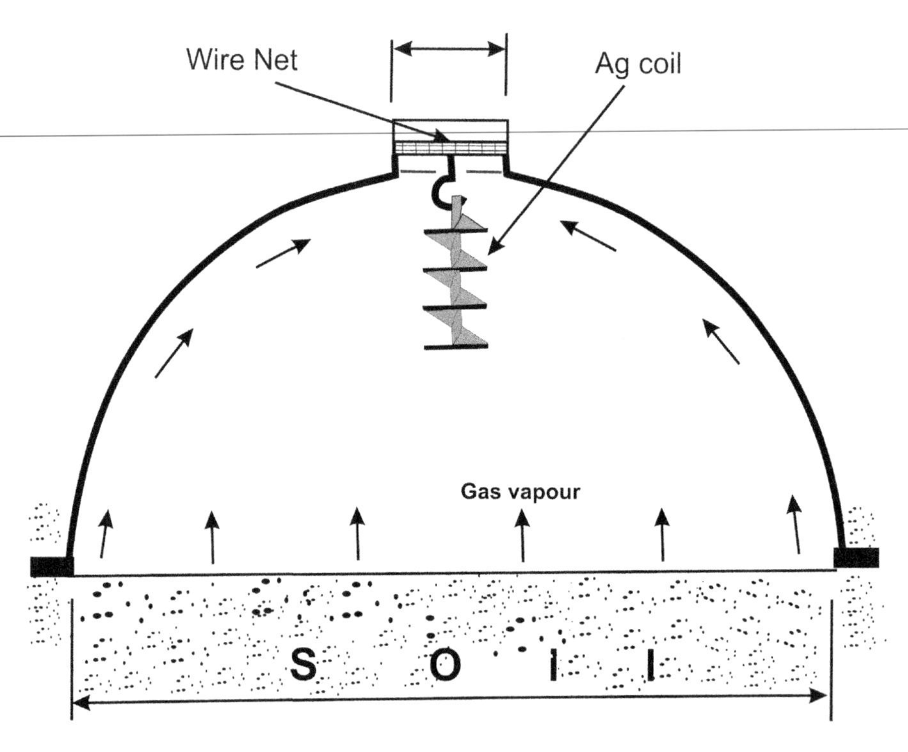

FIGURE 4.15 Vapor sampling instrument (Credit: AK Talapatra, 2006) [70].

"He" anomalies are produced by radioactive decay and found over oil reservoirs, hot spring, porphyry copper and uranium deposits. Samples from shallow depth of soil are collected and analyzed by mass spectrometry.

Nitrogen anomalies are based on the reason that N_2 concentrations increase toward the center of hydrocarbon-bearing basins. Methane (CH_4), nitrogen, other natural gases and asphalt (a sticky black and highly viscous liquid or semi-liquid) are present in most crude petroleum basins and coal deposits. The bubbles of natural gases, crude oil and hot asphalt from the underlying oil and gas field escape with high pressure to the Earth's surface through fissures. The gas causes bubbles that make the asphalt appear to boil. The release of gas and asphalt to the surface acts as an excellent guide to locate and develop hydrocarbon (oil and gas) basin and coalfields. These features (Figs 4.16 and 4.17) can be seen today at Rancho La Brea Lake Pits in front of the Page Museum in the heart of the Los Angeles City. The site represents one of the world's most famous fossil locations, trapped approximately between 110,000 and 10,000 years from now, above the renowned petroleum-producing basin of twentieth century at Los Angeles County. Asphalt and methane appear under surrounding buildings and require special operations for removal to prevent weakening building foundations.

SO_2, CO_2 and CS_2 are commonly found in soil over copper deposits and their wall rocks.

4.5.10. Electro-geochemical Survey

Electro-geochemical survey, also known as CHIM (CHastichnoe Izvlechennye Metallov) survey in China, became well accepted during 1970s for landscape geochemical prospecting around arid to semi-arid region with deeply weathered terrain. The vertical dispersion of metal ions from the deep-seated orebody to the surface by electrochemically mass transport through rock capillaries can be identified either in the soil or by vapor sampling. However, the ionic concentrations are too feeble to be detected by traditional geochemical methods. The electro-geochemical technique is capable of collecting larger volumes of mobilized metal ions on electrodes placed in the soil and applying small currents for a sustained period. The survey line set up over the expected hidden orebody is comprised of a series of specially coated carbon electrode pairs. The electrodes are placed at about 20 cm depth, about 60 cm apart and covered by soil. The electrodes are connected with 9 V DC battery and left for about 48–72 h. The electrode units are unearthed; the absorbent coatings scaled out and digested in concentrated nitric acid for elemental analysis by ICPMS. Soil samples are also collected from each electrode pit and analyzed by ICPMS for comparison. The survey scan lines are shifted on either side along the strike direction of expected mineralization at interval anywhere between 20 and 500 m depending on stages of application.

FIGURE 4.16 Oozing of hot methane and nitrogen gas bubbles from Rancho La Brea Tar Pits, Los Angeles County, one of the world's most famous fossil locations. The pits are on the top of crude oil basin discovered in 1900 followed by exploration and production from 1907.

FIGURE 4.17 Oozing of hot crude oil and shiny black asphalt to the surface (foreground) over the abandoned petroleum basin in front of Page Museum, Rancho La Brea, in the heart of Los Angeles County. The background is the lass green grass park for public recreation around the Museum (Author, Srishti and Srishta, reconnaissance tour, July 2010).

This technique is often recommended to validate targets indicated by geophysical conductivity, prominence of Hg concentration, nonpedogenic calcrete anomalies and shallow soil sampling before venturing for costly diamond drilling. It has successfully been tested for probing strike and depth extension for the concealed part of Challenger gold deposit in Australia and other deposits in United States, Canada and China. This method, together with Hg-soil geochemistry, is effective for selection of prospecting targets during reconnaissance for concealed ore beneath thick weathered overburden. The method has the advantage of being simple and rapid technique with high efficiency and reproducibility at relatively low costs.

4.5.11. Radiogenic Isotope Geochemistry

The study of radiogenic isotopes geochemistry plays a significant role in modern-day scientific research for determination of chronology of rock-forming events. Isotope geochemistry is an attribute of the relative and absolute concentrations of the elements and their isotopes in the Earth. Variations in the abundance of these isotopes can be measured by "isotope ratio mass spectrometer." The information can reveal the age of the rocks and minerals or the source of air and water. The study of isotope is divided into (a) stable isotope and (b) radiogenic isotope geochemistry. The most stable isotopes are comprised of carbon (stable ^{12}C, ^{13}C and radioactive ^{14}C), nitrogen (stable ^{14}N and ^{15}N), oxygen (stable ^{16}O, ^{17}O and ^{18}O) and sulfur (stable ^{32}S, ^{33}S, ^{34}S and ^{36}S).

The radiogenic isotope is comprised of lead-lead isotope geochemistry. Lead has four stable isotopes – ^{204}Pb, ^{206}Pb, ^{207}Pb, ^{208}Pb and one common radioactive isotope ^{202}Pb with a half-life of ~53,000 years. Lead is created in the Earth via decay of primarily uranium and thorium. The most important ratio pertains to daughter Pb isotope (^{206}Pb and ^{207}Pb) derived from the decay of radiogenic parent uranium (^{238}U and ^{235}U) and thorium isotopes (^{232}Th). The other radiogenic ratios are among (Sm-Nd), Rb-Sr and K-Ar system. Samarium-neodymium isotope system can be utilized to provide a date or isotope signature or fingerprint of geological and archaeological finds (pots, ceramics). ^{147}Sm decays to produce ^{143}Nd with a half-life of 1.06×10^{11} years.

4.5.12. Heavy Mineral Survey

Heavy minerals like ilmenite, sillimanite, garnet, zircon, rutile, monazite, magnetite, titanium, chromite, cassiterite, diamond, gold and platinum have a tendency to form onshore beach and offshore placer deposits. Prominent deposits occurring along coastline of countries like India (Fig. 4.18), Indonesia, Malaysia, and Australia bordering Indian Ocean are the largest marine resources of the world. The beaches in the east and west coast of India are enriched with on and offshore heavy minerals placer deposits and are exploited economically. The sampling is conducted by collecting regular interval vertical column of layered placer deposits. The isograde and isopach contours are drawn for computation of reserves.

4.5.13. Polymetallic-Polynodule Survey

Polymetallic-polynodules, popularly known as manganese nodules, are rock concretions on the sea bottom formed by concentric layers of iron, manganese and other high value metals around a tiny core. The size of a fully developed nodule vary from fraction of millimeter to as large as 20 cm with an average size between 5 and 10 cm. Nodules are formed by precipitation of metals from sea water over several million years. Polymetallic nodules occur in most oceans of the world with greatest abundance at vast abyssal floor at depth between 4000 and 6000 m. The areas of economic interest have been identified at north central Pacific Ocean, the Peru basin in the southeast Pacific and the center of the north Indian Ocean. The most promising deposits with respect to resource and metal content occur between Hawaii Island and Central America in equatorial Pacific Ocean.

The sample collection from prospective area of sea bottom is discussed at Chapter 7 and Fig. 7.31. The worldwide resource has been estimated at 5,00,000 million tonnes. The nodules are of greatest economic interest metal contents varying between nickel (1.25 and 1.50%), copper (1.00 and 1.40%), cobalt (0.20 and 0.25%), manganese (~30%), iron (~6%), silicon (~5%), aluminum (~3%) and with lesser amounts of Ca, Na, Mg, K, Ti and Ba.

Since 1970, research and development works initiated to identify the best ocean bed nodule deposits, establishing mining and process route by prospective mining consortia comprised of Federal and Private Companies from UK, United States, Germany, Belgium, Netherlands, Italy, Japan, France, Soviet Union, India and China. As a result, multiton of nodules from the abyssal plains (~5500 m depth) of the eastern equatorial Pacific Ocean were collected. Significant quantities of Ni, Cu, and Co extracted from this ore using both pyro and hydro methods. However, the activities could not be commercialized due to excess availability of onshore nickel metal, anticipated ecological imbalance and finally conservation of the natural resources for the future. The research continues.

4.6. REVIEW

Geochemistry plays vital role in mineral exploration long before conceptualization and continues much beyond the closure of mine operation. Baseline geochemical maps, at national and international levels, are generated by multi-elemental study of soil and rocks at interval on a scale of tens to thousands of kilometers covering the Earth's surface. The samples are analyzed for total range of trace element (Ag, As, Au, Bi, Ca, Co, Cr, Cu, F, Fe, Hg, K, Mn, Mo, Na, Ni, Pb, Pd, Pt, Rn, Sb, Si, Te, Zn, etc.). The baseline maps, showing the regional spatial variation in chemical composition, are aimed at mineral exploration and broad-based environmental issues. An entire range of geochemical techniques have been focused with decision-making criteria based on site conditions, judgment and previous experience during reconnaissance to closure of operations.

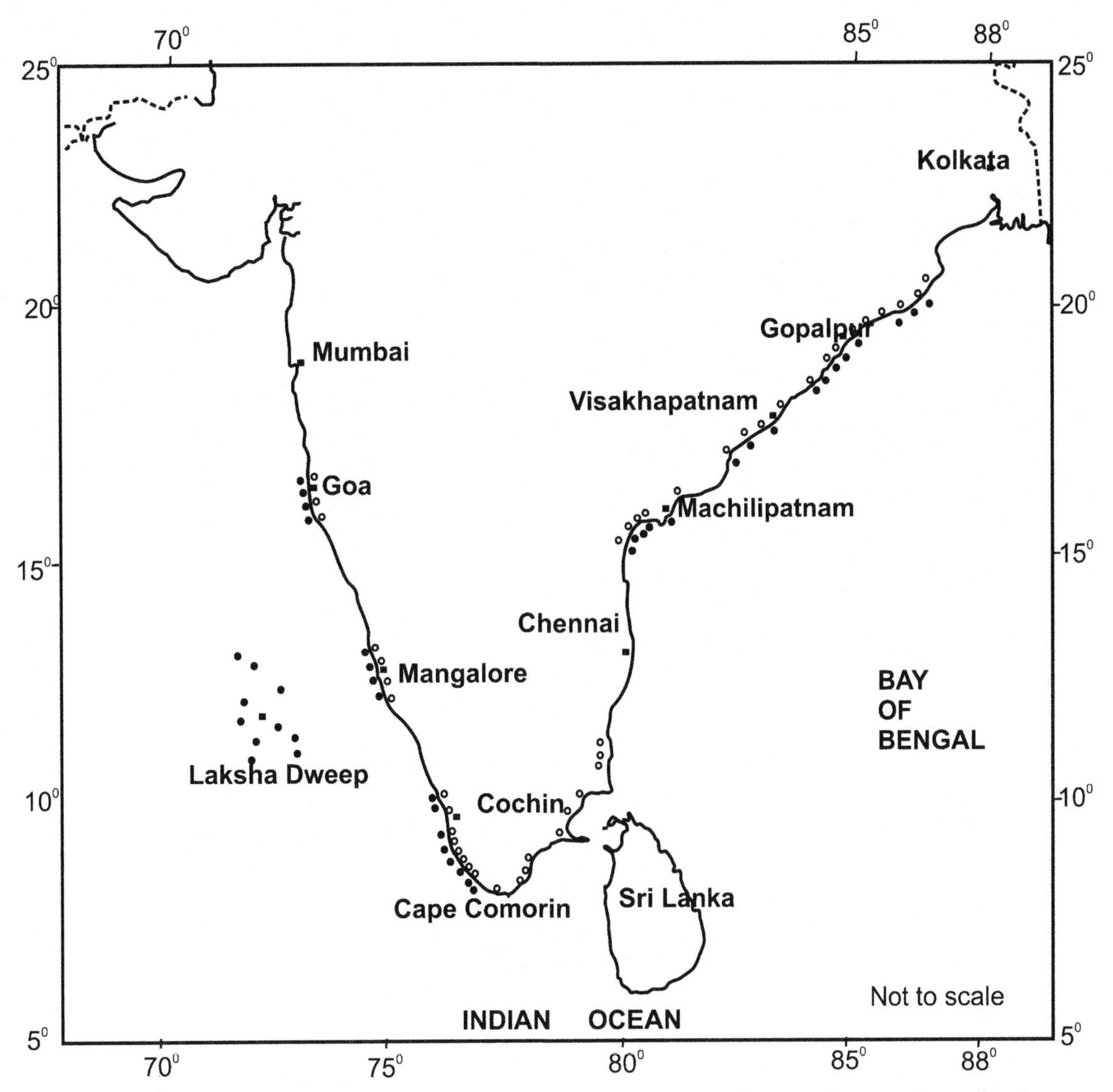

FIGURE 4.18 Heavy mineral placer deposits containing titanium, chromite, diamond, ilmenite, magnetite, monazite, rutile, zircon, garnet, and sillimanite along Indian coastline.

Important parameters to consider when designing or evaluating a geochemical survey are:

- Orientation survey,
- Areal extent of the survey,
- Sampling density,
- Type of samples to collect (soil, bedrock, stream water, vegetation, gas, etc.),
- Postcollection treatment of the samples (anomaly enhancement, sieving of soil samples into different particle size fractions),
- Choice of chemical analysis.

Geochemical exploration eventually works on real-time modeling — ability to determine new sampling, drilling locations before moving drill rig for target testing, infill drilling, need to redrill and step-out decisions. It reduces lead times, which is especially important, when exploration season is short.

FURTHER READING

Govett (1983) [28] was the premier textbook writer of exploration geochemistry. Bateman (1950) [4], Hawkes et al. (1962) [36], Levinson (1974) [51] and Beus et al. (1975) [3] described a good account of geochemical application in mineral exploration. Govett, 1983, gave a detailed analysis of rock geochemistry in mineral exploration in third volume of a series on "Handbook of Exploration Geochemistry." Rose et al. (1983) [60] dealt with advance methods in geochemical exploration. Talapatra (2006) [70] in his modeling approach elaborated nonconventional techniques for concealed land and offshore deposits supported by Indian case studies. Recent publications of Colin Dunn (2007) [14] and Carranza (2008) [8] can be referred for advance studies in geochemistry.

Chapter 5

Exploration Geophysics

Chapter Outline

Neves-Corvo multi-metal deposit, Portugal, hidden between 330 and 1000 m below the surface is an outstanding example of geophysical success in mineral exploration.

—Author.

5.1. INTRODUCTION

The science of geophysics is based on the principles of physics to the study of the whole of earth — from the deepest interior to the surface. The subject of geology, geography, geochemistry and geophysics together play a key role in earth science. The former two disciplines involve direct observations of rock exposures on the surface or subsurface workings or borehole cores. These are more often descriptive and qualitative. Geochemistry is partly descriptive and mainly quantitative. The geophysical tools operate much above the ground from aircraft or helicopter platform fitted with multisensors or on the ground in general.

Mineral Exploration. http://dx.doi.org/10.1016/B978-0-12-416005-7.00005-2

Geophysical studies are always quantitative and involve real measurements based on the variation of response pattern or contrast of propagating waves passing through nonhomogeneous medium. The propagation parameters are seismicity, density, magnetic susceptibility, electrical conductivity, resistivity, electromagnetic (EM) and radiometric radiance. The propagating wave reflects and refracts at the interface of rock types, structure, stratigraphic formation and presence of mineralization, water, oil and gas. The measures of variation are with respect to either position of the object such as strength of magnetic field along a profile or function of time such as propagation of seismic waves. The data in both the cases are presented as graphical form (Fig. 5.1). The graph in waveform will represent physical variations in the underlying structure ("signal" of mineral body) superimpose on undesired variation from nongeological features ("noise" due to presence of electrical cables, rail lines, traffic vibrations, factories, workshops). The "signal" is that part of the waveform that relates to the messages sought from geological features under investigation. The "noise," random or coherent, is other part of the waveform due to extraneous effect. "Anomaly" is a significant departure from the normal pattern of background values. Anomalies must be explained in terms of geologic conditions indicating possible occurrences of oil and mineral deposits.

Data collection, processing, interpolation and interpretation of complex geophysical field are cumbersome with computing limitations. Advent of digital computing during second half of the twentieth century enhanced the computing skill manyfold. The original data, generally smooth continuous function of time or distance, need to be expressed in digital form by sampling the function at a fixed interval and recording the instantaneous value of the function at each sampling point. The analog function of time $f(t)$ or distance $f(d)$ in original form (magnetic, seismic, gravity, etc.) can be converted as digital function $g(t)$ at Fig. 5.2. Data in digital form can easily be processed by high-power computer software. The transformed digital data are represented by only specified discrete values at a series of points at fixed interval. The new waveform is also complex type. Various mathematical techniques, such as Fourier analysis, convolution, cross-correlation, digital filtering, are applied to maximize the signal content useful for geological interpretation. The digital data processing and interpretation simulate an image or model of the subsurface structure.

The execution of geophysical data collection is both from airborne and ground base. Airborne techniques operate from aircraft and helicopters equipped with versatile and distinct multiple sensors including a range of digital cameras covering the entire Infrared (IR) to ultraviolet spectrum, gravity, magnetic, EM and radiometric antenna. A notebook computer is interfaced as proprietary software for flight preparation, digital acquisition and automatic data quality control. The application of geophysics in mineral prospecting begins in the reconnaissance and large area prospecting stage. The airborne methods serve the purpose to outline broad geological features from a vast area of thousands of kilometers for test drilling. The survey continues to the much-detail

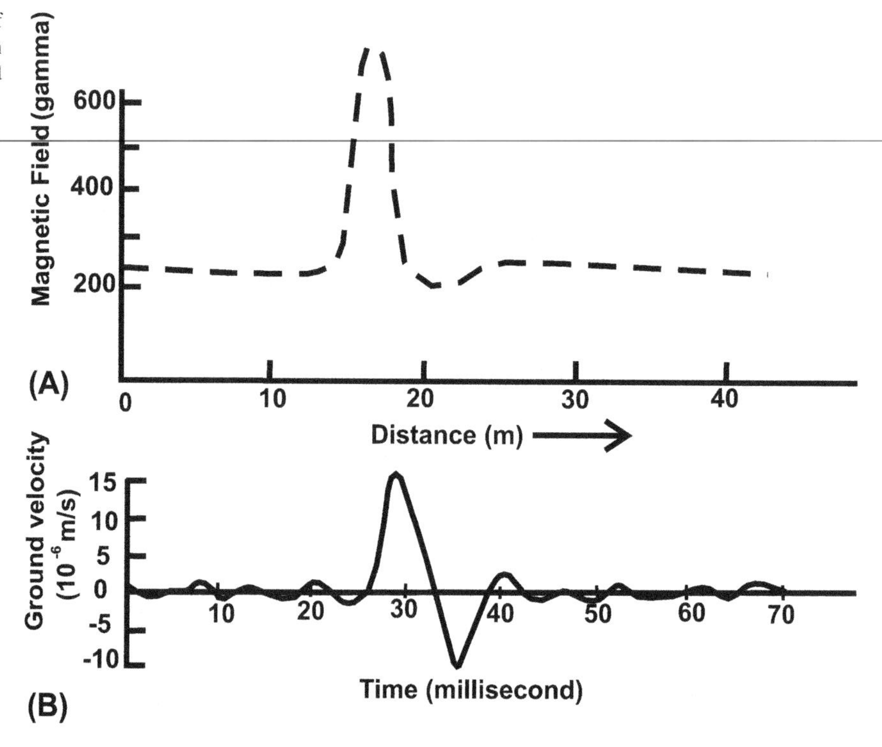

FIGURE 5.1 Graphical presentations of geophysical responses in waveform as a function of distance (A) or time (B) (Source: modified after Kearey et al. (2002) [44]).

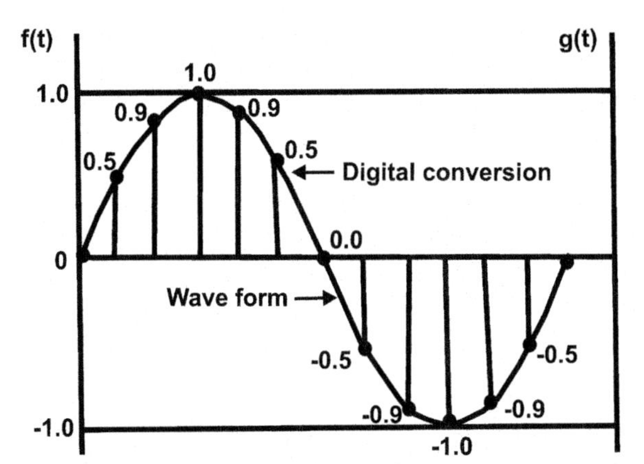

FIGURE 5.2 Digital transformation in waveform of geophysical responses for high-end computer processing (Source: modified after Kearey et al. (2002) [44]).

prospecting stages. The ground geophysical methods will be appropriate to further delineate probable shape-size during prospecting. The findings will prioritize the anomalies for target drilling. The advance geophysical borehole logging can establish the continuity of orebody in all direction during pre-feasibility and mining operation.

Geophysical surveying is capable to provide a relatively quick and cost-effective way of locating subsurface hidden mineral deposits. However, this survey does not dispense with the necessity of drilling. Many a time geophysical surveying is prone to major ambiguities or uncertainties of interpretation and misleading. Specialized training of instrumentation, mathematical interpretative skill, capacity of depth penetration of the method and experience of practicing exploration geologist will be important for successful investigation program. It can optimize exploration programs by maximizing the rate of ground coverage and minimizing the expensive drilling requirements.

The geophysical methods cover a wide domain of applications. This chapter will restrict to the application for mineral investigations only. Geophysical methods are often used in combination with geological and geochemical surveys for superior result. The various key geophysical methods are listed in Table 5.1.

5.2. SEISMIC SURVEY

5.2.1. Concept

Seismology is the science of earthquakes to study the causes and effects of minute pulsation to most catastrophic natural phenomenon inside the earth. The method can be classified broadly into two major divisions depending on the energy source of the seismic waves. If the source is due to the natural shock waves of earthquakes to derive information on the physical properties, composition and gross internal structure of the earth, it is called as "Earthquake seismology." If the seismic waves are generated by artificial explosions, such as detonating a charge of dynamite (land)

TABLE 5.1 Geophysical Surveying Methods with Parameters and Properties Suitable for Type Deposits

Method	Measured parameters	Operative physical property	Suitable deposit type
Seismic	Travel time of reflected and refracted seismic waves.	Density and elastic mode velocity	Oil and gas, layered sedimentary basin.
Gravity	Spatial variation in the strength of Earth's gravitational field.	Density contrast between the surrounding host rocks	Massive sulfides, chromite, salt dome, barite, kimberlite, concealed basin.
Magnetic	Spatial variation in the strength of geomagnetic field.	Magnetic susceptibility	Magnetite, ilmenite, pyrrhotite-rich sulfides.
Electrical			
1. Resistivity	Earth's resistance	Electrical conductivity	Groundwater, sulfide ore.
2. Induced Potential (IP)	Polarization voltage/frequency development of ground resistance	Electrical capacitance	Large sulfide dissemination, graphite.
3. Self-Potential (SP)	Electrical potential	Electrical conductivity	Sulfide veins, graphite, groundwater.
EM	Response to EM radiation	Electrical conductivity and inductance	Sulfide ore, graphite deposits.
Radiometric	Gamma radiation	Gamma ray	Thorium, uranium, radium.
Borehole geophysics and Mise-á-la-Masse	Down-hole probe	All types	Continuity of mineralization in strike and dip.

and/or nonexplosives such as vibroseis or compressed air (marine) at selected sites to infer information about regional or local structure, it is called "Explosion seismology." This is extensively being applied to interpret the interfaces of rock boundaries, particularly for layered sedimentary sequence, location of water table, and in the search of oil and gas. The survey can be carried out on land or at sea for exploration of offshore resources.

The fundamental principle of the survey is the mode of propagation of waves in elastic media or more precisely travel in rock media. The subsurface unit is assumed to be homogeneous and isotropic to simplify the wave propagation resulting in basic interpretation of the measured effects at the plane of discontinuity.

5.2.2. Stress and Strain

The propagation of seismic waves causes redistribution of internal forces and deformation of geometrical shape within a rock mass. The concepts of stress and strain are resultant of these changes.

When a body is deformed or strained, internal forces appear inside the body that tries to recover the original configuration. This balancing internal force or restoring force per unit area across a surface element "*A*" within the material created due to deformation is known as "stress." It can be expressed as:

$$\text{Stress} = (\text{Internal or restoring force}) / (\text{Area}) = F / A$$

The unit is dyne/cm^2.

The stress is termed as normal stress when F is perpendicular to the plane of the area element. It can be tensile or compressive stress depending on whether it is directed from or into the material on which it acts. The stress will be shearing stress when F is tangential to the area element.

"Strain" produced in a body by deforming forces is defined as the fractional changes in dimension being deformed per unit original dimension. Therefore, it can be expressed as:

$$\text{Strain} = (\text{Change in dimension}) / (\text{Original dimension})$$

The dimension can be length, area or volume. It is unitless. Strain that causes only a change in shape with no change in volume is called a shear or distortion strain. A change in volume without change in shape will be called a dilatation or contraction strain. Strains that are associated with relative changes in length in the directions of respective stresses are called normal stress.

5.2.3. Elastic Moduli

The elastic properties of a material are described by certain clastic constants, which quantitatively specify the relationships between different types of stress and strains. The velocity of elastic wave traversing in homogeneous medium depends on number of factors like Young's modulus, bulk modulus, rigidity or shear modulus and Poisson's ratio. Young's modulus (E) is the ratio between longitudinal stress: longitudinal strain. Bulk modulus (K) is defined as the ratio of the uniform compressive stress to the fractional change in volume. Rigidity stress (μ) is the measure of the stress/strain ratio in case of a simple tangential stress. Poisson's ration (σ) is a measure of the geometrical change in the shape of a clastic body.

5.2.4. Seismic Waves

Seismic waves from natural or artificial sources propagate outward as pulses. There are two groups of seismic waves such as body and surface waves.

Body waves propagate through the internal volume of an elastic solid medium and are of two types: "compression waves" and "shear waves."

Compressional waves (the longitudinal, primary, P-waves of earthquake seismology) are the fastest of all seismic waves. It propagates by compressional and dilatational uniaxial strains in the direction of wave travel through solid, liquid and gas medium. It is a body wave that causes the compression of rocks when its energy acts upon them. When the P-waves move past upon a rock, the rock expands beyond its original volume, only to be compressed again by the next P-wave (Fig. 5.3A).

Shear waves (the transverse, secondary, S-waves of earthquake seismology) carry energy through the earth in very complex pattern. The particles of the medium vibrate about their fixed mean position in a plane perpendicular to the direction of wave propagation. It moves slowly than P-waves and cannot travel through

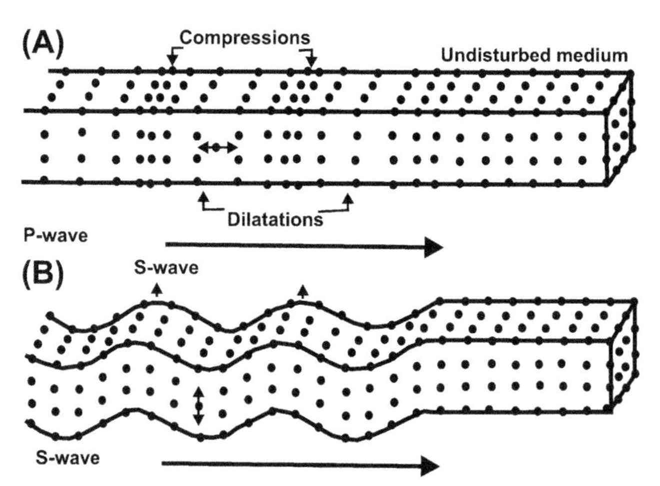

FIGURE 5.3 Propagation of (A) compression or primary and (B) shear or secondary waves showing type of elastic deformation of ground particles (Source: conceived after Kearey et al. (2002) [44].

outer core as it cannot exist in fluids, e.g. air, water and molten rocks (Fig. 5.3B).

Surface waves (Rayleigh and Love waves) travel only along a free surface or along the boundary between two dissimilar solid media. When the particle motion is a combination of both longitudinal and transverse vibration giving rise to an elliptical retrograde motion in vertical plane along the direction of travel, it is called *Rayleigh wave. Love wave* is a major type of surface waves having a horizontal motion that is shear or transverse to the direction of propagation.

The velocity of propagation of any body wave in any homogeneous, isotropic material is determined by the elastic moduli and densities of the material through which they pass. The traditional seismic survey uses only compress ional waves due to easy detection of the vertical ground motion in the detectors that reaches fast because of high-speed wave velocity. Recording stress and surface waves provides greater information about the subsurface, but at a cost of greater data acquisition and consequent complex processing. Multicomponent survey is becoming more popular and useful.

5.2.5. Seismic Reflection and Refraction Method

It is the frequently practiced method for mapping underground structure in sedimentary formation in connection with oil exploration. The principle behind this method follows laws of reflection and refraction of optical waves at the contact of two different media. Similarly, P and S seismic waves move uniformly from the source and reflect and refract on the boundary of a second medium with a different elastic velocity. The energy is partly reflected and partly transmitted in the second medium.

Method of reflection profiling can be explained by Fig. 5.4. A wave travels from source "*S*" and reflects at a point "*R*" of the interface at a thickness of h_1 and arrives at the geophone "G" at time interval of *Tx*. The velocity V_1 of the upper layer and depth h_1 to the interface can be obtained mathematically by recording the reflection times at two distances (*x*, *x'*).

The information obtained by a single reflected pulse at one detector position is not enough to establish the existence of a reflecting horizon. In practice, stepwise shifting of the entire shot-geophone with a series of multitrack geophone placed at short interval is used. A continuous mapping of the reflecting horizon is possible in this way as depicted at Fig. 5.5.

A schematic seismic profile of subsurface geological formation is given at Fig. 5.6.

A schematic illustration of marine seismic profiling system is given in Fig. 5.7.

Refraction prospecting uses refracted waves from near-surface explosions and subsurface layering. It is detailed on a smaller scale, particularly for unknown geology. It is a powerful tool of petroleum seismology and theoretical seismology for investigating the basement, depth and crustal structure.

The generation of seismic waves involves explosion of a dynamite charge in a hole or weight dropping. Truck mounted mechanical vibrators (Vibroseis) to pass an extended vibration of low amplitude into the ground and

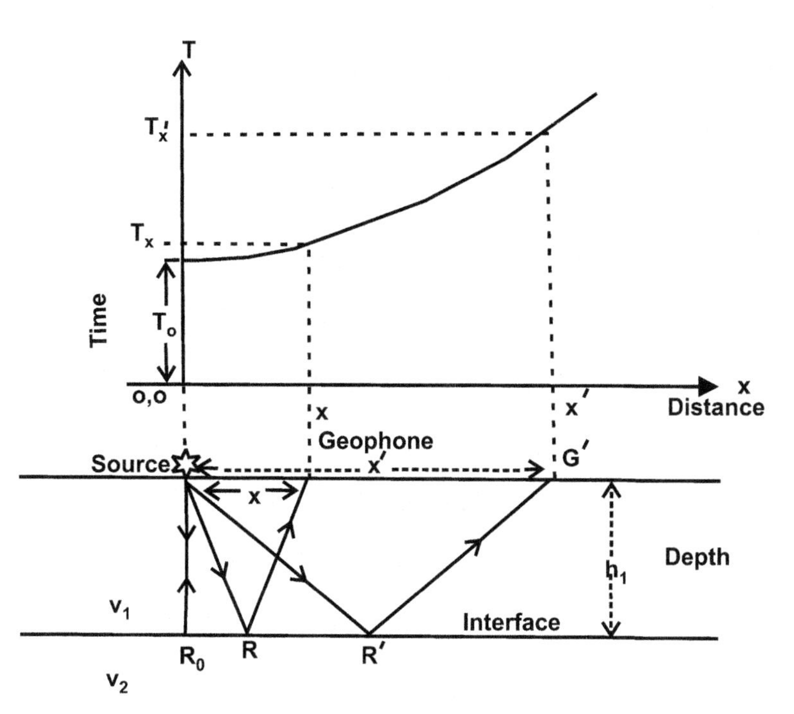

FIGURE 5.4 Method of seismic reflection profiling by time versus distance curve at media interface.

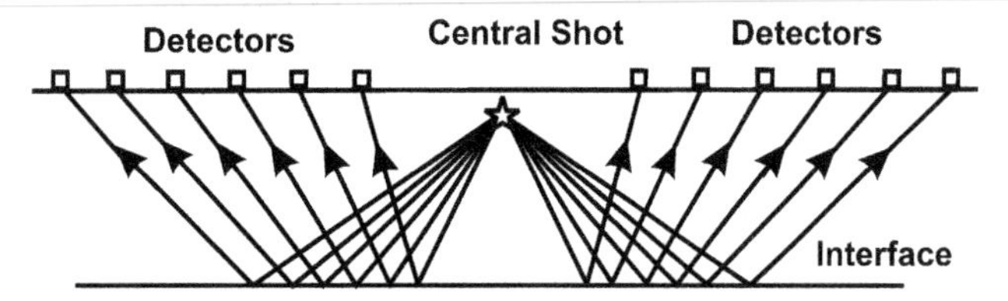

FIGURE 5.5 Multichannel seismic profiling between central shot and multiple detectors on either side (Source: modified after Kearey et al. (2002) [44]).

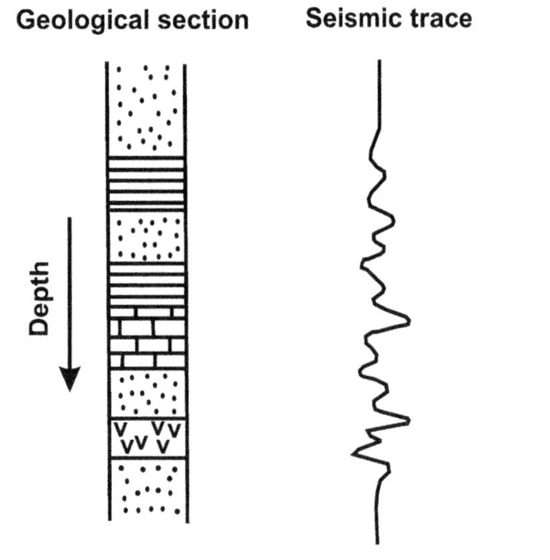

FIGURE 5.6 Seismic reflection profiling of subsurface geological formation (Source: modified after Kearey et al. (2002) [44]).

continuously varying frequency between 10 and 80 Hz. In marine seismic study, an electric or gas spark or air gun shot is used as an energy source. Device used to detect and receive seismic ground motion are called "Seismometer" or "Geophone." It is an electromechanical device used to convert mechanical input (seismic pulse) into electrical output and ultimately produce a continuous graph "Seismograph." The modern seismic survey simultaneously records ground motions receiving from all directions due to combinations of transverse and longitudinal waves by a three-component geophone (Fig. 5.8).

Seismic survey could explain the principal discontinuities, layering and probable materials in the interior of the earth. The technique is most suitable for investigation of oil and gas and to a lesser extent massive metallic deposits. With the introduction of marine seismic survey in the mid-twentieth century, the ocean floor, otherwise unknown, could be precisely mapped discovering the Mid-Atlantic Ridge at an average water depth of 5 km and deep oceanic trenches in the Western Pacific.

5.3. GRAVITY SURVEY

5.3.1. Concept

The basic concept behind Gravity survey is to investigate variation (gravity anomalies) in the Earth's gravitational field generated by differences of density between subsurface rocks. The variation in density is induced by presence of causative body such as salt domes, granite pluton, sedimentary basins, heavy mineral like chromite, manganese, faults and folds, etc., within the surrounding subsurface rocks. The size of the anomalies mainly depends on difference in density between host rocks and causative body, their geometrical form and depth of occurrence. The method is capable to carry survey on ground, air and marine environment.

5.3.2. Theory

The Newton's law of gravitation states that the force of attraction F between two masses m_1 and m_2, whose dimensions are small with respect to the distance r between them then, F is directly proportional to the product of the masses and inversely proportional to square of the distance between them. Therefore,

$$F \propto m_1 m_2, \quad F \propto 1/r^2$$

$$F = (G\, m_1 m_2)/(r^2)$$

$$G = (F\, r^2)/(m_1 m_2)$$

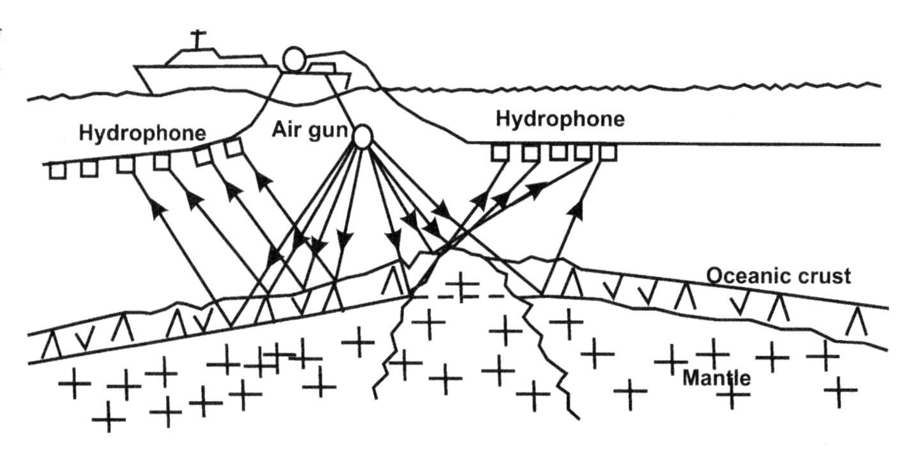

FIGURE 5.7 Schematic diagram of marine base under water seismic reflection profiling.

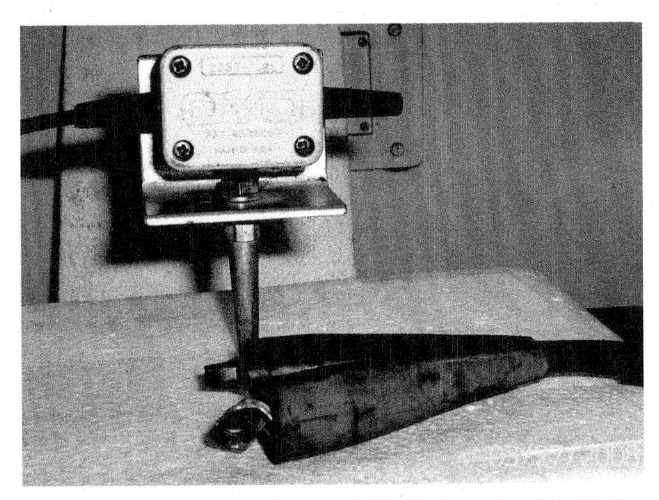

FIGURE 5.8 A three-component vertical 14-Hz "geophone" device for detection and recording of seismic reflection and refraction from subsurface interface.

Where, G is the universal gravitational constant (6.67×10^{-8} dyne cm^2/gm^2 in c.g.s. system) which is numerically equal to the force in dyne that will be exerted between two masses of 1 g each with centers 1 cm apart.

5.3.3. Unit of Gravity

The gravitational field is most usefully defined in terms of the gravitational potential U. The potential U due to a point mass m at a given point P, at a distance r from m, is defined as the work needed by the gravitational force in moving a unit mass from infinity to the final position P. The measuring unit is Gal (cm/s^2) that is strength of a gravitational field or unit of acceleration that will act upon a mass of 1 g with a force of 1 dyne. The value of gravity (g) has a worldwide average of about 980 Gal on the Earth's surface with a variation of 5.17 Gal from equator to pole. The field measuring unit commonly used for the measurement is milliGal (mGal) and gravity unit = 0.1 mGal; 1 Gal is precisely equal to 0.01 m/s^2.

Gravity anomalies result from the difference in density, or density contrast, between a body of rock and its surroundings. For a body of density ρ_1 embedded in material of density ρ_2, the density contrast $\Delta\rho$ is given by $\Delta\rho = \rho_1 - \rho_2$.

5.3.4. Rock Density

The bulk density of a rock depends on its mineral composition, porosity and fluid in the pore spaces. Variation of one or all will change the density locally. Density is commonly determined by direct measurement of samples in air and water. The difference in weight in air and water and the volume of water displaced by immersing the sample provide the volume of the sample. Thus, the sample density can be computed in the following way that is weight of the sample in air divided by the weight of an equal volume of water. The density of common rocks and ore minerals are given Table 5.2.

5.3.5. Measurement of Gravity

"Gravimeter" is the measuring instrument of gravitational field of the Earth at specific location. The instrument works on the principle of measuring the constant downward acceleration of gravity. There are two types of gravimeters: absolute and relative. Absolute gravimeters measure the local gravity in absolute units ("Gal" after "Galileo"). Absolute gravimeters are compact (Autograv CG-5 model) and used in the field. It works by directly measuring the acceleration of a mass during free fall in a vacuum. The accelerometer is rigidly attached to the ground.

Relative gravimeters are spring based. It is a specially assembled extremely sensitive spring balance carrying a fixed mass. The basic principle is that the changes in gravity will result change in weight of fixed mass with change of location. Thus, the length of the spring will differ a tiny little bit (Fig 5.9) with change of location on earth. The spring extension is recorded by suitable optical, mechanical or electrical amplifications with high precision. During any gravity survey, the gravimeter is calibrated at regular interval at a base station where the absolute value of gravity is known. Absolute gravity values at survey stations are obtained by reference to the International Gravity Standardization Network (IGSN).

The progressive spacing of the instrument in a gravity profile (Fig. 5.10) changes from few meters for detailed mineral investigation to several kilometers in regional reconnaissance survey. In case of rapid change in gravity field, the station density must be increased for precise interpretation of gravity gradient. The data sheet contains date, time, location, elevation, water depth and gravimeter reading for each survey station. They measure the ratio of the gravity at the two points.

A commonly used instrument for measurement is Worden gravimeter. It is a compact temperature-compensated gravimeter with precision level < 0.1 mGal. The system is held in unstable equilibrium about an axis. Any increase in gravitational pull on a mass at that end of weight arm is caused by a rotation opposed through a sensitive spring.

Ship-borne system is widely used in marine gravity survey for mapping the ocean floor. Satisfactory tests of airborne helicopter gravity survey are getting prominence. The principal factors that limit the accuracy of the measurement due to acceleration caused by motion of the carrier is about 1−1.5 mGal.

TABLE 5.2 Densities of Common Rocks and Minerals

Rocks	Density (10^3 kg/m^3)	Minerals	Density (10^3 kg/m^3)
Alluvium	1.96–2.00	Cassiterite	6.80–7.10
Amphibolites	2.79–3.14	Chalcopyrite	3.90–4.10
Anorthosite	2.61–2.75	Chromite	4.30–4.80
Basalt	2.70–3.20	Coal	1.11–1.51
Clay	1.63–2.60	Galena	7.40–7.60
Dolomite	2.28–2.90	Gold Native	19.30
Gabbro	2.85–3.12	Gypsum	2.30–2.80
Gneiss	2.61–2.99	Hematite	5.10
Granite	2.52–2.75	Magnetite	4.90–5.20
Limestone	2.60–2.80	Mercury Native	13.60
Peridotite	3.10–3.40	Platinum Native	14.00–19.00
Quartzite	2.60–2.70	Pyrite	4.90–5.20
Rhyolite	2.40–2.60	Pyrrhotite	4.50–4.80
Sandstone	2.05–2.55	Silver Native	10.10–11.10
Schist	2.50–2.90	Sphalerite	4.10–4.30
Shale	2.06–2.66	Quartz	2.59–2.65

5.3.6. Gravity Reduction

"Gravity reduction" is the process of routine correction of field gravity contrast data between an arbitrary reference point and a series of stations that are influenced by various extraneous effects, which are not related to subsurface geology. No interpretation should be attempted before correcting the field data. The reduction processes are due to the following.

5.3.6.1. Drift Correction

Instrumental drift is corrected by repeated readings at a base station at fixed time interval through out the day. The drift correction at time t is d. Positive drift requires negative correction and vice versa. It is due to deficient quality of the spring and change of temperature during the recording period.

5.3.6.2. Latitude Correction

The earth is a nonspherical body with equatorial bulging. Points near the equator are farther from the center of mass of the earth than those near the poles causing gravity to increase from the equator to the poles. The correction must be subtracted from, or added to, the measured gravity contrast, depending on whether the station is on higher or lower latitude than the base station.

5.3.6.3. Elevation Correction (Free Air)

The inverse square dependence on the distance of Newton's law causes the vertical decrease of gravity of the survey point with increase of elevation in the air from the center of the earth. An amendment factor is used in gravitational surveys that take into account the decrease in the force of gravity with increasing altitude. It assumes interference of air between the observation point and sea level. It is equal to −0.3086 mGal/m above sea level. This correction must be added to a measured gravity difference if the survey station lies above the datum plane and vice versa.

5.3.6.4. Elevation Correction (Bouguer)

The Bouguer correction removes the effect of gravitational pull as a function of change in elevation by approximating the horizontal rock layer present below the observation point. The extra mass beneath an object located at a high elevation (say on top of a mountain) causes a higher amount of gravitational force. If the strength of gravity at sea level is known, then its actual value at higher elevations can be approximated using the Bouguer correction. The Bouguer correction is mathematically expressed as (0.4186 "g" "h"), where "g" is the assumed average density of the earth and "h" is the difference in altitude between the place where the

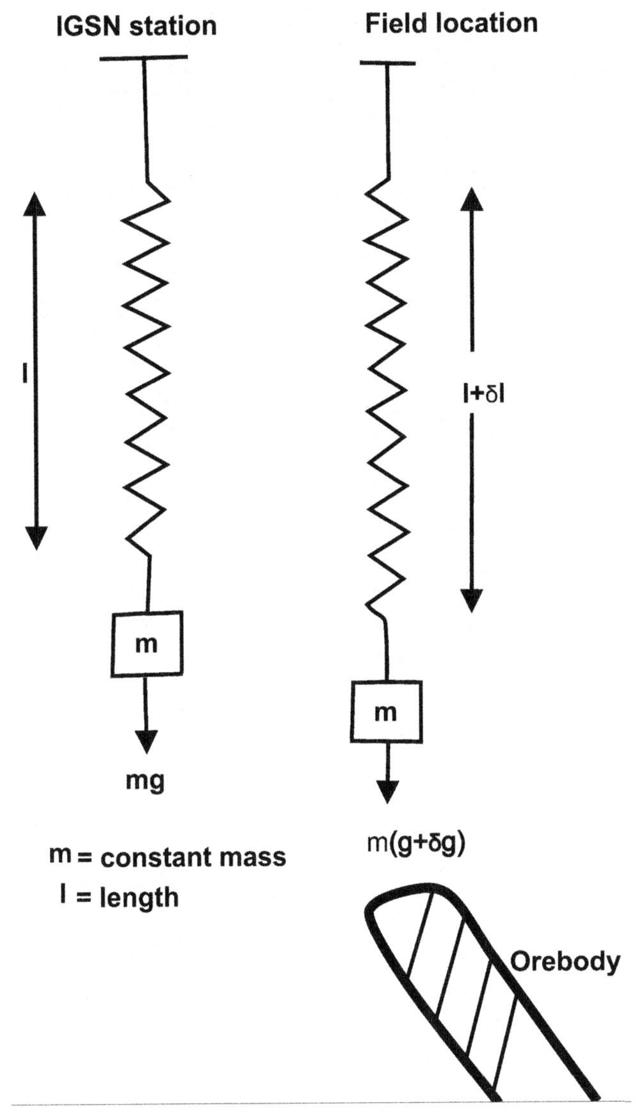

FIGURE 5.9 Schematic diagram showing the principle of stable relative gravimeter survey in the field.

gravity is known and where it is being measured. This value is added to the gravity at the lower location to yield an approximation of its value at the higher elevation.

5.3.6.5. Elevation Correction (Terrain)

The "elevation correction" or "terrain correction" accounts for the topographic relief in the vicinity of the gravity station. The effects are always positive. It is low in the flat areas, but high in steep sided valleys, at top of the cliffs and summits of the mountains.

5.3.6.6. Tidal Correction

The periodical gravitational effects of sun and moon create Earth's tides in the same way as ocean tides behave due to celestial attraction. It is considerably smaller than oceanic tides. Therefore, the distance between observation point and center of the earth will vary periodically caused by the combined effect of the Sun and Moon by few centimeters. This is known as tidal variation and needs correction for high precision survey.

5.3.7. Applications

One of the remarkable applications of gravity survey is the discovery of deep-seated Neves-Corvo massive multimetal sulfide orebodies. There was no surface signature. The ground gravity survey during 1973 over the Portuguese part of Iberian pyrite extension indicated strong positive anomalies that rapidly vary at regional background. The first few target drilling was discouraging. The operation was abundant. The exploration was renewed due to convincing Bouguer anomalies after a time gap of year and half. The fifth drill hole brought the success to intersect four concealed Cu-Zn-Pb-Sn orebodies (Neves, Corvo, Graca, and Jambual) at a vertical depth of 330 m from surface.

5.4. MAGNETIC SURVEY

Magnetic survey is widely used in mineral exploration for investigation of iron ore, magnetite, ilmenite, pyrite and pyrrhotite-rich sulfide deposits.

5.4.1. Concept

Investigation of subsurface geology on the basis of anomalies in the geomagnetic field resulting from the varying magnetic properties of underlying rocks and minerals is the basic principle of magnetic survey. The directional properties of magnetite-rich rock (lodestone) were discovered centuries before Christ. This was modified with the knowledge of Earth's magnetic field or geomagnetism and its directional behavior between the twelfth and sixteenth century. The quantification of the directional properties of geomagnetism and local anomalies with growing sophisticated instrumentation became increasingly important with respect to mineral prospecting during the eighteenth century onward.

Magnetic surveying for mineral investigation with high precision instruments can be operable in the air (airborne), at sea (marine) and on land (ground). The airborne survey is very attractive for scanning large area during reconnaissance to delimit target areas for detailed ground survey during prospecting stage. The process is rapid and cost effective. "Bird" is a magnetic sensor fixed in a string in the tail of the aircraft. Marine surveying is similar to airborne. The sensor is towed in a "fish" kept away behind the ship to remove the magnetic effect of the vessel. It is effective for investigation of polymetallic nodules on the ocean floor. Ground magnetic surveying is performed during prospecting and exploration over relatively small areas on previously defined target by airborne.

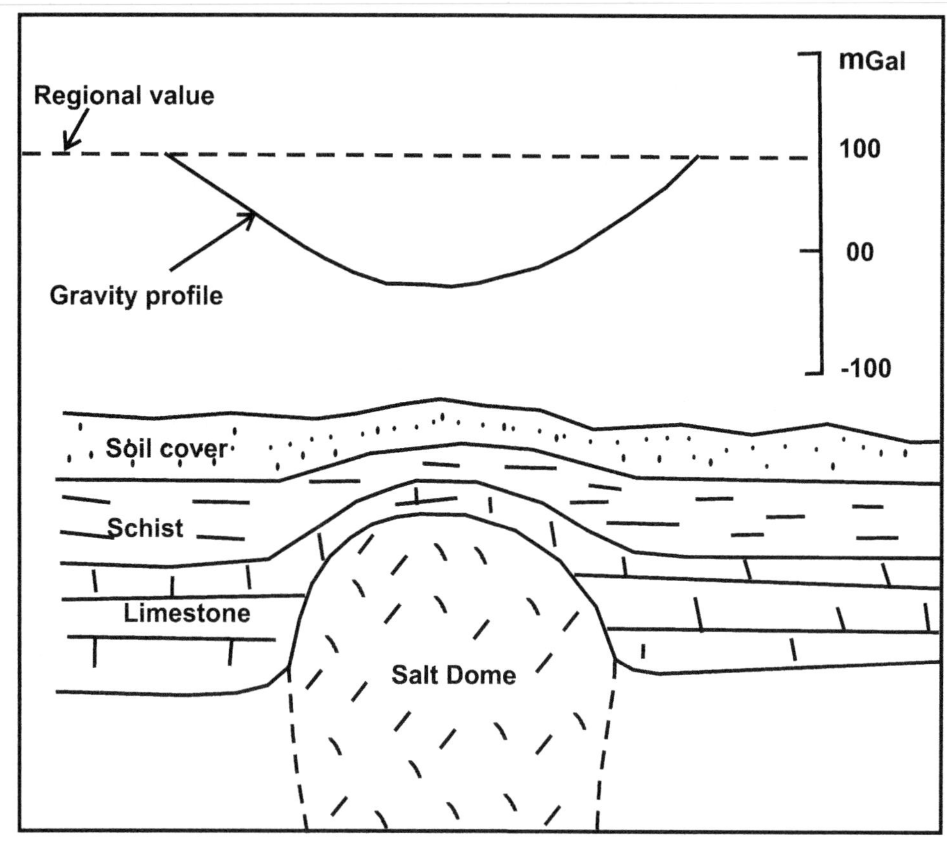

FIGURE 5.10 Gravity profile across a buried salt dome showing lower density compared to the surrounding country rocks.

5.4.2. Theory

A magnetic field or flux density is developed around a bar magnet. It flows from one end of the magnet to the other. The flux can be mapped by sprinkling iron filings over a thin transparent sheet over a bar magnet or by a small compass needle suspended within it. The curve orientations of the iron filings or magnetic needle are called "lines of force." The lines of force converge to points at both ends of the long bar magnet. These points are located slightly inside the magnet and referred as "poles" of the magnet. It always occurs in pair. A freely suspended bar magnet similarly assumes a position in the flux of the Earth's magnetic field. The pole of the magnet that aligns to point in the direction of the geomagnetic north pole is called "north-seeking or positive pole." It is balanced by a "south-seeking or negative pole" of identical strength at the opposite end of the magnet. The lines of force always diverge from north or positive pole and converge to south or negative pole (Fig. 5.11).

The magnetic field B or flux density due to a magnetic pole of strength m at a distance r from the pole is expressed as the force exerted on a unit positive pole at that point P. It is defined as

$$B = (\mu_0 m) / (4\pi \mu_R r)$$

Where, μ_0 is the constant corresponding to the magnetic permeability of vacuum and μ_R is the relative magnetic permeability of the medium separating the poles.

In geophysical applications, the unit of measurement of magnetic intensity is gamma (γ) which equals to 10^{-9} T or nanotesla (nT). The total magnetic intensity of the earth in polar region is of the order of 60,000 γ or 60,000 nT and at equator is 30,000 γ.

5.4.3. Earth's Magnetic Field

The planet Earth posses the property of huge magnet with north and south geomagnetic poles aligned 11.5° away from geographical north (to west) and south pole (to east). The orientation of a freely oscillating magnetic needle at any point on the Earth's surface depends on the direction of the geomagnetic field at that point. The geomagnetic field, "F," at any point has few elements to represent its magnitude and direction. The components are a vertical (Z), horizontal (H), declination (D) and inclination (I) as shown in Fig. 5.12. Declination is the angle between magnetic north and true or geographic north. Inclination (I) is the

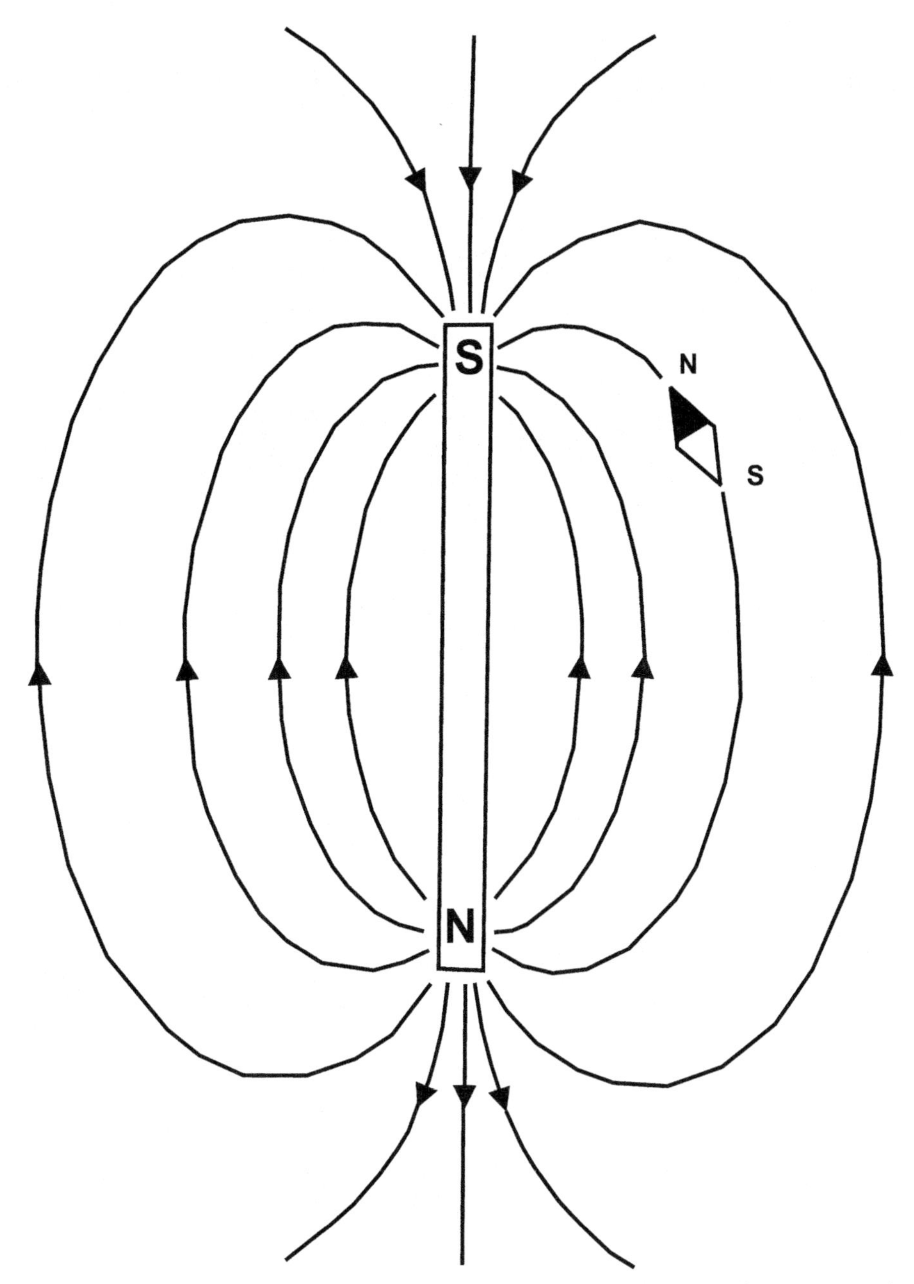

FIGURE 5.11 Lines of forces caused by a bar magnet always diverge from north or positive pole and converge to south or negative pole.

angle of F with respect to the horizontal component H. Magnetic anomaly is caused by superimposed presence of magnetic minerals and rocks on the normal geomagnetic field at that location.

5.4.4. Rock Magnetism

Magnetic susceptibility of rocks depends mainly on the proportion of Rock Forming Minerals (RFM). Most of the common rock types are either nonmagnetic or very feebly magnetic. When the rocks are composed of higher proportion of magnetic minerals like magnetite, ilmenite, pyrrhotite, etc., it becomes susceptible to magnetism. Basic igneous rocks are usually more magnetic, due to higher content of magnetite, than the acid igneous rocks. Metamorphic rocks vary in magnetic property on the similar analogy. Sedimentary rocks in general are nonmagnetic unless locally enriched with magnetite, ilmenite and pyrrhotite-magnetite-bearing sulfide deposits.

Common causes of magnetic anomalies are due to intrusion of basic and ultrabasic dykes, sills, lava flows and magnetic orebodies. The amplitude varies between as

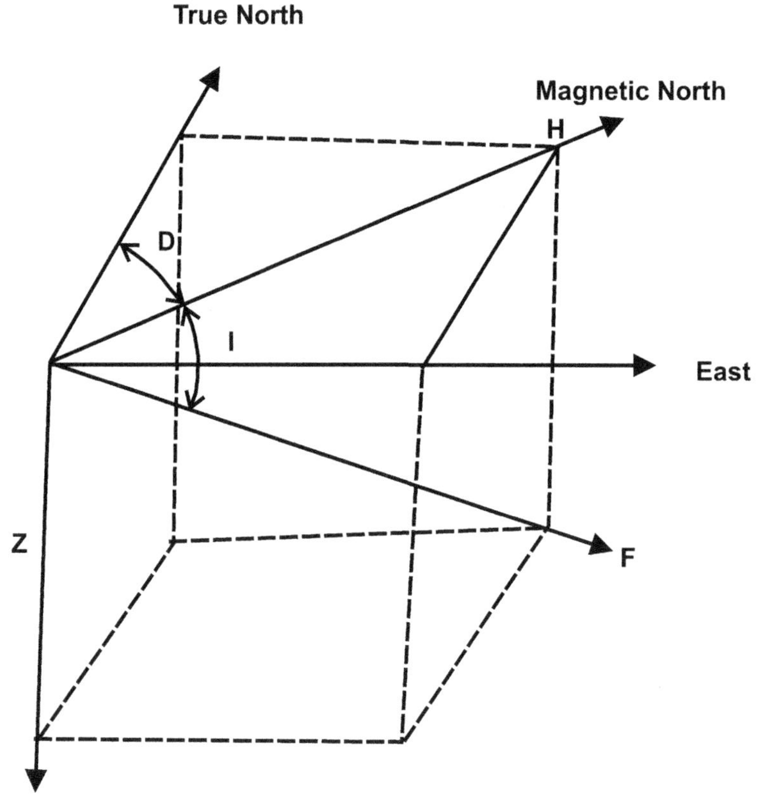

FIGURE 5.12 Schematic diagram of geomagnetic elements showing the declination (D) and inclination (I) of the total field vector F.

low as 20 nT in limestone and 800 nT in basic igneous rocks to more than 6000 nT over sulfide orebodies. Magnetic susceptibility caused by variation of rocks or orebodies is superimposed on the geomagnetic field at that location. Magnetic anomaly is the response signaled by the causative body over the regional trend of the country rocks.

5.4.5. Survey Instruments

Surveying instruments have been in use during early 1900s to measure the geomagnetic elements by magnetic variometer. It is essentially based on principles of suspended bar magnet in the Earth's field and cumbersome to handle in the field. Since then, the instruments are updated to be user friendly and compatible to computer-based processing for easy interpretation. The precision of the instruments is narrowed down to ±0.1 nT.

Fluxgate magnetometer was developed during 1940s employing two identical ferromagnetic cores of high permeability that provides instantaneous measurements. The instrument developed is "nuclear precision" or "proton magnetometer" (Fig. 5.13), consisting of a container filled with liquid rich in hydrogen atoms surrounded by a coil.

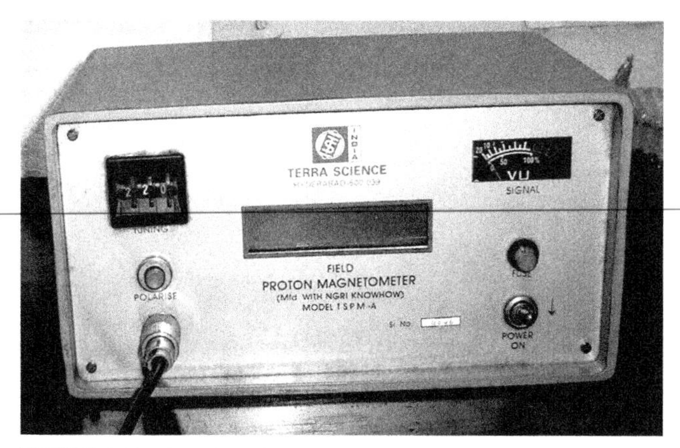

FIGURE 5.13 User-friendly proton magnetometer device compatible to high-end processing with precision.

The next generation instruments with higher precision are optically pumped or alkali vapor magnetometer and magnetic gradiometer.

5.4.6. Data Reduction

Reduction of magnetic data is essential to remove the noises caused by other elements not related to subsurface magnetism. The effect of *"diurnal variation"* on ground surveying can be removed by periodic calibration of the instrument at fixed base station. The same for aeromagnetic survey can alternatively be assessed by arranging numerous crossover points in the survey path (refer Fig. 5.20). Geometric correction is computed by using International Geomagnetic Reference Field (IGRF) which defines the theoretical undisturbed magnetic field at any point on the Earth's surface. In magnetic surveying, terrain correction is rarely applied due to negligible effect of vertical gradient of the geomagnetic field.

5.4.7. Applications

Magnetic surveying is extensively used for metallic mineral investigation particularly for iron ore with higher ratio of magnetite to hematite. It is capable of locating massive sulfide deposits especially in conjunction with EM methods, discussed later in this chapter. Aeromagnetic surveying should preferably be programmed at low-level flight path (100–200 m above ground) avoiding excessive monsoon, peak summer and foggy weather. The depth of penetration of airborne survey will depend on the capacity of instruments. In ground surveying the traverse is designed across the strike of the formations at line interval less than the length of the expected causative body. Magnetic anomalies caused by shallow objective are easily detectable than deep-seated target. The airborne magnetic and electromagnetic survey with advanced configuration system, both in frequency and time domain, with high penetration capacity can identify deep-seated metallic bodies. The application of the system has considerable increase of bandwidth of both helicopter-borne frequency-domain electromagnetic (FDEM) and fixed wing time-domain electromagnetic (TDEM) system.

The visual interpretation of isocontour maps of corrected magnetic data provides a qualitative existence of orebody. The approximate location and horizontal extent of the causative body can be determined by study of relative spreads of the maxima and minima of anomaly. A comparison of gravity and magnetic interpretation of manganese orebody is given in Fig. 5.14.

5.5. ELECTRICAL SURVEY

5.5.1. Concept

Geo-electrical methods of mineral investigation depend on the properties of conductivity and resistivity of subsurface rock mass to passing electric current. The methods are exercised either by introduction of artificially generated current through the ground or make use of the naturally occurring electrical field within the earth. The current is driven through a pair of electrodes connected to the terminal of a transmitter and the resulting distribution of potential in the ground is mapped by using another pair of electrodes connected to a sensitive voltmeter. In case of a homogeneous subsurface, the potential distribution and the lines of electrical flux can be measured by the magnitude of current introduced and

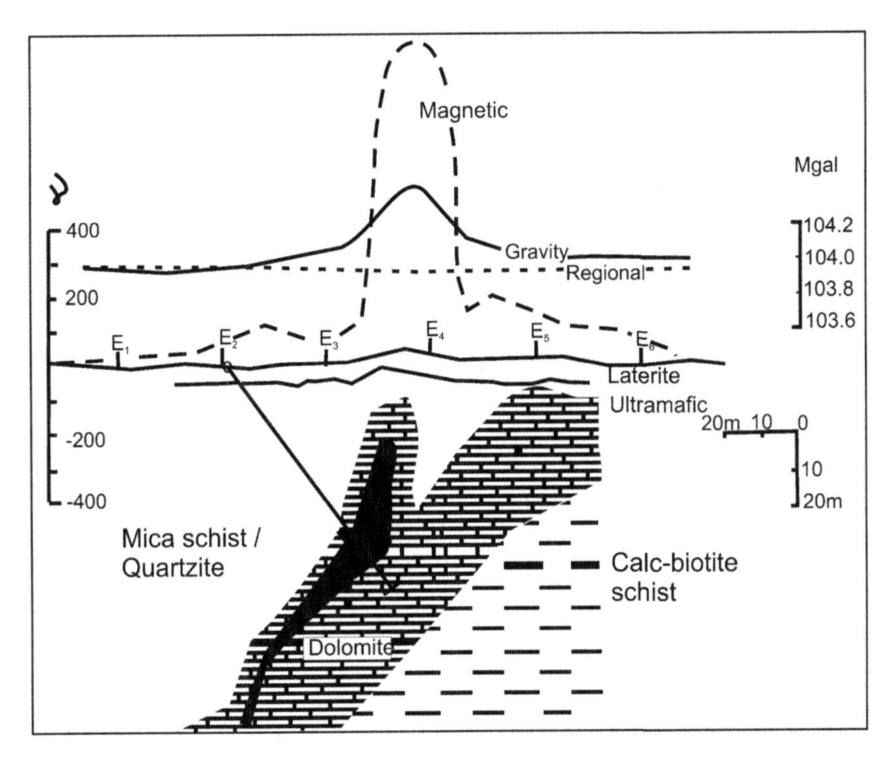

FIGURE 5.14 Geophysical interpretation of gravity and magnetic anomalies and confirmed by drill testing of rich sulfide orebody in Rajasthan.

variation in the receiving electrodes. The current deflects and distorts the normal potential in inhomogeneous condition in case of electrically conductive or resistive objects are present in the ground. The better conductive causative mineral bodies are massive sulfide deposits, graphite-rich bed, fractured or altered underground zones with confined water containing dissolve salts, at times even clay beds. Massive quartz veins are highly resistive to current flow.

Three types of geo-electrical methods are in use based on the type of current sources and their response to the subsurface rocks under investigation. The surveying methods are (1) resistivity, (2) induced potential (IP) and (3) self-potential (SP).

5.5.2. Resistivity Method

5.5.2.1. Definition

The property of the electrical resistance of a material is expressed in terms of its resistivity. It is defined as the resistance between opposite faces of a unit cube of the material. Resistivity is one of the extremely variable physical properties of rocks and minerals. Certain minerals, native metals and graphite, conduct electricity via passage of electron. Most RFM are insulator. Hard compact rocks are generally bad conductor of electricity. The electric current is carried through a rock mainly by passage of ions in pore waters. It follows that porosity and the degree of saturation governs the resistivity of rocks. Resistivity generally increases as porosity decreases.

In this method, artificially generated electric currents are introduced in to the ground and the resulting potential differences (volt) are measured at the surface. Deviations from the background pattern of potential differences indicate inhomogeneities and presence of anomalous object in the subsurface.

5.5.2.2. Electrode Configuration

A number of different electrode configurations have been designed for field practices using linear array type arrangements.

In the Wenner array, the four electrodes (current, $C_1 - C_2$ and potential, $P_1 - P_2$) are kept along a straight line at equal array spacing, a. In the Schlumberger array, the distance between the potential electrodes ($2l$) is small compared to the distance between the outer current electrodes ($2L$). In case of dipole-dipole array configuration, the potential probes, $P_1 - P_2$, are kept outside the current electrodes, $C_1 - C_2$, each pair having a constant mutual separation, 'a'. The current source will be treated as an electric dipole if the distance between the two pairs, 'na', is relatively large (Fig. 5.15).

5.5.2.3. Field Procedure

There are two common types of resistivity surveying being practiced.

When the current and potential electrodes are maintained along a straight line at same relative spacing around a fixed central point, it is termed as "vertical electrical sounding" (electrical drilling). The procedure is presumed on the analogy that the current penetrates continuously deeper with increasing separation of the current electrodes. The purpose of electrical sounding is to deduce the variation of resistivity with depth below a given point on the ground for near-horizontal layers of formation below. The method is useful for determining loose horizontal overburden thickness over hard rocks in river valley projects and groundwater.

(A)

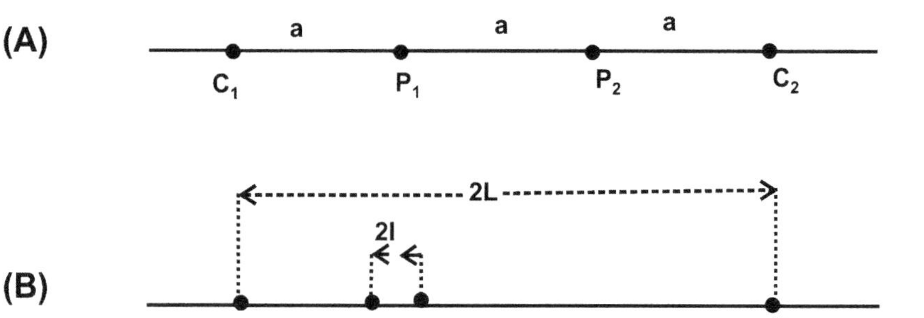

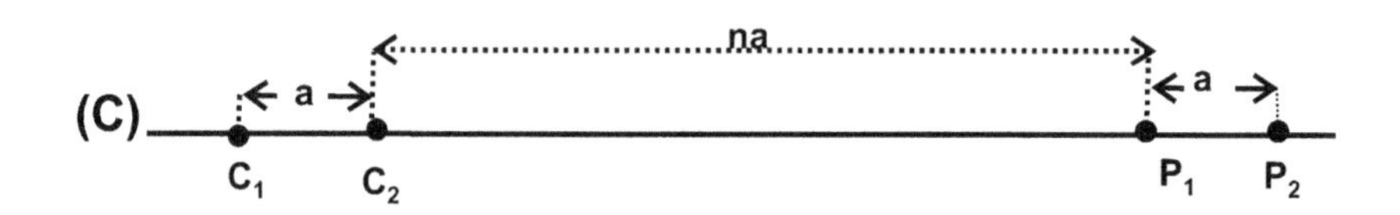

FIGURE 5.15 Schematic diagram showing common types of electrode configuration in resistivity surveying: (A) Wenner, (B) Schlumberger and (C) dipole-dipole.

"Constant separation traversing" is obtained by progressively moving an electrode spread with fixed electrode separation along a traverse line, the configuration of the electrodes being aligned either in the direction of the traverse (longitudinal) or at right angles to it (transverse).

5.5.2.4. Resistivity Survey Instruments

Resistivity survey instruments are designed to measure very low levels of resistance to a very high accuracy. Most resistivity meters employ microprocessor controlled low-frequency alternating current source (between 100 Hz for shallow probes around 10 m to less than10 Hz for 100 m penetration). Direct current source along with non-polarizing electrodes must be used for greater depth penetration of some hundreds of meters in favorable ground condition.

The unit of resistivity is the ohmmeter (ohm m or $'\Omega$m)

5.5.2.5. Applications

The resistivity surveying was not initially favored during reconnaissance survey because of slow process due to manual planting of the electrodes prior to each measurement. This restriction was no longer hold with increasing availability of noncontacting conductivity measuring device discussed in EM surveying. The techniques are widely used in hydrogeological investigations covering subsurface structure, rock types and water resources for drilling, engineering geological investigation sites before construction and for exploration of sulfide deposits (Fig. 5.16).

5.5.3. IP Method

5.5.3.1. Definition

When the externally applied direct current, connected through a standard four-electrode resistivity spread, is switched off abruptly (t'_0), the electrochemical voltage does not drop to zero instantly. The voltage dissipated gradually to zero at t'_3 after many seconds between the potential electrodes after a large initial decrease (Fig. 5.17). Similarly, when the current is switched on the initial voltage jump at t_0 and a slow increment takes place over a time interval (t_1-t_2) before the steady-state value is reached (t_3). The ground acts as a capacitor, storing electrical charge, and thereby becomes electrically polarized in state ($t'_1-t'_2$). The measurement of a decaying voltage over

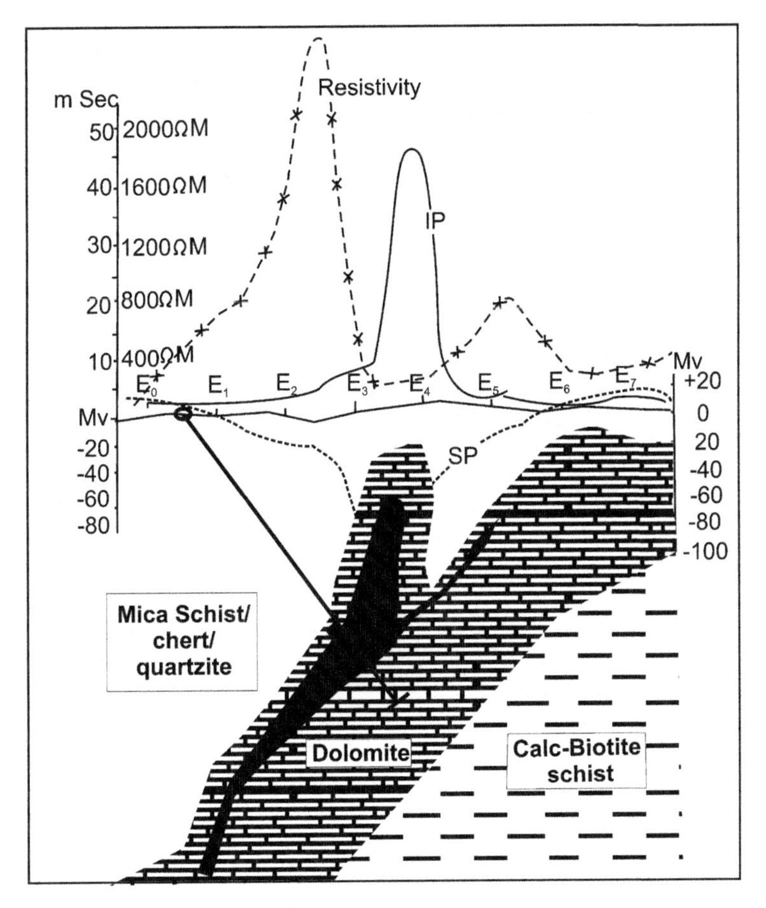

FIGURE 5.16 Geophysical interpretation of SP, IP and resistivity survey and confirmed by drill testing of rich sulfide orebody in Rajasthan. E_0, E_1, E_2...are measurement points on survey lines across the expected causative body.

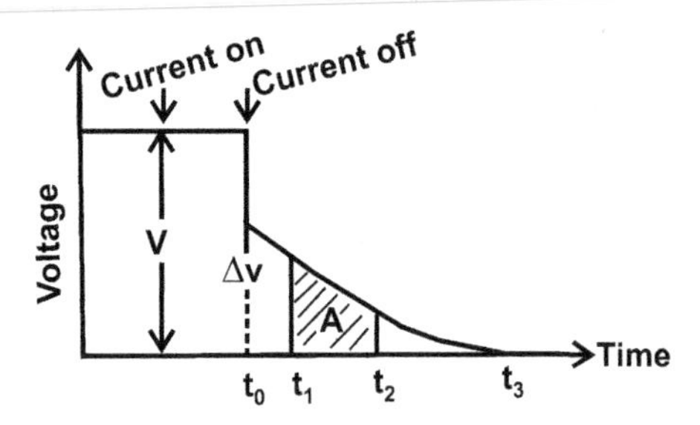

FIGURE 5.17 Schematic diagram elucidating the principles of IP method of survey by means of voltage versus time curve passing through conductive mass.

certain time interval is known as time-domain IP survey. Alternatively, if a variable low-frequency alternate current source is used for the measurement of resistivity, it is observed that the measured apparent resistivity of subsurface rocks decreases as the AC frequency is increased. This is known as frequency-domain IP surveying.

5.5.3.2. IP Measurement

The mechanism of IP method works on externally imposed voltage causing electrolytic flow in the pore fluid of rocks. Thus, negative and positive ions build up on either side of the minerals and charged during the imposed voltage. On removal of the impressed voltage, the minerals return to their original state over a finite period of time causing a gradual decay voltage. In the same analogy, metallic minerals, if present, became additionally charged during external current flow and voltage gradually decays on switching off the source. The effect is known as membrane or electrolytic polarization in case of nonmetallic minerals and electrode polarization or overvoltage in case of metallic minerals. Metallic sulfides and oxides and graphite are good conductor and respond better to this effect. The magnitude of electrode polarization depends both upon intensity of impressed voltage and concentration of conductive minerals. The ubiquitous disseminated sulfide orebody provides larger surface area available for maximum ionic-electronic interchange and hence extensively used for IP surveying.

The measurement of the decaying voltage (M), shortly after switching off the polarization current, is the area "A" of the decay curve over a specific time interval ($t'_1-t'_2$). The area "A" is determined within the measurement instrument by analogue integration. The measured parameter is called chargeability. The unit is milliseconds (ms). Different minerals are distinguished by unique chargeability such as pyrite (13.4 ms) and magnetite (2.2 ms).

$$M = A / V \text{ or } \Delta V / V$$

5.5.3.3. Applications

IP method is extensively and effectively used in base metal exploration uniquely for locating low-grade ore deposits such as disseminated sulfides. Example of rich zinc-lead deposit of Rajasthan is given at Fig. 5.16.

5.5.4. SP Method

5.5.4.1. Definition and Mechanism

The SP or in other words spontaneous polarization method based on natural potential differences results from electrochemical reactions in the subsurface. The method is unique as no current is artificially introduced in to the ground. The causative body has to lie partially close to the water table to form a zone of oxidation. Electrolytes in the pore fluids undergo oxidation and release electrons which are conducted upward through the orebody. The released electrons cause reduction of the electrolytes at the top of the orebody. An electronic circuit is thus created in the body so that the top of the body acts as negative terminal. The SP anomaly is always and invariably centrally negative over metallic ore deposits (Fig. 5.16). The bulk of the orebody exists below the water table and undergoes no chemical reaction and merely serves to transport electrons from depth generating stable potential differences over long periods at the surface.

5.5.4.2. Equipment and Field Procedure

Field equipments consist of a pair of nonpolarized electrodes connected by insulated cable via a millivoltmeter. Electrodes are composed of simple metal spikes immersed in saturated solution of its own salt (as Cu in $CuSO_4$) in a porous pot that allows slow leakage of solution into the ground. Measurements of potential differences are carried out by shifting successive electrodes or more commonly by fixing one electrode in barren ground (up to 50 mV) and the other is moved in steps across expected target area (1000 mV). Spacing between electrodes is generally 10, 20 and 30 m apart. Typical SP anomalies may have amplitude of several hundred millivolts with respect to barren ground and exceeds several thousand millivolts over deposits of metallic sulfides, magnetite and graphite.

5.6. EM SURVEY

5.6.1. Definition

Electromagnetic or EM surveying is based on the response of ground to the propagation of EM fields which are composed of an alternating electric intensity and magnetizing force. Primary or inducing field is generated by passing an alternating current through a coil (loop of wire called transmitter coil) placed over the ground. The primary field spreads out in space, both above and below the ground, and can be detected with minor reduction in amplitude by a suitable receiving coil in case of homogeneous subsurface. However, in the presence of a conducting body, the magnetic component of the EM field penetrating the ground induce alternating currents or 'Eddy currents' to flow within the conductor. The eddy currents generate their own secondary EM field distorting the primary field. The receiver then responds to the resultant of the arriving primary and secondary fields so that the response differs in phase, amplitude and direction from the response to the primary field alone. These differences between the transmitted and received EM field reveal the presence of the conductor and provide information on its geometry and electrical properties. The induction of current flow results from the magnetic component of the EM field. The whole process of EM induction has been generalized in Fig 5.18. Surface EM, using surface loops and receiver, or a down-hole tool lowered into a bore-hole, acts on the body of mineralization. Regional EM uses airborne, either fixed-wing aircraft or helicopter-borne EM tools. Both methods can map 3-D perspective of mineral bodies, deep in the earth, and guide further exploratory drilling for verification.

The depth of penetration of an EM field depends upon its frequency and electrical conductivity of the medium through which it is propagating.

5.6.2. Detection of EM Field

Two-frame compensator and Turam systems are employed to measure the phase and amplitude relationship between the primary, secondary and resultant fields. The spacing and orientation of the coils are critical. A long straight conductor is laid on the ground and supplied with alternating current. Field traverse is made by two coils at fixed distance apart and at right angles to the conductor. Changes of phase and amplitude are measured between the induced voltages in each coil.

The qualitative response of the EM field can easily be obtained by very low frequency (VLF) or Audio-frequency magnetic (AFMAG) methods. VLF uses powerful radio transmitters as used in long-range communication. The system needs a transmitter with magnetic vector across the strike of the geological formation and a receiver tuned to the particular frequencies of the transmitter. The depth penetration is low. AFMAG field method uses natural EM field induced by thunderstorms and known as "sferics." This method enables deeper penetration into the ground. The high-frequency waves propagate between the ground surface and the ionosphere and act as an efficient EM waveguide.

5.6.3. Time-Domain Electromagnetic Survey

The measurement of comparatively small secondary and resultant field is difficult in the presence of strong primary field. The solution came with the development of TDEM procedure. The primary field is not continuous and consists of a series of intermittent pulses separated by periods. The secondary field is only measured when the primary field is absent. The eddy currents defuse around the boundary of a highly conductive subsurface body and decay slowly when the inducing field is removed. The rate of decay of fading eddy currents can be measured to locate anomalous conducting bodies.

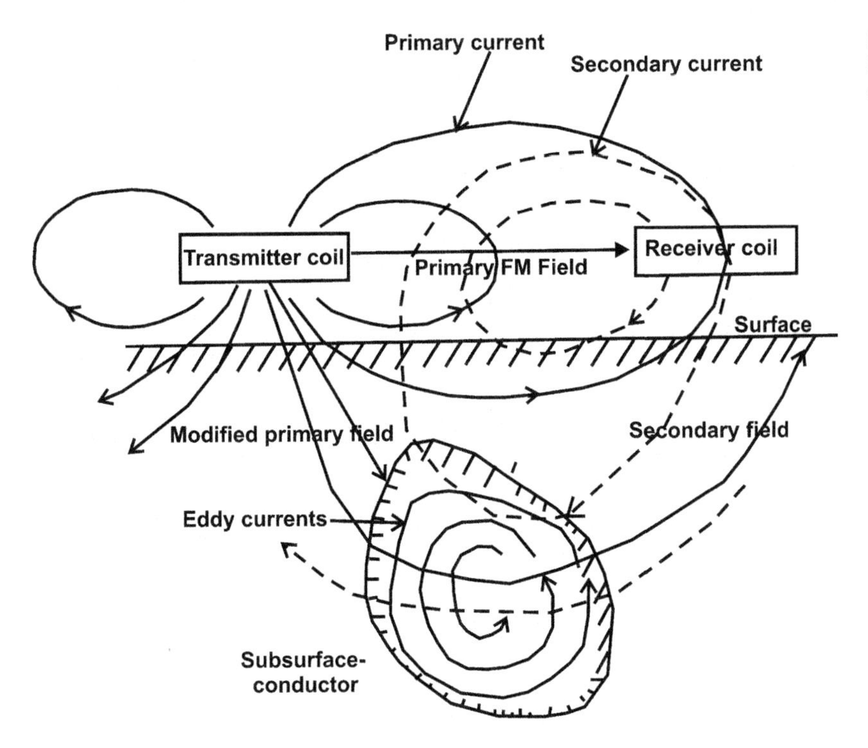

FIGURE 5.18 Conceptual diagram of EM induction processing system generating eddy currents in subsurface conductive mass.

5.6.4. Noncontacting Conductivity Measurement

The electrical methods such as resistivity, SP and IP require physical introduction of current into the ground through electrodes making it labor intensive, slow and costly. To overcome these drawbacks recently, noncontacting conductivity meters using magnetic component of the EM field has been developed. Therefore, there is no need for physical contact of either transmitter or receiver with the ground and can be used from aircraft at much faster speed than other electrical methods.

5.6.5. Airborne EM Survey

Airborne EM method is most widely used during reconnaissance survey for target detection. It is popular because its high speed helps to cover vast area in shortest time thereby making the technique cost effective. The operation is broadly divided into two types – passive and active. In case of passive system, only the receiver is airborne using very-low-frequency (VLF) and audio-frequency magnetic (AFMAG) field methods. In active system, both transmitter and receiver are mobile. The later system is more frequently used as it can be performed in areas where ground access is difficult. Active systems are two types – quadrature and fixed separation.

In quadrature systems, a large transmitter slings between tail and wings and the receiver is towed from a cable of around 150 m behind the aircraft. The receiver bird swings with the long hanging cable causing uncontrolled height and direction during the flight movement. In this system, only the phase difference between the primary and resultant field can be measured which is caused by a conductor.

In fixed separation system, the transmitter and receiver coils are mounted either on the wings of the aircraft flown at a ground clearance of 100–200 m or on a beam fixed beneath the helicopter flown at an elevation of 20 m maintaining fixed separation and height of the flight. In single aircraft or helicopter bound operation, an extremely small separation of the transmitter-receiver is used to generate and detect an EM field over a relatively large distance. This may cause significant signal distortion. The problem can be conquered by use of two planes flying in tandem, both flying at strictly regulated speed, altitude and separation. The rear craft carry the transmitter and the forward plane tow the receiver mounted in a bird (Fig. 5.19). Airborne TDEM methods INPUT® (Induced PULse Transient) is used to enhance secondary field measurement. This survey is relatively expensive and requires advance data processing facility. However, the cost is compensated by increased depth of penetration, minimized orientation errors and diagnostic of the type of subsurface conductor present.

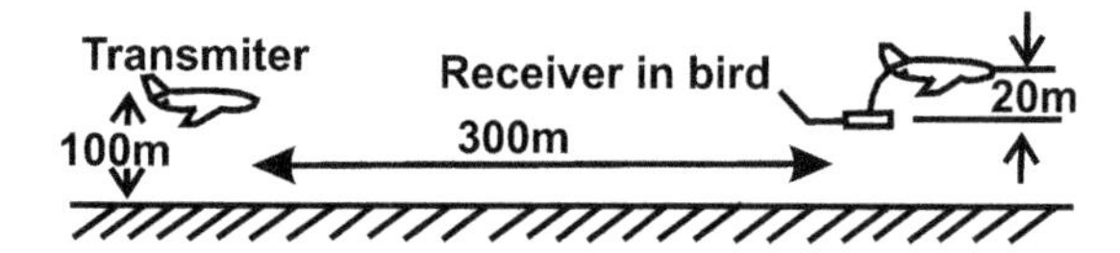

FIGURE 5.19 Schematic diagram showing EM aerial survey employing two airplanes.

A typical airborne survey path is designed to fly the craft in a regular to-and-fro grid interval (say alternate E-W/WE direction) covering the entire length of the license area. The flight continues in a similar pattern (say N-S/S-N) so that the survey covers along and across the total strike of the possible unknown mineralization (Fig. 5.20). The possibility of missing the target will be minimum.

5.6.6. Applications

The EM methods are most suitable for detecting bodies of high electrical conductivity such as metallic ores, particularly massive sulfides and hydrogeological studies. The methods are also in use for delineating faults, shear zone, and thin conducting veins. EM surveys can be used to detect palaeochannel-hosted uranium deposits associated with shallow aquifers, which often response to EM surveys in conductive overburden.

5.7. RADIOMETRIC SURVEY

5.7.1. Concept

The radiometric surveys detect and map natural radioactive emanations (γ ray) from rocks and soils. The gamma radiation takes place from the natural decay of elements like U, Th and K. The radiometric method is capable of detecting these elements at the surface of the ground. The common radioactive minerals are uraninite (^{238}U), monazite, thorianite (^{232}Th), rubidium (^{87}Rb) in granite-pegmatite, feldspar (^{40}K), muscovite, sylvite in acid igneous rocks and radiocarbon (^{14}C). Heat generated by radioactive disintegration controls the thermal conditions within the earth. Rate of radioactive decay of certain natural elements acts as powerful means of dating geological time of rock formation that is geochronology. Exploration for these minerals by radiometric survey became important because of the demand for nuclear fuels and also for detection of associated nonradioactive deposits such as titanium and zircon. Isotopes are elements whose atomic nuclei contain the same number of proton but different number of neutron. Certain isotopes are unstable. They disintegrate spontaneously to generate other elements. Radioactivity means disintegration of atomic nuclei by emission of energy and particles of mass. The by-

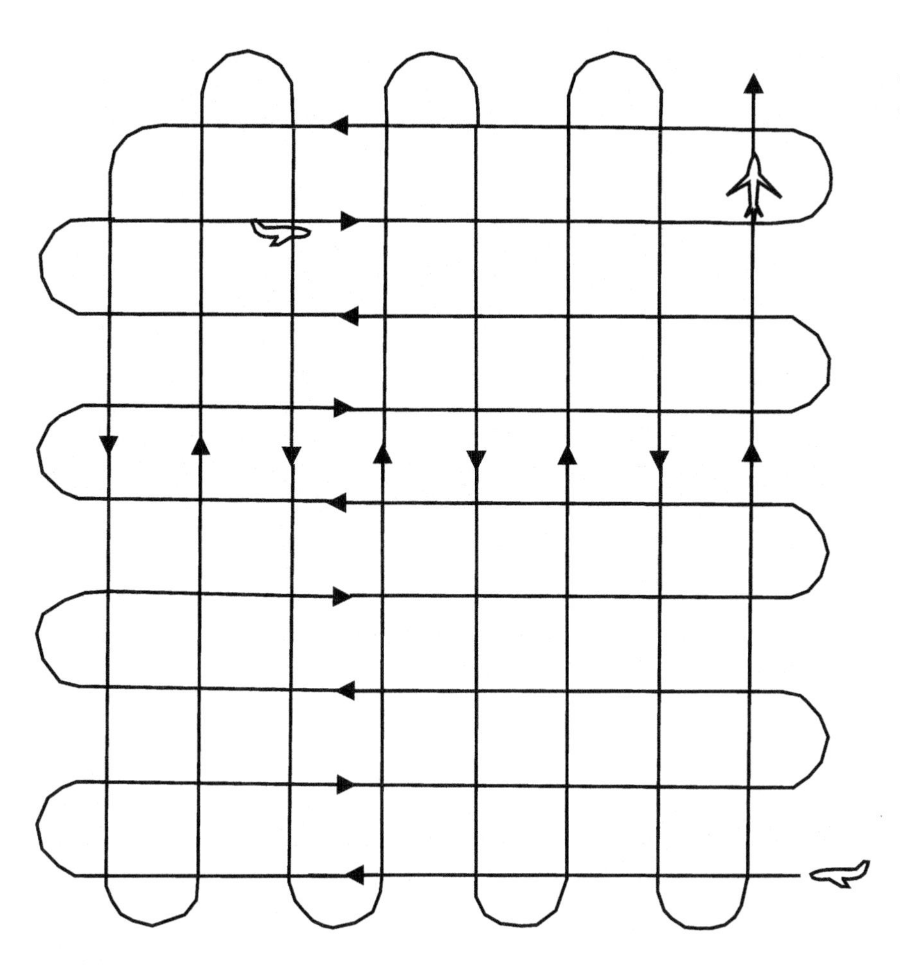

FIGURE 5.20 A typical airborne flight plan to fly (E-W/W-E) followed by (N-S/S-N) to scan the total area under investigation.

products of radioactive disintegrations are in various combinations of alpha (α) particles of helium nuclei, beta particles (β) of electrons emitted by splitting of neutrons and gamma (γ) ray of pure EM radiation.

5.7.2. Instrument

Radioactivity measurement is the detection of number of counts of emission over a fixed period of time. Standard unit of gamma radiation is the Roentgen (R). Several types of detectors are in use. Geiger or Geiger-Müller counter, equipped with meter and earphone, responds primarily to beta particles in the proximity of radioactive source. The scintillation counter is capable of detecting emissions of flash of lights by photomultiplier. It is a more efficient type capable of almost 100% detection of γ radiation. Geiger counter method is limited to ground surveys. Airborne radiometric surveys are exclusively carried out by Scintillation counter in combination with airborne magnetic and EM surveys for mineral investigation. Advance Scintillation detector is designed with enhanced detection sensitivity to differentiate the counts due to γ radiations of different energy levels, say, between uranium and thorium deposits.

5.7.3. Radiometric Dating

The disintegration, since crystallization, of a radioactive parent isotope (say uranium, thorium, rubidium, potassium and carbon) produces a stable daughter isotope (^{206}Pb, ^{207}Pb, ^{208}Pb, ^{87}Sr, ^{40}Ar, ^{40}Ca and ^{14}N, respectively) at a known rate of decay. Atomic proportion of parent and daughter nuclei of a mineral or rock sample can be determined by mass spectrometry. Radiometric age of the events can be computed mathematically from the father-daughter relationship and experimentally determined decay constant (λ). The precision of result depends on the collection of unaltered specimen, exact location, petrological origin, relation to surroundings, reliable chemical and isotope analyses and expertise interpretation. Radiometric methods applied in geochronology can be performed by several father-daughter isotope products such as potassium-argon, rubidium-strontium, uranium-lead, fission track, radiocarbon and tritium methods.

5.7.4. Applications

Radiometric surveying is becoming more and more significant for investigation of nuclear fuel and associated minerals, determination of geochronology of rock formations to unravel many untold stories, demarcation of geological structures, and intrusive. Survey can be conducted on the ground as well as airborne.

5.8. BOREHOLE LOGGING

5.8.1. Principles of Well Logging

Various types of boreholes are often drilled to study the subsurface geology along the depth of the hole. The information helps in interpretation of three-dimensional picture of the area. Shallow noncore holes are excavated by percussion drills in which rock fragments are blown out of the holes by air pressure. Deeper holes are sunk by rotary drills with rock cutting bits, tungsten-carbide or diamond bit. These provide noncoring flushed by drilling fluids (*mud*) and core type using core barrel and water circulation, respectively. In case of core-type drilling, the recovered cores are logged for lithology, mineralization (excluding fluid content), and structure and samples are chemically analyzed. In noncore-type excavation and high core loss section of core drilling, information along the depth can be obtained by geophysical methods known as down-hole geophysical surveying or wire-line logging using various sensors. The general objective of well logging is the evaluation of formations by measuring certain rock properties. The inside of the holes are protected from wall collapsing by casing to lower the geophysical probes and high-density drilling fluids/mud with cement. Most widely applied geophysical methods use gravity, electrical resistivity, SP, magnetic, EM induction, radioactivity, sonic velocity and temperature.

The instrument is housed in a cylindrical metal tube (Sonde) connected to an unbreakable multicore cable fixed in a rotating drum fitted with winches and a recorder. The probe is lowered to the bottom of the hole and logging is carried out while hoisting the instrument ups though the section. Logging data are automatically recorded on a paper strip and simultaneously on magnetic tape in analogue or digital form for subsequent computer processing. Geological properties obtained from well logging are formation thickness, lithology, porosity, permeability, proportion of water and hydrocarbon saturation and temperature.

5.8.2. Mise-á-la-Masse

Mise-á-la-Masse method, meaning "excitation of the mass" in French, is a variation of electrical resistivity survey by enhancement of sufficient resistivity contrast between the host rock and the sulfide ore compared to conventional survey. This is in fact a postdiscovery method with definite knowledge of existing sulfide mineralization. The interpreted equipotential maps are good indicators of continuity of mineralization in strike, dip, shape and interconnectivity between numbers of mineralized intersections.

One of the current electrodes (C_1, the positive electrode) is planted into the conductive mass under investigation, i.e. mineral body exposed in a pit or drill holes. The second current electrode (C_2, the negative electrode) is placed a large distance away, say 5–10 times the width of the mass on the surface or within another mineralized part of the drill hole. The current electrode (C_1) in the drill hole consists of lead metal rod attached to an insulated flexible copper wire. One of the potential electrodes P_1 is placed on the ground close to the mass and between C_1 and C_2. The second potential electrode P_2 is placed at a large distance from C_1 and in the opposite direction of C_2 (Fig. 5.21). Surface potentials are measured with respect to the distant electrode preferably over a grid around the drill hole or pit. The potential distribution from these two current electrodes reflects the geometry of the conductive mass. Thus, it can indicate the continuity of mineralization in strike and dip direction.

5.8.3. Applications

Well logging techniques are very widely used for electrical imaging, mine mapping, hydrocarbon and hydrological exploration for getting important in situ properties for possible reservoir rocks. Electric logs are considered the most useful for evaluating formation fluid properties. Modern well logging techniques are being used in locating deep-seated metallic ore deposits, extension of orebodies in all directions, freshwater resources and as a part engineering test drilling operation. Cross-hole and Up-hole tomography involves sending seismic, electrical, and EM or radar signals between transmitter in one borehole and a receiver in

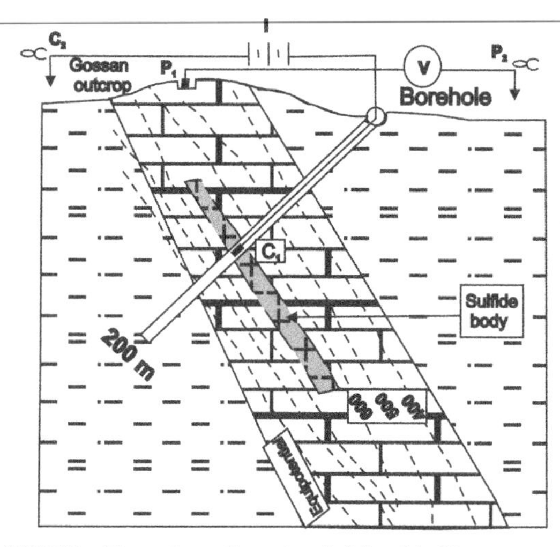

FIGURE 5.21 Electrode configurations in Mise-á-la-Masse survey when one current electrode located in borehole passing through sulfide orebody and the other on the surface/another borehole to complete the circuit. Both the potential electrodes are located on the surface (Credit: Prof. Shalivan).

another borehole. Mise-á-la-Masse survey conducted successfully for Mataloko geothermal field, Flores, East Nusa Tenggara, Indonesia, lead-zinc-copper exploration at Sawar belt, zinc-lead deposit at Kayar, Ajmer district, and gold-pyrite exploration at Bhukia-Jagpura, Rajasthan, India.

5.9. REVIEW

Exploration geophysics is the applied science of measuring physical properties of rocks and minerals and more specifically, to detect the measurable physical contrast between them. The physical properties, under reference, are seismic, gravity/density, magnetic, electrical, electro-magnetic and radiometric. Exploration geophysics is competent to detect and map the subsurface special distribution of rock units, structures like faults, shears, folds and intrusive, target style of mineralization including 3D perspectives, hydrocarbon accumulations and groundwater reservoirs. The methods can also be used in other related areas such as monitoring environmental impact, ancient mining voids, buried archaeological spots, subsurface salinity mapping, and civil engineering sites.

Exploration geophysics is efficient in large area reconnaissance/prospecting for rapid delineation of targets. It is comprehensive using multidisciplinary methods to identify complex mineral environment. Geophysics is cost-effective in reducing the vast volume of earth with decision to initiate drilling or abandoned the area. The techniques are time tested and proven for discovery of deep seated mineral deposits all over the world.

FURTHER READING

Sharma (1986) [68], Robinson et al. (1988) [59] and Arnaud (1989) [1] are worth reading for concept development in exploration geophysics. The work of Kearey et al. (2002) [44] is suggested for further enhancing the knowledge in the subject.

Chapter 6

Photogeology, Remote Sensing and Geographic Information System in Mineral Exploration

Chapter Outline

Mineral Exploration. http://dx.doi.org/10.1016/B978-0-12-416005-7.00006-4

To see a world in a grain of sand
And a heaven in a wildflower:
Hold infinity in the palm of your hand,
And eternity in an hour.

—William Blake.

6.1. INTRODUCTION

In the ancient days shepherds, hunters and travelers used to cover vast land areas. In the process many deposits were discovered out of inquisitiveness. Most of the base metal deposits of the Princely states in India and Saudi Arabia were discovered by them some 3000 years ago. Broken Hill zinc-lead deposit was an accidental discovery in New South Wales Province of Australia by shepherds looking for tin in gossans in 1883. The Sudbury basin, known for the largest resources of Ni-Cu-PGE, was reported by one blacksmith in 1883 during construction of the first transcontinental Canadian Pacific Railway. The Bushveld Igneous Complex, the World's largest Cr-PGE resources, was discovered in 1897 on a routine geological mapping. The Sukinda deposit, the largest chromium resources in India, was a chance discovery by one local tribal villager working for Tata Steel in early 1940s. The knowledge of earth science i.e. rocks, minerals and its occurrences matured with the time. But still a century ago the explorers took traverses by walking or on camel/elephant/horse back for geological studies. It involved physical touching, hammering, breaking and examination of the rocks. It was all together difficult missions to approach remote hazardous terrain. It often imposed restriction on accurate location and detail mapping.

This process of physical approach was replaced by Remote Sensing (RS) techniques over nine decades ago. Since 1920 use of aerial photographic interpretation in the field of earth sciences became a fast and effective tool for exploration of natural resources. The science further progressed with the launching of Landsat-1 satellite in 1972. This made remotely sensed high-resolution digital imagery of electromagnetic spectrum (EMS) available for interpretation and use in commercial exploration of mineral or petroleum in shortest possible time. The use of Geographic Information System (GIS) in mineral exploration was the application aspect by allowing the integration of dissimilar digital data sets into a single, unified database. The recommended approach was to compile all type of available geoscientific data within the GIS envelope in the context of an exploration model to produce a mineral potential map.

Geologists had always been interested in bird's eye view of Earth's surface from a height. It helped him to understand and overview the geomorphology, lithology, vegetation and structures. Geomorphology represents all facets of the landform-related features. Lithology refers to fundamental and broad distinction between soils, igneous, sedimentary and metamorphic rocks. Vegetation focuses upon the plant cover and the underlying soils and rocks on which it grows. The structures identify the kind of deformation the rocks had undergone such as fractures, shears, folds, faults and lineaments. These attributes contributed several dimensions for the geological events generated during millions of years. The understanding guided him to search for minerals and fuels.

Data collection in RS technology is recording data about an object, area or phenomenon under investigation without coming in direct physical contact. There are two types of information collection system.

(1) Still photographs snapped from space flights or airborne cameras, and
(2) digital recording by multispectral electronic scanners or sensors from airplanes or satellites.

6.2. PHOTOGEOLOGY

Photogeology is the simplest approach in RS techniques and its applications. It is the derivation of geological information from interpretation of aerial photographs. Nadar, a French photographer, was the first to suggest the use of aerial photos taken from a captive balloon in 1858 for preparation of topographic and cadastral maps. Albert Heim made a balloon flight over the mountains in 1898. He expressed that the structures were more clearly defined in the aerial view. Wilbur Wright took the first photo from an airplane in 1909 and opened the door to photogeology. World War I (1914-1918) had a tremendous influence on the development of aerial photography, its adaptation to common reconnaissance and needs to surveillance. The science of photointerpretation was born. Many of its basic techniques were developed during the1920s. The expertise improved during World War II (1940-1945). The development continued through 1940s and 1960s approaching to the highest capabilities. The growth finally assimilated into the newly developing geological remote sensing.

6.2.1. Evolution of Camera

Cameras had come a long way since it was first invented around early 18th century. It used direct sunlight that penetrates through a pinhole forming conical shape reverse to the object on the opposite wall of the dark room. This was modified as pinhole camera replacing the hole by a lens. The image was recorded on glass plate of the box. The first Black and White (B&W) photograph reported in 1814. The crucial starting point in the history of the camera started from 1837 by permanent photograph using visible light or rays. The first color photograph was produced on glass plate in 1907. In early 1940s commercially successful color photographs were produced on film. The quality of photographs was improved by introducing wide-angle lenses and filters. The digital single-lens reflex camera (SLR/DSLR) turned out in 1981. This is the most recent addition to the world of high-resolution cameras, which provide more features than any other camera ever produced. Digital cameras hold photos on a memory card, which allows one camera to hold over hundred photographs. It became increasingly popular in next few years. Digital cameras are continuously modified to smaller (mounted on cellular phone) and faster with 12.4 million pixels mounted in space shuttle of NASA in 2008.

The successive journey of aerial photography started with camera fitted on hydrogen-filled balloon, pigeons, kites, parachutes, helicopter, fixed-wing aircraft and finally in space shuttle. The camera takes photographs of ground from higher horizon without any ground-based support. Cameras may be hand held or firmly mounted on stand or board. The images may be taken by photographer triggered remotely or automatically. Aerial photographs are usually taken between mid-morning and mid-afternoon when the sun is high with minimal shadow effect.

6.2.2. Classification of Aerial Photographs

Aerial photographs are classified on the basis of camera axis i.e. oblique and vertical, type of film emulsion and scale.

6.2.2.1. *Oblique Photographs*

Photographs can be snapped either at high- or low-angle oblique camera position to the objects. High-angle oblique photographs include a horizon. Oblique photographs are handy to obtain permanent records of inaccessible mountain cliff, canyon, gorge, steep-angle quarry faces, dam sites and similar features. These photographs can be studied to identify stratigraphy, rock color and texture, erosion, fold, fault and linear structures. The information is valuable during mineral search and to attain corrective measures at some point of structural failures.

6.2.2.2. *Vertical Photographs*

Vertical photographs are taken by a camera pointing vertically downward and the axis of the camera/lens is perpendicular to the ground. Aerial photographs show a perspective view due to effect of distortion on account of image motion, displacement, change in topography and effect of parallax. The principal point (PP)/plumb point/nadir or geometric center has no image displacement. It is the point on the photograph that lies on the optical axis of the camera. It is determined by joining the fiducial marks recorded on the photograph (Fig. 6.1).

Flight lines are the paths that an aircraft takes in order to ensure complete coverage of the area to be photographed. Flight lines are arranged to give a succession of overlapping photos (Fig. 6.1) to minimize distortion. The photos overlap within and between flight lines. The overlaps in these two directions are called forward overlap (end-lap) and side lap. Forward overlap along the flight line between two adjacent photographs (stereo pairs) is about 60% to provide complete coverage and stereoscopic view of the area. The forward overlaps between first, second and third adjacent photographs are 60 and 30% respectively. Side lap between flight lines is usually about 30% to ensure that no areas are left uncovered. A nadir line is a line traced on the ground directly beneath an aircraft while taking photographs of the ground from above. This line connects the image center of

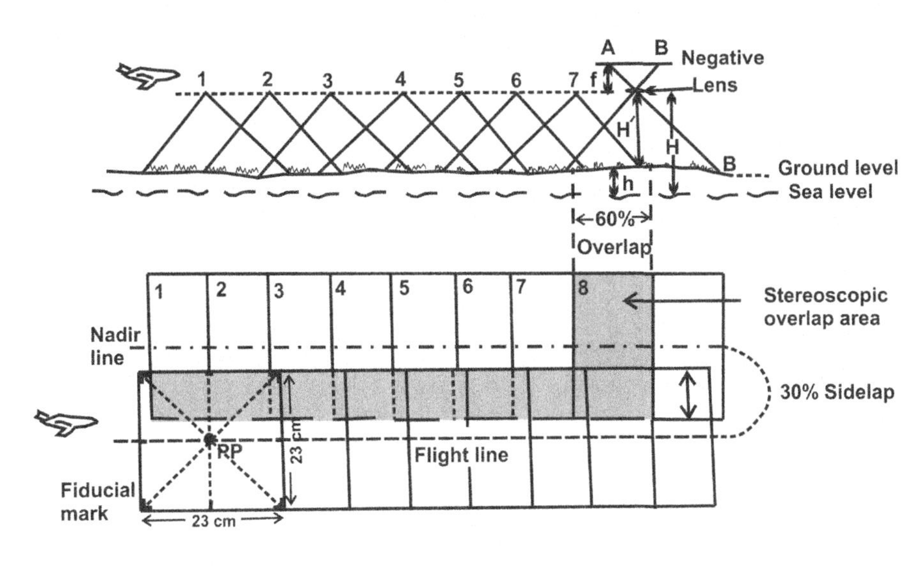

FIGURE 6.1 Schematic typical photograph coverage over flat terrain showing forward overlap and side lap, nadir and flight line, PP, fiducial mark (Source: conceived after many).

the successive vertical photographs. Nadir line is rarely in straight-line format due to changes in flight travel course and needs necessary correction. The title strip of each photo frame includes flight, line and photograph number, date and time of the exposure, bubble balance, sun elevation, flight height and focal length of the camera.

6.2.2.3. Film Emulsion

The common types of aerial photos are based on film emulsion and can be grouped as:

(1) Panchromatic B&W
(2) Infrared (IR) B&W
(3) Color (True)
(4) IR color or false-color composite (FCC).

A true-color image of an object is the same color as it appears to the human eyes—a green tree appears green, blue water as blue etc. True color for B&W images perceived lightness of the object as original depiction. A false-color IR image of an object depicts complementary that differs from original color as appears in human eyes, say vegetation, forest and agricultural land depict red in lieu of original green color.

6.2.2.4. Scale

There are four broad group of scale to distinguish the aerial photographs: small, medium, large and very large. A large-scale map indicates that the Representative Fraction (RF) is large, i.e. the RF's denominator is small. 1:10,000 and 1:2500 maps are large scale. Small-scale maps have a small RF. 1:250,000 and 1:50,000 maps are small scale.

(1) Small scale– $\geq$1:50,000 to 1:250,000: Reconnaissance
(2) Medium scale– 1:10,000 to 1:50,000: Prospecting
(3) Large scale– 1:2000 to 10,000: Detail exploration
(4) Very large scale– $\leq$1:2000: Mine exploration.

Therefore, the thumb rule will be "the larger the scale of map larger the objects with more detail features and better resolution".

6.2.2.5. Determination of Scale

Scale is the ratio between a distance of two points on the aerial photograph and the distance of the same points on the actual ground. The unit of scale is expressed as equivalent (1 = 1000 mm) or dimensionless fraction (1/1000) or dimensional ratio (1:1000). In vertical aerial photograph the Scale is a function of the focal length (f) of the camera and the flying height above the average ground level (H') of the aircraft. The aircraft flies at a nearly constant height. When a plane flies over a flat terrain the scale will be constant. In case of flying over undulated mountainous terrain the scale will vary rapidly across the adjacent photographs. Therefore, the scale of aerial photographs is a function of terrain elevation.

$$\text{Scale}(S) = f/H'$$

where H' is the difference between the terrain elevation (h) and height of the aircraft above a datum (H), usually the mean sea-level value available from the altimeter in the aircraft.

Following Fig. 6.1,
Focal length (f) = 50 cm
Craft height above datum (H) = 5500 m
Terrain height (h) = 50 m
Scale (S) = 50 cm/(5500 − 500) m = 50 cm/500,000 cm = 1/10,000 or 1:10,000.

6.2.3. Parallax

Parallax is an apparent displacement or difference of orientation of an object viewed at two different locations during vertical aerial photography. Objects at a higher height lie closer to the camera and appear relatively larger than similar objects at a lower elevation. The tops of the objects are always displaced relative to their bases. It can be measured by the angle of inclination between those two lines. Nearby objects have larger parallax than distant objects when observed from different positions. This difference in parallax gives a 3D effect when stereo pairs are viewed stereoscopically.

6.2.4. Photographic Resolution

The resolution of aerial photographs will depend on various factors. The effect of scale is closely related to ground distance from the camera i.e. closer the distance higher the resolution. Correct exposure time will give higher resolution while using slow and fast film. Higher resolving power of the camera lens will give better results. Movement of the camera lens during exposure must be minimized. The vibration of the camera and the aircraft should be minimal for better resolution. The resolution will also change depending on atmospheric condition at the time of filming and quality of film processing.

6.2.5. Problems of Aerial Photograph

There are few inherent problems associated with aerial photography and particularly related to aircraft movement. The aircraft usually deviates from the line of flight, altitude and tilting of wings. These will cause drifting of the photographs, change of scale between adjacent frame, distortions and resolution of photographs.

6.2.6. Photo interpretation

There are two commonly used equipments for study of aerial photographs. One is called stereoscope. A set of two photograph shots from two successive stations can be viewed using stereoscope. The left and right photographs are viewed commonly by left and right eyes respectively. When the two images merge into each other a 3D mental model is perceived. It enables a viewer to see 2D image that is actually two separate images printed side by side with common overlapping areas in 3D. The equipment is low-end field stereoscope. It can easily be carried to the field for quick checking of features in the photographs with field observations. The other one is mirror stereoscope and used in the process laboratory. The equipment is capable to view full 23 × 23 cm photograph frame.

The modern instruments used in mineral exploration program are color additive viewer and electronic image analyzer. In case of color additive viewer multispectral photographs are taken simultaneously using three or four cameras in narrow spectral band of 0.4-0.5 (blue), 0.5-0.6 (green), 0.6-0.7 (red) and 0.7-0.9 (IR) μm. An FCC is thereby generated by superimposition of multi-spectral photographs. Human eye will differentiate and interpret color composite than gray tones. The electronic image analyzer scans B&W aerial photographs and produces a close circuit video digital images for interpretation.

6.2.7. Application in Mineral Exploration

A systematic evaluation, interpretation and identification of key elements from aerial photographs have multidimensional applications for society in general and mineral exploration in particular. The key information include topography, land distribution and land use pattern, soil and rock types, drainage system and texture, surface erosion, structures like fold closure, faults, shears and lineaments, and surface indicators like presence of weathering, gossans and old mining-smelting remnants. The first few parameters like topography, land distribution and land use pattern help in urban and agricultural planning. The remaining information contributes toward conceptualizing existence of minerals and possible exploration targets. The approach is more significant particularly for mineral deposits occurring at remote inaccessible areas. Soils and rock types indicate the presence of mineral hosting environment suitable for type minerals. Surface erosion, structures like fold closure, faults, shears, and lineaments are favorable indicators for deep-seated deposits. Vegetation including anomalous colors and toxic effects along with drainage pattern guides formulation of geochemical stream sediment sampling. The last but not the least, surface indicators like presence of gossans and old mining-smelting remnants is a sure catch for mineral exploration program. A typical aerial photograph is given in Fig. 6.2 for general familiarity.

FIGURE 6.2 A typical aerial view of Colorado Main River in the background from 7000 feet height of Southern Rim, The Grand Canyon, Natural Wonders of the World, Arizona.

6.3. REMOTE SENSING

6.3.1. Definition and Concept

RS is the emerging technology that has undergone phenomenal development over aerial photography and photo interpretation. It is the comparable way of collecting information about an object, area and phenomenon without coming in direct contact or touching itself. The information is acquired by a remotely placed sensor far away from the source object. In totality, RS implies data acquisition of electromagnetic radiation (EM radiation or EMR) from sensors flying on aerial or space platform and its interpretation for physical attributes of the ground objects. The fundamental and simple difference between photogeology and RS is as between photographs and images. The data in photograph are the reflection of natural light first recorded on to a light-sensitive emulsion coated base film (negative). These are then developed in laboratory and printed on similarly light-sensitive emulsion coated paper (positive/paper print/hard copy) for further processing and interpretation. In case of an image the data are the reflected and emitted multispectral electromagnetic energy (EME) initially recorded directly in digital form on to a magnetic tape or disc. The soft copy is processed and interpreted. The later phenomena are the basis of RS having the maximum capability, liberty and flexibility for manipulation of multispectral response over photogeology. It is young and growing science.

The essential principle of RS involves that each type of object reflects or emits a certain intensity of light when in contact with different range of wavelength of EMS, depending on the physicochemical attributes of the object. It can be clearly demonstrated that multispectral images in green, red and near-IR (NIR) bands can distinguish between different types of objects like water, soil, rocks, surface weathering and vegetation.

6.3.2. Energy Sources and Radiation

6.3.2.1. Electromagnetic Energy

Visible light, one of the many forms of EME, with wavelength (λ) varying between 0.4 and 0.7 μm is detected by human eye. Human brain receives color impulses of visible objects from the eye via three separate light receptors in the retina. These receptors respond to blue, green and red light respectively and known as additive primaries or primary colors. The eye receptor systems stimulate equally to a white visual effects if these three color beams overlap. A relative quantity of mixing of the just three primary colors i.e. blue, green and red lights reflected from the objects change to all possible colors representing the full range of "Rainbow" i.e. Violet, Indigo, Brown, Green, Yellow, Orange and Red (VIBGYOR) that the human eye perceives (a bowl of fruits) as visible light by synthesizing.

Other wavelengths (0.7-300 μm), especially longer than visible ray, are known as IR. The IR is divided into various components with respect to increasing wavelength. These are NIR (λ 0.7-1.1 μm), Middle Infrared (MIR λ 1.1-3.0 μm), Far Infrared (FIR λ 7-17 μm) and Thermal Infrared (TIR λ 3.0-5.0 μm). Color NIR beams generate three supplementary colors namely cyan, magenta and yellow.

There are two principal types of energy (EME) such as light energy as reflected and heat energy as emitted from the objects after partial absorption. The data acquisition in RS technology works on this concept of transmitting rays/energy at various wavelengths and receiving the relative reflectance and emitted energy to distinguish the characteristic attributes of objects under investigation.

Ultraviolet ray (λ 3 nm-0.4 μm) transmits heat energy that burns human skin and affects the eyes. Transmission of IR energy at λ 8-14 μm emits less than 200 °C and suitable for thermal spring. Similarly, IR energy at λ 3-5 μm emits more than 200 °C and suitable for volcanic study. Microwave with 0.8-100 cm wavelength can penetrate into the cloud and even subsurface. Various rays/energy and its characteristic wavelengths are tabulated at Table 6.1.

6.3.2.2. Electromagnetic Radiation

EM radiation is a phenomenon that takes the form of self-propagating energy waves as it travels through space (vacuum or matter). It consists of both electric and magnetic field components. The energy waves oscillate in phase perpendicular to each other and perpendicular to the direction of energy propagation. It is observed that longer the wavelength involved lower would be the frequency as well as energy. EM radiation is classified into several types according to the frequency of its wave. These types include (in order of decreasing frequency and increasing wavelength) such as: cosmic radiation, gamma radiation, X-ray radiation, ultraviolet radiation, visible radiation, IR radiation, terahertz radiation, microwave radiation and radio waves. A small and variable window of frequencies is sensed by the eyes of various organisms. This is known as the visible spectrum (λ 0.4-0.7 μm) or light. EM radiation

TABLE 6.1 **Various Rays/Energy and Its Characteristics Wavelength**

Ray	Wavelength	Ray	Wavelength
Gamma ray	<0.03 nm	Photographic IR	0.7-3 μm
X-ray	0.03-3 nm	NIR (magenta, cyan and yellow)	0.7-1.3 μm
Ultraviolet ray	3 nm-0.4 μm	MIR	1.3-3.0 μm
Visible ray	0.4-0.7 μm	FIR (thermal, emissive)	7-17 μm
Blue	0.4-0.5 μm	TIR (forest fire)	3.0-15.0 μm
Green	0.5-0.6 μm	Microwave	0.3-300 cm
Red	0.6-0.7 μm	Television/radio waves	1.5 km
IR	0.7-300 μm	EM radiation	0.74-300 μm

1 nm = 10^{-9} m, 1 μm = 10^{-6} m.

carries energy and momentum that may be imparted to matter with which it interacts.

A blackbody is an idealized theoretical radiator that absorbs 100% of all EM radiation that hits it. No EM radiation passes through it and none even reflect. The object appears perfectly black when it is cold because no light (visible EM radiation) is reflected or transmitted. There is no material in nature that completely absorbs all incoming radiation. However, graphitic carbon absorbs 97% incoming radiation and is the perfect emitter of radiation. A blackbody emits a temperature-dependent spectrum (thermal radiation) of light and is termed as blackbody radiation. The blackbody emits the maximum amount of energy possible at that particular temperature.

The EM propagation travels the path length twice between the source, object and sensor through the total thickness of the atmosphere. The compositional nature of the atmosphere affects the propagating energy by partial absorption and scattering. Atmospheric absorption results in the effective loss of energy to atmospheric constituents. The most efficient atmospheric absorptions are water vapor, carbon monoxide, carbon dioxide and ozone. The unpredictable diffusion of radiation by particles within the atmosphere is called the atmospheric scattering. Atmospheric windows are the ranges of wavelength in which the atmosphere is particularly transmissive.

6.3.2.3. Electromagnetic Spectrum

The EMS of EM radiation is the range of all possible frequencies of EM radiation. The behavior of radiation depends on its wavelength with inversely proportional to frequency and wavelength i.e. higher frequencies have shorter wavelengths and vice versa. The EMS of an object is the characteristic distribution of EM radiation emitted or absorbed by that particular object. A complete range of EM spectrum is elaborated in Fig. 6.3.

6.3.2.4. Spectral Reflectance/Response Pattern

The difference between the intensity of EM radiation reflected or emitted by an object at different wavelength is called spectral response or signature. The curve generated by intensity of energy versus wavelength is spectral response curve. A single feature or a response pattern can be diagnostic in identifying an object.

6.3.2.5. Data Acquisition

The data detection and acquisition is performed either photographically or electronically. The process of photography relies on chemical reaction on light-sensitive film. On the other hand, the electronic process administers electromagnetic signals to the objects. The EM signals are received back to the sensors with broader spectral range of sensitivity. It is capable to store and transmit as and when required. Most of the data acquisition of RS technology is synonymous to multispectral satellite imagery as detailed in Table 6.2. These images received from these satellites are readily available at cost for the exploration agencies.

6.3.3. RS System

An ideal and typical RS system would require the following components for better understanding of general operation and utility.

6.3.3.1. Platform

Platforms are vehicles or carriers to carry the remote sensor. The typical platforms for RS data acquisition are terrestrial (ladder, trucks for ground investigations), aerial (kites, balloons, helicopters and aircrafts for low-altitude RS) and space borne (manned or unmanned rockets, satellites from high altitude). The key factor for the selection of a platform is the altitude that determines the

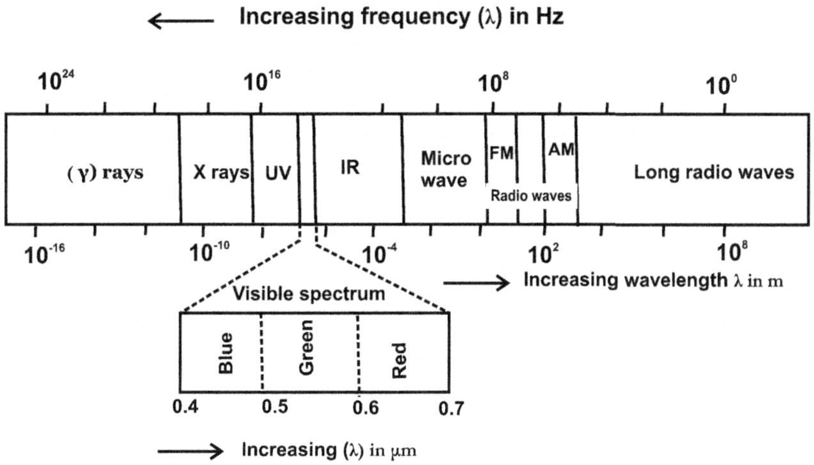

FIGURE 6.3 A complete possible range of EMS with increasing frequency (Source: modified after Wikipedia).

TABLE 6.2 Salient Features and Chronological Development of Major Landsat-Type Earth Resources Satellite Platforms of the 20th Century

Satellite	Country	Year	Nature	Altitude (km)	Sensor
Landsat-1	USA-NASA	1972	Sun Sys	919	MSS, RBV
Landsat-2	USA-NASA	1975	Sun Sys	919	MSS, RBV
Landsat-3	USA-NASA	1978	Sun Sys	919	MSS, RBV
Landsat-4	USA-NASA	1982	Sun Sys	705	MSS, TM
Landsat-5	USA-NASA	1984	Sun Sys	705	MSS, TM
SPOT-1	France	1986	Sun Sys	832	HRV
IRS-1A	India	1988	Sun Sys	904	LISS-1
SPOT-2	France	1990	Sun Sys	832	HRV
IRS-1B	India	1991	Sun Sys	817	LISS-2
Landsat-6	USA (EOSAT)	1993	Sun Sys	705	ETM
SPOT-3	France	1993	Sun Sys	832	HRV
IRS P1	India	1993	Sun Sys	817	LISS-2, MOS
IRS P2	India	1994	Sun Sys	817	LISS-2, MOS
IRS-1C	India	1995	Sun Sys	817	LISS-3
IRS-1D	India	1997	Sun Sys	817	LISS-3
SPOT-4	France	1998	Sun Sys	832	HRV IR
IRS P5	India	1998	Sun Sys	832	LISS-4
Landsat-7	USA-NASA	1999	Sun Sys	705	ETM+
IKOCOS	Commercial	1999	Sun Sys	–	High-resolution MSS
Terra	USA-Japan	1999	Sun Sys	705	ASTER, etc.

best possible ground resolution and which is also dependent on the instantaneous field of view (IFOV) of the sensor on board the platform.

The first Landsat-1 satellite was launched in July 1972 by NASA, USA. This was originally named Earth Resources Technology Satellite (ERTS) to provide multispectral imagery for the study of renewable and nonrenewable resources. Landsat-4 (1982) incorporated Thematic Mapper (TM) which scans in seven bands, two of which (five and seven) were specifically opted for geological purposes. A great variety of satellites were built for monitoring various environmental conditions on land and at sea. It can view the Earth in vertical, side, or limb modes. The new methodology of Earth science is based on satellite data that allow a whole Earth approach to study the environment. Remotely sensed satellite data and images of the Earth have four important advantages compared to ground observations such as synoptic view, repetitive coverage, multispectral and low-cost data.

India began development of an indigenous Indian Remote Sensing Satellite (IRS) program to support the national economy in the areas of agriculture, water resources, forestry and ecology, geology, water sheds, marine fisheries, coastal management, weather forecast, natural calamities and disaster management. IRS satellites are the mainstay of National Natural Resources Management System (NNRMS), for which Department of Space (DOS) is the nodal agency, providing operational RS data services. Data from the IRS satellites are received and disseminated by several countries all over the world. New applicatins in the areas of urban sprawl, infrastructure planning and other large-scale applications for mapping have been identified with the advent of high-resolution satellites. Salient features of some important Landsat-type Earth resources satellite platforms of the 20th century are given in Table 6.2.

The path of a celestial body or an artificial satellite as it revolves usually in an elliptical path around another body is

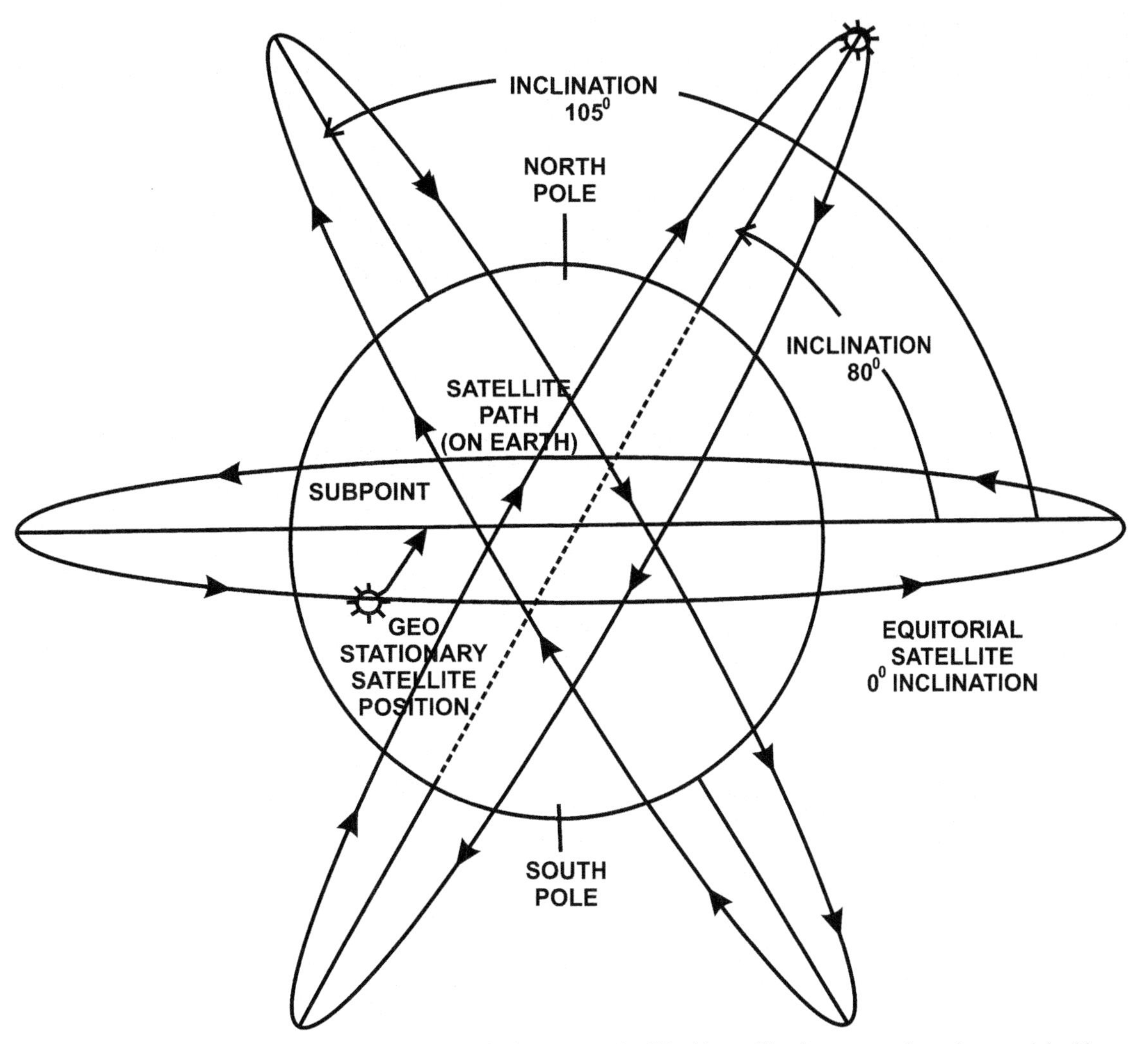

FIGURE 6.4 Diagram showing the typical elliptical orientation path of Earth's satellites known as polar and equatorial orbit.

called an orbit. In case of Earth's satellites it makes one complete revolution in 12 h. There are two types of orbit i.e. polar and equatorial orbits.

In case of a polar orbit the satellite travels over both north and south poles at about 850 km altitude above the Earth at an angle of 80° and 105° from equatorial plane (Fig. 6.4). N to S rotation is called descending and S to N is ascending. Satellites take about 100 min for a complete revolution in polar orbit. It can see small portion at a time covering the whole globe at high resolution. Polar orbit is essentially sun-synchronous and geostationary. A satellite in such an orbit can observe all points on the Earth during a 12-h day. This type of orbit is useful for spacecraft that performs mapping or surveillance.

In case of equatorial orbit the satellite flies along the line of the Earth's equator (Fig. 6.4). A satellite must be launched from a place on Earth close to the equator to get into equatorial orbit. Equatorial orbits are useful for satellites observing tropical weather patterns, as they can monitor cloud conditions around the globe.

Sun-synchronous orbit (heliosynchronous or dawn-to-dusk orbit) is a geocentric orbit which combines altitude and inclination in such a way that an object on that orbit ascends or descends over any given point of the Earth's surface at the same local mean solar time. The surface illumination angle will be nearly the same every time. This consistent lighting is a useful characteristic for satellites that image the Earth's surface in visible or IR wavelengths and for other RS satellites (e.g. those carrying ocean and atmospheric RS instruments that require sunlight).

Geostationary orbit is a geosynchronous orbit directly above the Earth's equatorial orbit (0° latitude and 36,000 km altitude) stays over the same spot with a period equal to the Earth's rotational period. Geostationary objects appear motionless in the sky from Earth's surface, making

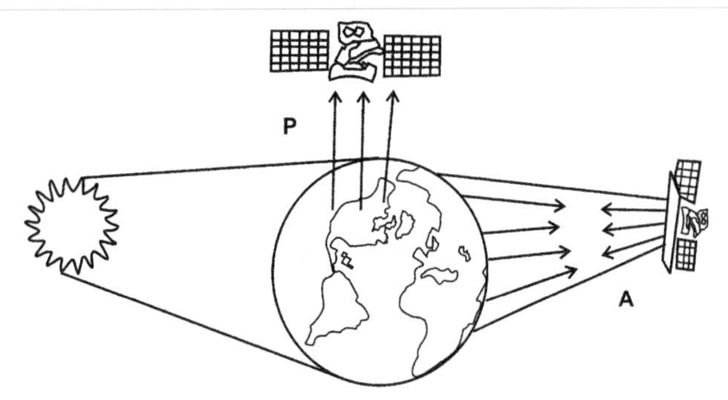

FIGURE 6.5 Schematic diagram showing the position of data collecting device with respect to Sun and Earth to designate it as Passive (P) and Active (A) sensors.

the geostationary orbit of great interest for communication purposes and weather forecasts. It observes an evolving system with lower spatial resolution. Satellites in geostationary condition differ in location by longitude only due to the constant 0° latitude and circularity of geostationary orbits.

6.3.3.2. Sensors

Sensors are device like photographic cameras, scanners, radiometers mounted on suitable platform to detect and record the intensities of electromagnetic radiation in various spectral channels. Sensors are of two types: Passive and Active.

Passive sensors are designed to record data using available naturally occurring energy source reflected, emitted and transmitted parts of the EMS. It relies on the solar illumination side of the Earth or natural thermal radiation for their source of energy (Fig. 6.5). Detection of reflected solar energy can only proceed when the target is illuminated by the sun. This restricts visible light sensors on satellites from being used during a nighttime pass. The examples of passive sensors are Landsat Multispectral Scanner (MSS), Landsat Thematic Scanner (TM) using additional wavelengths to produce superior spectral and spatial resolution, airborne scanning system SPOT with stereoscopic capabilities and space shuttle.

Active sensors use their own illumination as source of energy and can make observations on both the sun-lit as well as the dark side of the Earth anytime, regardless of the time of day or season (Fig. 6.5). The sensor emits radiation which is directed toward the target to be investigated. The radiation reflected from the target object is received and recorded by the sensor. An active system requires the generation of a fairly large amount of energy to adequately illuminate the targets. Some examples of active sensors are a Synthetic Aperture Radar (SAR) and laser fluorosensor.

6.3.3.3. Sensor Resolution

RS sensors have four types of resolution, viz., spatial, spectral, radiometric and temporal.

(1) Spatial resolution includes the geometric properties of the ground covered under IFOV of the sensor. IFOV is defined as the maximum angle of view in which a sensor can effectively detect electromagnetic energy (imaging).
(2) Spectral resolution is the span of the wavelength over which a spectral channel operates by the sensor. It is defined by the bandwidth of the EM radiation of the channels.
(3) Radiometric resolution is the degree of intensities of radiation the sensor is able to distinguish.
(4) Temporal resolution involves repetitive coverage over an area by the sensor and is equal to the time interval between successive observations. The repeated coverage will identify changes in the objects under study.

An ideal RS system should fulfill the following:

(1) Uniform energy source of all wavelengths at a constant high level of output, irrespective of time and place.
(2) Atmosphere between the source, target and receiver expected to be clean for to and fro energy radiation.
(3) Sensitive super sensor for acquisition of data.
(4) Real-time data processing and interpretation system.
(5) Multidisciplinary users having adequate knowledge, skill and experience of RS-GIS data acquisition, analysis and extract their own information.

6.3.4. Characteristics of Digital Images

6.3.4.1. Pixel Parameters

In digital imaging, a pixel (picture element) is the smallest item of information in an image. Each pixel is represented by a number equivalent to average radiance or brightness of that very small area. Pixels are normally arranged in a 2D grid (x, y) and are often represented using dots or squares. The "z" value represents the grayscale value of 256 different brightness levels between 0 (black) and 255 (white). Each pixel is a sample of an original image, where more samples provide more accurate representations of the original object. Pixel size determines the spatial resolution. The intensity of each pixel is variable in color system and each pixel has typically three or four components such as red-green-blue or cyan-magenta-yellow and black. An image is built up of a series of rows and columns of pixels.

6.3.4.2. Mosaics

Each image has a uniform of scale and resolution for a scan path with forward and side overlap. A mosaic is a set of images (photographs) arranged to facilitate a bird's eye view of the entire area. This can be done by cutting and merging of each overlapping scene of B&W or color images or prints.

6.3.5. Digital Image Processing

Multispectral satellite sensor data are gathered and stored in digital form on computer compatible magnetic tapes (CCTs) at ground station. It is further processed at main computing center. The main functions of the geological exploration data for this total system are image restoration, enhancement and information extraction.

6.3.5.1. Image Restoration

Image restoration is the process of correcting the defects in images during data collection and subsequent transfer to ground station. The process involves replacing lost data (pixel and line), filtering of atmospheric and other noises and geometrical correction.

6.3.5.2. Image Enhancement

Image enhancement is the procedure of improving the quality and the information content of original data before processing. The common practices include contrast enhancement, spatial filtering, density slicing and FCC. Contrast enhancement or stretching is performed by linear transformation expanding the original range of gray level. Spatial filtering improves the naturally occurring linear features like fault, shear zones and lineaments. Density slicing converts the continuous gray tone range into a series of density intervals marked by separate color or symbol to represent different features.

FCC is commonly used in RS compared to true colors because of the absence of pure blue color band as further scattering is dominant in blue wavelength. FCC is standardized as it gives maximum identical information of the objects on Earth and satisfies all users. In standard FCC vegetation looks red (Fig. 6.6) because vegetation reflects very high in NIR and the color applied is red. Water bodies look dark if it is clear or deep because IR is an absorption band for water. Water bodies give shades of blue depending on its turbidity or shallowness because such water bodies reflect in green wavelength and color applied is blue.

6.3.5.3. Information Extraction

Information extraction is carried out online by ratioing, multispectral classification and principal component analysis to enhance specific geological features. Ratio is prepared by dividing gray level of a pixel in one band by that in another to recognize ferruginous and limonitic capping (gossans) useful for identifying sulfide deposits.

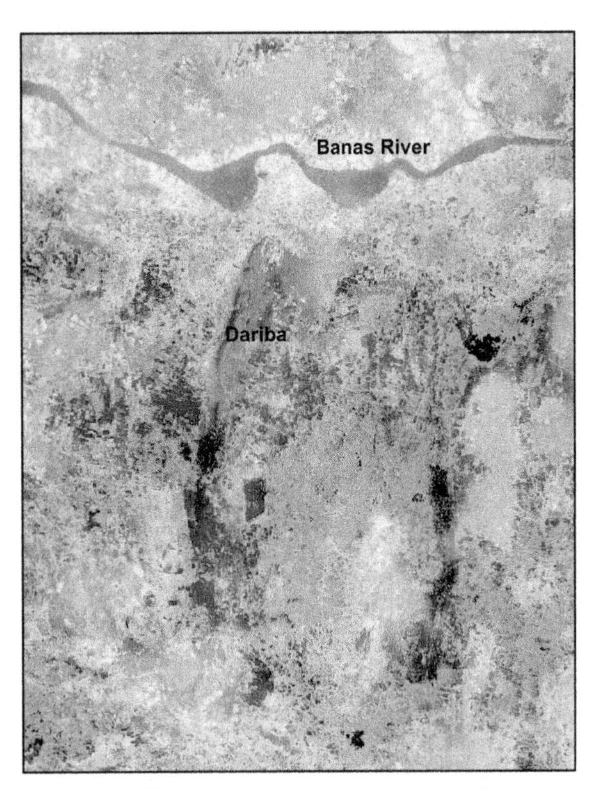

FIGURE 6.6 FCC image of aeromagnetic and satellite data over Rajpura-Dariba sulfide belt, Rajasthan, India (Credit: Dr A. Bhattacharya and Haldar (2007) [33]).

Multispectral classification generates small groups of pixel of different reflectance and marked by different colors or symbols to represent same kind of surface signature. Principal component analysis is a commonly used method to improve the spread of reflectance by redistributing them. It is used to enhance or distinguish difference in geological features (Fig. 6.7) e.g. elevation, land cover, lineaments, rock types, vegetation index, turbidity index, forest fire, flood and archaeological features.

6.3.6. Interpretation

Interpretation or extraction of information from processed satellite images are usually done by photogeological or spectral approaches. Each frame of Earth cover image has its own spectral reflectance characteristics. The characteristics ("signature") are unique and make possible to distinguish the objects of interest from its intermixed background. The final process is completed by the analysis of the data or images using image interpretation techniques. The techniques of RS and image interpretation yield valuable information on Earth resources.

6.3.7. RS Application in Natural Resources

Multispectral RS techniques have significant potential for multipurpose applications in all branches of Earth

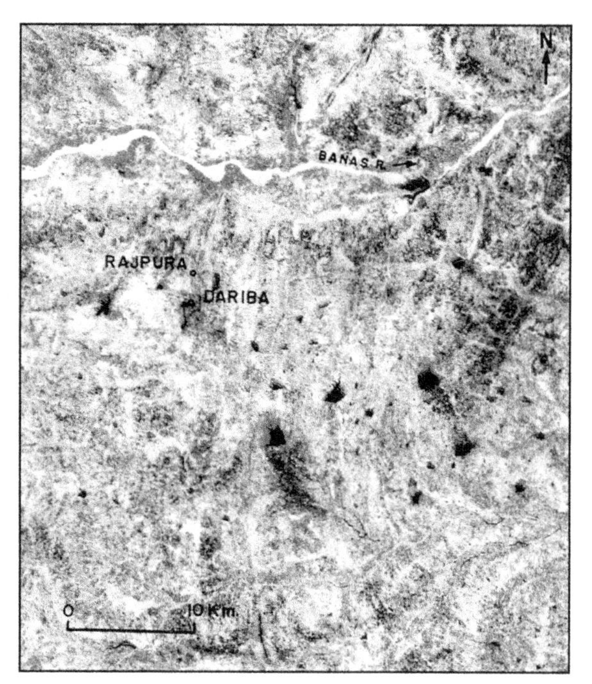

FIGURE 6.7 Principal component image of Rajpura-Dariba sulfide belt, Rajasthan, India (Credit: Dr A Bhattacharya and Haldar (2007) [33]).

science such as geomorphology, structure, litho-mapping, stratigraphic delineation, geotechnical and geo-environmental studies. Many applications are closely related to mineral exploration and resource evaluation. It forms the concept base by virtue of its synoptic overview to locate and delineate metallogenic provinces and mineral deposits including hydrocarbon and water at reduced time and cost. RS applications play a significant role at all the sequential stages of exploration starting from reconnaissance, large area prospecting, prospecting, detail exploration, active mining and geo-environment to mine closure.

The most powerful data sources at reconnaissance or preliminary stage of mineral search are satellite base images at small scale (1:50,000 or 250,000). The objective is to identify metallogenic provinces out of extremely large license area. The targets can be checked by limited test drilling. The explorer then moves to prospecting stage using photogeology (1:25,000 or 50,000) and supplemented by aerial geophysics to identify anomalies representing possible target(s) for systematic drilling. At this juncture thematic map generation is useful for prioritization of exploration targets. The prospecting activities lead to detailed exploration by detail mapping (1:10,000, 5000, 2000), ground geochemistry, geophysics and close space diamond or reverse circulation drilling to estimate the reserves and resources. Environmental baseline maps generated at the initiation of exploration can be useful and compared with the mine closure plan for environmental restoration of ecosystem.

RS study of geomorphology reveals various types of landforms (tectonic, volcanic, fluvial, coastal and deltaic, aeolian and glacial). The salient guides are predominantly surface indications like sustained weathering and erosion (residual and supergene enrichment), oxidation (gossans), drainage pattern (stream sediment sampling), placer deposits (diamond, gold, ilmenite, monazite) formed as a result of mechanical concentration of fluvial, aeolian, alluvial, eluvial and marine process, and finally remnants of ancient mining-smelting activities. A long continuous belt of placer deposits around the East and West Coast of India, Indonesia and Australia was identified from remote sense data.

The aerial and space base acquired data provide a complete new dimension to the mineral exploration by integration of structural (rings, folds, faults, shear zones, lineaments) data to a composite aerial view. These structures, in many cases, are the governing factors in localization of economic mineral deposits. The identification of rings, shear zones and lineaments using RADARSAT images help to find the areas with the probability of diamond pipe and base metal mineralization. The final structure layer is prepared using visual interpretation and software processing.

The remote sense data generate broadscale litho-maps including mineral assemblages and formation of a stratigraphic succession model. Mineral deposits have certain affinity to particular group of host rocks e.g. base metals with dolomite, calc-silicate and black-schist, phosphorites with dolomite, iron with BHQ and coal with shale and sandstone. Similarly some minerals are closely related to the certain stratigraphic age group e.g. gold with Archean greenstone/greenschist horizon (>2500 Ma), coal with Permo-carboniferous (248-360 Ma) and hydrocarbon with Cretaceous (65-144 Ma) age. Some mineral deposits are preferentially confined to rocks of type genetic aspect e.g. >60% of zinc-lead deposits is related to Proterozoic SEDEX type. The interpretation of RS data serves as useful guide during mineral exploration by identification of these features.

The role of geobotany in mineral exploration is dealt at Chapter 4. Dense vegetation, masking the surface, is inaccessible and considered as hindrance to mineral exploration. The information collected from RS platforms can open the reality below the ground and add to resource. The relative geobotanical abnormalities with respect to the vast area can easily be detected and mapped from aerial view. Two types of anomalies are common in the growth of vegetation. Morphological features include changes in color of leaves and flowers and dwarfing due to toxic effect of metals in the soil. Taxonomic differences refer to relative abundance or absence of certain species.

Exploration for ground water requires identification of aquifers that exists few meters to 100s of meters from the surface. EM radiation and even microwave can barely penetrate few meters in the ground. This causes limitation of data acquisition and use of RS as direct guide for deep-seated ground water exploration. However, many of the surface features, responsible for subsurface water condition, can be mapped by RS leading to regional and local ground water maps. The regional ground water survey can be interpreted from second-order indirect indicator namely landforms, rock types, soil moisture, rock fractures, drainage characteristics, and vegetation. The local exploration guides are from first-order direct indications of recharge and discharge zone, soil moisture and anomalous vegetation.

Hydrocarbons (oil and gas pools), on the other hand, exist few kilometers from the surface confined to either suitable stratigraphic or structural traps. Hydrocarbon exploration by multispectral RS data acquisition primarily depends on second-order indirect evidences namely striking circular drainage anomaly, geobotanical and tonal anomaly due to seepage of underlying hydrocarbons, regional lineaments in the oil-bearing region and films of oil slicks on ocean and sea water surface.

A remote sense data interpreter has to rely on direct or indirect clues such as general stratigraphic setting, alteration and oxidation zones, favorable host rock assemblages, rings-folds-faults-shears and lineaments, morphology, drainage pattern and effect on vegetation to guide the exploration rapidly. The alteration and structure along with other information layers i.e. geo-maps, geophysics and geochemistry are used to produce the primary exploration model in GIS. The best results are achieved by giving higher weights to the RS layers. The quality of results is evaluated by field checking. RS interpretations play high reliability in mineral exploration.

To conclude, mineral potentiality mapping is systematic plan to collect, manage and integrate various geospatial data from different sources and scales during multistage activity. GIS can describe and analyze and interact to make predictions with models and provides support for decision makers.

6.4. GEOGRAPHIC INFORMATION SYSTEM

6.4.1. Definition

GIS is composed of three critical words to define the module GIS. "Geographic" refers to known location of the primary database consisting of observations of features, activities or events defined in space as point, lines or area, assay value, in terms of geographic coordinates: latitude, longitude and elevation. The measuring units are either in degree-minutes-seconds or in Universal Transverse Mercator (UTM) coordinate System. The various types of data are captured under Database Management System (DBMS) or Relational Database Management System (RDBMS) in different layers. Information means that the data are processed within GIS using high-speed powerful software tools for the analysis of spatial data to yield useful knowledge, making maps into dynamic objects, models or output as required by the user. System implies a group of interacting, interrelated, or interdependent function to reach the end objectives of different users.

The sequences of activities in GIS function are:

(1) Data collection: measurement aspects of geographic phenomena and processes;
(2) Storing: measurement stored to digital database to emphasize spatial themes, entities, and relationships;
(3) Retrieving: operate to create more measurements and to discover new relationships by integrating;
(4) Transformation: convert new representations conform to uniform frameworks of entities and relationships;
(5) Processing: system for capture, storage, retrieval, analysis and display;
(6) Modeling: system to store and maneuver geographic information; and
(7) Display: maps and reports.

A GIS can be defined as an organized assemblage of hardware, software, geographic data, and personnel designed to efficiently capture, store, update, maneuver, analyze, and display all forms of geographically referenced information.

6.4.2. Components of GIS

The main components of GIS to perform the activities involve five subsystems (Fig. 6.8):

(1) Hardware
(2) Software
(3) Geographic-referenced data
(4) Method and
(5) People—professionals (multiuser from same database).

The hardware component of the GIS is the main input-output system and consists of computers or Central Processing Unit (CPU) for storage of data and software. A high-capacity disk drive of CPU is the storage unit of data and program. The digitizer and scanner are attached for converting the maps and documents into digital form. The output units consist of monitor for online display, plotter and printer for results as hard copy prints of maps, images and other documents. Temporary and permanent devices for storing and retrieving of digital information are pen drive, CD, magnetic tape, and external Hard Disk Drive (HDD) or hard disk with high capacity.

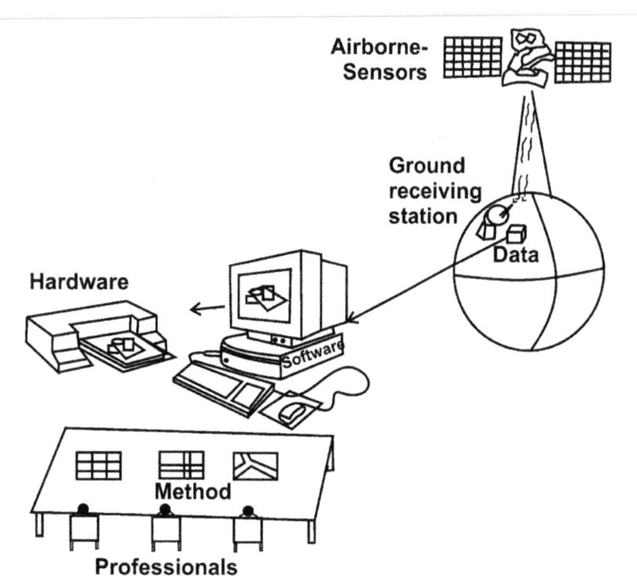

FIGURE 6.8 A typical functional aspects of GIS interfacing between major components from data collection to multiuser end results.

Software is the key subsystem that includes the programs and interface for driving the hardware. It is responsible for total data management, storing, analyzing, maneuver and displaying the data or geographic information. Efficient quality software must be user friendly, compatible, well documented and cost effective. There are many commercial software available which can be tailor made for specific uses. Some of the popular names are ArcInfo, ArcView, MapInfo, etc.

Data, more precisely georeferenced data, are the most significant component of the GIS system. It can be purchased from a commercial data provider or collected in house and compiled to custom specification. The key functionality of GIS is integration of spatial and tabular data stored in DBMS, RDBMS and Structured Query Language (SQL).

The right method is a key for successful operation of GIS technology. The well-designed implementation plan with decision support and business rules are unique to each organization.

The benefits of GIS technology cannot be completely utilized for a meaningful end result without competent people. The specialists design the total system for wide range of end users. The users belong to multidiscipline e.g. computer science, agriculture, forest, town planner, industry, geology, geography etc. Each one can share a common database and generate results as required by them. The identification of specialists and end users is significant for proper implementation of GIS technology.

Generally, GIS is considered to be expensive and difficult. But with the advent of new technology like graphical user interface, powerful and affordable hardware and software it is gaining ground and included in the mainstream.

6.4.3. Capabilities

GIS has capability of multiuser multipurpose function:

(1) GIS is a high-tech equivalent of a quicker and efficient map generator. It can access and store data in digital format, enables complex analysis and modeling.
(2) Capable to conduct location analysis of various attributes stored at different layers, linking them to explain the causes and effects that yield a result. The presence of surface oxidation at one layer supported by geobotanical evidences in other can lead to discovery of massive sulfide deposit underneath.
(3) Responds on query and displays results after satisfying certain spatial condition. A well can be planned within a preset distance from the township by satisfying spatial conditions of possible aquifers, township and existing pipeline.
(4) Capable of performing temporal analyses at time interval over many years to derive relationship between changing land use practices and future requirements.
(5) Evaluate different scenarios applying sensitivity analysis and forecast the best one. It can continuously monitor and revise decision with changing assumption and additional input.

6.4.4. Data Input

DBMS, RDBMS and SQL software with integral error checking facility incorporate or import data from outside sources, update and alter data, if necessary. Data can be directly entered to GIS platform manually or created outside in standard ASCII (American Standard Code for Information Interchange) files. GIS is also capable of importing data files that are created outside in other formats. GIS must provide the capability to export data to other systems in common format (e.g. ASCII). Maneuvering the database to answer specific data-related questions is organized through a process known as database analysis. GIS must provide ways to modify, refine, revise and update the database.

Two basic types of data are normally incorporated into GIS applications. The first type consists of real-world phenomena and features that have some kind of spatial dimension such as geological map (Fig. 6.9). These data elements are depicted mathematically in the GIS as points, lines, or polygons that are referenced geographically or geocoded to some type of coordinate system. It is entered into the GIS by devices like scanners, digitizers, Global Positioning System (GPS), air photos, and satellite imagery. The other type is sometimes referred to as an attribute or

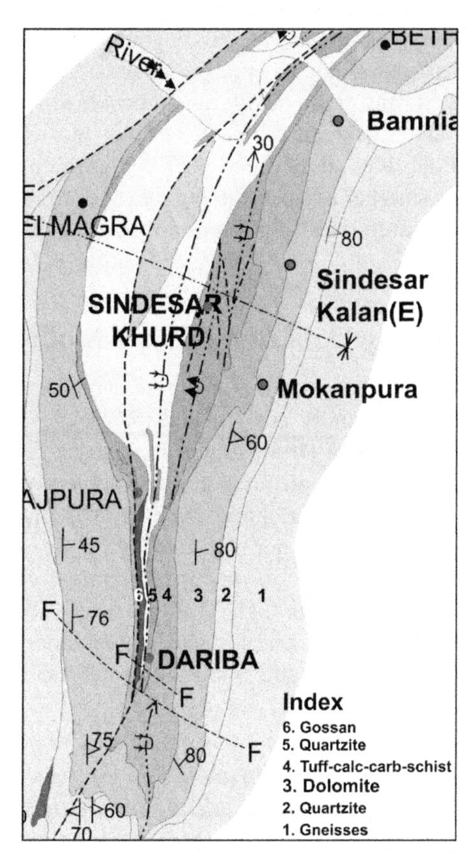

FIGURE 6.9 Data model of real-world phenomena and features having spatial dimension (e.g. geological map of Rajpura-Dariba base metal belt) usually incorporated into GIS applications.

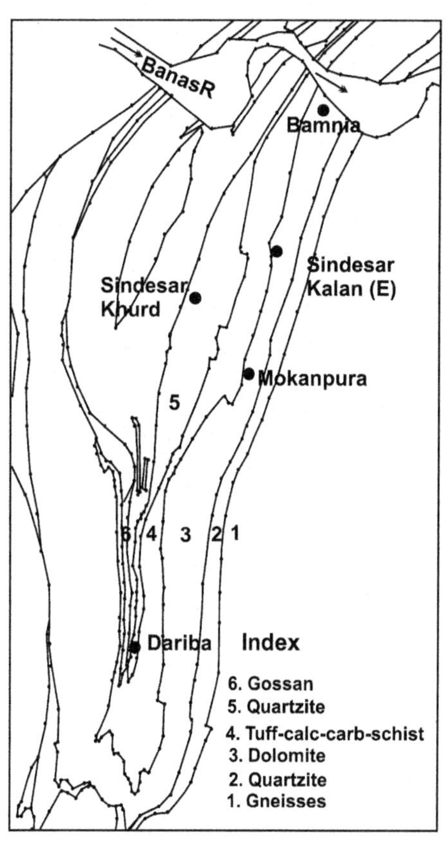

FIGURE 6.10 Data model by transformation of geographic phenomena of Rajpura-Dariba base metal belt into vector mode.

point data such as soil profile. Attributes are pieces of data that are connected or related to the points, lines and polygons mapped in the GIS. These attribute data can be analyzed to determine patterns of importance. Attribute data are entered directly into a database where it is associated with element data. The GIS uses one of the two primary types of spatial data, i.e. vector and raster, to represent the location component of the geographic information.

The vector data model of geographic phenomena is directional and represented by points, lines and areas. The points are zero-dimensional objects (nodes) and can be represented by a single coordinate (x, y), e.g. sample, mine, house and city location. The lines are 1D objects (line) and represented by a set of points or pair(s) of coordinates e.g. roads, streams, faults, shears and lineaments. The polygons (2D areas underlain by dolomite, orebody) are bounded by a closed loop, joined by a set of line segments. 3D objects are represented (x, y, z), where z is the value of elevation e.g. hills. The vector data model is particularly useful for representing discrete objects e.g. sample point, roads, streams, faults and rock boundaries in the form of points, lines and polygons and stored as set of coordinates. The data in vector format are geometrically more precise and compact (Fig. 6.10).

In raster data model the surface is divided into regular grid of cells represented by rows and columns or pixels of identical size and shape (Fig. 6.11). The location of geographic objects or conditions is defined by the coordinates of the cell position. Each cell indicates the type of objects, class, category, measurements, condition or an interpreted value that is found at that location over the entire cell. The smaller the cell size higher is the resolution with accurate and detail information. But the data storage will be enormous. On the contrary larger the cell size less data space will be required with approximate or less accurate information. In case of very large grid, several data types may occur in each cell and will be treated as homogeneous unit during analysis to generate inaccurate results.

The advantages and disadvantages between vector and raster data are modified after David J. Buckley and given in Table 6.3

6.4.5. Projection and Registration

One often comes across maps of the same area containing different features (geology, geochemistry and geophysics) and in different scales. GIS has capability to maneuver the

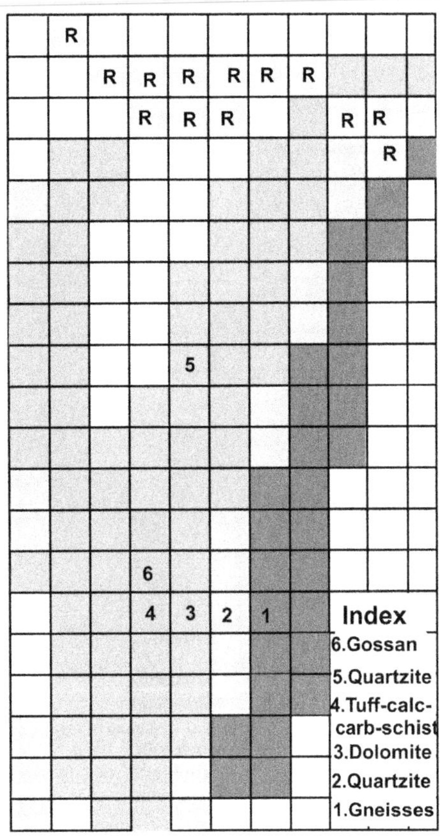

FIGURE 6.11 Data model by conversion of geological information of Rajpura-Dariba base metal belt to raster mode for GIS applications in mineral exploration.

geocoded map information from different sources to common scale so that it can fit and register. Projection is a fundamental component of mapmaking in GIS. It is a mathematical function of transferring information from 3D surface to fit in 2D medium. The digital data may have to undergo other transformation like projection and coordinate conversions to integrate them into a GIS before it can be analyzed. This process inevitably distorts at least one of the following properties like shape, area, distance and direction. Different projections are used for different types of map for specific uses. GIS has the processing power to transform digital information gathered from sources (digitized data, aerial photographs, satellite and GPS) with different projections to a common frame.

6.4.6. Topology Building

Topology defines the mathematical representation of spatial relationship between geographic features. It describes the relationships between connecting or adjacent coverage attributes. Topological relationships are built from simple elements into complex elements such as points (simplest elements), arcs (sets of connected points) and area (sets of connected arcs). Three types of relationship exist in topology: connectivity, area definition and contiguity. Storing connectivity makes coverage useful for modeling and tracing in linear networks. Storing information about area definition and contiguity makes it possible to merge adjacent polygons and to combine geographic features from different coverage with overlay operations.

6.4.7. Overlay Data Analysis and Modeling

Overlay analysis or spatial data analysis is a function in GIS applications to spatially analyze and interpolate multiple types of data streaming from a range of sources. Each type of data pertaining to same area or similar objects is registered in vector or raster mode in individual files. New information can be created by overlaying or stacking (Fig. 6.12) of the related layers with common georeferences and analyzing through GIS function. There are several different spatial overlay and manipulation operations to arrive at a specific model that can be used on features of user's interest.

6.4.7.1. Digital Evaluation Model, Digital Terrain Model, Terrain Evaluation Model and Triangulated Irregular Network Model

A Digital Evaluation Model (DEM) is a 3D representation of a surface topography using a raster or vector data structure. Any digital representation of a continuous surface of relief over space is known as DEM. It describes the elevation of any point in a given area in digital format. A Digital Terrain Model (DTM) or Terrain Evaluation Model (TEM) exhibits 3D spatial distribution of terrain attributes. It is a topographic map in digital format, consisting not only of a DEM but also types of land use, settlements, drainage, road network and related features. Triangulated Irregular Network (TIN) is a 3D surface model derived from irregularly spaced points and break line features. The basic unit is a triangle consisting of three lines connecting three nodes. Each triangle will have three neighbors except outer periphery. Each node has an x, y coordinate and a z value or surface value. These models are created by digitizing contour maps along with points, lines and polygon data of related objects followed by vector to raster conversion and interpolation to derive the desired results. These are useful for road and rail line and dam site planning, reservoir capacity estimation, identifying ridgelines and valleys, visibility studies, cut and filling problems.

6.4.7.2. Mineral Exploration Model

Overlay analysis is useful in mineral exploration for identifying targets. The example (Fig. 6.12) contains conceptual

TABLE 6.3 Salient Advantages and Disadvantages of Vector and Raster Data Model

Vector data model	Raster data model
Advantages	
Data storage at original georeferenced coordinates at higher spatial resolution, maintain and form without generalization.	Geographic location of each cell is identified by its position in cell matrix. Best resolution at smallest cell size.
Graphic output is more accurate and realistic like cartographic representation.	Data structure is simple and compact, storage in flat ASCII format for easy to program and quick to analysis.
Most data, e.g. hard copy maps, are in vector form and no data conversion is required.	Retrieval, updating and generation of data. Grid cell systems compatible with raster-based output devices—plotters, graphic terminals.
Efficient encoding of topology and operations by network analysis.	Topology can be completely described with network linkage.
Data are less voluminous and technology is less expensive.	Discrete data enable integrating two data types. Computational efficiency in overlay analysis.
Disadvantages	
Location of each vertex needs to be stored unambiguously.	Cell size decides resolution. Smaller the cell size complex the data structure and more expensive technology.
Vector data conversion to topological structure is processing intensive and requires extensive data cleaning.	Difficult to adequately represent linear features depending on the cell resolution. Network linkages are difficult to establish.
Algorithms for analysis functions are complex inherently limiting functionality for large data.	Processing of associated attribute data is complex with large volume of data.
Elevation data are not effectively represented in vector form. Often requires sizeable data generalization or interpolation of data layers.	Most input data are in vector form and need conversion from vector to raster by escalating processing, generalization and unsuitable cell size.
Spatial analysis and filtering within polygons is not possible.	Most output maps from grid cell systems do not conform to high-quality cartographic needs.

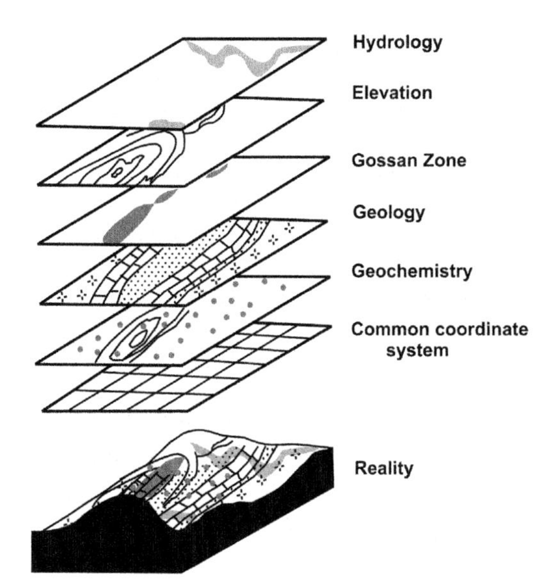

FIGURE 6.12 Concept of overlay analysis and integration of multilayer data of Rajpura-Dariba base metal belt for identifying drill targets: an example of GIS application in mineral exploration.

data from a range of source maps of Rajpura-Dariba base metal belt, Rajasthan, India. The files contain hydrology, elevation, surface signature (gossan), geology and geochemistry. All source data have been geocoded and registered on a cell-by-cell basis from a series of land data files in individual layer. The analyst module of GIS manipulates and overlays simultaneously the information derived from various data files. The analyst uses the system to interrelate the geocoded source data files. The interpreted result is expected to define target area for exploration of possible sulfide deposits. The example is a concept base model attempted to generate thematic maps. Addition of data of remote sense EM and surface geophysics will certainly strengthen the model to forecast exploration targets.

6.4.8. GIS Application in Mineral Exploration

GIS applications in mineral exploration are being widely used internationally. GIS platform allows in establishing

Mineral Deposit Database (MDD) for a region or country by integration of all available geoscientific data (even dissimilar) into digital single and unified database. The recommended approach is to compile all the available geoscientific data within the GIS in the context of an exploration model. It will produce a brief report and mineral potential maps of a province, region, district, belt, deposit and block. Careful consideration must be given in developing the model so that all the relevant aspects of the deposit being sought are represented. The model is important in deciding logical weight to apply by a geologist having adequate knowledge of the model and the deposit related to each of these aspects. The final map indicates the ranks and priority for exploration target in the study area.

An extension of the GIS database is incorporation of exploration data consisting of deposit name, location, infrastructure available, parks and reserve forests, historic and current exploration details, drill hole location, summary logs, drill sample assay value, geochemistry, EM and gravity images, mineral occurrences, reserve and resource detail and finally the lease status like RP/PL/ML. This Mineral Resource Information System will act as an exploration guide to new targets and decision base for free areas. It will be in SQL base system so that any investor can design his objectives and search online for desired results.

6.5. GLOBAL POSITIONING SYSTEM

The global positioning system (GPS) is a universal satellite-based navigation system developed, replaced, monitored and maintained by United States Department of Defense originally for military applications. Its official name was Navigation Satellite Timing and Ranging (NAVSTAR). The first global positioning space vehicle was launched in 1978. Thc total nctwork satellite launching was completed in 1994. The system became fully operational in 1995. Since then the system is available for civilian use that works worldwide under any weather conditions, 24 h a day without paying any routine subscription or setup charges. The total number of satellite in the constellation today is 60 (16 for civilian use and remaining for military and standby). A GPS satellite weighs about 1000 kg. The precision for civilian uses is in centimeter scale and the same for military purposes is in millimeter scale. The GPS consists of three major segments (Fig. 6.13). These are the space segment (SS), a ground control segment (CS) and a user segment (US).

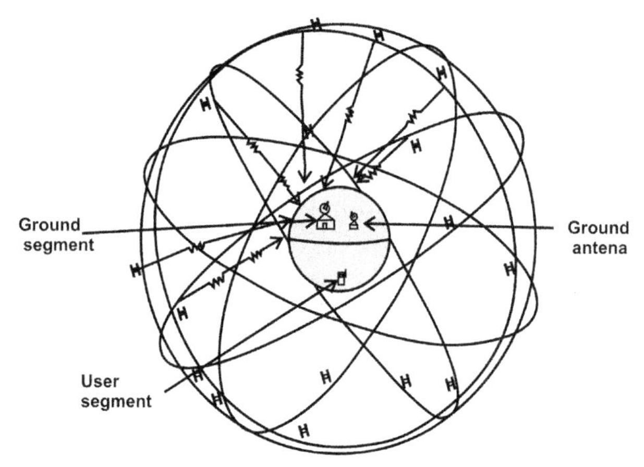

FIGURE 6.13 Conceptual overview of constellation of GPS satellites in space, ground control and user segments.

6.5.1. Space Segment

The Space Segment originally comprises of the 24 orbiting satellites (21 active and 3 standby), in 6 circular orbital planes with 4 satellites in each plane. The orbital planes are centered on the Earth and have 55° inclination relative to Earth's equatorial plane. The planes are equally spaced separated by 60° apart along the equator from a reference point to the orbit's intersection. The satellites are orbiting at an altitude of approximately 20,200 km. Each satellite makes two complete orbits each day i.e. a complete rotation around the Earth in 12 h. The orbits are so arranged that at least six satellites are always within line of sight from almost everywhere on Earth's surface.

6.5.2. Ground CS

The flight paths of the satellites are tracked by ground CS located at various monitoring stations of respective participative countries. In the event of any deviation of the space vehicle from its designed orbit the ground control station transmits the tracking information to the master control station. The master control in turn uploads orbital and clock data to each GPS satellite regularly with a navigational update using the ground antennas. These updates synchronize the atomic clocks on board of the satellites within a few nanoseconds of each other.

6.5.3. User Segment

The US uses various types of receivers to compute the coordinates (X, Y), elevation (Z), velocity and time estimates. GPS receivers are composed of an antenna, tuned to the frequencies transmitted by the satellites, receiver-processors, and a highly stable atomic clock. The receiver computes its position and time by making simultaneous measurements to number of satellites. A 2D position i.e. latitude and longitude can be computed by signals of three satellites. Signals from at least four satellites are required for determination of 3D location including elevation and clock bias. The receiver displays information comprised of number of visible satellites, location and speed to the user. The location works on both in LONG/LAT i.e. degree/minutes/seconds and UTM in metric grid system. There are 60 zones to cover the entire Earth's surface. The receivers can include an input for differential corrections and relay position data to a PC. It can interface with other devices using methods including a serial connection, USB or Bluetooth.

6.5.4. Signals

GPS satellites transmit low-power time-coded radio signals of 1575.42 MHz frequency in Ultra-high frequency (UHF) band. The signals are recorded by ground-based GPS receiver. The signals travel by line of sight through semi-transparent objects and not solids, metals and EM field. The sources of error and interference of signal transmission consist of ionosphere and troposphere delay, signal multi-path error, clock error, orbital error, number of visible satellites, satellite geometry and shading, and international degradation of the signals.

6.5.5. Types of GPS

The types of GPS receiver can broadly be classified on the basis of accuracy required and associated cost of the unit.

6.5.5.1. Handheld GPS

Handheld GPS is the simplest, cheapest and easiest unit consisting of a single receiver. The technology is simple and therefore least accurate. The location position, computed from the signals, can be distorted by 10-30 m. Most of the mobile cell phones provide name of the place while traveling by receiving signals from local antenna.

6.5.5.2. Differential Code Phase GPS

GPS measurements are prone to multiple errors on account of uncertainties in satellite ephemeris, atmospheric condition, quality of receiver and multipath situation. Differential GPS works on simultaneous measurements by receiver at a reference station with precisely known location, clock-time and number of roving receivers moving from point to point. The positional introduced noises measured at the base station are used to compensate instantly the position measured by the rovers to attain greater accuracy. The base station may be a ground base facility or a geosynchronous satellite. In either case, the base station is known with precise location value. This known value can be compared with the signals received from the GPS satellites and thus find the international error introduced by each GPS signals. The correction can be immediately transmitted to the mobile GPS receivers i.e. real-time Differential Global Positioning System (DGPS) to compute a much more accurate position. The receiver position can also be corrected at a later time during processing by GPS software. The use of DGPS can greatly increase positional accuracy within 1 m.

6.5.5.3. Carrier Phase Tracking GPS

Carrier phase tracking GPS signals have resulted in development of land surveying, geological mapping and as a guide to reach the target destination. The positions can be measured up to a distance of 30 km from a reference location without any intermediate point. A small handheld unit allows positions and traverse routes to be downloaded to GIS software for geological mapping. A small unit can store more than 300 sample positions and tens of routes.

6.5.5.4. Electronic Total Station

An ETS is an electronic instrument used in modern surveying. The total station is an electronic theodolite integrated with an electronic distance meter (EDM) to read distances from the instrument to a particular spatial entity. Some models include internal electronic data storage to record surveyed points (x-northing, y-easting and z-elevation), distance, horizontal and vertical angles. Data can be downloaded from the total station to a computer. Application software is used to process results and generate maps of the surveyed area.

A total station is used to record the absolute location, geological contacts (maps), results of geological, geochemical and geophysical survey, borehole program and even underground working layout and stopes. The recorded data are downloaded into a computer, processed and compared to the designed layout. Control survey stations at regular intervals are installed in the underground to facilitate the survey by ETS.

6.5.6. Applications

GPS is widely used for military and civilian purposes. The military applications of GPS have many purposes like reconnaissance and route map creation, navigation of soldiers to locate them in the darkness or in unfamiliar territory, and to coordinate the movement of troops and supplies. It helps in missile and projectile guidance for accurate targeting of various military weapons. GPS satellites also carry a set of nuclear detonation detectors consisting of an optical, X-ray and Electromagnetic Pulse sensor. This forms a major portion of the United States Nuclear Detonation Detection System.

The civilian uses for obtaining instant and precise location (latitude, longitude and altitude) on land and sea during geological fieldwork, mineral mapmaking and checking, land surveying, commerce and scientific applications, tracking and surveillance. It provides precise time reference including the scientific study of earthquake and as a required time synchronization method for cellular network protocols.

6.6. SOFTWARE IN RS-GIS

RS-GIS software includes the programs and an interface. It is required for driving the hardware. The software modules represent a total system and responsible for generating, storing, analyzing, maneuvering and displaying the results or geographic information. The strength of suitable software module is to maintain user friendliness, compatibilities, documentation and cost effectiveness. Exploration and mining companies use GIS to identify prospective areas for exploration, 3D orebody modeling, mining, infrastructure layout and environmental system management. The progress of GIS into three dimensions is a revolutionary change for the utility of the technology in mineral exploration. The producers and their main products of GIS software are listed without any discrimination of superiority:

6.6.1. ArcGIS

The commercial software ArcGIS is group of GIS software produced by Environmental Systems Research Institutes (ESRI). The name of the software is arrived by combining two words, i.e. Arc and Info making ArcInfo, Map and Info making MapInfo. Arc/Map mean the graphical entities and Info means attribute. Arc and Map having Information, make them intelligent, which can be used for querying and analysis.

ArcGIS is a high-performance, dynamic software family that can produce significantly better looking accurate maps at shortest time. It provides review and responds to errors, unsupported content. It can preview map document and estimated rendering time, save map document to a map service definition (MSD) format, combine layers (referencing feature or raster data) into a single layer package comprising both the layer file and data. It has the facility to easily share these layer packages with other users via files, email, or the ArcGIS Online sharing facility. ArcGIS Online shares work with own or global groups, and access maps from around the world. The main components of ArcGIS are ArcInfo, ArcView and ArcReader.

ArcInfo is comprehensive GIS within ArcGIS software family. It includes all the capabilities of ArcView and ArcEditor. It also adds the advance geoprocessing and data conversion functionality. ArcInfo provides all the tools to build and manage a complete intelligent GIS including maps, data, metadata, geo-data sets and workflow models. This functionality is accessible via user-friendly interface and customizable through model, scripting, and applications. The key features include advance spatial analysis, extensive database management, multiuser data editing and integration, deployment and high-end cartography in exploration program.

ArcView is full-featured GIS software for visualizing, analyzing, creating, and managing data with a geographic component tied to a place. It consists of an address, postcode, GPS location, city, local government area, or other location. ArcView allows one to visualize, explore and analyze these data, revealing patterns, relationships and trends.

ArcReader is user-friendly desktop mapping application. It allows viewing, exploring, and printing maps and globes. ArcReader possesses interactive mapping capabilities by accessing wide variety of dynamic geographic information and view high-quality maps.

6.6.2. AutoCAD

AutoCAD Map, patented by Autodesk, is a powerful drafting tool used mostly for engineering drawings for its accuracy and 3D viewing. It works on 2D and 3D coordinate system. It is user friendly and available at reasonable cost. It is extensively used by geologists for preparation of geological maps, subsurface plans, sections for estimation of resources and reserves by conventional methods.

6.6.3. IDRISI

IDRISI Selva is an integrated raster-based GIS and image processing broad application base software solution developed by Clark Labs. It provides many modules for the analysis and display of digital spatial information. The Land Change Modular (LCM), for example, provides land cover change analysis and prediction software with tools to analyze, measure and project the impacts of changes on habitat and biodiversity.

6.6.4. Integrated Land and Water Information System

Integrated Land and Water Information System (ILWIS) are developed by International Institute for Aerospace Survey and Earth Sciences in the Netherlands. It is GIS-RS software for both vector and raster processing. The main features include digitizing, editing, analysis and display of data as well as production of quality maps.

6.6.5. MapInfo

MapInfo, developed by MapInfo Corporation, is a natural resources software solution for mineral exploration, mining, engineering, infrastructure and environment. MapInfo Vertical Mapper module has a wide range of analysis tools to reveal trends in data set. It has unique prediction capabilities to specify a test location and identify areas with statistically similar attributes. The software executes 3D orebody modeling and estimation by all standard interpolation principles from existing point files or tables, regardless of data type. It performs TIN with smoothing, Inverse-weighted distance function (ISD),

Natural Neighbor, Rectangular (Bilinear) interpolation, Kriging and Custom Point Estimation. The advance module includes modeling options by overlaying multiple layers and applying a mathematical function, calculate steepness of slope or the direction the slopes are facing in a grid, show cross sections, 3D perspective view of the terrain with optional overlay of imagery and Natural Neighbor analysis.

6.6.6. Microstation

Microstation developed by Bentley Systems is easy to access powerful and interoperable 3D-CAD software platform for the design, construction, operation and dynamic view.

6.6.7. PAMAP

PAMAP, developed by PCI Geomatics, is an electronic map of Pennsylvania as a seamless, consistent and high-resolution set of digital geospatial data product.

6.6.8. SPANS

SPatial ANalysis System (SPANS) is a quadtree/raster base software solution for spatial analysis and modeling like topology and contouring. It uses quadtrees as the unique medium for storing all map data. The software is user friendly and menu-driven installed on desktop/personal computer. The solution software is developed by Tydac Technologies Inc., Canada.

FURTHER READING

Evans (1999) [24] has introduced the subject on photogeology and RS. GIS and its role in geo-environmental issues have been illustrated by Sarkar (2003) [65]. Further detail can be available from recent development in the subject by Lillesand and Kiefer (2003) [52], David (1997) [18] and Gupta (2008) [30]. A good introduction to RS is given by Campbell (2007) [7]. GIS applications have been very thoroughly discussed by Bonham Carter (1995) [5].

Chapter 7

Sampling Methods

Chapter Outline

Sampling requirements are set by the orebody, not by the engineer.

—Author.

7.1. DEFINITION

Sampling is the process of taking a small portion of an object such that the consistency of the part shall symbolize the whole—entire property or only of adjacent portion of the object under assessment. The objects in geological aspect are granite hill, limestone deposit, alluvial soil, weathered profile, beach sand, poly-metallic nodules, mineral occurrences, drill core, well water, gas. The sample interval and quantity will depend on the homogeneity or complexity of the mineral under search. The "unit" of the sample size i.e. millimeter (mm), centimeter (cm), meter (m), feet (f), gram (gm), kilogram (kg), pound (lb), liter (l) etc. must be specified in particular to make it significant. The unit is needed for precise computation of the average grade by conventional and statistical method. There are various sampling practices as suitable to the situation.

7.2. SAMPLING EQUIPMENT

The samples are collected by various means and methods as suitable and convenient to the situations without compromising the quality and reproducibility.

7.2.1. Conventional Equipment

The most frequently used sampling tool, an indispensable companion of a geologist in the field, is a hammer and chisel. These handy tools are used for the collection of rock samples, chip samples and channel samples. In case of grab, muck, car and bulk samples, when large volume of samples is required, spade, shovel and mechanized loader are preferred. However, the most important sampling technique required for establishing the continuity of mineralization in all directions, 3D visualization and assessment of a mineral deposit is drilling holes. The technology needs elaboration.

Mineral Exploration. http://dx.doi.org/10.1016/B978-0-12-416005-7.00007-6

7.2.2. Drilling Techniques

The principle of drilling appears from the remote antiquity used for boring short holes in the construction of pyramids by the Egyptians (2630 BCE). The first machine was manufactured in 1862-1863. It was operated by hand rotation. The technology became popular for digging holes in different media with diverse applications. The drilling is extensively employed in mineral exploration. The increase of demand progressively stimulated the manufacturers for effective design of machines and accessories with an aim to improve efficiency and lower the cost. The search for deep-seated orebody, petroleum and gas reservoir is unthinkable without availability of efficient drill machines which can collect samples at depth more than 2000 m.

There are essentially three methods of drilling: percussive, rotary, and combined effect of percussive and rotary movement. Rotary drilling methods include auger drilling, diamond drilling and its variations. Jackhammer and wagon drills fall in the percussive cum rotary drill type.

7.2.2.1. Percussion Drilling

The percussion or churn drill digs a vertical hole. It employs the principle of freely falling chisel bit hung on a cable to which percussive motion is imparted by one of the various types of power units. The power units are manual lift and drop, compressed air and electrically driven winches. The tungsten carbide bit fitted in a hammer is lifted few meters and allowed to drop (Fig. 7.1) to hit the bottom of the hole. The process continues in succession. The churning motion of the bit crushes and scraps the ground, and so a hole is dug. The cutting of rocks thus produces mud or slurry by lowering water. The crushed material is removed from the bottom of the hole at a regular interval to make a sample. Churn drilling is suitable for soft and medium formation. In harder formation resharpening of cutting bit is required frequently resulting in lowering of progress. The capacity of the churn drill is limited to relatively short holes say 10-50 m.

The cost of sampling is comparatively much lower than diamond drilling under similar conditions. But the chances of contamination between sample depths are high. This makes the method inefficient to demarcate correct orebody contacts and assessment of average grade. It can provide information regarding presence of mineralization that can be precisely explored by diamond drilling. The system has been modified for collection of regular sample by connecting water pump (Fig. 7.2). The return water brings the rock cutting to the surface and a sample can be prepared. Another modification is by introducing a slit in the bit

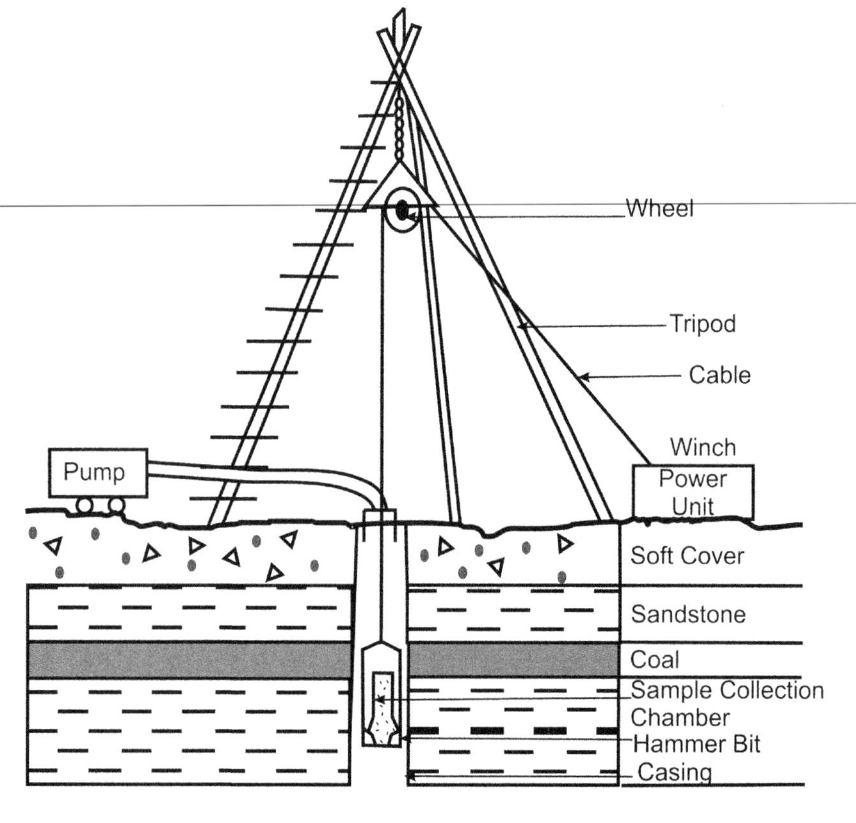

FIGURE 7.1 Schematic conceptual diagram of percussion drilling procedure often used in engineering geology for foundation testing and occasionally in the initial stage of mineral exploration and particularly for low-profile budget.

FIGURE 7.2 Percussion drilling in operation for rock quality and quick target test for mineral occurrences at low cost and time. The hammer is detailed in the inset.

which holds the rock cuttings for a sample at standard interval. The method is regularly being used for tube wells and foundation testing.

7.2.2.2. Percussive Cum Rotary Drilling

When the penetration of drill bit (integral or detachable tungsten carbide) takes place due to the resultant action of both percussive and rotary movements it is known as percussive cum rotary drill such as Jackhammer (Fig. 7.3) and Wagon drills. The percussive action produces longitudinal impact on the rod to break the rock particles. The rotational movement exerts force on the bit head to penetrate into the rocks. The depth of drilling is limited to 6 m. The drills are compressed air driven. Water is injected through hollow steel drill rods to cool the bit head from excessive heating. The return water flushes out the crushed material from the hole for free movement of bit and rods. If collected properly, it will serve as samples for assaying to know the metal content of the advancing face. These types of drills are used primarily for development of tunnels, advance mining faces and breaking big boulders in road and other construction areas. It has limited use in mineral exploration. The samples are used to estimate the approximate metal content of big rock exposures, mine blast quality in advance for grade control and scheduling purposes. It can provide information of roof and floor of coal seam including thickness of coal bands within.

FIGURE 7.3 Jackhammer drilling in rich sulfide mineralization for multipurpose use as underground mine face development and sample source for ore continuity and grade control.

7.2.2.3. Auger Drilling

Auger drills have limited use, but it plays a significant role in sampling and evaluation of soft and loose ground like soil, beach sand, mine dump, mill concentrate and tailings. Auger drills can be hand operated (Fig. 7.4A) or mechanically powered. Its advantages are low cost, speed and mobility. Hand-operated augers can operate up to 3 m with hole diameter between 10 and 15 cm depending on softness of the material to cut. Mechanically driven augers with efficiently designed cutter heads (Fig. 7.4B) can drill up to 30 m or more depending on the subsoil condition. Samples out of multipurpose auger drilling are useful to provide grade and other specifications mentioned for above material quickly at low cost. However, samples may often fail to provide accurate information due to wall collapse and related contamination.

7.2.2.4. Diamond Drilling

The surface and underground diamond drills are the most versatile tools. It is extensively used for mineral exploration, dam site and other foundation test works, information on advance mine development face, drainage of mine workings, mine ventilation, oil structure investigations and oil-gas well drilling. Extreme hardness of diamond enables it to cut all types of rocks and minerals found in the Earth's crust.

The diamond drill unit (Fig. 7.5) consists of an engine (motor), attached to a drilling head and hoisting units, water pump together with drill rods, core barrel, casing pipes, cutting tools and a tripod stand. The engine is powered by diesel, electricity or compressed air. The motor, mounted on a cemented platform or truck, transmits revolving power through a transmission and clutch to a set of gears to the drilling head. There are usually three to four set of feed gears in the swivel head, with capacity ranging between 100 and 1000 revolutions per inch of advance of the feed screw. The chuck equipped with jaws is placed at the

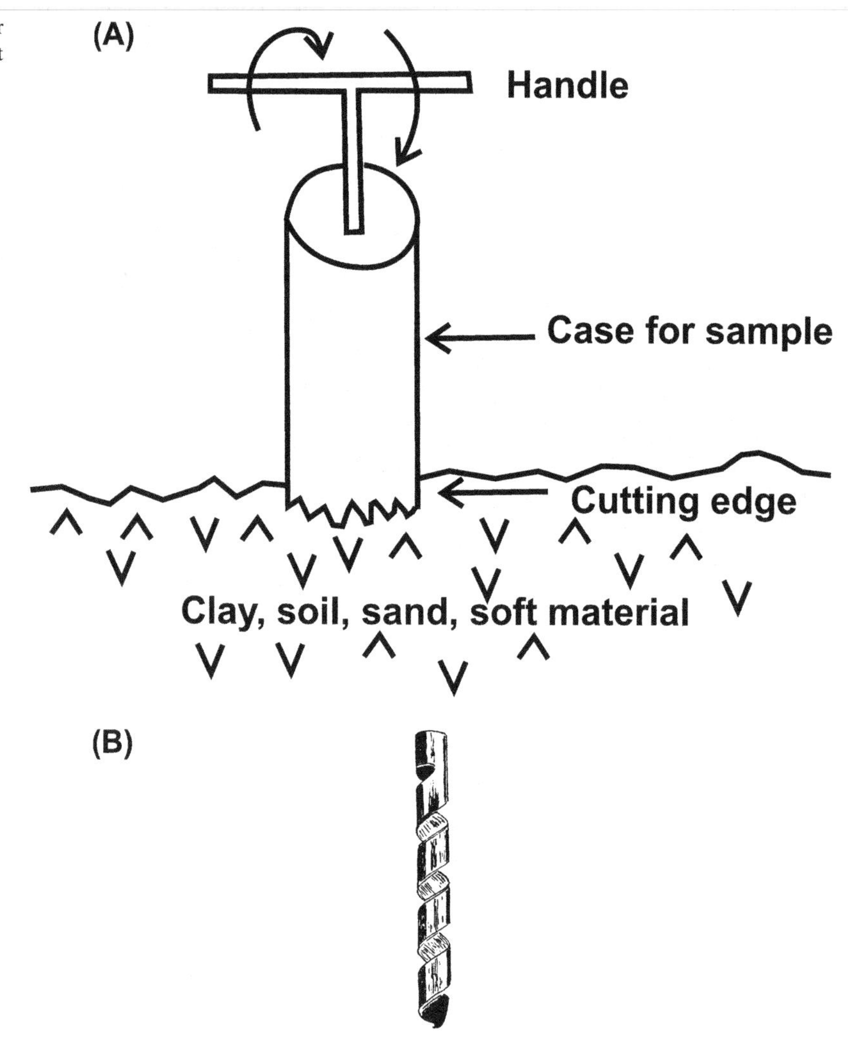

FIGURE 7.4 Sketch diagram of (A) auger drilling—a simplified easy to hand operate low-cost sampling unit and (B) cutter head.

bottom of the feed screw through which drill rods pass. The drill rods are attached to core barrel and diamond bit. The total drill string is forced downward with the high-speed revolution of the chuck by tightening jaws. This results in cutting of core and making a hole.

A "tripod" is commissioned by erecting three poles of about 30 feet long around the drill unit. The function of tripod is to raise and lower 10 or 20 feet rods during drilling operation with the help of "hoist". A "scaffold" is nailed and chained to the tripod where the helper can stand safely and operates the rod hoisting. The screwing and unscrewing of rods is done efficiently by automatic mechanized means. The rods are withdrawn at intervals of 10 feet or less depending upon the drilling condition. The core is removed from the barrel for geological studies (logging) and storage.

The modern diamond drill rig has capacity to sink +2000 m depth through fully mechanized operation with two to three crews (Fig. 7.6).

The hollow steel, flush jointed or coupled "drill rods" are usually 10 feet (3.05 m) long through which water is pumped to cool the bit and flush the rock cuttings. "Diamond core bit", the main cutting tool, is a cylindrical hollow tube made of special alloys with a crown at one end (Fig. 7.7). The crown is composed of superior diamond holding qualities of powered metal alloys on which diamonds of different sizes, quantities and design are set depending on the rock types to be drilled. The fragment sizes are expressed as stones per carat (spc), say 80/120 spc i.e. between 80 and 120 spc (1 carat = 200 mg).

"Reaming shell" is mounted between the cutting bit and core barrel. It is an annular bit with diamonds set only on the outer surface or periphery (Fig. 7.8). The reamer shell widens the borehole diameter drilled by the diamond bit by about 0.30-0.40 mm. It maintains an uniform hole diameter, reduces the wear and tear of the core bit and barrel, and improves flow of return water.

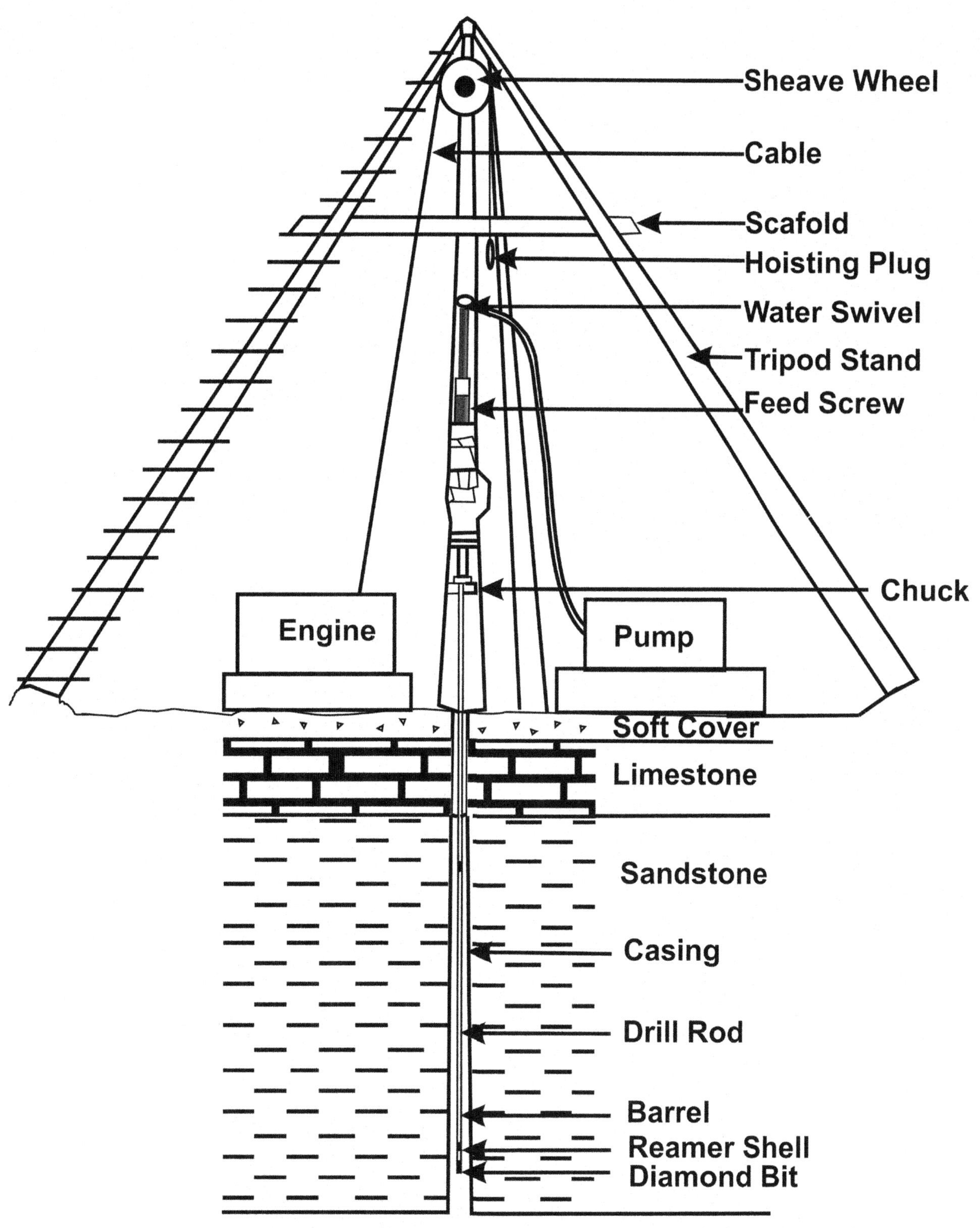

FIGURE 7.5 Schematic diagram of surface diamond drilling unit showing various components and functions to sink a borehole for recovery of core samples of all rock types passing through including structural features and mineralization.

A "core barrel" is attached between the lower end of the drill rods and the reamer shell. The core barrel holds the core inside while drilling and brought to the surface. Core barrels are "single tube" used for core drilling under best core recovery condition or noncoring bits in blast hole drilling.

FIGURE 7.6 Truck-mounted diamond drill rig in operation for base metal exploration in Australia and capable of drilling 50-100 m a day.

"Double tube" core barrels are suitable for average core recovery, where the inner and outer tubes are connected and rotate simultaneously. "Triple tube" core barrels are complex and expensive type used in broken, friable and sheared formations where assay value of the samples is necessary. "Core lifter" or core spring (Fig. 7.9) is placed at the lower end of the barrel that holds the core from dropping out of the barrel.

A cylindrical "core" (Fig. 7.10) of rock is cut with the advancement of the bit. The core represents as sample and provides physical and accurate records of formations through which the drilling continues.

Water is essential for the drilling operation. Sometimes drill sites are located at remote places. Therefore water is stored to a tank dug close to the borehole collar. Water is either pumped from a nearby source or periodically supplied by tanker. Care is taken to recirculate the water after settling of rock cutting in the tank. Water loss in the hole can be prevented by "casing" the hole or localized cement grouting of the fractured areas. In case of highly fractured ground condition water can be substituted by use of "drill mud"

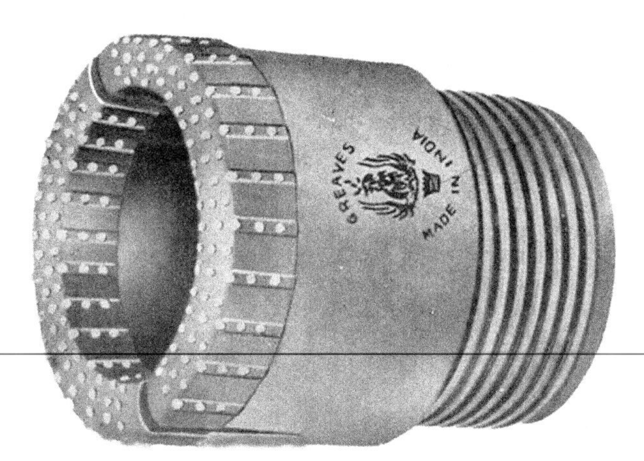

FIGURE 7.7 Standard diamond core bit studded with technically designed tiny diamonds at the crown composed of superior quality alloy grooved with channels to cool the bit head and flash the fine cuttings.

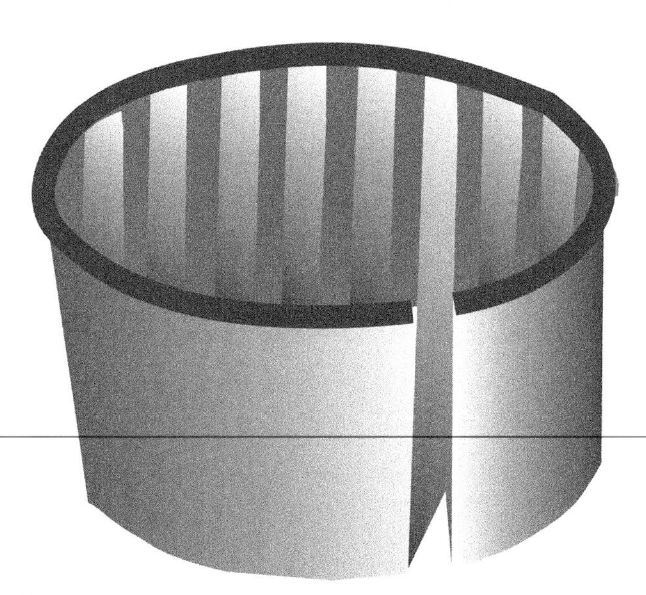

FIGURE 7.9 Standard spring-type core lifter that protects the core from slipping into the hole and regrind.

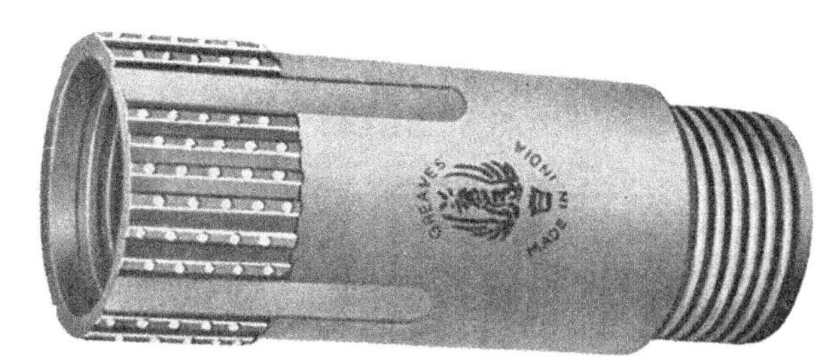

FIGURE 7.8 Standard reamer shell coupled behind the core bit embedded with diamonds at the outer surface to increase the hole diameter for easy flow of return water and fine cuttings.

FIGURE 7.10 Standard drill core showing stratiform sphalerite, galena in calc-silicate host rock with red pencil line marked for splitting into two near-identical halves at Rajpura-Dariba Mine, India.

(bentonite clay, polymer). It will significantly reduce fluid loss, hole collapse and improve drilling efficiency.

When diamond bits and reaming shells are worn out beyond further drilling the remaining diamonds are "salvaged" from the matrix by acid bath. The recovered stones are sorted for size and condition. It is then mixed proportionately with fresh diamond for setting new bits. The value of the recovered diamonds is credited to the purchaser.

The underground drills are usually a lightweight compressed air or electric-driven machine mounted on a single or double drill column (Fig. 7.11). The drill rods are 5 or 10 feet pull due to narrow space of mine workings. The drill units can work from horizontal to 90° up and down. But normal practice is to drill between 0 and ±45° to avoid excessive load of the rods and better safety of the drill crew. The capacity is around 300 m.

The capacity of drilling depth depends on the power and condition of the machine, terrain, angle of the hole and type of rocks to be penetrated. The efficiency will reduce with steep-angle holes and extensive length of hole depth. Drilling can be done over 2000 m depth. The size of core is decided by the exploration agency based on type of minerals under investigation and prevailing rock condition. The size of bit, reamer shell, barrel, and drill rods are arranged accordingly. Generally, the drilling starts with HX size and progressively reduces to NX, BX, AX, and rarely to EX (Table 7.1). The cost of drilling will be higher with larger core diameter and more reliable sample representation, and vice versa.

FIGURE 7.11 Typical compressed air-operated underground drill unit that can operate at 0-360° rotations.

The "collar", starting point, of a surface exploration drill hole must be closed after completion with wooden plug to prevent pebbles and soil being dropped down the hole, thus making it difficult to move back in the hole again if required. It is also desirable to fill the entire hole with cement at a later date to prevent inundation of underground mine by water gushing during heavy rains or interconnected with water channels. The top of the collar should be covered with a cemented platform marking the project and hole number, coordinates, angle and direction, depth, start and end of drilling date etc., for future references. The collars of underground boreholes must be plugged to avoid accident at upper or lower levels by explosive gases during mine blasts.

7.2.2.5. *Wire Line Drilling*

The "wire line drilling" performs by withdrawing the core and inner tube assembly from the hole without pulling out the hollow drill rods by a separate hoisting unit fixed at a different pulley. The inner tube assembly is lowered down inside the barrel after taking the core out and drilling continues. Therefore lowering and hoisting of drill string, barrel and drilling head is not required after every run drilled. It saves considerable time and energy.

"Continuous core drilling" works on the principle of reverse circulation (RC). RC refers to circulating the water down to the bit head outside the drill rods and returning it up through inside the bit, core barrel and drill rods. The effect of reverse circulating water is to continuously float the core back to the surface through bit, barrel and rods. In this case, the drill rods and barrel are hollow enough so that the core can move upward as the drilling continues. The cores are collected at the surface, placed in the core box with proper orientation and depth. This type, being

TABLE 7.1 Standard Drilling Type, Hole Diameter and Core Diameter

Drilling type	Hole diameter (mm)	Core diameter (mm)
HX	99.20	76.20
NX	75.60	54.70
BX	59.90	42.00
AX	48.00	30.00
EX	37.70	21.40

a continuous core drilling process, can operate for long duration till the bit is effective. It saves the time of lowering and hoisting of drill string at every 3 m, which was required in case of conventional diamond drilling. It avoids over-grinding of core. Thus, the efficiency of drilling performance increases many times resulting in substantial cost reduction and improved core recovery.

7.2.2.6. RC Drilling

In conventional diamond drilling, water is pumped through the drill string in the borehole to the bit head. The pressure of the return water takes the rock cuttings and sludge through the annular space between drill rods and outer wall of the hole. The sludge gets continuously intermixed during the drilling operation.

RC drilling commonly prefers compressed air produced by a hydraulic top drive motor under dry drilling condition. The compressed air is introduced into the drill through dual concentric air pipe (between the outer and inner pipe) and flows to the "down-the-hole or DTH" tungsten carbide hammer bit. The compressed air and hammering initiate the generation of chips at the drill head depth. The rock cuttings created at the bit face are set in continuous motion upward by high-pressure return air through the inner sample tube. The return air is supplied by installing high-pressure vacuum pump. The entire rock cuttings (no core) are removed to the surface and collected as inverted sequence in the cyclone above ground level (Fig. 7.12). The samples are separated at regular intervals representing particular depths.

The RC drills (Fig. 7.13) are commonly used for open pit excavation of iron ore, bauxite, limestone, rock phosphate, and coal seam. It drills 10- to 15-cm large diameter vertical blast holes at high speed and low cost. The modified RC drills can rotate the drill string from vertical to any steep angle and became very popular for mineral prospecting throughout the world. The RC drill sample enables quick testing of drill targets during Reconnaissance to initial stages of exploration. It creates a rational base for further diamond drilling program for obtaining more precise sample location and high-quality core samples.

The advantage of RC drilling is its capability of drilling in the hardest formation with easy penetration at high speed and recovery of uncontaminated samples even in a broken and fractured formation. The fast rate of drilling thereby promotes planning of future core drilling program and mining operation. It is exceptionally cost effective and provides easy mobility of the rig from one location to other for both prospecting and mining.

7.2.2.7. Borehole Survey

The boreholes have a tendency to deviate both in inclination angle and direction from the original setting either due to drilling through rock types with different hardness,

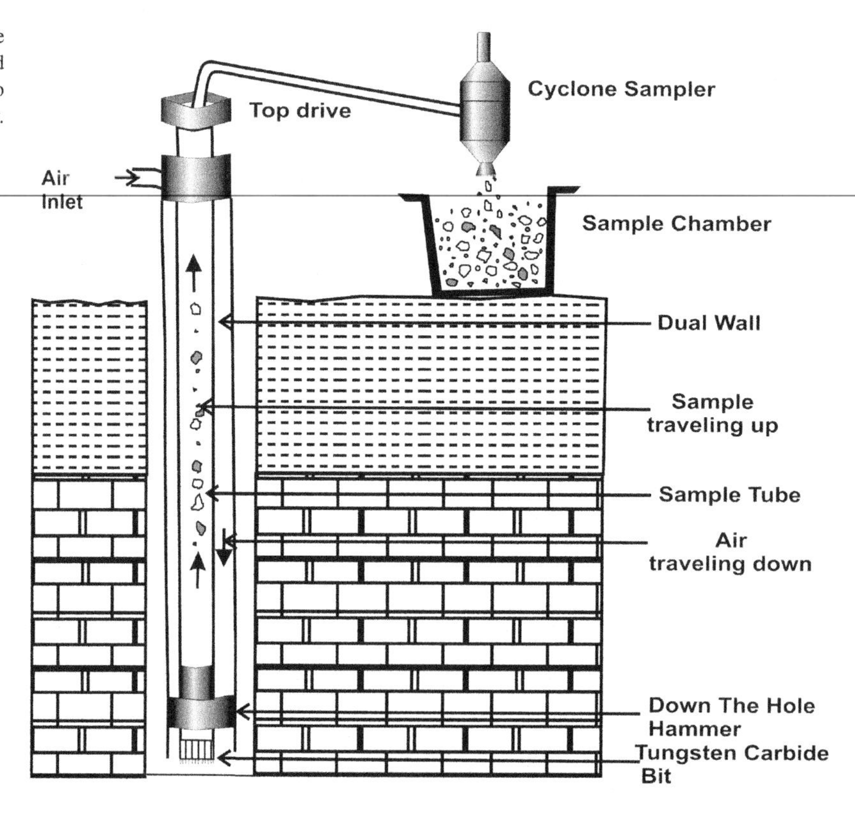

FIGURE 7.12 Conceptual framework of noncore RC drilling and sample collection, widely adopted by exploration companies all over the World due to fast sampling at low cost with reasonable reliability.

FIGURE 7.13 RC drilling in operation and sample collection for iron ore deposits (Credit: Biplab Mukherjee).

structures or due to unwanted over speed drilling, defective drill rods, barrel, bits etc. The deviation at certain interval of depth (say 30-50 m) is recommended and measured by various ways.

The simplest way to survey the angle of a hole is by "etch testing". It consists of a hollow container fitted at the lowering end of the drill rod in place of the barrel and bit. A special type of glass culture test tube of about 13 cm long, partially filled with hydrofluoric (HF) acid and corked with a rubber stopper, is placed in the container. The container is lowered to the desired depth of survey and kept stationary for about 45 min. During the period the HF acid reacts with the glass tube and forms a horizontal etch in the inner surface. The container is then withdrawn from the drill hole, the tube is washed and an etching line is marked with ink. The angle of the hole at that point is measured as shown in Fig. 7.14. The method could provide the deviation of angle only and used in case of low-budget exploration for limestone etc.

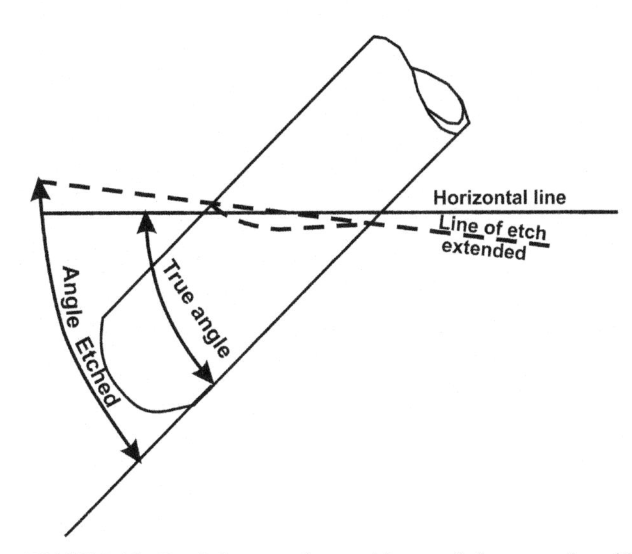

FIGURE 7.14 Borehole survey by special type of glass test tube acid etching method at lowest cost for measuring the deviation of angle only and not the direction.

The "Tropari" is a single-shot, micro-mechanical borehole surveying instrument operated by a timing device. The unit provides both direction and inclination at high precision which can be used to define the attitude of the borehole at the desired survey depth. The instrument is essentially a pivoted ring-mounted compass (Fig. 7.15). It is attached with a clock mechanism to lock the compass needle and dip indicator within a preset delay within 5-90 min from lowering the unit. The unit is hoisted to the surface after automatic locking of the system to record azimuth direction and inclination. The device can record one set of reading i.e. inclination and direction at a time. It has to be lowered repetitively to all desired depth resulting loss of drill shifts. The time loss will be considerable in case of drill hole with extended depth. It is not cost effective for long holes. Regular surveys at intervals as the borehole progresses will allow a plot of the borehole to be drawn from the data.

This deficiency has been substituted by introducing "reflex multishot borehole camera". It is an electronic multishot instrument, ideal for carrying out surveys in a nonmagnetic environment as the borehole progresses. The features and benefits include capabilities like robust, reliable and fully integrated, perform on accurate electronic measuring principle, measures in all directions, fast and user friendly. The unit is comprised of a multishot camera, a tiny magnetic compass, flashlight and auto-locking clock device (Fig. 7.16). The shot and flash are synchronized with the clock. The angle and directional data are recorded on a 16-mm film. The string is lowered slowly and kept stationary for 2 min at desired depth of survey. Then the process continues lowering the unit till end of the hole. The entire drilling depth can be surveyed at one single run. The survey film is developed in dark room for total set of inclination and bearing at respective depths. The contemporary models are capable to conduct borehole orientation survey and onsite access to survey data for incorporating corrective measures.

FIGURE 7.15 Borehole survey by Tropari unit capable of measuring both angle and direction for one reading at a time (Credit: Vinod Jhanjari).

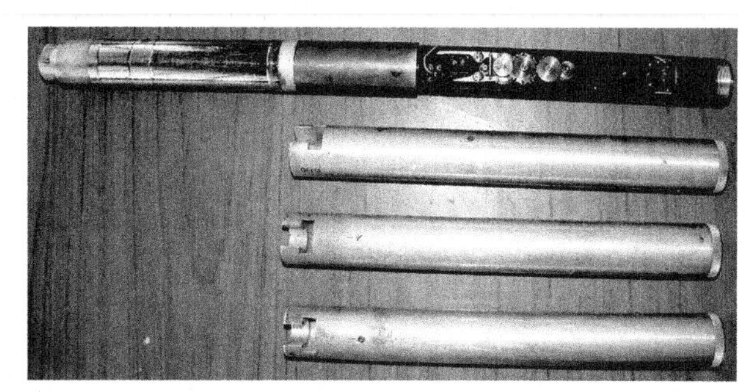

FIGURE 7.16 Borehole survey by reflex multishot camera competent to measure both angle and azimuth with highest precision for several depth of hole at one lowering—a total borehole survey solution (Credit: Vinod Jhanjari).

The device is useful to the geologists, drillers, minerals, tunnellers and ground engineers all over the World.

FIGURE 7.17 Standard wooden, plastic and metallic (aluminum) core box essential for drill core preservation.

7.2.2.8. Core Recovery

Recovery of drill core is an important parameter for efficient mineral exploration program. Good core recovery is the responsibility of the drill crews. Careless crew can ruin core by drilling too fast, overdrilling a run beyond the core barrel capacity, using undesired type of barrel and faulty core lifter. It can cause core drop in hole and regrinding. The recovery of core is considered necessary to be 100% except drilling through highly fractured, sheared and caved zones. Minimum recovery through mineralized area must be more than 90%. Driller should use double or triple tube barrel in mineralized formation. He may even change to short run length with dry drilling, if necessary. Core recovery dictates the reliability and precision of estimation of a deposit. The existing hole should be rejected in worse core recovery condition and a fresh hole be drilled with utmost care.

7.2.2.9. Core Preservation

Drill hole costs millions of dollars in mineral exploration. Valuable drill cores are collected carefully and kept in wooden or aluminum core boxes marking the direction of drill depth with→. Each run length is separated by peg mark indicating the depth of drilling. Boxes are generally1.5 m long and can hold about 8 m of core in five channels separated by thin plates (Fig. 7.17). Box may have hinged or nailed cover to protect the core while shifting. Drill core is a vital resource that needs to be accessed easily and kept in good condition. Many exploration companies around the World create core library facilities to store for 15-30 years. Core boxes are stacked in order in uniquely designed shacks. Each box in the library can be identified by project name, box and borehole number, drill run and date of drilling. During the mining stages of a deposit core of couple of standard sections is preserved for future study and rest can be destroyed. Special lighting is provided to view the core. Online core library data repository is new technology under development.

7.2.2.10. Core Logging

"Diamond drilling" and "drill core" play key role in mineral exploration. The quality of drilling by best possible recovery, proper core placement in preservation box with correct arrow marking and shifting to the core shade from drill-site need extreme care before study of the core and sampling. Any slip in between will add uncertainty of accurate ore boundaries in space. The reason may be on account of misplacing or missing core during the drilling, collection, placement and shifting. Apparently, it looks insignificant in comparison to shifting of contacts by fraction of meter, change of reserve by few 1000 tonnes and metal grade in second decimal. But maintaining seriousness at each small and big issue is important. Therefore, the core is spread initially on extra long conical plastic tray in the core laboratory. The arrow direction on each piece is checked and the edges of two successive core pieces be perfectly matched. Any discrepancy in identical matching must be sorted out without any compromise and recorded. Now the core is washed thoroughly by water spray (Fig. 7.18) and ready for geological study and sampling. The author experienced and appreciates the unique sincerity of the geologists at Lennard Shelf exploration camp of Meridian Minerals Limited, Australia.

"Core logging" is the geological study and recording of drill cores. Records are made on printed sheets (Table 7.2). It covers the general description of core i.e. from and to, core size (NX, BX, AX, BQ), run and core length, percent recovery, color, grain size, textures, structures, foliation with core axis, fractures, shears, folds, faults, mineral composition, alterations, visual estimates of metal values, sample number and finally rock name. Lines are marked on

FIGURE 7.18 Checking of core in perfect order and thorough washing before logging and sampling point to the sincerity and reliability of data collection norm in mineral exploration at Lennard Shelf camp, Australia (Alaik Kubra, Exploration Geologist, Meridian Minerals Limited).

mineralized core for splitting into two halves of identical mirror images. Visual estimates are made before sample preparation and assaying. Each visual estimate compares with assay value, rechecks the core in case of major difference and sends duplicate for reanalysis. "Tough-book core logger" (Fig. 7.19), a portable rugged laptop, is extensively used for recording at site. It is designed with multiple Excel sheets for database like collar, survey, rocks and assay files. It can be interfaced to process computer by both wireless and hardware transmission for 3D modeling and reserve estimate. Many exploration companies conduct continuous color photographs of entire length of borehole for permanent records.

7.3. SAMPLING METHODS

There are a variety of sampling practices as appropriate to the specific situation. The sampling aims to generate the best representative of the object under search. Pitting, trenching, stack and placer sampling are practiced for surface exploration. Diamond drill core, RC drill cuttings, sludge, channel and chip sampling are suitable for both surface and underground exploration. Grab, muck, car, bulk sampling are carried out for quick estimation of run-of-mine grade. This facilitates in grade monitoring and control for blending purposes.

7.3.1. Soil Sampling

Soil samples, from residual or transported material, are collected (Fig. 7.20) on a relatively closely spaced rectangular or square grid pattern at specified interval designed during orientation survey. Sieve analysis of samples indicates that minus 80 mesh fraction of the soil represents sufficient material for further processing as well as provides maximum contrast between background and threshold value. It becomes standard practice to obtain minus 80 mesh fraction for analysis of various elements of interest.

7.3.2. Pitting

It is commonly practiced during the initial stage of surface geochemical exploration. The sampling is carried out by

TABLE 7.2 A Sample Borehole Core Log Sheet

Project: BH NO: Date of Logging: Logged by:

From (m)	To (m)	Core size	Run length (m)	Core length (m)	% Core recovery	Color	Grain size	Structure	RQD	RFM	OFM	VE	Remarks

RQD: Rock Quality Designation RFM: Rock forming minerals OFM: Ore forming minerals VE, Visual estimate.

FIGURE 7.19 "Tough book"—a high-tech Internet interfaced data sharing drill core logger used at Lennard Shelf Exploration Camp, Australia.

FIGURE 7.20 Collection of soil sample during Reconnaissance/orientation survey for Pt-Pd target search around Tagadur chromite-magnesite open pit mine at Nuggihalli Schist belt, Karnataka, India, (Finn Barrett, Consultation Geologist, Goldstream Mining NL, Australia).

excavating about $1 \times 1\ m^2$ pits in rectangular or square grid pattern covering the entire target area. The depth of the pits varies depending on the extent of weathering and nature of the rocks. The material from each meter of vertical depth is kept in separate low-height rectangular flat stacks to determine the variation in grade and other distinctive features. Each stack represents a sample. The pits showing the presence of mineralization can be contoured (Fig. 4.2) to identify the strike and depth continuity of the orebody for drill testing.

7.3.3. Trenching

Trenches are often cut across the orebody (Fig. 7.21) after the probable configuration of mineralization is outlined either by pitting or by rock/soil sampling. The material recovered from each meter of trench is stacked separately as a sample for analysis to identify the variations across the mineralization. The walls of the trenches

FIGURE 7.21 Trench sampling for platinum group of elements and chromite during Reconnaissance survey at Sitampundi layered igneous complex, Tamil Nadu, India.

can be sampled by channel cut or chipping for comparing with results of stack sample to fix the mineral boundary.

7.3.4. Stack Sampling

Stack sampling is the collection of representative broken material generated by pitting, trenching, mine production or any other process. The samples are collected by auger drilling principle by inserting 10- to 40-cm diameter cylinder down to the base of the stack. It can also be done by collecting bucket full of sample in the same way. A number of collection points from a stack are selected. A composite sample is prepared by combination of one from the central part and four more from halfway between the center and corners of the stack (Fig. 7.22). Alternatively, particularly in case of small pits and narrow trenches, the total recovered material can be reduced by successive crushing, cone and quartering and treated for chemical analysis.

FIGURE 7.22 Stack sampling of chromite mine production ore at Sukinda belt, India.

7.3.5. Alluvial Placer Sampling

Alluvial placer deposits are formed by weathering, transportation and deposition of valuable minerals. The large alluvial placer platinum deposit at Ural Mountains in Russia is an example of such deposits discovered way back in 1823. In general these deposits are less consolidated, loose and soft materials. Scooping by hand spade or by auger drills is employed to collect the loose sandy samples at certain grid interval up to certain depth. Whereever necessary a casing is driven into the deposit to protect wall collapse and contamination during sample collection.2

FIGURE 7.24 Channel sample cut by pneumatic cutter for exploration of platinum group of elements at open pit bench face, Boula-Nausahi chromite mine, Orissa, India.

7.3.6. Channel Sampling

Channel sampling is suitable for uniformly distributed mineralization in the form of veins, stringers and disseminations. The sampling is performed by cutting of channels across mineralized body in fresh surface exposures or underground mine workings, such as face, walls and roof. The area is cleaned to remove the dusts, dirt, slimes and soluble salts by any of the three processes. These are washing with hosepipe (air/water) or scrubbing with a stiff brush or by chipping of outer part of the rocks to smoothen the sampling face. A linear horizontal channel is cut between two marked lines at a uniform width and depth (Fig. 7.23). The width is generally 5-10 cm at a depth of 1-2 mm. The length of sample varies depending on the variation in mineralization across the orebody. But in practice, the length is kept at uniform unit between 1 and 2 m within the mineralization. The uniform length of sample eased the statistical applications.

The tools are a hammer and sharp chisel (a pointed stub of drill steel) or a pneumatic hammer with pointed or chisel bit (Fig. 7.24). While the sampler cuts the channel, a second person collects the chips, fragments and fines in a clean box, sack or on a canvas sheet spread on the floor. A sample of 1 m length will weigh around 1-2 kg.

7.3.7. Chip Sampling

If the mineralization is irregularly distributed or disseminated and not easily recognized by necked eye, channel sampling may not be representative. The better alternative practice is to take samples by chipping off fragments of about 1-2 by 1-2 cm size covering the entire surface exposure, underground mine face, wall and roof in a regular grid interval, say 25 × 25 cm (Fig. 7.25). The area is cleaned before sample cutting. The sampler chips off the fragments with the help of hammer and a pointed chisel. The chips are collected in a clean box or satchel or on a canvas sheet spread on the floor. The weight of samples from a 3 × 3 m area is generally between 1 and 2 kg. Channel sampling is laborious, tedious, time consuming and expensive compared to chip sampling. Chip sampling is preferred for low-cost fast sampling, identification of mineralized contacts and quick evaluation of grade of the area.

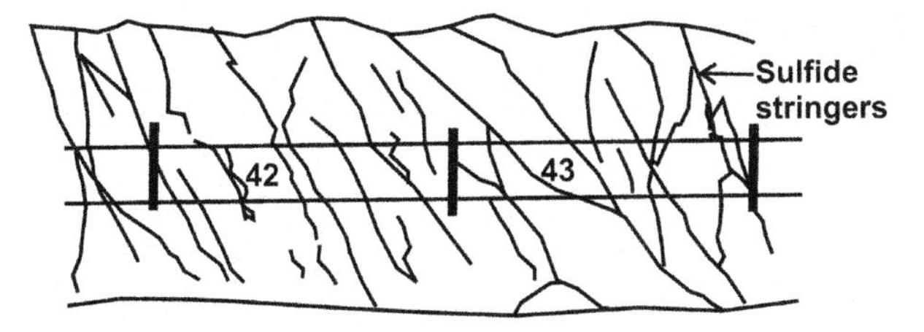

FIGURE 7.23 Schematic presentation of channel sampling of mineral exposure at surface and underground mine crosscut wall at 1 m interval.

7.3.8. Diamond Drill Core Sampling

The mineralized portion of core is sampled and analyzed for metal content. The core is spitted into two identical halves lengthwise with respect to the mineral distribution as

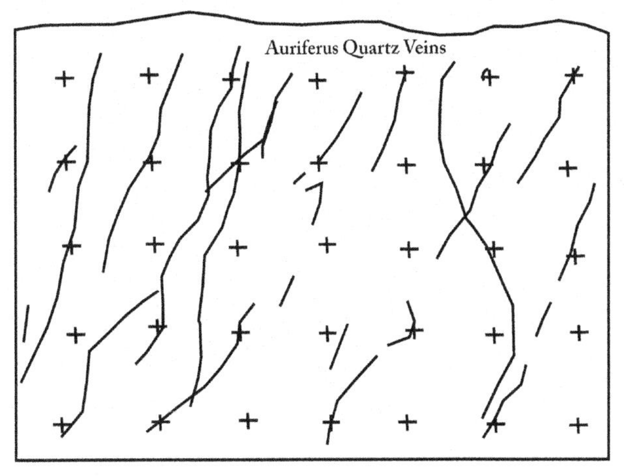

FIGURE 7.25 Chip sampling (+) of wall/face in irregular vein-type deposits such as auriferous quartz veins.

observed during logging. One half is grinded, reduced and sent to laboratory for chemical analysis. The other half is preserved in the core boxes as original record for future check studies. The second halves can also be used as composite sample for metallurgical test works during the initial stage of exploration to develop the laboratory scale beneficiation process flow diagram. The metallurgical test work will indicate the amenability, optimum grinding, liberation and recovery pattern leading to producing clean marketable concentrates.

A simple type of core splitter operates manually by framing a splitting unit by a short piece of rail foundation, with a matching chisel fitted in the top and a hammer as illustrated in Fig. 7.26. Core is placed tightly between the rail and chisel and hammered from the top to split the core as sample. The hammering can be powered by compressed air. This is handy at low cost. The unit can be used in the camp site located at remote places.

FIGURE 7.26 Semi-mechanical core splitter used at Khetri copper mine during 1970s.

The modified version works by electric power. The cutting head is either diamond core cutting unit or a blade made of hard metal alloy (Fig. 7.27) that cuts core into two smooth identical halves (Fig. 7.28).

7.3.9. Sludge Sampling

Sludge is the finer coproduct particles of diamond drilling generated by cuttings of rocks between core and outer hole diameter. The core recovery often becomes poor in case of drilling through fractured mineralized zones. In such situation the sludge forms an integral part of sample while passing through the concerned mineralized horizon. It is relevant to recover the maximum portion of sludge in such circumstances.

Sludge collection can be done in various ways as suits to the operator. The simplest way is to put a plastic or metallic tub and allow the return water to pass. The cuttings are settled and can form a sample corresponding to drilling interval. The method can be modified by using a large sludge box with three to four longitudinal partitions. The return water can flow in zigzag pattern between the successive partitions so that

FIGURE 7.27 Fully mechanized electric core cutter being used at Lennard Shelf Exploration Camp, Australia.

FIGURE 7.28 Stratiform sulfide drill core split into two identical halves by electric cure cutter at Rajpura-Dariba mine, India.

settling of materials becomes better. Commercially designed sludge cutters with mechanical operation are available. Sludge samples are not incredibly authentic due to contamination between drill runs. But it serves the purpose of existence and to some extent the quality of mineralization in the absence of good core recovery.

7.3.10. RC Drill Sampling

Sample from RC drilling is the collection of rock cuttings with respect to the drill depth. The entire rock cuttings are raised to the surface by return air pressure and collected in glass chambers in inverted sequence of depth (refer Fig. 7.12 and Fig. 7.13). The samples are used in valuation of the intersections/deposit. RC drill samples are not the exclusive and ultimate solution of mineral deposit evaluation. These samples are supported, complemented and balanced by some amount of diamond drilling for global perspective. RC drill samples are extensively used in grade control system of mine production in advance and particularly for large open pit mines.

7.3.11. Grab Sampling

Grab sampling is done at any stage of exploration and more so during the mine production phases for quick approximation of the grade. Hand specimen and smaller size are randomly picked up from loose broken material on outcrop, pits, trenches, mine workings, stope draw points, mine cars, train load shipments, and all shorts of stock piles. Care should be taken to avoid inclusion of any foreign objects like wood, iron pieces, nails and plastics.

7.3.12. Muck Sampling

Muck sample is composed of few handheld spade or mechanized shovels full of mineralized fragments and fines collected from the mine face or stope draw points (Fig. 7.29). These samples collected from actual hoisting of mine face are useful for comparing with predicted value of the stope or face blasts. These samples are also compared with cuttings/sludge samples of the Jackhammer and long hole drills. The grade may not match on day-to-day basis. But the average over a period of week, fortnight, month, quarter and annual can be comparable depending on the heterogeneity of the deposit. This also helps to indicate the intrinsic external dilution during mining.

FIGURE 7.29 Muck sampling collected from all sides using a handheld spade or mechanized shovel depending on the volume of the sample.

7.3.13. Car Sampling

Car sample is composed of a hand full of broken pieces picked up randomly from every 5th/10th/15th moving car coming out from the underground mine (Fig. 7.30) or dumpers/trucks of surface mine or aerial ropeway tubs which transport ore to integrated or third party beneficiation plant and smelters. This is compared between the run-of-mine grade and mill head grade for valuation and grade control. The same sampling method is followed for valuation of metal grade, deleterious components and moisture content of concentrate being shipped for the integrated or third party smelter.

7.3.14. Bulk Sampling

Bulk sample is composed of large volume of material (100-1000 tonnes) representing all the distinctive characters of entire orebody with respect to metal grade and mineral distribution. The samples are collected from different parts of the stock pile of trial pit of surface mine, trial crosscuts of underground mine and Run of Mine (ROM) ore of regular production. The best collection equipment will be shovels due to handling of large volume. The total material is mixed thoroughly to reduce the heterogeneity. The sample is used for developing beneficiation flow sheet for optimum uses of reagents and maximizing the recovery efficiency.

7.3.15. Ocean Bed Sampling

The deep ocean floor mineral resources are comprised of poly-metallic nodules, manganese crusts and active or extinct hydrothermal sulfide vents. It covers large areas

FIGURE 7.30 Car sample in underground mine by collecting handful of ore randomly from moving mine cars.

FIGURE 7.31 Bucket-in-pipe poly-metallic nodule sample lifting system collected from deep ocean floor (Credit: Arpita De).

between 4000 and 6000 m below the ocean's surface. The poly-metallic nodules contain mainly nickel, copper, cobalt and manganese. The manganese crusts include primarily manganese, copper, vanadium, molybdenum and platinum. The sulfide vents contribute largely copper, zinc, lead, gold and silver. These raw materials are found in various forms on the ocean floor, usually in higher concentrations than land base mines.

The sample collection from prospective area of sea bottom is conducted by progressive reduction of sample grid through 100 × 100, 50 × 50, 25 × 25 and 12.5 × 12.5 km^2. The collection unit is designed as bucket-in-pipe nodule lifting system (Fig. 7.31) and tested successively. The quantity of materials collected in a trip is about 10 tonnes which is reduced to 200-500 kg to represent a sample.

FIGURE 7.32 Sample size reductions by small portable laboratory Jaw crusher.

7.4. SAMPLE REDUCTION FOR CHEMICAL ANALYSIS

Few grams (~5 gm) of homogeneous fines at ~100 mesh size are required by the laboratories for chemical analysis. Therefore, the samples collected by different methods are reduced without sacrificing the property of the mass being sampled. This is done by progressive grinding of fragment size and gradual reduction of quantity at stages. The samples can be prepared manually by mortar and pestle. Manual processing is extremely slow and impractical to cope up large number of project samples. Therefore, the size is reduced successively through a course of crushing (Jaw crusher, Fig. 7.32), grinding (Roll crusher, Fig. 7.33), and pulverizing (disc/ball/rod pulverizer/mill, Fig. 7.34).

After each stage of crushing, grinding and pulverizing the samples turn into relatively homogeneous. The sample quantity is reduced after every stage by Cone and Quartering as shown in Fig. 7.35 for procedural concept and Fig. 7.36 for technical application.

FIGURE 7.33 Sample grinding by laboratory scale small Roll crusher.

FIGURE 7.34 Sample size reductions by laboratory scale disc, ball or rod pulverizer.

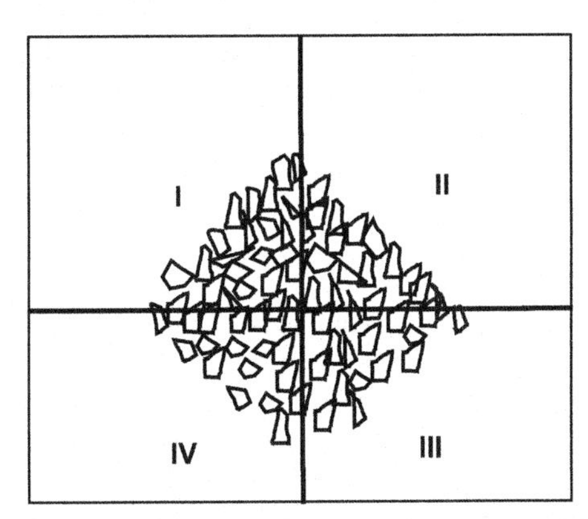

FIGURE 7.35 Sample quantity reductions by Cone-and-Quartering principles after every stage of crushing, grinding and pulverizing.

FIGURE 7.36 Sample quantity reductions for chromite production grade at mine head by Cone-and-Quartering practices. The quarters II and IV will be mixed for further reduction.

FIGURE 7.37 Sample quantity reductions by Jones refill splitter.

The reduction can be done by automatic Jones refill splitter (Fig. 7.37). The crusher-grinder-pulverizers are small in size and of low cost. The units can easily be moved to other exploration camp and installed. The units are suitable for processing samples in laboratory.

Generally two samples of 50 gm each are prepared. One part is sent to laboratory and the duplicate is preserved at the exploration department for future references. The laboratory technicians further grind the material, mix and reduce to about 5 gm for analysis and preserve the remaining for references.

7.5. ACCURACY AND DUE DILIGENCE IN SAMPLING

Sampling due diligence, that makes an authentic geological resource assessment, needs validation of six principal components.

(1) Sample representation, integrity and security,
(2) accuracy of laboratory assays,
(3) insertion of blank and standard at industry-accepted interval in the sample string within mineralization,
(4) adequacy of samples,
(5) quality assurance and quality control (QA/QC) protocols,
(6) mineral resource continuity.

The sampling errors generally occur due to the following reasons:

(a) If the sample area is represented by rich stringers and veins, the unconscious natural tendency of the supervising staff is to collect more samples from the richer part to represent it as an attractive property. This will unduly increase the metal content of the sample. One has to maintain religiously uniform depth and width throughout each length of sample.
(b) If the orebody consists of a mix of hard and soft material, there will be a tendency of the sampler to chisel disproportionate amount of soft rock mass to ease the work. He is tempted to complete more numbers of samples in a shift and thus eligible for incentive bonus. Same caution as at (a) is to be observed.
(c) In case of a drill core sample there will be a tendency to split the core along the foliation and bedding plane for easy cut without considering the distribution of mineralization. The geologist must make color line pencil mark on the core surface (Fig. 10.2) for equal representation of two identical halves (Fig. 7.10).
(d) In a mineral exploration camp, analytical laboratory, date recording in register and data entry in computer, the possibility of interchange of sample numbers is very likely incidence. This cannot be ruled out in the event of handling 100s of samples in a day from large exploration camp, mine production unit and beneficiation plant. Check and recheck at every stage have to be cautiously introduced into the system.

7.6. QUALITY ASSURANCE AND QUALITY CONTROL

It is necessary to test the bias likely to be associated during the sample preparation and analysis. The QA/QC measures are mandatory in modern exploration program for creating a reliable database which is free of any prejudice. The QA/QC pass database assures a trustworthy quantity (tonnes) and quality (percent grade) within acceptable confidence of reserve and resource base of the deposit under feasibility study and mining investment. This is more relevant to upload the investigation data and reports in standard stock exchange for commercial trading in a competitive manner (Refer Chapter 9, Section 15.9). Some of the control measures are:

(a) During exploration of a deposit some of the sampling methods in combination, such as the drill core, channel and chip, may be conducted. Therefore, before the resource estimation a sampling campaign should be conducted by creating drill, face, muck, grab channel and chip samples over the same location and length of mineralization. The results can be compared and statistically tested before accepting the best suitable sampling technique for the evaluation of the deposit.
(b) In case of drill core sampling normally one half of the core is processed for analysis without confirming the equal representation of the other half in the system. The second half of core should be processed, analyzed for a certain length of intersection, statistically verified and accepted (Refer Table 15.7 and Fig. 15.17).
(c) There can be an inherent human and process error in a laboratory while analyzing a sample. Duplicate samples of known value are inserted at industry-accepted intervals of 10th or 20th position in the sample run and analyzed throughout the exploration phase at the same laboratory without disclosing the identity of the sample (Refer Table 15.7).
(d) The standard (certified reference material of known value) and blank (certified reference material of zero value) samples are dispatched to the laboratory together with the routine samples for quality control purposes. These samples are inserted at the start, end and every 10th or 20th position in the sample string. It is desirable to change the sequence of insertion of blank and standard from time to time for quality assurance. In case of major differences the samples are sent for repeat reanalysis. The QA/QC can be repeated in batches.
(e) If the samples from the same deposit are analyzed at different laboratories there is likely to be some bias due to different laboratory personnel (analysts) and analytical procedures (AAS, Volumetric, XRD etc.). A set of same sample should be analyzed at all the concerned laboratories as well as in a Referee Lab of international repute (Refer Table 15.7).
(f) The "data error" between paired set can be tested by various statistical tests as discussed in Chapter 9. The simplest one will be scatter plot of paired data (Refer Fig. 15.17). The scatter diagram will easily identify the presence of extreme erratic high or low sample values with respect to each other. The erratic sample pair must be sorted, identified, isolated and their authenticities investigated along with probable source of errors. The samples can be verified and rejected if it do not satisfy

and fit into the geological condition. The filtered data set will be suitable for QA/QC analysis.

(g) QA/QC program is comprised of the assay data pertaining to various sampling methods, duplicate, standard, blank, both half-core and interlaboratory analysis from activities (a) to (f) above. The data sets are statistically compared and tested for mean, variance, scatter plot, correlation coefficient, "*f*" and "*t*" tests (refer Chapter 9 on Statistical and Geo-statistical Application in Geology). The original versus respective new assay values received from the laboratory are plotted on a graph (scatter plot, refer Fig. 15.17). The plot must show a remarkable degree of correlation at high confidence (r^2) level. If the check assay results are lying in the acceptable range of the standard deviation or within less than 5% variation from the mean value at 95% level of significance, then the assay results are captured in the main assay database. The file turns into "Stable Database". It is a continuous process with incoming addition assay input till the exploration ends. In the process some of the results are unacceptable and the rejected values are not included in the database. Once technically convinced by QA/QC protocols the total sample database can only be used for estimation of reserve and grade parameters.

(h) The confirmation of correlation between two complementary assay data strings can be performed by (refer Chapter 9):

(1) a percentile-percentile (PP) plot,
(2) a quartile-quartile (QQ) plot and
(3) a cumulative frequency (CF) plot.

7.7. OPTIMIZATION OF SAMPLES

Most of the sampling methods are necessary and is a continuous process during mine planning, production scheduling and grade control in a working mine. The drilling and particularly the diamond drilling is the most authentic and most costly sampling tool. So the hardest question is often raised in exploration program as: "when to start drilling?" and even harder is "when to stop?"

The first step is governed by the evidences identified from surface signature, supported by airborne and ground geophysical and geochemical anomalies indicating subsurface continuity of mineralization during Reconnaissance. The interpretative skill and experience of the geologists act as the prime factor at this juncture. Once the evidences potentially indicate an existence of mineralization, then decision for drilling is taken. Main purpose of drilling is to delineate the deposit and to establish the continuity in strike and depth for purpose of resource estimation. The success story may begin with a bit of luck.

The drilling program continues in sequential manner to achieve the defined objectives. The drilling interval and associated quantity will depend on the complexity and value of the commodity. The drilling quantity should be adequate to establish 60% of the total reserve as demonstrated (proved + probable) category for feasibility study and investment decision. The drilling can be stopped for the time being at that stage until additional reserve is required.

The precision of width (tonnage) and grade (percent metal) with increasing number of boreholes in an ongoing base metal exploration project is discussed in detail and determined by applying statistical tools (Chapter 15, Section 15.9.15, Table 15.12 and Fig. 15.23). The curves become steady after drilling at 100 × 50 m spaced grid interval. This provides adequate confidence for definitive feasibility report preparation. The information is adequate and suggests for an investment decision for mine development. Further drilling improves the confidence marginally.

Therefore, the drill sampling is optimized and exploration stops at that point. The mine development initiates and regular production continues. The exploration drilling will renew again in future for enhancement of ore reserves along dip and strike. The cycle of exploration repeats and finally stops when the deposit is fully exhausted in all respect.

FURTHER READING

Bremner et al. (1996) [6] highlighted Trends in Deep Drilling. Sampling methods have elaborately been discussed by Banerjee (1997) [2] and Evans (1999) [24].

Chapter 8

Mineral Resource and Ore Reserve Estimation

Chapter Outline

Any due diligence investigation of a reserve/resource requires a geologist to do the audit and to prepare the data to be audited
—L. A. Wrigglesworth.

8.1. DEFINITION

Mineral resources and ore reserves are defined by the quantity (tonnage) and quality (grade of elements) of in situ concentration of material in or on the Earth's crust. The resources and reserves exist within well-defined 3D mineralized envelopes. The boundaries are drawn between ore and waste or between several grades of ore of all possible bodies within overall framework of mineralized horizon. The evaluation is based upon the information generated during various stages of exploration from inception to date. Data are collected from all types of sampling program, validated with due diligence and captured in main database as discussed in Chapter 7. In situ geological resources are generally higher than minable ore reserves. This is due to availability of detailed sampling input providing higher confidence in estimation, fixing of firm mining boundaries and rejection of mineralization around irregular shapes and tail ends of the body. The firm knowledge of reserves and resources are needed for investment decision of any property.

Mineral Exploration. http://dx.doi.org/10.1016/B978-0-12-416005-7.00008-8

8.1.1. Estimation of Resource and Reserve

The mineral resource and ore reserve potential of mineral deposit is estimated principally by one straightforward formula with minor variation. The unit of measurement is tonne (unit in metric system i.e. 1,000 kg).

$$t = V \times \text{Sp. Gr.}$$

$$V = A \times \text{influence of third dimension}$$

$$\text{Total } T = \sum_{i=1}^{n} (t_1 + t_2 + t_3 \ldots t_n)$$

where,

t or T = measured quantity in tonne.
V = volume in cubic meter (m^3).
A = area in square meter (m^2) is derived by measurement from plans or sections of the geologically defined mineralized area of the deposit.

"Influence" of third dimension is the thickness of horizontal deposit like coal seam, bauxite, placer deposits or drill section interval for base metal deposits.

Sp. Gr. = specific gravity, bulk density and tonnage factor, though not truly synonymous, is used in computation of tonnes by including likely volume of the void and pore spaces. Measurement of number of undisturbed drill cores or bulk samples are the most reliable means of establishing a tonnage factor.

8.1.2. Mineral Resource

The mineral resource is the in situ natural concentration or occurrence of mineralization within a geologically defined envelope. The geological characteristics (quantity, grade, and continuity) are partly known, estimated, or interpreted from broad base evidences and regional knowledge. The presence of mineralization is inferred without programmed framework of verification and cutoff concept. The main emphasis is the estimation of resource inventory of low confidence made during early stage of exploration or around the outer periphery of economic concentration. The evidences are based on wide space sampling program discussed at Chapter 7. The economic viability is premature and intends to establish after the advance stages of exploration. The form, quantity and grade indicate intrinsic future interest and reasonable prospects for eventual profitable extraction.

8.1.3. Ore Reserve

The mineral reserve or precisely ore reserve is that well-defined part of the deposit at specific cutoff after completion of detailed exploration. The reserve is estimated with a high level of confidence based on detail and reliable information. The sample locations are spaced closely enough to confirm geological and/or grade continuity. This reserve must be techno-economically viable. The geological characteristics must be so well established to support production planning. The deposit can be mined and marketed at a profit. The metallurgical tests show optimum recovery. A pre-feasibility/scoping study or feasibility report is prepared to make an investment decision. It includes mine planning, financial analysis including losses associated with mine dilution and metallurgical processing.

8.1.4. Minable Reserve

The minable reserve is very précised accounting of the quantity and grade present within the stope boundary and finally sum total of all stopes. Minable reserve includes all three types of planned and unplanned dilution associated during large-scale mining (Fig. 8.1). The "internal dilution" is comprised of narrow low-grade or barren rocks that exist within mineral body (Fig. 8.1) or between rich mineral bodies (Fig. 8.3). The "external or planned dilution" is the intended addition of extraneous barren rocks outside the ore contacts for uniform designing of blast holes. The "unplanned and wall dilutions" are on account of over drilling and blasting beyond designed design, deviation of blast holes, weak and sheared formation at ore contacts. The wall dilution can be anticipated based on past experience with similar mining method, the type and structure of wall rock and rock mechanic studies. It is pragmatic to consider all dilution waste at "0" grade to produce a conservative estimate. A margin of 5-10% mining loss of ore is expected at the contacts depending on mining methods. The total waste dilution can be expressed as:

$$\% \text{ Waste dilution} = \text{Total waste}/\text{Ore}$$

Some part of ore reserve is blocked in vertical pillars around the shafts/inclines and horizontal Crown/Sill pillars

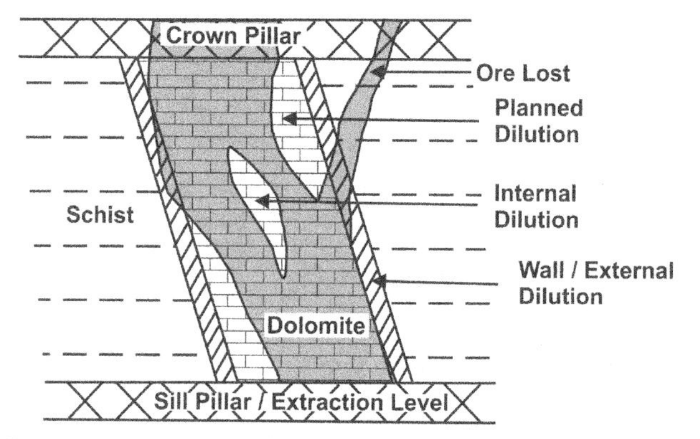

FIGURE 8.1 Schematic cross-section showing smoothening of stope boundaries and expected planned, internal and wall/external dilution during mine planning. The actual will vary to certain extent depending on many factors.

between mining blocks. The pillars act as mine support systems. It is not considered as ore reserve until and unless it becomes recoverable at a later phase of mine life. The category of pillar reserve is then upgraded and merged with minable reserve.

8.2. ESTIMATION OF QUALITY

The quality or grade of mineral resources and ore reserves is the relative concentration of minerals and metals. The unit of measurement is expressed as percent (say 10% Zn, 15% Ash content in coal and 45% CaO in limestone). The various features associated with grade estimation are as follows:

8.2.1. Cutoff Grade

"Cutoff" is the most significant relative economic factor for computation of resource and reserve from exploration data. It is an artificial boundary demarcating between low-grade mineralization and techno-economically viable ore (Fig. 8.2) that can be exploited at a profit. The cutoff boundaries change with the complexity of mineral distribution, method of mining, rate of production, metallurgical recovery, cost of production, royalty, taxes and finally the commodity price in international market. Change of any one criterion or in combination of more gives rise to different cutoff and average grade of the deposit. Cutoff never changes on short-term basis. Market trend is continuously monitored over long-term perspective and situation may compel to change the cutoff or close the mining operation. The concept works well in case of deposits with disseminated grade gradually changing from outer limits to core of the mineralization.

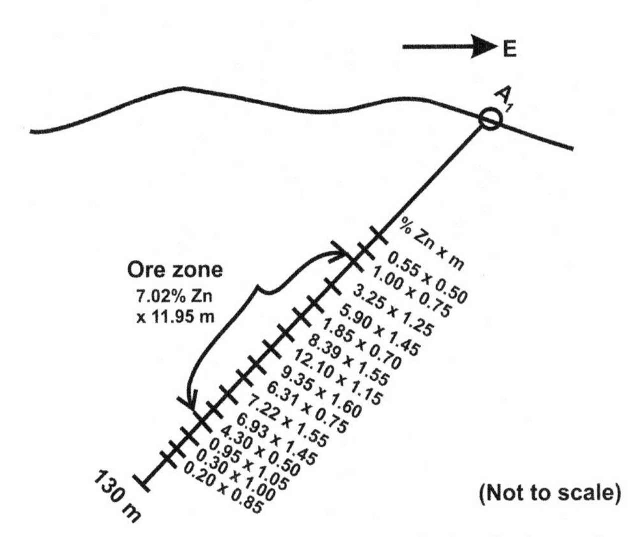

FIGURE 8.2 Average grade computation of mineralized zone from borehole samples leaving low values on either side of ore boundaries at 3% Zn cutoff and 2 m minimum mining width.

In heterogeneous vein-type deposits with rich mineral/metal at the contacts the cutoff has little application in defining the ore limits. In large-scale mechanized mining operations the internal waste partings are unavoidable. The minimum acceptable average grade, defined by combination of alternate layers of ore and waste is the basic criterion of decision making. In this situation an even run-of-mine (ROM) grade is obtained by scheduling ore from a number of operating stopes with variable grades. Combination of ore veins and waste partings with marginal cost analysis will define the shape of orebody. The ore veins at the margins along with the internal waste must satisfy the cost of production by itself, otherwise the marginal vein should be excluded while mine planning. This is known as variable or dynamic cutoff concept (Fig. 8.3).

Cutoff grade perceptibly denotes as simple issue, but it is probably the most misunderstood or misused factor in resource/reserve estimation. The selection of cutoff must be critically reviewed before acceptance. Cutoff grades are normally expressed in percentages (%) of metals for base, ferrous and nonferrous metals (Cu, Pb, Zn, Fe, Al, Cr etc.), and in grams per metric tonne (g/t) or parts per million (ppm) or ounces per dry short tonne for precious metals (Au, Ag, Pt, Pd etc.). It can be given as a percentage equivalent of the predominant mineral commodity in case of multi-metal deposits (% Eq. Cu means % equivalent of Cu).

$$\% \text{ Eq. Cu} = \% \text{ Cu} + \{(\text{Ni price} \times \% \text{ Ni})/\text{Cu price}\} + \{(\text{Au price} \times \% \text{ Au})/\text{Cu price}\} + \cdots$$

$$\% \text{ Eq. Zn} = \% \text{ Zn} + \{(\text{Pb price} \times \% \text{ Pb})/\text{Zn price}\} + \{(\text{Ag price} \times \% \text{ Ag})/\text{Zn price}\} + \cdots$$

8.2.2. Minimum Width

The ultimate use of reserves and grades are related to mine the orebody economically. Mining of ore, by open-pit and underground methods, requires minimum width of the orebody for technical reasons. Narrow width of orebody restricts the vertical limit of open-pit mining due to increase of ore to waste ratio with depth. A minimum of 3 m is suitable for semi-mechanized ore extraction in underground mining. However, greater the width of the orebody larger will be the volume of ore production, higher the mechanization and ore man shift (OMS) leading to low-cost production. Therefore, cutoff base mineralized zone computation is performed keeping in view the minimum width.

8.2.3. Cutting Factors

Many of the base metal (Cu, Pb) and the majority of the precious metal (Au, Ag, Pt, Pd) deposits show occasional or

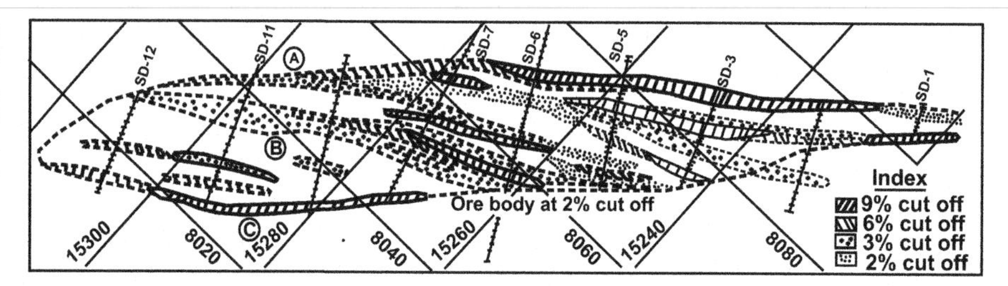

FIGURE 8.3 Concept of dynamic or variable cutoff for vein-type deposit at Balaria zinc-lead mine, Rajasthan, India, Haldar (2007) [33]. The material between "B" and "C" has been excluded for mining. The area is not economically payable by itself.

frequent high sample values. These values are considered to be erratic and designated as nugget value. Some of the estimators of exploration and mining companies prefer to introduce a cutting factor i.e. an arbitrary upper limit marker value in the ore reserve estimation. Any individual assay value, greater than the cutting factor, is reduced to the later before computation of average grade. Some group of estimators practice logarithmic transformation of all sample values for average grade estimation to reduce the nugget effect. This thumb rule applications can significantly understate or overstate the average grades of resources and reserves. The equal length samples should be statistically analyzed. If the higher values together represent 10-20% of the population it should be considered as natural phenomenon which coexists with lower values. This phenomenon can be further supported by volume-variance relationship. It means that bigger the volumes of sample (say 1 day mine production) smaller will the grade variation. Therefore, cutting factor is not suitable without conducting statistical studies of the distribution pattern based on adequate number of samples.

8.2.4. Average Grade

The average grade of an intersection along a trench, borehole, underground workings, cross- and long section, level plan, individual orebody, total deposit, national and global resources and reserves is computed by the formula:

(a) Composite grade of channel, borehole intersection:

$$\text{Grade}(g) = \sum(l_1 \times g_1 + l_2 \times g_2 \ldots + l_n \times g_n) / \sum_{i=1}^{n}(l_1 + l_2 \ldots + l_n)$$

where,

l = length of sample

g = grade of sample.

Exercise: A diamond drill hole has been sampled (Fig. 8.2 and Table 8.1) as given below. Calculate the average grade of the intersection at 3% Zn cutoff.

The mineralized zone has been demarcated between 31.35 and 43.30 m along the borehole at 3% zinc cutoff.

TABLE 8.1 Sample Assay Value of Borehole A1

Sample no.	From (m)	To (m)	Sample length (m)	% Zn
A1/1	30.10	30.60	0.50	0.55
A1/2	30.60	31.35	0.75	1.00
A1/3	31.35	32.60	1.25	**3.25**
A1/4	32.60	34.05	1.45	**5.90**
A1/5	34.05	34.75	0.70	**1.85**
A1/6	34.75	36.30	1.55	**8.39**
A1/7	36.30	37.45	1.15	**12.10**
A1/8	37.45	39.05	1.60	**9.35**
A1/9	39.05	39.80	0.75	**6.31**
A1/10	39.80	41.35	1.55	**7.22**
A1/11	41.35	42.80	1.45	**6.93**
A1/12	42.80	43.30	0.50	**4.30**
A1/13	43.30	44.35	1.05	0.95
A1/14	44.35	45.35	1.00	0.30
A1/15	45.35	46.20	0.85	0.20

Figures in bold signify metal values at 3% Zn cutoff and 2 m minimum mining width for average grade computation.

$$\sum(l_1 + l_2 \ldots + l_n) = 11.95 \text{ m},$$

$$\sum(l_1 \times g_1 + l_2 \times g_2 \ldots + l_n \times g_n) = 83.914$$

$$\text{Average Grade}(g) = 83.914/11.95$$
$$= 7.02\% \text{ Zn for } 11.95 \text{ m}.$$

(b) Average grade of section, plan, orebody, deposit, national and global:

$$\text{Grade}(g) = \sum(t_1 \times g_1 + t_2 \times g_2 \ldots + t_n \times g_n) / \sum(t_1 + t_2 \ldots + t_n)$$

where,

t = tonnes of subblock
g = grade of subblock.

(c) Average grade by statistical and geostatistical method:

The estimation of average grade by statistical and geostatistical method is discussed in Chapter 9.

8.2.5. Minable Grade

The minable grade is the average grade of the stope/mine after taking into consideration of internal and external waste inclusion and loss of ore at the irregular mineralized boundary during stope planning (Section 8.1.4). This is different from average grade of the deposit and generally of higher than the cutoff grade.

8.2.6. ROM Grade

The run of mine (ROM) grade is the final quality of ore coming out of the mine or mine head. The mine production grade is advised to be lowered by 5-10% for estimation and forecast plan. This is on account of inherent internal and unavoidable external dilution and mining losses due to change in blast hole orientation and length, improper blasting, extra dilution along sheared contacts and incomplete recovery from stopping area.

8.2.7. Mill Feed and Tailing Grade

All samples discussed so far are composed of heterogeneous fragment size and do not represent in realistic sense. On the other hand, the mill feed ore is sampled after a continuous process of systematic mixing, crushing, grinding and pulverizing in the beneficiation plant. The fragment size is nearly uniform, may be at (−) 100 mesh size and collected by automatic sampler at 15-30 min interval at the discharge point of ball/rod mills. The modern day plants are equipped with advance microprocessor-based sampling probe installed in the conditioner. The multiple assay values are displayed and monitored at every 2-5 min interval in the centralized computer control room. The reagents in the floatation cells are adjusted accordingly by auto-control system. The online trend of the mill feed grade, average grade of the shift and day is available. The total scheme of microprobe metal analysis, screen display, data capture and reagent control is a centralized integrated system. The mill feed sample grades are considered as final value of the ore production to reconcile the grade of mine for the hour/shift/day/month/quarter/annual and life of the mine as the case may be.

Similarly the tailing outflow is also continuously sampled by manual method or electronic probe at 15-30 min interval, displayed and captured in the centralized circuit panel. The mill feed and tailing grades facilitate in computation of metal balance in the plant with respect to recovery parameters and overall performance.

8.3. CONVENTIONAL RESOURCE/ RESERVE ESTIMATION

In general, different conventional estimation methods are employed depending on the shape, dimension, complexity of the mineral deposit and sample type and interval during exploration. The procedures turn into complex with intricacy of deposit having large volume of sample information. The examples of simple deposits are seam, horizontal layers and placer type of fuel and industrial minerals. The exploration is carried out mostly by short vertical holes. The complex type encompasses base and noble metals. The sampling is primarily by large number of fan shape diamond or reverse circulation (RC) drill holes in various angles. The number of samples are usually very high. The in-between type will be variation of ferrous and nonferrous metal deposits. The area of influence is used to assume the continuity of mineralization between sampled data. It must be judged critically to minimize the error in estimation of tonnage and grade of the deposit. The various times-tested traditional estimation procedures are:

(1) Old style
(2) Triangular
(3) Square and Rectangle
(4) Polygonal
(5) Isopach and Isograde
(6) Cross-section
(7) Longitudinal Vertical Section
(8) Level plan
(9) Inverse power of distance

The triangular, square, rectangle and polygonal methods are point estimates by de-clustering of cells around the samples. The de-clustering methods divide the entire section and plan area into representative polygon around samples and called a cell. It is always safe to follow two to three complementary methods of estimation for any deposit. The outcome of each procedure must be close to each other with respect to its tonnage and grade to accept the final result. Otherwise, the computational procedure must be checked. The example can be estimation of a mineral deposit by employing cross-section, long section, level plan, statistical methods and compared.

8.3.1. Old Style

The old style method was in practice during past for single vein-type deposit like gold-bearing quartz veins of Kolar and Hutti, Karnataka, India, and gold mines in South

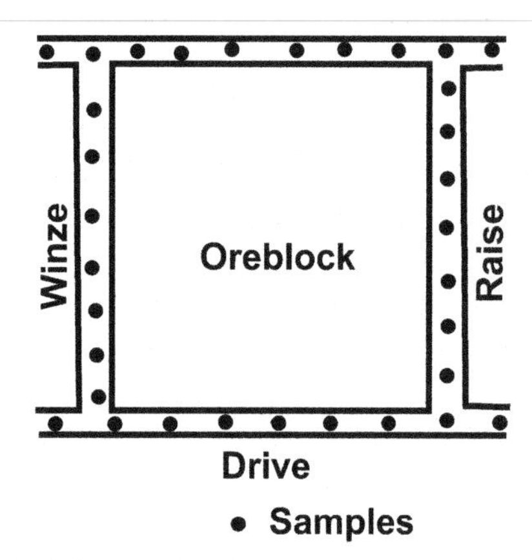

FIGURE 8.4 Old style estimations for vein-type deposits averaging the channel and chip sample values around the ore block.

Africa. The auriferous veins are usually either exposed or close to the surface. The geological and geochemical exploration is supported by few numbers of surface drill holes to establish the existence and continuity of mineralization down depth. The initial entry is by adit, incline and shaft. The mine levels are developed within the orebody at short vertical interval of around 30 feet. The extraction levels are suitably located on one side of the orebody based on dip of the lode. The levels are connected by raises and winzes passing through the mineralization. Channel and chip sampling is conducted at short interval of 3-6 feet all along the drives and raises (Fig. 8.4). The reserve is estimated by multiplication block area, thickness of vein and Sp. Gr. The grade is computed by averaging all the sample values generated within the block.

8.3.2. Triangular

Triangular method is employed for flat type of near surface deposits having better continuity such as laterite and bauxite. Triangles are formed by joining three adjacent positive intersections defining a block (Fig. 8.5). The horizontal area of each block is measured and multiplied by the thickness of the mineralization to get the volume. The reserve is obtained by multiplying the volume with bulk Sp. Gr. of ore. The grade is computed by averaging the three corner values of the triangle.

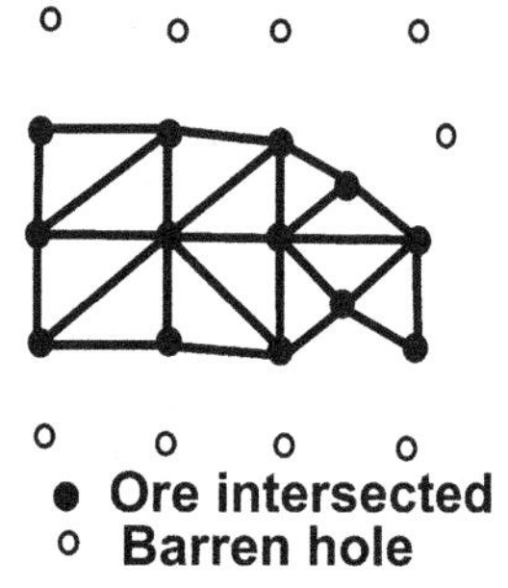

FIGURE 8.5 Reserve estimations by Triangular method for flat-type deposits considering area of the triangle for tonnage and average grade of the holes located at the three corners.

8.3.3. Square and Rectangle

The sampling for flat-type deposits can also be planned by drill location at center of square or rectangular grid (Fig. 8.6). The reserve and grade of the deposit will be estimated in the same way as at triangular method.

8.3.4. Polygonal

Polygons are drawn either by joining each positive borehole or by perpendicular bisectrix around each borehole (Fig. 8.6). Reserve and grade can be estimated as at 8.3.3.

8.3.5. Isograde/Isopach

The estimation by Isograde (Fig. 8.7 contours of identical grade) method is suitable for flat and low-dipping disseminated deposits with variable thickness and grade. The area between two successive contours is measured and multiplied by the difference between the lower and upper contour values to obtain the volume. The reserve is computed by further multiplying the volume and bulk specific gravity. The grade is computed by weighted

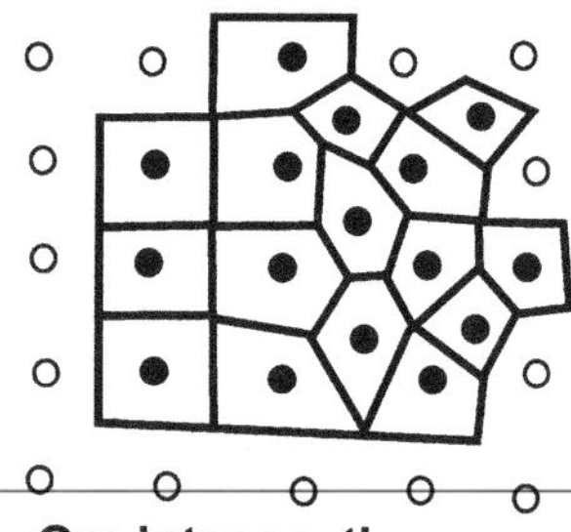

FIGURE 8.6 Reserve estimations by Square, Rectangle and Polygonal method keeping samples at the center of the square or polygon.

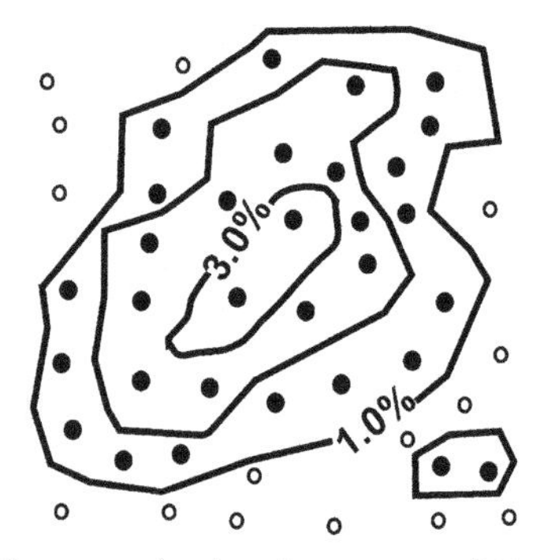

FIGURE 8.7 Reserve estimations by contours of identical grade and thickness of mineralization.

averaging the assay values falling within the two contours.

8.3.6. Cross-Section

Geological cross-section is a vertical image of the plane across the geological continuity of the area. The extent of section is limited by the available surface geological data and borehole information. The total surface features such as rock contacts, structures, mineralized signatures, weathering and gossan are plotted with local coordinate system along the surface profile. The scale is often selected as 1:2000, 1:1000, 1:500. Contours indicate elevation of the profile. All the boreholes falling on and around the section are plotted based on its collar coordinate, direction, angle of drilling, deviation and length of hole. The information of core recovery, rock contacts, structures, analytical results, individual or composite value, RQD from the log sheets are plotted along the trace of the hole. The geological correlation is made taking into consideration knowledge of the area and experience of the geologists. The orebody can be extended up to the surface if it is directly exposed such as depicted by chromite deposit or by indirect signature like presence of oxidation/gossan of base metal deposits. Otherwise, the orebody will be treated as concealed type and shape will be drawn by drill information. The orebody configuration can be very simple consisting of one vein or it can be multiple in numbers and giving a complex type by splitting and coalescing with each other.

The total mineralized area is divided into several subblocks around each borehole intersection by halfway influence principle (Fig. 8.8). The halfway demarcation is made by joining midpoints of hanging and footwall mineralization contacts between two adjacent boreholes. The area of each subblock is measured by geometrical formulas for rectangular, square and triangular orebody. A planimeter or an overlay of transparent graph sheet or AutoCAD software can be used for measuring area of irregular orebody. Planimeter is a drafting instrument used to measure the area of a graphically represented planar region by tracing the perimeter of the figure. The volume of the subblock is computed by multiplying the third dimension i.e. half of drilling interval on either side. The extremities of the orebody at both the end sections can be logically extended any distance less than equal to half of the drill interval. Halfway influence on either side, for volume computation between sections, may introduce significant errors in tonnage and grade if similar configuration does not exist in the adjacent sections. It is recommended to draw Longitudinal Vertical Section and Level Plan simultaneously to depict a reasonable 3D perspective. The tonnage and average grade of the section is computed by the formula at Sections 8.1.1 and 8.2.4.

Exercise: A zinc-lead deposit in Rajasthan was identified by gossan outcrop extending over 1500 m in NE-SW

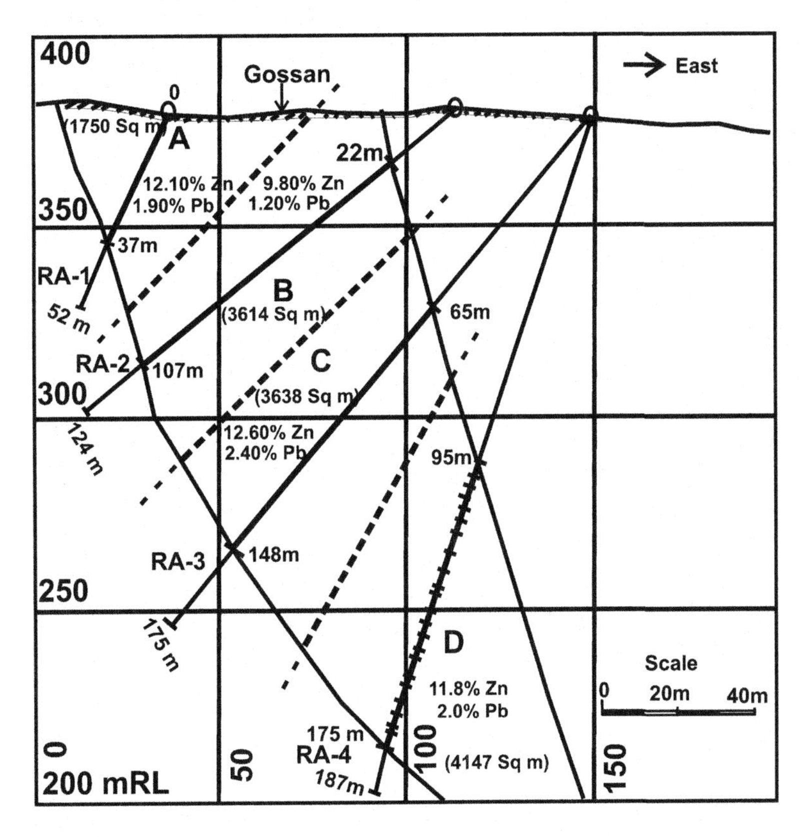

FIGURE 8.8 Reserve estimations by cross-section method—most popularly and widely adopted by all level of professionals since many decades.

TABLE 8.2 Details of Drill Hole Information Along Section A-A for Estimation of Reserve and Grade

Borehole	Block	Area (m^2)	Volume (m^3)	Tonnage (t)	% Zn	% Pb
A-1	A	1750	87,500	262,500	12.10	1.90
A-2	B	3614	180,700	542,100	9.80	1.20
A-3	C	3638	181,900	545,700	12.60	2.40
A-4	D	4147	207,350	622,050	11.80	2.00
Total		13,149	657,450	1,972,350 or 1.97 Mt	11.51	1.88

direction. Surface exploration was conducted by diamond drilling at 50 m section interval. Four boreholes have been drilled at section A-A (Fig. 8.8). The section subblock area around each borehole has been demarcated by halfway influence concept. Estimation of reserve and grade for the section with bulk specific gravity 3.00 would be (Table 8.2):

Where,
Area = measured by planimeter or superimposed graph sheet
Volume = Area × Halfway influence (50 m)
Block tonnage (t) = Volume × Bulk Sp. Gr. (3.00)
Total section tonnage (T) = Sum of all block tonnes

$$\Sigma\ (t_1 + t_2 + \cdots + t_n)$$

$$\text{Average grade} = \text{Grade}(g) = \Sigma(t_1 \times g_1 + t_2 \times g_2 + \cdots + t_n \times g_n / \sum_{i=1}^{n} \times (t_1 + t_2 + \cdots + t_n)$$

The reserves and grades of whole orebody are the cumulative tonnage and weighted average grades of all sections (Fig. 8.9).

8.3.7. Longitudinal Vertical Section

Longitudinal vertical section (more suitably projection) is the creation of a vertical image along the elongated direction presenting features like lithology, ore geometry,

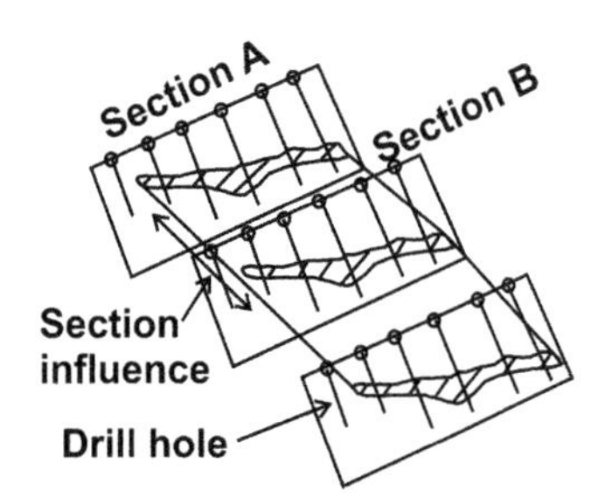

FIGURE 8.9 Concept of cross-section method for section and deposit reserve and grade estimation.

categorization and ore reserve. The trace of the surface profile and subsurface position of mineralized information as gathered by drill holes and underground workings are plotted in the vertical plane. The negative information of drill holes is considered to delimit the mineralization from barren rocks.

The total mineralized envelope on the longitudinal vertical section is divided into subblocks around the positive intersection with the principle of halfway influence (Fig. 8.10 and Table 8.3). The tonnage and average grade of individual subblock and total ore deposit is computed as discussed in the case of cross-section method at Section 8.3.6.

Exercise: A concealed silver-rich zinc-lead deposit in Rajasthan was identified at a depth of 120 m from the surface. Surface drilling was conducted at 100 m interval. All mineralized intersections of surface holes were projected on a longitudinal vertical section. The subblocks around each intersection was demarcated by halfway influence and individually measured. Estimation of reserve and grades for the section with specific gravity of 3.00 would be:

Where,
Area = measured by planimeter or superimposed graph sheet
Volume = Area × mineralization plan width (m)
Block tonnage (t) = Volume × Bulk Sp. Gr. (3.00)
Total long section tonnage (T) = Sum of all block tonnes

$$\Sigma\ (t_1 + t_2 + \cdots + t_n)$$

$$\text{Average grade} = \text{Grade}(g) = \Sigma(t_1 \times g_1 + t_2 \times g_2 + \cdots + t_n \times g_n / \sum_{i=1}^{n} \times (t_1 + t_2 + \cdots + t_n)$$

8.3.8. Level Plan

Level plan is the horizontal plan image of any subsurface datum plane. It is very similar to surface geological map to large extent. Plan view of a particular level is created taking

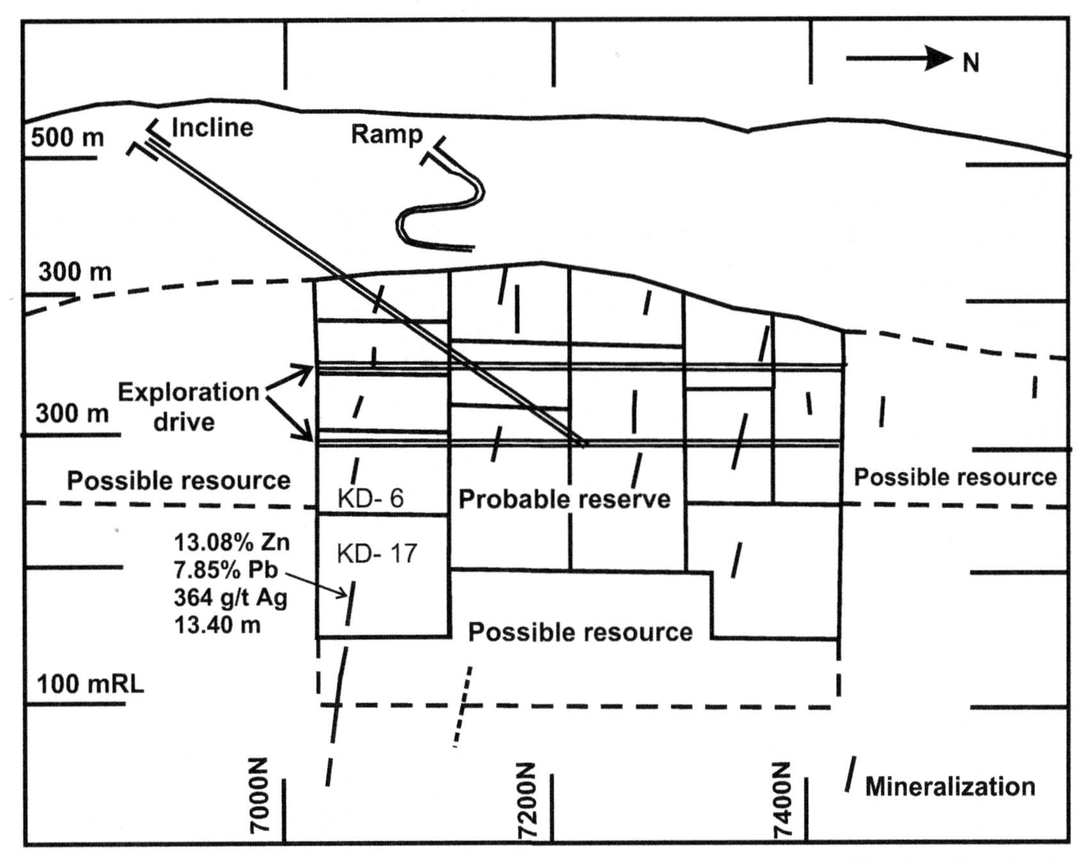

FIGURE 8.10 Estimation of reserve and grade by Longitudinal Vertical Section—an alternative process to validate the estimate by other techniques.

measurements from all the cross-sections and underground drill and development sampling. The reserve is computed by the same way as discussed in Sections 8.3.6 and 8.3.7 (Fig. 8.11 and Table 8.4).

Exercise: A concealed silver-rich lead-zinc deposit in Rajasthan was identified by routine drilling along the structural lineament at a depth of 120 m from surface. Surface drilling was conducted at 100 m interval followed by entry to the deposit by incline and development of footwall drive for close space underground drilling at 50 m interval and delineation of orebody. The reserve and grade is estimated by level plan area method as:

TABLE 8.3 Details of Boreholes Information along Longitudinal Vertical Section for Estimation of Reserve and Grades

Borehole	Area (m^2)	Volume (m^3)	Tonnage (t)	% Zn	% Pb	g/t Ag
KD-17	8000	96,000	282,000	13.08	7.85	364
KG-06	5525	112,157	336,472	5.15	7.75	266
Total	13,525	208,157	624,472	9.27	7.80	311

Where,

Area = measured by planimeter or superimposed graph sheet

Volume = Area × Halfway influence (50 m)

Block tonnage (t) = Volume × bulk Sp. Gr. (3.00)

Total section tonnage (T) = Sum of all subblock tonnes $\Sigma\,(t_1 + t_2 + \cdots + t_n)$

$$\text{Average grade} = \text{Grade}(g) = \Sigma(t_1 \times g_1 + t_2 \times g_2 + \cdots + t_n \times g_n / \sum_{i=1}^{n} \times (t_1 + t_2 + \cdots + t_n)$$

8.3.9. Inverse Power of Distance

The most accepted computerized extension functions applied in the mining industry for computation of mine production blocks and subblocks are based on the principle of gradual change for making value estimates. One of the common methods is generally referred to as the "Inverse Power of Distance" or ($1/D^n$) interpolation. The technique

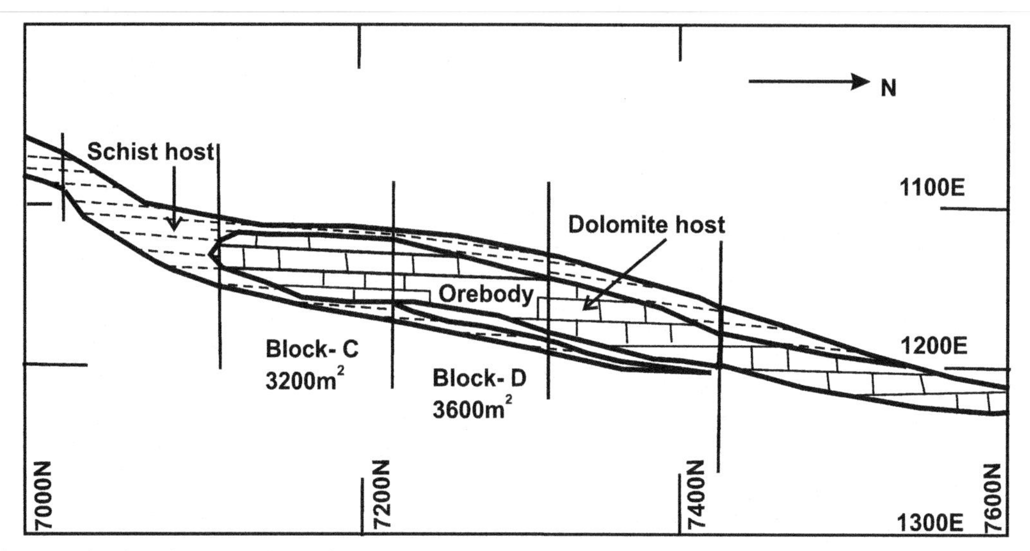

FIGURE 8.11 Estimation of reserve and grade by level plan method—an alternative technique to cross-check estimates by other procedures.

TABLE 8.4 Details Subblock Area for Estimation by Level Plan Method

Block	Area (m^2)	Volume (m^3)	Tonnage (t)	% Zn	% Pb	g/t Ag
C	3200	160,000	480,000	5.01	0.93	89
D	3600	180,000	540,000	7.89	1.88	86
Total	6800	340,000	1,020,000 1.02 Mt	6.53	1.43	87

uses straightforward mathematics for weighting the influence of all surrounding samples upon the block being estimated as depicted in Fig. 8.12. It is necessary to select only those samples falling within the influence zone relevant to the mineralogical behavior (continuity function) of the population. It is also important to reflect the anisotropic character within the deposit and vary the distance weighting function directionally with the help of semi-variogram function in various directions (refer Chapter 9).

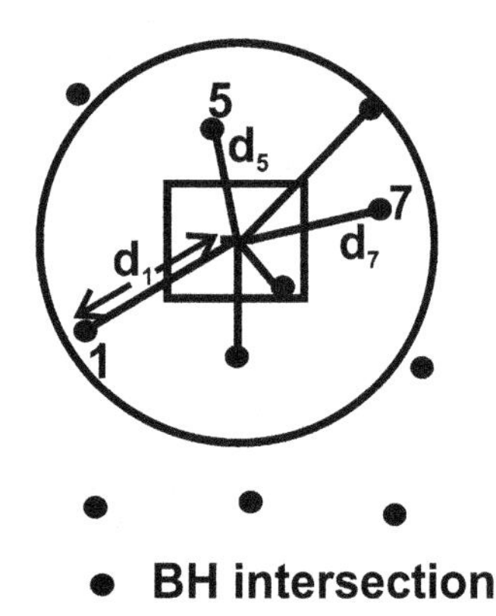

FIGURE 8.12 Principle of Inverse Power of Distance method considering samples falling within an optimum search circle or ellipse in two dimensions.

The mining block is divided into series of regular 2D or 3D slices within the planned boundary equivalent to blast hole of mine production. Figure 8.13 illustrates a 2D cross-section model or rock matrix. The block dimensions and approach for an open-pit mine are 12.5 m along strike (infill drill interval), 10 m vertically (bench height) and 5 m across the dip (face movement). Each cell is designated by a code number (say, −200, 17, 19) controlled by identification of section, bench and cell; e.g. −200 is south 200 section, 17 is the bench between 330 and 340 m level and 19 is the cell position between 40 and 45 east.

The samples along the boreholes are converted to uniform 5 m composite length. The selection of samples for computation of a panel is controlled by a search ellipse oriented with its major axis along the down-dip of the orebody (range = 90 m) and minor axis across (range = 30 m). In case of 3D computation the intermediate axis is oriented along the strike direction (range = 115 m). The ranges in various directions are obtained from semi-variogram study (Chapter 9). The ellipse moves on the plane of the cross-section, centering the next computational panel while doing the interpolation. The strong anisotropic nature, if observed in the semi-variogram, is further smoothened by differential weighting factors on samples selected through search ellipse screening. The samples located down-dip are assigned greater weighting factor than across the orebody. These factors were tested in various options near controlled cells. In this method the near sample points get greater weighting than points further away. The power factor is often employed as d^2.

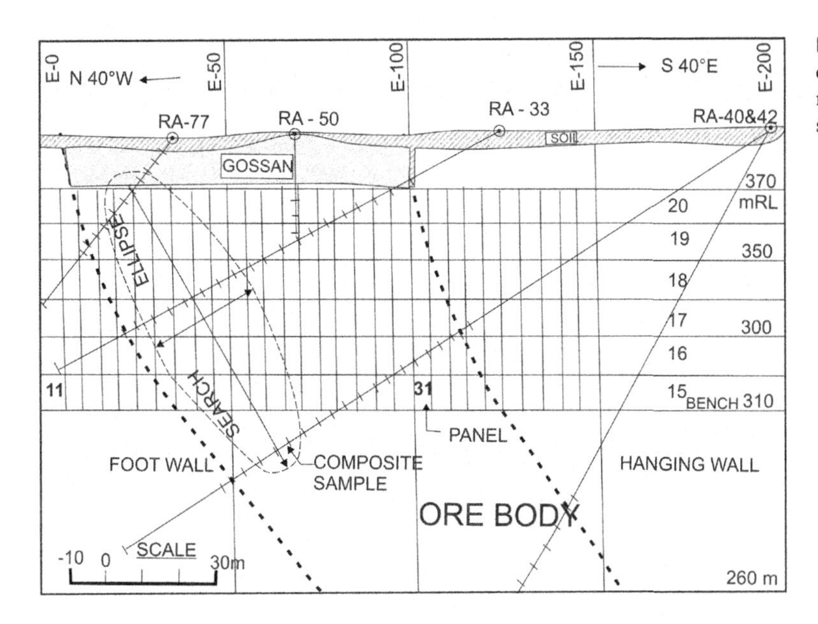

FIGURE 8.13 Computation of small block reserve on cross-section employing Inverse Squared Distance method—very significant information for production scheduling and grade control.

The tonnage of each panel is computed by block dimension and not affected by the variation of grade in all directions. The cell values (tonnage and grade) can be displayed as series of bench plan for production scheduling. Inverse Power of Distance computation is performed by using in-house or commercial software.

$$G_B = \sum_{i=1}^{n}\{g_i/(d_i)^k + \cdots + g_n/(d_n)^k\} / \sum_{i=1}^{n}(1/d_i{}^k + \cdots + 1/d_n{}^k + C)$$

Where,

G_B = estimated block grade
g_i = grade of the ith sample
d_i = distance between block center and ith sample
K = 1, 2, 3 (power and often = 2)
C = arbitrary constant

8.4. MINERAL RESOURCE AND ORE RESERVE CLASSIFICATION

The mineral resources and ore reserves are estimation of tonnage and grade of the deposit as outlined three dimensionally with variation in density of sampling and even with limited mine workings. The estimate stands on certain interpretations and assumptions of continuity, shape and grade. Therefore, it is always approximate and not certain until the entire ore is taken out by mining. Various types of sampling are conducted at different density or interval with associated uncertainties during exploration. One part of the deposit may have been so thoroughly sampled that we can be fairly accurate of the orebody interpretation with respect to tonnage and grade. In another part of the same deposit sampling may not be of intensely detail, but we have enough geological information to be reasonably secure in making a statement of the estimate of tonnage and grade. The knowledge may be based on very few scattered samples on the fringes of the orebody. But we have enough information from other parts of the orebody supported by geological evidences and our understanding of similar deposits elsewhere to say that a certain amount of ore with certain grade may exist. Increase of sampling in lower category region will certainly enhance the status as mining proceeds.

Mineral resource and ore reserve classification system and reporting code have been evolved over the years by different countries exclusively on the basis of geological confidence, convenience to use and investment need in mineral sector. Conventional or traditional classification system was in use during twentieth century. New development took place from third and fourth quarter of the same century satisfying statutes, regulations, economic functions, industry best practices, competitiveness, acceptability and internationality. There are several classification schemes and reporting codes worldwide such as USGS/USBM reserve classification scheme, USA, United Nations Framework Classification (UNFC) system, Joint Ore Reserve Committee (JORC) code, Australia and New Zealand, Canadian Institute of Mining, Metallurgy and Petroleum (CIM) classification, South African Code for the Reporting of Mineral Resources and Mineral Reserves (SAMREC) and The Reporting Code, UK.

The basic material and information for mineral resource and mineral reserve classification scheme and reporting code must be prepared by or under the supervision of "Qualified Persons or QP". The QP is a reputed professional with graduate or postgraduate degree in geosciences or mining engineering. The QP must possess sufficient experience (more than 5 years) in mineral exploration, mineral project assessment, mine development, mine operation or any combination of these. The QP may preferably be in good standing or affiliated with national and international professional associations or institutions. The QP is well informed with technical reports including exploration, sampling adequacy, QA/QC and analytical verification, discrepancy and limitations, estimation procedure, quantity, grade, level of confidence, categorization, and economic status (Order of magnitude, Prefeasibility, Scoping Study and Feasibility study) of the deposit concerned. The QP should be in a position to make the statements and vouches for the accuracy and completeness of the contained technical report including information and the manner in which it is presented, even he/she is not the author of the report. This is a matter of professional integrity and carries legal risk. The misleading statements can result in legal sanctions in the country and other jurisdictions.

8.4.1. Conventional Classification System

The degree of assurance in the estimates of tonnage and grade can subjectively be classified by using convenient terminology. In order of increasing geological exploration creating high confidence level and techno-economic viability the categorization has broadly been grouped as "Economic reserves" and "Sub-economic conditional resources". The economic ore reserves and sub-economic resources are further subdivided as Developed, Proved, Probable and Possible (Fig. 8.14). The classification system helps the investor in decision making for project formulation and activities required at different phases. These terms are supported by experience, time tested, and well accepted over years. The terminology is comparable with equivalent international nomenclature that is used by USGS or Russian systems as Measured, Indicated and Inferred.

8.4.1.1. Developed

The exposed parts of orebody represent "Developed" or "Positive" or "Blocked" reserves. Exposure can be by trenches or trial pit on the surface for open-pit mines or bounded on all sides by levels above and below, and connected by raises and winzes on the sides of the block for underground mines. Definition or delineation drilling at 30-15 m interval completed and all sides are sampled. The block is ready for stope preparation, blast hole drilling, blasting and ore draw. The draw point sampling is just carried out at this stage to assign stope production grade, blending ratio for the stockpile, reconciliation with respect to additional dilution and errors in estimation. The risk of error in tonnage and grade is minimal. The confidence of estimate is ~90%.

8.4.1.2. Proved

The "Proved" or "Measured" reserves are estimated based on samples from outcrops, trenches, development levels and diamond drilling. The drilling interval would be 200 or even 400 m for simple sedimentary bedded deposits (coal seam, iron ore) with expected continuity along strike, other than structural dislocation. The sample interval would be at 50 by 50 m for base metal deposits. The deposit is either exposed by trenches or trial pit for open-pit mines and by development of one or two levels for underground drilling. Further stope delineation drilling and sampling will

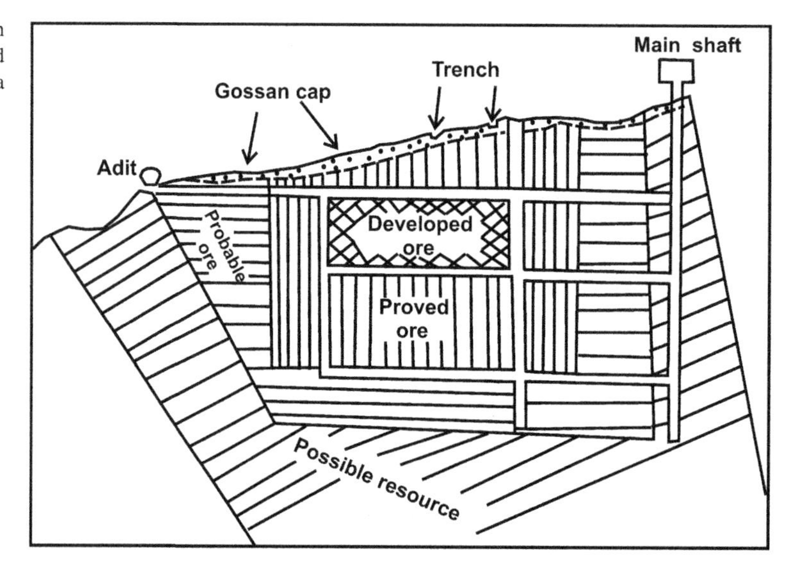

FIGURE 8.14 Conventional reserve classification systems showing various categories of reserves and resources based on enriched geological experience—a good option for small players in mining industry.

continue to upgrade the category to developed reserves. The confidence of estimate is ~80%.

8.4.1.3. *Probable*

The "Probable" or "Indicated" reserve estimate is essentially based on wide-spaced sampling, surface and underground drilling at 100-400 m interval depending on the complexity of the mineralization. The opening of the deposits by trial pit or underground levels is not mandatory to arrive at this category. The confidence of estimate is ~70%.

The sum total of Developed, Proved and Probable reserve is termed as "Demonstrated" category. The reserve of a project under investment decision should contain about 60% in the Demonstrated category.

8.4.1.4. *Other Ore*

Part of the ore reserve is blocked in Sill, Crown and Rib pillars for stability of the ground during mining operation and related impacts (Fig. 8.1). This blocked reserve is designated as "Other Ore" and is monitored as Proved category. As and when the other ore is likely to be recovered after completion of the nearby stopping blocks, it is elevated to Developed category.

8.4.1.5. *Possible*

"Possible" or "Inferred" resources are based on few scattered sample information in the strike and dip extension of the mineral deposit. There would be sufficient evidences of mineralized environment within broad geological framework having confidence of about 50%. The possible resource will act as sustainable replacement of mined out ore.

8.4.2. USGS/USBM Classification Scheme

The USGS through the years collects nationwide information about the mineral resources and reserves. In order to make a standard classification system Dr V. E. Mckelvey, Director, USGS, first conceptualized set of resource classification system in 1972 as indicated in Fig. 8.15.

The USGS and the USBM developed a common classification scheme in 1976. Additional modifications were incorporated to make it more workable in practice and more useful in long-term public and commercial planning. The success of future plan program will rely entirely on (1) precise knowledge of available reserves and resources for fixing priority, (2) developing existing unworkable deposits to economic proposition by cost cutting and technological breakthrough and (3) the probability of new discovery at regular basis. The resource base must be continuously reassessed in the light of new exploration input, advancement in mining and process technology and change in commodity price. The collaboration continued to revise the Bulletin 1450-A. The final document was published in 1980 as USGS Circular No. 831—"Principles of a Resource/Reserve Classification for Minerals."

	DISCOVERED	UNDISCOVERED
COMMERCIAL	RESERVES	RESOURCES
SUB-COMMERCIAL	RESOURCES	

FIGURE 8.15 The initial concept of resource classification system conceived as McKelvey Box in 1972.

The concept of classification and block diagram was developed as 2D representation as given in Fig. 8.16. The *X* and *Y*-axis represent the geological degree of assurance and the increasing economic feasibility, respectively. The geological axis is broadly divided into Identified and Undiscovered resources with further subdivision based on increasing exploration support. The economic feasibility axis is similarly divided into Economic and Sub-economic with further subdivision based on techno-economic viability on present market price.

The definition and specification of various identified resources have been described. The resource classification scheme gives emphasis to Identified Sub-economic resources for future target. It also initiated the concept of probability of existence of undiscovered resources simply on hypothetical and speculative ground.

8.4.2.1. *Paramarginal*

The portion of Sub-economic resources that either exists at the margin of economic-uneconomic commercial border, being nonrenewable asset, can be exploited at marginal profit with innovative mining and metallurgical techniques. The other type of Paramarginal resources is not commercially available solely because of safety, legal or political circumstances.

The example can be cited from Gorubathan multi-metal deposit, West Bengal, India, having high-grade metals (>10% Zn + Pb) on account of misbalancing Himalayan Ecosystem and extension of orebody below the railway line at Balaria base metal mine of Zawar Group, Rajasthan.

8.4.2.2. *Submarginal*

The portion of Sub-economic resources that would require much higher price at the time of mining or a major cost reduction advance R & D technology toward mining and metallurgical recovery.

An example can be cited to Sindesar Kalan base metal deposit, 6 km north of Rajpura-Dariba mining project, having 100 Mt of 2.50% Zn + Pb metal in graphite mica-schist host rock. The deposit is exposed to the flat surface

FIGURE 8.16 USGS resource classification scheme (Source: adopted from Mckelvey 1972).

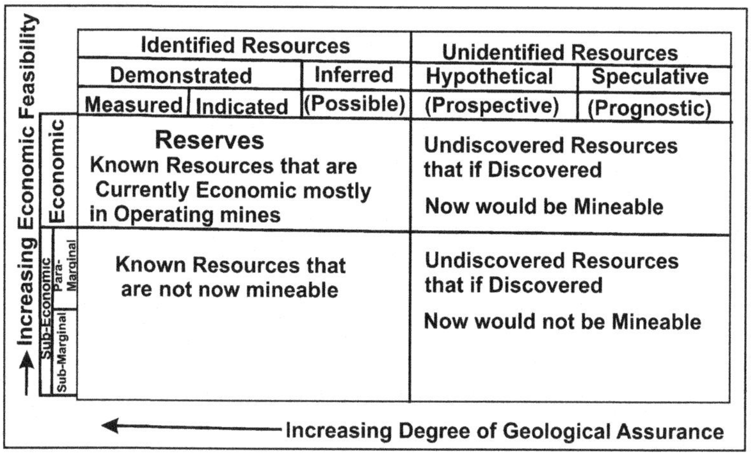

Identified Resources
Demonstrated | Inferred
Measured | Indicated | (Possible)
Unidentified Resources
Hypothetical | Speculative
(Prospective) | (Prognostic)
Increasing Economic Feasibility
Economic
Reserves
Known Resources that are Currently Economic mostly in Operating mines
Undiscovered Resources that if Discovered Now would be Mineable
Sub-Economic
Para-Marginal
Sub-Marginal
Known Resources that are not now mineable
Undiscovered Resources that if Discovered Now would not be Mineable
Increasing Degree of Geological Assurance

land and open-pit mining cost will be low with the support of major common infrastructure at Dariba mine. The technological breakthrough in metallurgical recovery of low-grade ore from graphite mica-schist host and increase in metal price will convert it to economic category in distant future.

8.4.2.3. Hypothetical

The "Hypothetical" or "Prospective" resources are undiscovered theoretical mineral bodies in nature that may logically be expected to exist in known mining district or region under favorable geological conditions. The existence, if confirmed by exploration and reveals quantity and quality assessment, would be reclassified as Reserves or Identified Sub-economic resources.

Neves Corvo poly-metallic deposit, Portugal, located ~250 km in the southeast extension of Iberian Pyrite Belt (IPB) in Spain, and Sindesar Khurd zinc-lead-silver deposit, located 6 km in the northeast extension of Rajpura-Dariba belt, India, were discovered at a depth of 330 and 120 m respectively under similar geological condition below barren surface cover as a routine exploration in the known belt.

8.4.2.4. Speculative

The "Speculative" or "Prognostic" resources are tentative mineral bodies in nature and are undiscovered so far that may occur either in known favorable geological setting where no discoveries have yet been made or unknown type of deposits that remain to be recognized. This is useful for long-term allocation of exploration budget. The existence, if confirmed by exploration and revealed quantity and quality assessment, would be reclassified as Reserves or Identified Sub-economic resources.

Uranium deposits worldwide are hosted by one of the geological settings as follows: unconformity related, conglomerate, sandstone, quartz-pebble, vein type, breccia complex, collapse breccia pipe, intrusive, phosphorite, volcanic, surficial, metasomatite, metamorphic, lignite and black shale. The search for uranium can be speculated for this favorable environment and tested.

8.4.3. UNFC Scheme

The UNFC system is a recent development in reserve categorization (E/2004/37—E/ECE/1416, February 2004). The scheme is formulated giving equal emphasis on all three criteria of exploration, investment and profitability of mineral deposits. The format provides (1) the stage of geological exploration and assessment, (2) the stage of feasibility appraisal and (3) degree of economic viability. The model is represented by multiple cubes ($4 \times 3 \times 3$ blocks) with geological (G) axis, feasibility (F) axis and economic (E) axis. The three decision making measures for resource estimation are further specified with descending order as follows:

Geological Axis (G) →
(1) Detailed Exploration
(2) General Exploration
(3) Prospecting
(4) Reconnaissance
Feasibility Axis (F) →
(1) Feasibility Study and Mining Report
(2) Pre-feasibility Study
(3) Geological Study
Economic Axis (E) →
(1) Economic
(2) Potentially Economic
(3) Intrinsically Economic.

The scheme is presented in 3D perspective (Fig. 8.17) with simplified numerical codification facilitating digital processing of information. Each codified class (Table 8.5) depicts a specific set of assessment stages with associated economic viability. The scheme is an internationally understandable, communicable and acceptable across national boundaries under economic globalization that makes easy for the investor to take correct decision.

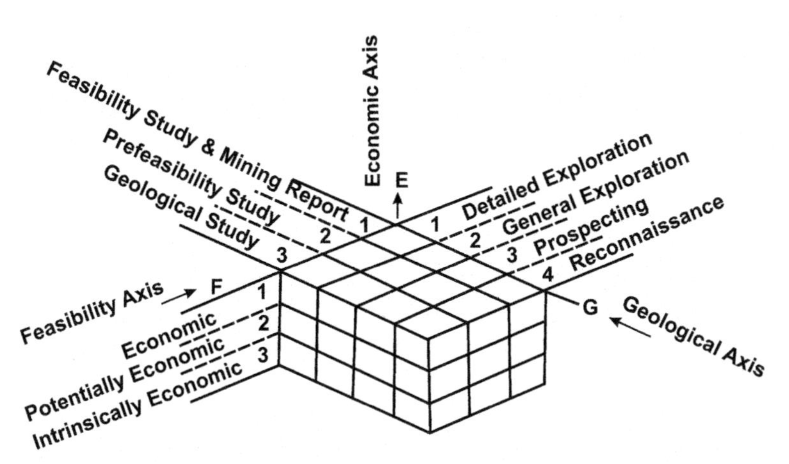

FIGURE 8.17 Resource and reserve scheme by UNFC system adopted by many countries including Government of India.

8.4.4. JORC Classification Code

The Minerals Council of Australia (MCA), The Australian Institute of Mining and Metallurgy (The AusIMM), and The Australian Institute of Geoscientists (AIG) established the Australian JORC for public reporting of Exploration Results, Mineral Resources and Ore Reserves. The scheme was formulated on the basic principles of transparency, materiality and competency. The other organizations represent on JORC are the Australian Stock Exchange (ASX), Securities Institute of Australia (SIA) and incorporated into the New Zealand Stock Exchange (NZX) listing rules. All exploration and mining companies listed in ASX and NZX are required to comply with JORC Code and regulate the publication of mineral exploration reports on the ASX. Since 1971 the Codes are being effectively updated for comparable reporting standards introduced internationally. The JORC Code applies essentially to all solid mineral commodities including diamond and other gemstones, energy resources, industrial minerals and coal. The general relation between Exploration Results, Mineral Resources and Ore Reserves classifies tonnage and grade estimates. The format reflects the increasing levels of geological knowledge and rising confidence. It takes due consideration of mining, metallurgical, technical, economic, marketing, legal, social, environmental and governmental factors. The scheme imparts a checklist for authenticity at each level.

TABLE 8.5 Example of UNFC Codification System

Economic axis	Feasibility axis	Geological axis	Code
Economic	Feasibility Study and Mining Report	Detailed Exploration	111
Economic	Pre-feasibility Study	Detailed Exploration	121
Economic	Pre-feasibility Study	General Exploration	122
Potentially Economic	Feasibility Study and Mining Report	Detailed Exploration	211
Potentially Economic	Pre-feasibility Study	Detailed Exploration	221
Potentially Economic	Pre-feasibility Study	General Exploration	222
Intrinsically Economic	Geological Study	Detailed Exploration	331
Intrinsically Economic	Geological Study	General Exploration	332
Intrinsically Economic	Geological Study	Prospecting	333
Intrinsically Economic	Geological Study	Reconnaissance	334

Mineral resources are concentration or occurrence of mineral prospects that eventually may become sources for economic extraction. It is placed in the Inferred category. Mineral Reserve on the other hand is the economically minable part of Measured and/or Indicated ore. It includes dilution and allowances on account of ore losses, likely to occur when the material is mined. The relationship between mineral resources and mineral reserves is presented in Fig. 8.18.

Reporting of Exploration Results include total database, sufficient information, clear, unambiguous and understandable non-misleading reports generated by exploration programs that may be useful to the investors. The report includes statements of regional and deposit geology, sampling and drilling techniques, location, orientation and spacing, core recovery, logging, assaying including reliability and cross-verification, 3D size and shape, diagrams, estimation methods employed, mineral tenements and land tenure status. It should also include exploration done by other agencies, baseline environmental reports, nature and scale of planned further work.

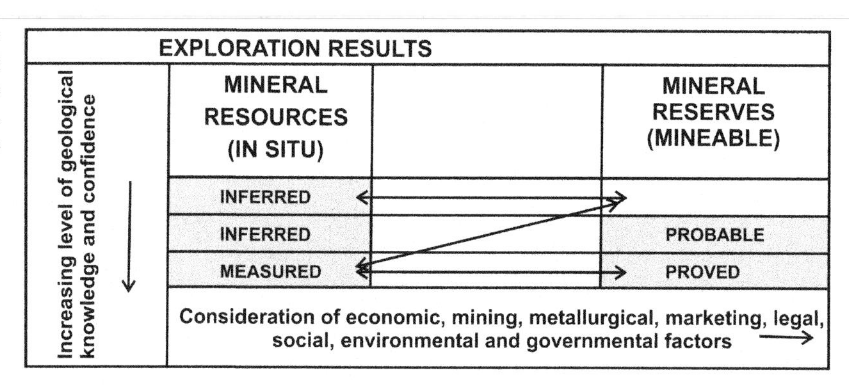

FIGURE 8.18 JORC CODE developed by professionals of Australian Institute of Mining and Metallurgy (The AusIMM) showing relationship between Mineral Resources and Mineral Reserves. JORC compliance organizations are registered with ASX (Source: modified after www.jorc.org).

Reporting of Mineral Resources and Ore Reserves would be comprised of database integrity, location, geological characteristics, continuity, dimension, cutoff parameters, bulk density, modeling techniques, quantity, grades, estimated or interpreted from specific geological evidence and knowledge, accuracy, confidence and reviews, mining and metallurgical factors and assumptions, cost and revenue factors and market assessment.

The Code applies to the reporting of all potentially economic mineralized material in the future. This includes mineralized-fill, remnants, pillars, low-grade mineralization, stockpiles, dumps and tailings where there are reasonable prospects for eventual economic extraction in the case of Mineral Resources and where extraction is reasonably justifiable in the case of Ore Reserves.

The JORC code is now well accepted in Australia and New Zealand. In recent years it has been used both as an international reporting standard by a number of major international exploration and mining companies and as a template for countries in the process of developing or revising their own reporting documents, including the United States of America, Canada, South Africa, and the United Kingdom/Europe, South America including Mexico, Argentina, Chile and Peru.

8.4.5. Canadian Resource Classification Scheme

The mineral resource classification scheme in Canada (Fig. 8.19) is known as National Instrument 43-101 (the "NI 43-101") used for standards of disclosure of scientific and technical information about mineral projects within the country. The NI covers metallic minerals, solid energy products, bulk minerals, dimension and precious stone, and mineral sands commodities. The NI is a codified set of rules and guidelines for reporting mineral properties owned or explored by national or foreign exploration and mining companies listed into Stock Exchanges: the Toronto Stock Exchange (TSX) Venture Exchange TSX, TSX, Canadian Securities Administrators (CSA), ASX, Johannesburg Stock Exchange (JSE) and London Stock Exchange. The NI is broadly comparable and

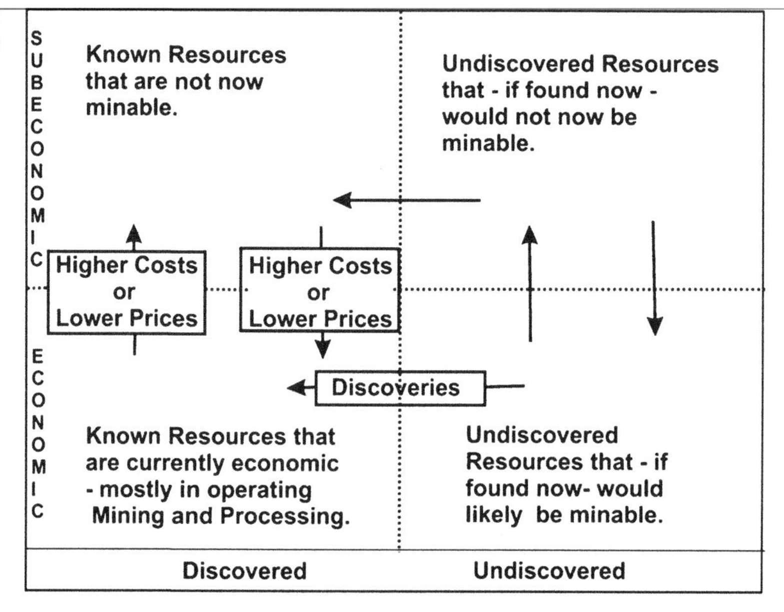

FIGURE 8.19 Schematic view of Canadian mineral resource classification scheme (Source: compiled after many).

interchangeable to JORC and the SAMREC Code. The NI 43-101 ensures that misleading, erroneous or fake information relating to mineral properties is not published and promoted to investors on the Stock Exchanges within the country overseen by the Canadian Securities Authority. The reporting format includes scientific or technical information on mineral resource or mineral reserve of the property.

8.4.6. Comparison of Reserve Classification

Conventional reserve classification system is plain and simple representation of the status of mining and other category of reserve and resources. It is more of qualitative depiction of reserves and easily understandable by small mine owners and common users without having advance knowledge of the trade. It has better applicability within the undeveloped countries and not exchangeable in "true sense of meaning of the term" with geologists, mining engineers, and others operating in the mineral field in developed and developing part of the globe.

The USGS/USBM mineral resource classification system conveys a common classification and nomenclature, more workable in practice and more useful in long-term public and commercial planning. The objectives are based on the probability of discovering new deposits, developing economic extraction processes for currently unworkable deposits and knowing immediate available resources. It believes in continuous resources reassessment with new geological knowledge, progress in Science and Technology (S & T) and Research and Development (R & D) and changes in economic and political conditions. The departments monitor the need, understanding and classification of mineral resources all over the world and expect it to be universally accepted system. The classification of mineral and energy resources is necessarily arbitrary, because the definitional criteria do not always coincide with natural mineralization boundaries. The system can be used to report the status of mineral and energy-fuel resources for the nation or for specific areas.

The UNFC is universally applicable scheme for classification/evaluation of mineral reserves and resources. Most importantly, it allows a common and necessary international understanding of these classification/evaluation. The system is designed to allow the incorporation of currently existing terms and definitions into this framework and thus to make them comparable and compatible. This approach has apparently been simplified by clearly indicating the essential characteristics of extractable mineral commodities in market economies, notably (i) increasing level of geological knowledge; (ii) field project status and feasibility and (iii) degree of economic/commercial viability. This resource classification system is unique and outstanding.

The UNFC is a flexible system that is capable of meeting the requirements for application at national, industrial and institutional level, as well as to be successfully used for international communication and global assessments. It meets the basic needs for an international standard required to support rational use of resources, improve efficiency in management, and enhance the security of both energy supplies and of the associated financial resources. The classification will assist countries with transition economies in reassessing their mineral resources according to the criteria used in market economies.

The classification has given maximum importance to commercial aspects suitable to planners, bankers and other financial institutions. But the small mine owners may find difficult to adopt the system and file the data in national mineral inventory. In the developed nations, very large mineral areas are exploited by fully mechanized method of mining, and sophisticated computerized equipments are used for data acquisitions at mine site. In under developed/developing countries, a vast majority of mining areas are relatively small and are exploited by manual methods. It will be unjustified to assume and expect from small entrepreneur to generate data in the format required as per UNFC. Exploration agencies, without adequate technical knowledge on economic investment decision, may find it difficult to classify resource and reserves. Varied nature of the mineral resource database available in countries across the world as such makes it difficult to evaluate the global mineral resources under a uniform matrix, since not all deposits are equally well known and the degree of exploration varies to a great extent. Any classification must meet first the local needs. Frequent changes and modification may not achieve the very objective.

The strengths of JORC classification are its clarity, transparency, materiality and competency. The reporting system for Exploration Results, Mineral Resources and Ore Reserves is exhaustive with checklist at each level. The database format includes the increasing levels of geological knowledge acquired during successive exploration phase attaining higher confidence. It also takes into account rational reflection of mining, metallurgy, technical, economic, marketing, legal, social, environment and governmental issues. The reporting domain is a complete documentation of exploration input, mineability, extraction recovery and economic viability supported by the essence of pre-feasibility or feasibility study, whichever is possible. It provides a clear vision and mission of the project under consideration for investment decision. All global exploration and mining companies listed in standard Stock Exchanges, particularly in Australia (ASX) and New Zealand (NZX), are required to comply resource-reserve reporting with JORC Code. The current project status with ongoing exploration and other test works are regularly communicated online through Stock Exchanges. The investors and the financial hubs can initiate and expedite mutual commercial transactions accordingly.

The JORC code is in full acceptance in Australia and New Zealand. In recent years it is being used as an international reporting standard by number of major international exploration, mining and financial companies from USA, Canada, South Africa, Europe, and South America including Mexico, Argentina, Chile and Peru.

The Canadian NI 43-101 code of reporting requires significantly more technical disclosure to the Stock Exchange by code originated from the Canadian Securities Authorities. The equivalent JORC is primarily a code for reporting the status of a mineral resource derived by an independent mineral industry body formed from industry professional associations.

Finally, constant efforts are to be made to simplify, harmonize and unify the USGS, UNFC, JORC, NI, Canadian Institute of Mining, Metallurgy and Petroleum (CIM) and Council of Mining and Metallurgical Institutions (CMMI) classification systems for an acceptable reporting standard code for all.

8.5. ORE MONITORING SYSTEM

The status of ore reserves and resources is revised at the end of each calendar or financial year as practiced in the country. The revision is made due to changes on account of annual depletion, status resources and reserves up gradation and addition of reserve with exploration activity during the year. The mine production at the end of each year is taken out of Developed ore. Similarly part of Proved ore is elevated to Developed category due to mine development. Part of Probable ore is added to Proved reserve based on exploration input. The part of Possible resource is enhanced to Probable reserve due to ongoing exploration in the project.

The revised resource-reserve table provides realistic status of minable ore to the management and managerial staff at all levels. This enables to draw short- and long-term strategy for mining operation toward planned production and exploration program for enhancing the resources to various category for sustainable mine life.

8.5.1. Representation of Mine Status

The ore reserve status of a producing mine can be depicted on a longitudinal vertical projection showing the block-wise reserves and status of individual stope/production center (Fig. 8.20). This type of projection will act as a management tool and to review the scheduling of production and monitoring activities.

8.5.2. Forecast, Grade Control and Monitoring System

In any producing mine the advance scheduling of ore production is made as information base for higher authority in the planning to technical person in the site to fulfill their

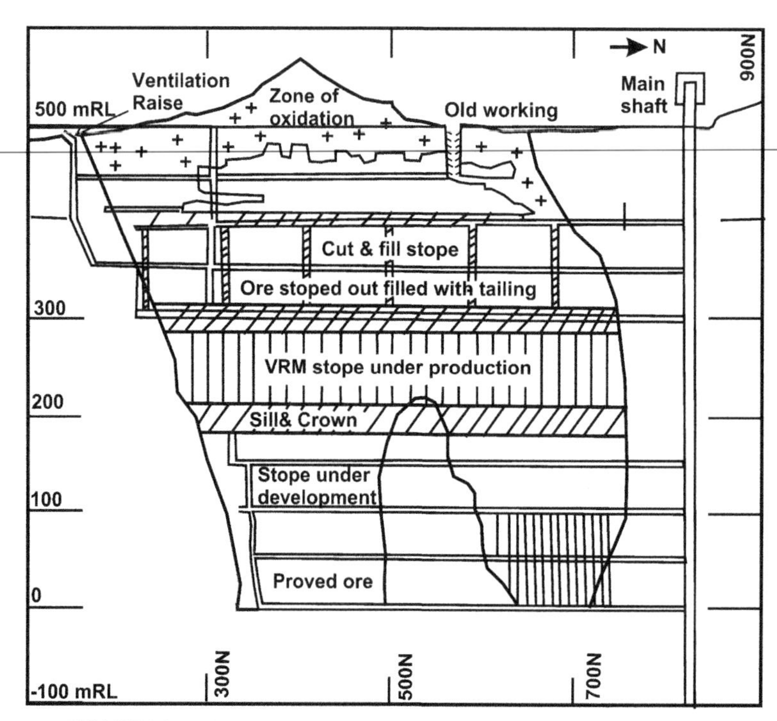

FIGURE 8.20 Schematic views of ore reserve monitoring and mining status.

respective objectives. It is made with respect to tonnes and grade for duration like 5 years/annual/quarterly/monthly/fortnightly/weekly and daily. This is based on optimized supply of ROM ore and to deliver the average mill feed grade over the life of the mine. The scheduling practices are evolved by simple mathematical calculation to complex Linear Programming tools using dynamic system.

The mine production grade can be anticipated by sampling the blast holes of the open-pit benches and sludge of the underground mine drill hole cuttings. Subsequently the grade can be further corroborated by draw point sample from stope face to schedule the grade control operation. The final mine production grade over stipulated period can be achieved by ROM sampling at mine head before it is diverted to various stockpiles.

All these mine samples are collected at coarse fragment sizes. Therefore, the final deposit or precisely the minable grade of the deposit is reconciled by back calculation of mill feed and tailing at −100 mesh sizes after balancing beneficiation recovery.

FURTHER READING

Popoff (1966) [57] elaborated principles and conventional methods of geological reserve classification. GSI, 1980, introduced the National Classification Scheme for the Indian mineral sector. USGS and USBM, 1976 [73] and 1980 [74], developed a common classification and nomenclature, which was published and revised as USGS Bulletin 1450-A—1976 [73], "Principles of the Mineral Resource Classification System of the USBM and USGS". UNFC system is a recent development in reserve categorization (E/2004/37—E/ECE/1416, February 2004). The Minerals Council of Australia (2004) [72] along with other institutions established the Australian JORC [42] for public reporting.

Chapter 9

Statistical and Geostatistical Applications in Geology

Chapter Outline

Statistical and Geostatistical study enhance analytical and interpretative skill of operator to visualize inside characteristics of orebody.

—Author.

9.1. WHY GEOSTATISTICS?

Geologists and mining engineers advocate that the grade of a mineral deposit depends on the location of subblock within the orebody and the sample value of the small block. Thus, conventional methods of resource and reserve estimation [polygonal, triangular, rectangular, cross- and longitudinal sections, level plan and inverse distance (refer Chapter 8)] methods are developed based on spatial position and value of the surrounding ground. There is no objective way to measure the reliability of the estimation. "Experience factors" are often applied to make corrections when the estimate goes wrong. Use of such methods may direct to write-off payable deposits, while uneconomic deposits may be overvalued and never deliver the predicted grades to the mill.

Workers in the field (Davis, 1973) [19] maneuvered this shortcoming by borrowing techniques from formal statistical theory. These include sample mean, range, standard deviation, variance, frequency distribution, histogram plot, correlation coefficients, analysis of variance, t-test, and

Mineral Exploration. http://dx.doi.org/10.1016/B978-0-12-416005-7.00009-X

f-test, trend analysis and distance inverse for univariate and multivariate elements. The global reserve potential with an overall confidence limit (CL) of the estimation (tonnage and grade) using such techniques can be computed. The classical statistical methods assume that sample values are randomly distributed and are independent of each other. It does not include the inherent geological variance within the deposit and ignores spatial relationships.

Krige (1951 [47], 1962 [48]) and subsequently Matheron (1971) [54] developed the theory of regionalized variables (RV) to resolve this problem. RV is a random quantity that assumes different values depending on its position within some region. In ore reserve estimation, the orebody is the region, and grade, thickness etc. are the RV. This technique provides the "best linear unbiased estimate" (BLUE) of the reserve in context of both local as well as global base. It also yields a direct quantitative measure of the reliability. The procedure involves creation of a semi-variogram, sample cross-validation and an estimation technique known by "Point" and "Block Kriging".

9.2. STATISTICAL APPLICATIONS

Statistics is a branch of applied mathematics. It deals with development and applications of procedures to collect, organize, analyze, interpret, present and summarize quantitative data for understanding and drawing conclusions. It uses probability theory to estimate population parameters—present including the future prediction. Statistics deal with various types of data. "Univariate" statistics describe the distribution of individual variable. In Earth Science data often show relationship and inter-dependencies between variables and called "Bivariate" or "Multivariate" statistics. These distributions are comparable. The important parameters of classical statistics are the mean ($\overline{X}$), median, mode, variance (σ^2 or S^2), standard deviation (σ or s), and coefficient of variation (CV). In case of statistical applications there should be at least 30 samples for drawing any inference.

The statistical procedures, test of hypothesis and applications involve different assumptions and accordingly grouped into "parametric" and "nonparametric". Parametric statistics fundamentally relies on the assumption that the parent populations are normally distributed and preserve homogeneity of variance within groups. Nonparametric statistics does not require normal distribution or variance assumptions about the populations from which the samples were drawn.

9.2.1. Universe

The "Universe" is the total mass of material in which one is interested and the total source of all possible data. Universe can be characterized by one or more attributes. It may be one-dimensional or multidimensional. The example of a geological Universe is the area of our interest—a stratigraphic province, a mineral deposit with defined boundary. The characteristics of Universe change with change of boundary limits.

9.2.2. Population, Sample and Sampling Unit

"Population" consists of a well-defined set of elements—either finite or infinite. A "sample" is a subset of elements taken from a population. A sampling unit is the part of the Universe on which the measurement is made. It can be a 5 kg of rock or ore sample, a mine-car load of muck sample and a 3 m piece of drill-core. When one makes statement of characteristics of a Universe, one must specify the "sampling unit" to recognize different population and treat them separately. The population of 1 m drill-core, channel or chip samples may be statistically very different from the population of $10 \times 10 \times 10\ m^3$ blocks of ore. The size of a sampling unit is very important. One should never make a statement about a population of grade, without specifying the grade of what—say 2% Cu.

9.2.3. Probability Distribution

Probability is the determination of the probable from the possible. The possible outcome of a random selection is described by the probability distribution of its parameter. When a coin is tossed the possible outcomes are either head or tail, each with an independent equal probability of ½. Similarly, when a perfect dice is thrown, the possible outcomes are either 1, 2, 3, 4, 5 or 6, each with equal probability of 1/6. This is a case of discrete variable, which can assume only integer values, the distribution will associate to each possible value x of probability $P(x)$. The value will be nonnegative and the summation of all possible $P(x)$ will be equal to one. Probability in an applied sense is a measure of the likeliness that a random event will occur.

Most geological events are not discrete. Instead, they have an infinite and continuous number of possible outcomes. The range of possible outcomes is finite. But the exact result that may appear cannot be predicted within the range. Such events are called continuous random variables. For example, one may wish to know the chance of obtaining a grade in the interval of 2-4 or 6-8% in a mineral deposit. In practice, such a distribution will never be known. All that can be done is to compute an experimental probability distribution and then try to infer which theoretical distribution may have produced such an experimental sampling distribution e.g. normal curve.

9.2.4. Frequency Distribution

The measurements of some attributes, say the analytical value of samples of a mineral deposit, can be classified into grade group of 0-1, 1-2, 2-3% and so forth. The frequency of occurrences (number of assay per grade interval) in the different groups is tabulated. This information would constitute a grouping or "frequency" or relative "frequency distribution". The grouped information is plotted with class intervals on the abscissa (x-axis) and frequency on the ordinate (y-axis). It allows proportionate area for occurrence of each frequency against grade interval and provides a visual picture of the representation of data. The plot is called "histogram" (Fig. 9.1). The distribution is a function of sample size. The sampling unit should be mentioned while defining the histogram. The population may be symmetric, asymmetric, normal, lognormal, bimodal, etc. If there are "n" number of samples (Table 9.1) each with a value of x_i, where $i = 1, 2, 3, \ldots, n$, then the samples can be grouped into a number of classes such as in Table 9.2.

TABLE 9.1 Percent Zinc Content of Exploration Drill Hole Samples at 1 m Uniform Length

% Zn	% Zn	% Zn	% Zn	% Zn	% Zn
5.70	3.90	4.50	3.30	3.60	4.60
1.50	2.60	5.50	3.40	4.50	2.20
4.20	4.40	2.40	2.60	7.50	4.70
4.40	3.50	1.40	3.70	3.60	1.20
2.50	0.50	5.50	4.70	0.70	2.40
0.50	5.60	2.60	3.30	1.80	4.30
3.70	2.70	3.40	4.30	1.60	3.60
3.30	5.10	4.60	3.50	2.70	4.50
5.90	3.10	4.80	1.50	2.80	3.50
2.30	1.30	3.60	2.40	5.30	5.40
2.90	4.90	2.10	4.10	3.70	6.50
0.30	3.40	3.20	3.40	3.80	2.30

9.2.5. Normal or Gaussian Distribution

The "normal" or "Gaussian" distribution is a continuous probability distribution that has a bell-shaped probability density function (Gaussian function) or informally the bell curve. The frequency distribution plot of Table 9.2 and Fig. 9.1 depicts the bell-shaped curve and stands for normal or Gaussian distribution of attributes with definite function of spread from the mean (μ). In case of a theoretical normal distribution curve the thumb rule is that 68% of the population or sample assays will be within plus or minus one standard deviation (δ or S) from the population or sample mean (μ or $\overline{X}$). Similarly, 95% of the population or sample values will stretch out between plus and minus two standard deviations (more precisely 1.96) on either side of the mean. The theoretical normal distributions show single mode, but in natural condition the mineral distribution can be of bimodal and multimodal character. Normal distribution is an approximation to describe real-valued random distribution that clusters around a single mean value.

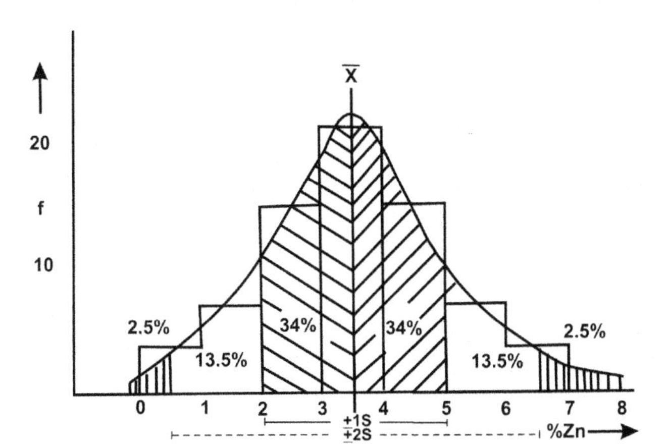

FIGURE 9.1 Relative frequency plot showing the percent area covered by 1 and 2 standard deviation on either side of central value or mean grade of 1 m sample population at Table 9.2 and represents a standard normal probability distribution.

A "cumulative" frequency is a process to understand collective information in a data set that is greater than equal to a particular value. The plot will display graphically the observations by number (n) or percentage (%) or proportion (1). The cumulative frequency curve provides the continuous information instead of discrete number of a particular group. Therefore, a cumulative frequency distribution obtained from "n" number of samples can be transformed into a continuous probability distribution (percentage or proportion) simply by dividing each frequency by "n", the total number of observations (Table 9.2, Fig. 9.2). The total area underneath a cumulative frequency distribution curve is transformed to continuous probability. This approach to probability is intuitively appealing to geologists, because the concept is closely akin to unification.

9.2.6. Minimum, Maximum, Range, Median, Mode and Mean

The "minimum" and "maximum" are the smallest and largest value in the data set respectively. The smallest and the largest number in Table 9.1 is 0.30 and 7.50% Zn.

The "range" (R) is simply the difference between the smallest and the largest values. It can be expressed that the

TABLE 9.2 Frequency Table of 1 m Borehole Assay Data of Table 9.1

Sample class of % Zn					
Lower limit	Upper limit	Midpoint	Number of samples	Cumulative frequency	Cumulative frequency %
0	1	0.5	4	4	5.6
1	2	1.5	7	11	15.3
2	3	2.5	15	26	36.1
3	4	3.5	21	47	65.3
4	5	4.5	15	62	86.1
5	6	5.5	8	70	97.2
6	7	6.5	1	71	98.6
7	8	7.5	1	72	100.0

percent Zn varies between 0.30 and 7.50 with a range of 7.20% Zn. The "mid-range" (MR) is half of the range $(x_{max} - x_{min})/2$.

The "median" (Md) is the middle value of a group of numbers (say, assay values) that have been arranged by size of ascending order or frequency distribution. Median splits the data set in two equal halves. When the total number of observations in the list is odd, the median is the middle entry in the list. In an odd series of 1, 2, 3, 4, 5, 6, 7, 8, 9 → 5 is the median. When the numbers of the list are even, the median is equal to the sum of the two middle (after sorting the list into increasing order) numbers divided by two. In an even series of 1, 2, 3, 4, 6, 7, 8, 9 → (4 + 6)/2 = 5 is the median. The odd and even rule is important. In an order data string half of the population is below the median value.

The "mode" (Mo) is the value which occurs most frequently in a set of numbers or the highest frequency in grouped data or in a sample population. In a series of 1, 2, 3, 4, 4, 4, 5, 5, 6 → 4 is the most frequently occurred number (three times). There may be more than one mode when two or more numbers have an equal or relatively highest number of instances (bi-mode or multimode). A mode does not exist if no number has more than one instance.

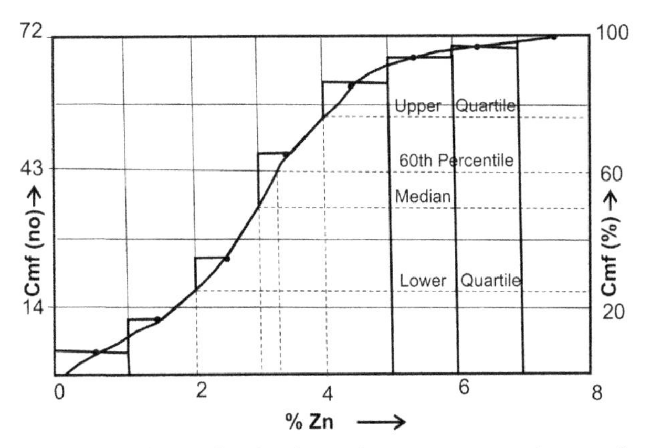

FIGURE 9.2 Cumulative plots by number, percent, proportion, quartile, deciles and percentiles of 1 m sample grades at Table 9.2.

"Quartiles" split the ascending or descending order data set into quarters. A quarter (≤25%) of the data falls below the lower or first quartile (Q_1). Similarly, a quarter (≥75%) of the data lies above the upper or third quartile (Q_3). "Deciles" split the order data set into 10 divisions. One tenth of the data falls below the lower or the first deciles and 1/50th is the fifth deciles and equal to median. "Percentiles" split the data set into 100th divisions and 50th percentile is equal to median. Twenty fifth and 75th percentiles are equal as first and third quartile respectively. The order data set is plotted in the x-axis in normal scale and cumulative frequency percent in y-axis in normal (Fig. 9.2) or logarithmic scale. "Quantiles" are generalization of the data splitting concept. It represents the points taken at regular intervals from the cumulative distribution function of a random variable. A "quantile-quantile" or "q-q" plot is a graphical presentation on which the quantiles from two distributions are plotted versus one another. A q-q plot visualizes and compared two distributions. This is particularly useful when there are sufficient reasons to expect that the two distributions are similar.

The sample "mean", arithmetic mean or average value $(\overline{X})$ is conceptualized on the central tendency of distribution parameters around which it is distributed. It is computed by sum of the values of all observations within the population divided by the number of samples. It is expressed as:

$$\text{Sample mean } (\overline{X}) = (1/n) \times \left\{ \sum_{i=1}^{n} X_i \right\}$$

From Table 9.1,

$$\sum_{i=1}^{n} x_i = 248.80$$

$$n = 72$$

$$\text{Mean}\ (\overline{X}) = 248.80/72 = 3.45\%\ \text{Zn}$$

The computation of arithmetic mean assumes that all the observed values are composed of the same size of sample length. Therefore, unequal sampling units cannot be utilized together to calculate valid statistical parameters. Transformation of raw data to equal length sampling unit is prerequisite in statistical applications.

In asymmetric curve the three measures of central tendency i.e. mode, median and mean will be different (Fig. 9.3A). In case of symmetric curves, such as normal distribution curve, the three measures will coincide (Fig. 9.3B).

9.2.7. Sample Variance and Standard Deviation

A reasonable way to measure sample variation about the mean value is to subtract the mean from the observed value and then sum as follows:

$$\sum_{i=1}^{n} \{X_i - (\overline{X})\}$$

As the algebraic sum of the deviation of a set of numbers from its mean is always zero, the above formula becomes meaningless as a measure of "variance". The average variance is thus expressed as the sum of the squared deviation of the mean from each observed values divided by the number of observations minus 1, representing "degree of freedom". The population sample variance estimator is expressed as:

$$\text{Sample variance}\ (S^2) = \sum_{i=1}^{n} \{X_i - (\overline{X})\}^2/(n-1)$$

$$\text{or,} = \left\{ \sum_{i=1}^{n} X_i^2 - \left(\sum_{i=1}^{n} X_i \right)^2 \right\} \Big/ (n-1)$$

$$\therefore\ S^2\ \text{for Zn distribution} = 154.638/71 = 2.178$$

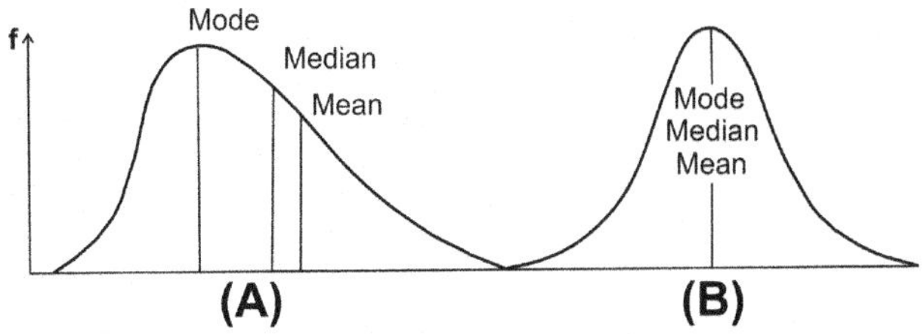

FIGURE 9.3 Schematic diagram showing (A) asymmetric and (B) symmetric probability distribution.

The square root of sample variance is the sample "standard deviation" estimator:

$$\text{Sample standard deviation}\ (S) = \sqrt{S^2} = \sqrt{2.178}$$
$$= 1.476$$

$$\text{Variance of the mean} = S^2/\sqrt{n} = 2.178/\sqrt{72}$$
$$= 0.256$$

$$\text{Standard error of the mean} = S/\sqrt{n} = 1.476/\sqrt{72}$$
$$= \pm 0.174$$

$$\text{Or, CL at 68\% probability level} = S/\sqrt{n} = 1.476/\sqrt{72}$$
$$= \pm 0.174$$

$$\text{CL at 95\% probability level} = S/\sqrt{n} \times 1.96$$
$$= \pm 0.34\%\ \text{Zn.}$$

The CV or relative standard deviation is the ratio between standard deviation and mean. CV often represents the alternative measure of skewness to describe the shape of the distribution. CV is generally ≤ 1 for normal distribution. A CV ≥ 1 indicates the presence of some erratic high sample values that have a significant impact on mean grade of the deposit. The erratic high values may be due to sampling error. These samples are sorted, located and rejected if do not fit in the geological system.

$$\text{CV} = S/(\overline{X}) = 1.476/3.455 = 0.427$$

The computation of mean from grouped data:

$$\text{Mean}\ (\overline{X}) = \sum (F \times \text{mp}) / \sum F = (249/72)$$
$$= 3.458\%\ \text{Zn}$$

where, F = frequency and mp = midpoint.

$$\text{Sample variance}\ (S^2) = \{\sum (F \times \text{mp}^2)/N\} - (\overline{X})^2$$
$$= (1012/72) - (3.458)^2$$
$$= 14.05 - 11.95 = 2.10$$

9.2.8. Probability Plot

The mean and standard deviation can be obtained from the plot of the upper class value limit (grade) versus the cumulative frequency percentages on linear probability paper (Fig. 9.4). A normal distribution will create a straight line on normal probability paper. The grade corresponding to 50% probability line gives the mean. The difference in grade between 84% and 50% probability gives an estimation of standard deviation.

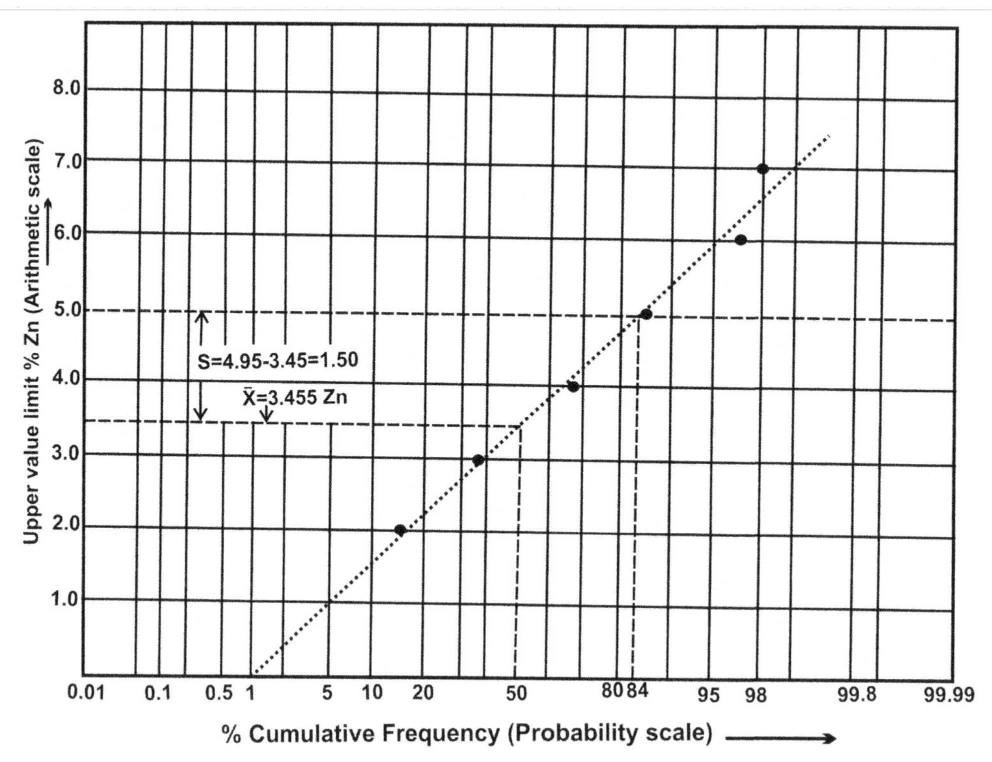

FIGURE 9.4 Probability plot of Table 9.2 for computation of sample mean and standard deviation of Zn grade.

9.2.9. Lognormal or Pareto Distribution

When the data behave exponentially the population often exhibits skewed to the right or left and better described as lognormal or Pareto distribution. The skewed lognormal distribution corresponds to spotty or low-grade mineral deposits. The high values are occasionally encountered while the majority of the samples are of lower grade. A lognormal distribution generally corresponds to normal distribution by logarithmic transformation of these values. The plot of group values often depicts a straight line on log probability paper. Otherwise a three-parameter lognormal paper is used to obtain a straight line. A constant can be added to each class for better result. The mean value for a logarithmic distribution is estimated by the following relationship from Table 9.3, Figs 9.5 and 9.6.

$$\text{Mean} = e^{\alpha} + \beta^2/2 = e^{\text{geometric mean}} \times e^{\beta^2/2}$$

$$\text{Standard deviation } (S) = e^{\alpha} + \beta^2/2 \times \sqrt{(e^{\beta^2} - 1)}$$

$$= \text{Mean} \times \sqrt{(e^{\beta^2} - 1)}$$

$$\text{CL} = (S \times 1.96)/\sqrt{n} \text{ at 95\% probability.}$$

The geometric mean (α) is the logarithm (Ln) of mean grade at 50% point frequency. The geometric standard deviation of the log value (β) is the difference of natural logarithm of grade at 50 and 84% cumulative frequency ordinate. It corresponds to one standard deviation.

9.2.9.1. Computation

A total of 2229 samples from surface borehole at 0.50 m uniform length from Baroi lead-zinc deposit, Rajasthan, India, have been tabulated at Table 9.3.

The frequency diagram and three-cycles logarithmic plot have been depicted at Figs 9.5 and 9.6 respectively.

$$N = 2229$$

$$\text{Ln at } 50\% = \text{Ln}(5.00) = 1.6094$$

$$\text{Ln at } 84\% = \text{Ln}(11.05) = 2.4024$$

$$\beta = \text{Ln } 84\% - \text{Ln } 50\% = 0.79299$$

$$\beta^2 = 0.6288$$

$$\beta^2/2 = 0.3144$$

$$\text{Mean} = e^{(1.6094 + 0.3144)}$$

$$= e^{(1.9238)}$$

$$= 6.848 - 3.0 \text{ (constant added)}$$

$$= 3.85\% \text{ Pb}$$

TABLE 9.3 Grouped Borehole Sample Data of Baroi Lead-Zinc Mine, India

Class interval (% Pb)	f	Cumulative frequency	Cumulative frequency %	Class interval (% Pb)	f	Cumulative frequency	Cumulative frequency %
0-1	811	811	36.38	15-16	54	2183	97.94
1-2	296	1107	49.66	16-17	2	2185	98.03
2-3	229	1336	59.94	17-18	5	2190	98.25
3-4	202	1538	69.00	18-19	3	2193	98.38
4-5	135	1673	75.06	19-20	4	2197	98.56
5-6	102	1775	79.63	20-21	4	2201	98.74
6-7	82	1857	83.31	21-22	—	22.01	98.74
7-8	59	1916	85.96	22-23	5	2206	98.97
8-9	49	1965	88.16	23-24	—	2206	98.97
9-10	44	2009	90.13	24-25	1	2207	99.01
10-11	38	2047	91.83	25-26	2	2209	99.10
11-12	31	2078	93.23	26-27	—	2209	99.10
12-13	20	2098	94.12	27-28	2	2211	99.19
13-14	21	2119	95.07	28-29	2	2213	99.28
14-15	10	2129	95.51	29-30	2	2215	99.37
				>30	14	2229	100

$$\text{Standard deviation} = 3.85 \times 0.932$$
$$= 3.58\%$$
$$\text{CL} = (3.58 \times 1.96)/\sqrt{2229}$$
$$= \pm 0.149\%\ \text{Pb}$$

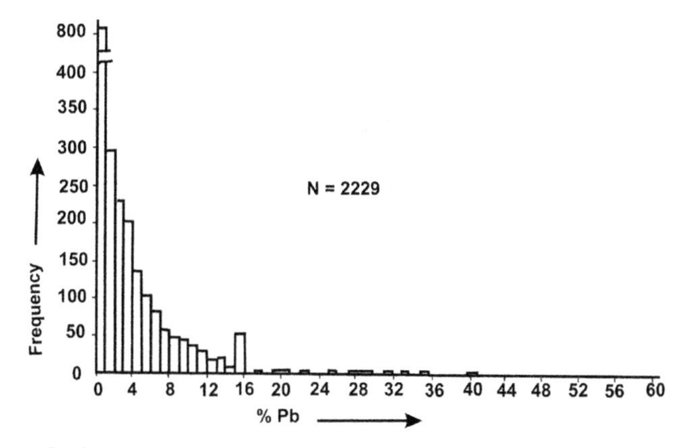

FIGURE 9.5 Frequency plot of percent lead grade at 0.50 m uniform sample length showing lognormal probability with positively skewed distribution.

9.2.9.2. Sichel's "t" Estimate

Sichel formulated tables to simplify the estimation procedure, especially in the case when few samples are available such as:

$$N = 2229$$
$$\alpha = \text{Ln at } 50\% = \text{Ln}(5.00) = 1.6094$$
$$\text{Ln at } 84\% = \text{Ln}(11.05) = 2.4024$$

$$\beta = \text{Ln } 84\% - \text{Ln } 50\% = 2.4024 - 1.6094$$
$$= 0.7929$$
$$\text{Log variance} = \beta^2 = 0.6288$$

Read variance β^2 for equivalent "t" value from mean of lognormal population at standard Sichel's Table B at specified "n". If the variance is in between the table value interpolated value can be computed between higher and lower table values.

$$\text{Mean} = \text{Log value at } 50\% \times t - \text{constant (if any)}$$
$$= 5 \times 1.3638 - 3$$
$$= 6.82 - 3$$
$$= 3.82\%\ \text{Pb}$$

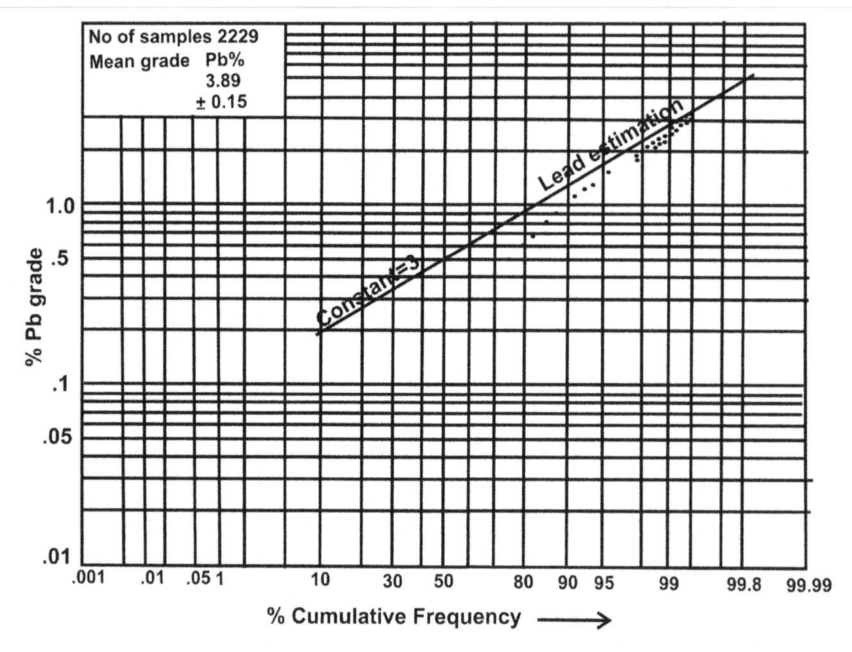

FIGURE 9.6 Plot and computation of $\overline{X}$, S^2 and CL of lead sample values by double logarithmic paper.

9.2.9.3. Confidence Limits

Read lower and upper limit of "*t*" of log variance equivalent at specified "*n*" from standard Sichel's "*t*" Table.

Lower limit = Mean × "*t*" lower value = 3.82 × 0.554 = 3.649

Upper limit = Mean × "*t*" upper value = 3.82 × 1.048 = 4.00

9.2.9.4. Coefficient of Skewness

The quantity of symmetry of the sample distribution is the "coefficient of skewness" or "skewness". The statistics is plotted as histogram and defined as:

$$\text{Coefficient of skewness} = \left[1/n\sum_{i=1}^{n}(X_i - \overline{X})^3/\delta^3\right]$$

9.2.10. Covariance

In typical geological situations, more than one variable is measured on each observational point e.g. percent Pb, percent Zn, percent Fe, g/t Ag and g/t Cd content of mill feed. The data can conveniently be arranged in an $n \times m$ array, where n is the number of observations (months) and m is the number of variables (percent or g/t metals). The statistical parameters such as $\overline{X}$, S^2, S, CL, CV can be computed. However, different variables measured on the same object usually tend to change together in some manner. Variables which have no relation to each other are said to be mutually independent i.e. an increase or decrease in one variable is not governed by a predictable change in another variable. As variable may not be totally or always independent, some measure of their mutual interface or relationship is expected. The procedure employed to compute the variance of a single parameter can be extended to calculate the measure of mutual variability for a pair of parameters. This measure, known as "covariance" (COV), is the joint variation of two variables about their common mean

$$\begin{aligned}\mathrm{COV}(X,Y) &= \left\{\sum_{i=1}^{n}(x_i-\overline{x})\times(y_i-\overline{y})\right\}\Big/(n-1)\\ &= \left\{\sum_{i=1}^{n}x_i\,y_i - \left(\sum_{i=1}^{n}x_i \times \sum_{i=1}^{n}y_i\right)\Big/n\right\}\Big/(n-1)\end{aligned}$$

The mill feed data by months during 2010-2011 of a process plant are tabulated at Table 9.4 for computation of COV between lead and silver content during the year.

9.2.10.1. Computation of COV

	Pb	Ag
$\sum X$	16.32	470
$\sum X^2$	23.749	19000
$\overline{X}$	1.36	39
S	0.3758	7.334

$$\sum_{i=1}^{n} \cdot \mathrm{Pb}_i \times \mathrm{Ag}_i = 663.92$$

$$\begin{aligned}\mathrm{COV(Pb, Ag)} &= \{663.92 - (16.32 \times 470)/12\}/11\\ &= 2.247\end{aligned}$$

TABLE 9.4 Monthly Average Mill Feed Grade of a Zinc-Lead Beneficiation Plant during 2010–2011

Month	% Zn	% Pb	% Fe	g/t Ag
April	3.67	1.67	5.50	49
May	3.34	1.12	4.60	38
June	3.30	0.86	4.30	33
July	3.55	1.28	4.40	41
August	3.76	1.40	3.60	43
September	3.50	2.25	4.30	52
October	3.80	1.40	4.30	45
November	4.12	1.50	4.50	41
December	3.97	1.60	3.70	36
January	3.32	1.12	3.20	33
February	2.92	0.93	3.60	30
March	3.37	1.19	3.80	29

9.2.11. Correlation Coefficient

In order to estimate the degree of interrelationship between variables in a manner not influenced by unit of measurements, the correlation coefficient "r" is used. Correlation is the ratio of the COV of two variables to the product of their standard deviation.

$$r(j,\ k) = \mathrm{COV}_{JK}/S_J S_K$$

Correlation coefficient is a ratio and expressed as unitless number. COV can be equal but cannot exceed the product of the standard deviations of its variables. So the correlation coefficient will range between $+1$ (perfect direct relationship) and -1 (perfect inverse relationship). The spectrum of the "r" exists between these two extremities of perfect positive or negative relationships and passing through zero (no relation) in linear function. The correlation between Pb and Ag from data of Table 9.4 will be:

$$r(\mathrm{Pb, Ag}) = \{\mathrm{COV}_{\mathrm{Pb,\ Ag}}/(S_{\mathrm{Pb}} \times S_{\mathrm{Ag}})\}$$
$$= 2.247\ /\ (0.3758 \times 7.334) = 0.815$$

$$\text{Goodness of fit} = r^2 = 0.66 \text{ or } 66\%$$

9.2.12. Scatter Diagram

The interrelation between two variables can be visualized by plotting independent variable along x-axis and its corresponding dependent variable along y-axis. The plot is known as scatter diagram which provides an instant sense of the expected relationship (Fig. 9.7).

9.2.13. Regression

Once the inter-dependability of two variables is established it will be possible to express the linear mathematical relationship between them such as Y is a function of X i.e. $Y = f(X)$. The linear equation can be stated as:

$$\tilde{Y}_i = b_0 + b_1 X_i$$

where, $\tilde{Y}_i$ is an estimated value of Y_i at specified value of X_i,

Y_i is the dependent or regressed variable,
X_i is the independent or regressor variable.

The fitted line will intercept the y-axis at a point b_0 at a slope of b_1. The equation must satisfy the condition that the deviation of Y_i from the line is minimal such as:

$$\sum_{i=1}^{n} (\tilde{Y}_i - Y_i)^2 = \text{Minimum.}$$

$$b_1 = \left\{ \sum_{i=1}^{n} X_i Y_i - \left(\sum_{i=1}^{n} X_i \times \sum_{i=1}^{n} Y_i \right) \Big/ n \right\} \Big/ \left\{ \left(\sum_{i=1}^{n} X_i^2 - \left(\sum_{i=1}^{n} X_i \right)^2 \right) \Big/ n \right\}$$

$$= \mathrm{SP}_{xy}/\mathrm{SS}_x$$

$$= (\text{Sum of products for COV})/(\text{Sum of squares for variance})$$

$$b_0 = \overline{Y} - b_1\,\overline{X}$$

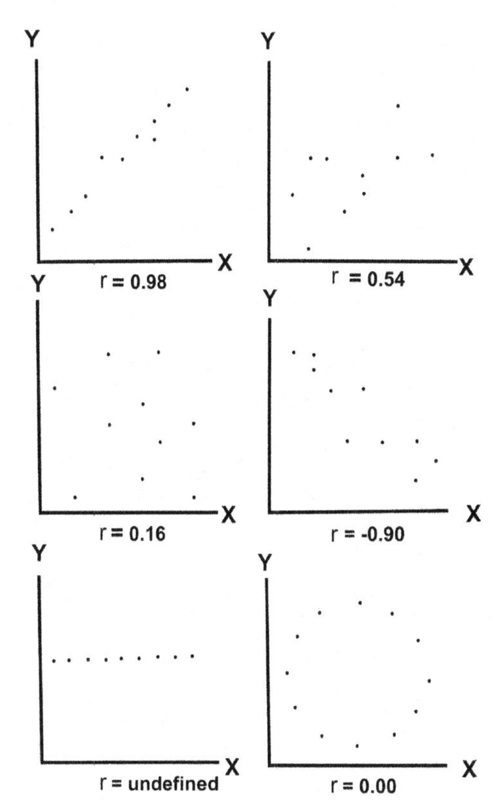

FIGURE 9.7 Plot of scatter diagrams of two variables showing different types of correlations between them depending on mutual inter-dependability.

The silver (dependent) and lead (independent) content from Table 9.4 are plotted in scatter diagram (Fig. 9.8). The regression equation can be computed as:

$$b_1 = (663.92 - (16.32 \times 470)/n)/\{23.749 - (16.32)^2/n\}$$
$$= 24.720/1.554 = 15.907$$
$$b_0 = 39.167 - (15.907 \times 1.36)$$
$$= 17.533$$

∴ the equation is:

$$\text{Ag(g/t)} = 17.533 + (15.907 \times \%\ \text{Pb})$$

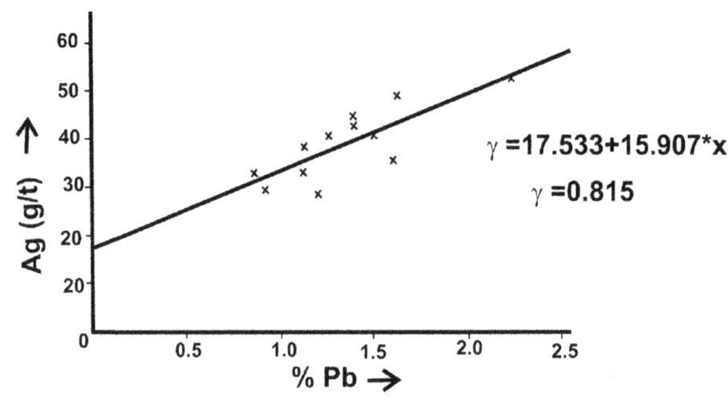

FIGURE 9.8 Plot of scatter diagram showing the regression line of Pb vs. Ag from Table 9.4.

9.2.14. The Null Hypothesis

The null hypothesis is a typical statistical theory which suggests that no statistical relationship and significance exists in a set of given single observed variable, between two sets of observed data and measured phenomena. The hypotheses play an important role in testing the significance of differences in experiments and between observations. H_0 symbolizes the null hypothesis of no difference. It is presumed to be true until statistical evidence nullifies it for an alternative hypothesis. Let us take two sets of mill feed silver samples from Table 9.5 and wish to compare the mean grade between set and population and between two sets. The null hypothesis presumes and will state as:

$$H_0 : \mu_1 = \mu_0$$

where,

H_0 = the null hypothesis of no difference
μ_1 = the mean of population 1, and
μ_0 = the mean of population 2.

The null hypothesis states that the mean μ_1 of the parent population from which the samples are drawn is equal to or not different from mean of the other population μ_0. The samples are drawn from the same population such that the

TABLE 9.5 Average Monthly Mill Feed Silver Grade (g/t) of a Zinc-Lead Mine

Sample	Set-I	Set-II	Difference (d) Set (I-II)	(Difference)^2
April' 2010	49	51	−2	4
May	38	31	7	49
June	33	33	0	0
July	41	41	0	0
August	43	40	3	9
September	52	43	9	81
October	45	51	−6	36
November	41	32	9	81
December	36	35	1	1
January' 2011	33	30	3	9
N	10	10	10	10
SUM ($\sum$)	411	387	24	270
AVG ($\overline{X}$)	41.1	38.7	2.4	27.0
VAR (S^2)	40.77	61.57	23.60	1078.67
STD (S)	6.38	7.85	4.86	32.84

variance and shape of the distributions are also equal. Alternative statistical tests as t, F and chi-square can only reject a null hypothesis or fail to reject it. The evidences can state that the mean of the population from which the samples are drawn does not equal to the specified population mean. It is expressed as:

$$H_0: \mu_1 \neq \mu_0$$

9.2.15. The *t*-Test

The "t"-test follows parametric statistics assuming certain conditions of normal probability distribution and equality of variance between groups and parent populations. The t-test or Student's t-test is based on t-probability distribution which is similar to normal distribution with widespread and dependent upon the size of sample taken. This is powerful to test the hypothesis and useful to establish the likelihood that a given sample could be a member of a population with specified characteristics (mean) or for testing hypothesis about the equivalency of two samples. The t-statistics can be computed by:

$$t = (\overline{X} - \mu_0)/(S/\sqrt{n})$$

where, $\overline{X}$ = mean of the sample
μ_0 = hypothetical mean of population
n = number of observation, and
S = standard deviation of observations.

If the computed value exceeds the table value of "t" at specific degree of freedom and level of significance and lies in the critical region (Fig. 9.9) or region of rejection the null hypothesis ($\mu_1 \leq \mu_0$) is rejected, leaving the alternative that $\mu_1 > \mu_0$. If the computed value of "t" is less than table value, it can only be stated that there is nothing in the sample to suggest that $\mu_1 > \mu_0$, but cannot specify that $\mu_1 < \mu_0$. This indecisiveness is a consequence of the manner in which statistical tests are formulated. They can demonstrate, with specified probabilities, what things are not. They cannot stipulate what they are.

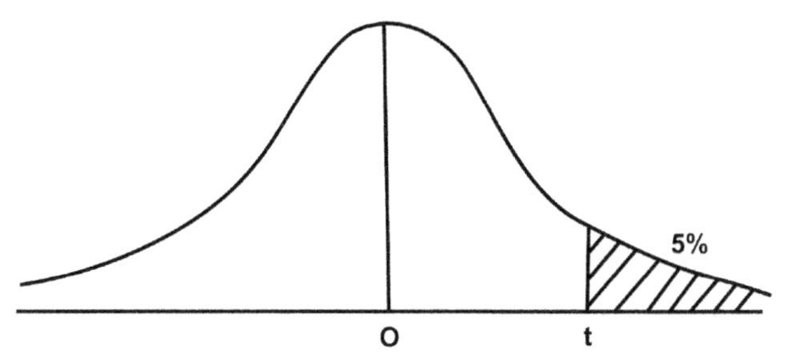

FIGURE 9.9 Schematic diagram showing analysis of t-test.

The average mill feed silver grade (g/t) of a zinc-lead mine over a period of 10 months is tabulated in Table 9.5. There are two sets of samples collected at different time intervals. The hypothetical mean grade (μ_0) of the deposit is 40 g/t. The "t" value of samples at Set-I is computed to test the hypothesis of likelihood of the means of the samples ($\overline{X}$) and population (μ_0).

$$t = (\overline{X} - \mu_0)/(S/\sqrt{n}) = (41.10 - 40.00)/(6.38/\sqrt{10})$$
$$= (1.10/2.02) = 0.54$$

The calculated value of 0.54 is much less than the table value of t (1.83) for nine degrees of freedom and the 5% ($\alpha = 0.05$ for one tail test) level of significance. The t computed value lies away from the region of rejection. The test is indecisive to suggest that the population mean is greater than 40 g/t.

Critical Values of "t" for various Degrees of Freedom and Selected Levels of Significance from Table-21, Penguin-Honeywell Book of Tables F.W. Kellaway (ed.), 1968, [45].

The two sets of samples (I and II) collected at different time intervals are unpaired and independent. The "pooled t" estimate tests the null hypothesis and the equality of the mean of two normally distributed populations. The pooled test statistical form stated as:

$$t = (\overline{X}_1 - \overline{X}_2)/\left\{\text{Sp} \times \sqrt{(1/n_1 + 1/n_2)}\right\}$$

where, Sp is the pooled estimate of the population standard deviation based on the both sample sets and given by:

$$\text{Sp}^2 = \{(n_1 - 1) \times S_1^2 + (n_2 - 1) \times S_2^2\}/(n_1 + n_2 - 2)$$

$$\text{Pooled } t = (41.10 - 38.70)/$$

$$\left[\left\{\sqrt{(9 \times 40.77) + (9 \times 61.57)/18}\right\} \times \sqrt{0.2}\right]$$
$$= 2.40/\sqrt{\{(366.93 + 544.13)/18\}} \times \sqrt{0.20}$$
$$= 2.40/\left(\sqrt{51.17} \times \sqrt{0.20}\right) = 2.40/(7.15 \times 0.44)$$
$$= 2.40/3.146 = 0.76$$

The table value of t for a two-tailed test with 18 degree of nine degrees of freedom representing each set, and the 10% (5% in each tail) level of significance are -1.73 and $+1.73$. The calculated t-value of 0.76 does not fall into either critical region. The null hypothesis cannot be rejected. There is no evidence to suggest that the two samples came from populations having different means.

The "paired t-test" provides a hypothesis test of the differences between population means for a pair of random samples whose differences are approximately normally distributed. Let us assume that the two sets of samples at Table 9.5 are repeat analysis of silver at two different laboratories. The paired t-value can be computed for testing equivalency of the means.

$$\text{Paired } t = \bar{d}/\overline{\text{Sd}}$$

where, d_i = difference of each paired sample

$\bar{d}$ = mean difference

$$\overline{\text{Sd}}^2 = \left\{n \times \sum_{i=1}^{n} d_i^2 - \left(\sum_{i=1}^{n} d_i\right)^2\right\} \Big/ n(n-1)$$

$$\overline{\text{Sd}} = \sqrt{(\text{Sd}^2/n)}$$

From Table 9.5:

$$\bar{d} = 2.4 \quad \sum d = 24 \quad \sum d^2 = 270$$

$$\text{Sd}^2 = \{(10 \times 270) - 576\}/(10 \times 9) = 23.60$$

$$\overline{\text{Sd}} = \sqrt{(23.60/10)} = \sqrt{2.36} = 1.536$$

$$\text{Paired } t = 2.40/1.536 = 1.56$$

The estimated "t" value of 1.56 is less than table value of 1.83 for nine degrees of freedom and the 5% ($\alpha = 0.05$ for one tail test) level of significance. It does not fall into critical region and it will be concluded that there is no evidence to suggest that the two samples came from populations having different means. Three assumptions are necessary to perform this test:

(1) Both samples were selected at random,
(2) Populations from which the samples were drawn are normally distributed, and
(3) Variances of the two populations are equal.

9.2.16. The *F*-Test

The F-test pertains to parametric statistics and follows the assumptions and conditions of normal probability distribution of the populations from which the samples are drawn. The equality of variances of two data sets can be found and tested based on a probability distribution known as the F distribution. "F" is the ratio between the variances of two populations. F distribution is dependent upon two values of one associated with each variance in the ratio. The samples are expected to be randomly collected from a normal distribution to satisfy the condition of F distribution. The equation is stated as:

$$F = (S_1^2/S_2^2)$$

where, $S_2 > S_1$

If the computed F value exceeds the table value it is concluded that the variance in the populations is not same in the two groups. If the calculated value is less than the table value we would have no evidence that the variances are different.

9.2.17. The Chi-Square Test

Chi-square (χ^2) test relates to nonparametric statistics, typically easy to compute and capable to process data measured on nominal scale (categorical variables). χ^2 is a statistical test frequently used to compare observed data with data expected to obtain according to a specific hypothesis. If a sample of size "n" is taken from a population having a normal distribution with mean "μ" and standard deviation "α" that may allow a test to be made of whether the variance of the population has a predetermined value. Each observation can be standardized to the standard normal form with "0" mean and unit variance following: $Z_i = (X_i - \bar{X})/S$ for samples or $Z = (X_i - \mu)/\alpha$ for population parameters. If the values of Z are squared to eliminate the sign and summed it formed a new statistic as: $\sum Z^2 = \sum_{i=1}^{n}\{X_i - \mu)/\alpha\}^2$.

The data processing, interpretation and testing the hypothesis are similar to parametric "t"- and F-tests. Chi-square test computes a value from the data using χ^2 procedure. The value is compared to a critical value from a χ^2 table with degree of freedom equivalent to that of the data. If the calculated value is ≥table value the null hypothesis is rejected. If the value is less than the critical value the null hypothesis is accepted. χ^2 is an important test in nonparametric statistics. It is used to test the difference between an actual sample and another hypothetical or previously established population which is expected due to chance or probability. The basic χ^2 equation:

$$\chi^2 = \sum_{i=1}^{n} (O_i - E_i)^2/E_i$$

where, O_i = observed frequency of ith class

E_i = expected frequency of ith class.

Example: A balanced coin tossed for 20 times and yielded 12 times head and 8 times tail against expected probability value of 10 and 10 (Table 9.6).

TABLE 9.6 Results of Coin Toss and χ^2 Calculation

	Observed (O_i)	Expected (E_i)	$(O_i - E_i)^2$	$(O_i - E_i)^2/E_i$
Head	12	10	2	0.40
Tail	8	10	−2	0.40
Total	20	20		0.80

Critical Values of "χ^2" for various Degrees of Freedom and Selected Levels of Significance from Table-24, Penguin-Honeywell Book of Tables F.W. Kellaway (ed.), 1968, [45].

The calculated value of χ^2 is 0.80 (without any correction factor) is much less than the table value of 3.84 at 1 degree of freedom is not statistically significant at 0.05 level. It can be concluded that the deviation between observed and expected frequency is due to sampling error or by chance alone.

9.2.18. Analysis of Variance

The analysis of statistical variance (ANOVA) uses comparison of group of samples. The observed variance in a particular variable is partitioned into components attributable to different sources of variation. There are several models and procedures to perform the analysis of variances due to effects of categorical factors. The models are One-Way ANOVA, Multifactor ANOVA, Variance Component Analysis and General Linear Models. One-Way ANOVA deals with single categorical factor equivalent to comparing multiple groups of data set. Multifactor ANOVA uses more than one categorical factor arranged in crossed pattern. Variance Component Analysis model utilizes multiple factors arranged in hierarchical manner. Linear model is a complex type uses categorical and quantitative, fixed and random factors arranged in crossed and nested pattern. The general format of the one-way model is presented at the Table 9.7.

Five ore samples are collected from a zinc-lead deposit. The samples are analyzed for percent Pb content at six standard laboratories (Table 9.8). Each assay value of one sample analyzed at different laboratories is called a replicate. In one-way analysis, total variance (TV) of the data set is broken into two parts: variance within each set of replicates and variance among the samples.

The TV of all observations (all replicates of all samples) can be computed as:

$$SS_T = \left(\sum_{j=1}^{m} \sum_{i=1}^{n} Xij^2 - \sum_{j=1}^{m} \sum_{i=1}^{n} Xij^2 \right) \Big/ N$$

$$= 96.66 - (52.77)^2/30 = 3.84$$

TABLE 9.7 Model of One-Way Analysis of Variance of Grouped Data

Source of variation	Sum of squares	Degree of freedom	F-test
Among samples	SS_A	$m-1$	MS_A/MS_W
Within replicates	SS_W	$N-m$	
Total variation	SS_T	$N-1$	

where, m = number of samples, n = number of replicates and $N = m \times n$.

TABLE 9.8 Percent Pb Assay Value of Five Samples Analyzed at Six Standard Laboratories

		$m=5$					
	Replicate (Lab)	Sample-1	Sample-2	Sample-3	Sample-4	Sample-5	
	1	1.92	1.87	1.25	2.03	1.99	$N=5\times6=30$
	2	1.87	1.43	1.43	2.25	2.43	
$n=6$	3	2.13	2.02	0.87	1.76	1.76	
	4	1.65	1.76	1.14	1.84	2.02	
	5	1.73	1.93	0.95	1.59	1.84	
	6	2.24	1.61	1.65	1.90	1.91	
	$\sum x$	11.54	10.62	7.29	11.37	11.95	$=52.77$
	$\sum x^2$	22.45	19.03	9.29	21.80	24.07	$=96.66$

The variance among the samples and within replicates:

$$SS_A = \left\{ \sum_{j=1}^{m} \sum_{i=1}^{n} Xij^2/n \right\} - \left(\sum_{j=1}^{m} \sum_{i=1}^{n} Xij \right)^2 /N$$

$$= 95.20 - (52.77)^2/30 = 2.37$$

$$SS_W = SS_T - SS_A = 3.84 - 2.37 = 1.47$$

The results of the analysis of variances as derived from data set at Table 9.8 are given in Table 9.9.

The computed F value of 10.17 is greater than the table value of 2.76 at 5% level of confidence (Fig. 9.10). Therefore, it can be concluded that the variation is not the same in the two groups.

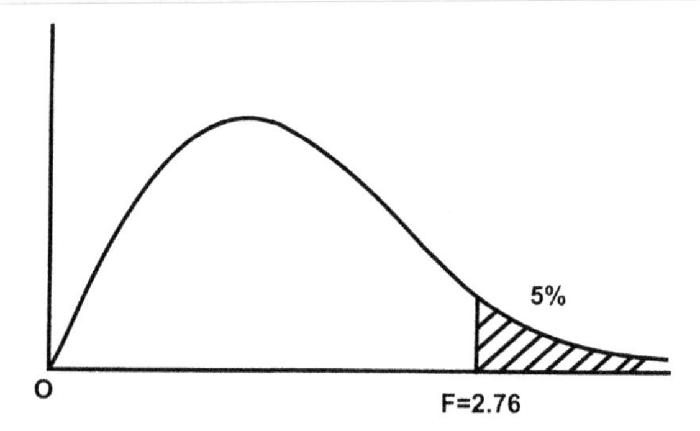

FIGURE 9.10 Analysis from F-test from table value at 5% level of confidence.

9.2.19. Trend Surfaces Analysis

The trend surface mapping is a mathematical technique of computing an empirical 2D "plane" or 3D "curves", "contours" and "wire mesh" derived by regression method. It provides the general structure of spatial variation present in the data set. It is a linear function of the geographic coordinates of a set of scattered observations and estimated new values on a regular grid points/cells. The estimated values must minimize the deviations from the trend. It is based on the utilization of total data set to construct the global functional relation and hence it is a wide range operator. Local extreme values will affect the trends. It is a global pattern recognition technique in geology, geochemistry, geophysics, environmental sciences. It is not fit for local grade estimation. The mathematical equation expressing the functional relationship can be stated as:

$$Zn = a_0 + a_1x + a_2y + e \quad \text{(first degree equation)}$$

$$Zn = a_0 + a_1x + a_2y + a_3x^2 + a_4xy$$
$$+a_5y^2 + e \quad \text{(second degree equation)}$$

and by similar types of equations, where,

Zn = dependent variable (sample value)

x and y = independent variables (location of sample coordinates)

e = random error component,

$a_0, \ldots, a_n$ = unknown coefficients.

The number of coefficients are dependent on the degree (NDEG) of the polynomial used and can be computed as {(NDEG + 1) × (NDEG + 2)}/2. The usage of higher degree of polynomial attempts to reduce the noise level in the data or in other words minimizing the sum of squares of the residuals and allow for the definition of the trend in a unique manner. However, it is generally observed that after reaching a certain degree, the linear fit becomes static. This is because of error propagation and the noisy nature of the large matrices.

In a first-degree equation we need three normal equations to find the three coefficients:

$$\Sigma Zn = a_0 n + a_1 \Sigma x + a_2 \Sigma y$$

$$\Sigma x\, Zn = a_0 \Sigma x + a_1 \Sigma x^2 + a_2 \Sigma xy$$

$$\Sigma y\, Zn = a_0 \Sigma y + a_1 \Sigma xy + a_2 \Sigma y^2$$

The equation can be written into matrix form such as:

$$\begin{bmatrix} n & \sum x & \sum y \\ \sum x & \sum x^2 & \sum xy \\ \sum y & \sum xy & \sum y^2 \end{bmatrix} \times \begin{bmatrix} a_0 \\ a_1 \\ a_2 \end{bmatrix} = \begin{bmatrix} \sum Zn \\ \sum x\,Zn \\ \sum y\,Zn \end{bmatrix}$$

Solving this set of simultaneous equations we obtain the coefficients for the best fitting linear trend surface.

The trend surface technique is illustrated by a 2D case study. A zinc-lead orebody extends along N42°W-S42°E over a strike length of 90 m. It is lens shaped and widest at the center. Width varies between 2 and 16 m with an average of 7 m. It pinches at both ends. The mineralization dips to west and plunges 55° northwesterly. The orebody has been drilled from underground at four levels at 15 m interval in addition to channel sampling at levels. The mineralized intersections through diamond drill holes indicated that it is predominantly zinc rich with moderate iron and subordinate amount of lead. The grades vary

TABLE 9.9 Results of Analysis of Variance

Source of variation	Sum of squares	Degree of freedom	Mean squares	F-test
Among samples	2.37	4	0.59	10.17
Within replicates	1.47	25	0.058	
Total variation	3.84	29		

between 3.96 and 11.18 with average of 6.87% Zn, 0.00 and 5.98 with an average of 0.40% Pb, and 2.44 and 15.25 with an average of 6.05% Fe. The drill and channel intersection values are plotted along a longitudinal section. The dependent variables e.g. Pb, Zn, Fe and width are assumed as a numerical functional relation of criteria variables of x (strike) and y (level) coordinates. Trend surface maps of lead and zinc at fourth degree are generated on longitudinal vertical section (Figs 9.11 and 9.12 respectively) considering local geographic coordinates x and y.

The zinc value has been computed at every observed drill intersection point of the orebody using fourth degree trend surface equation. The comparison between actual sampled and computed values is given in Table 9.10.

Analysis and observations: From the table we obtain the following:

TV = 71.80

Unexplained variance (UV) = 12.10

Explained variance (EV) = (TV − UV) = 59.70

Goodness of fit (GF) = (EV/TV × 100 = 83.15%.

The GF is a measure of the efficiency of the trend equation applied and depends on the input data and inherent structure. Even with large database, extreme values will lower the GF. It may be necessary to point out that GF is a useful measure for a homogeneous population and not appropriate for skewed or mixed one. Hence, lack of significance or having high significance needs to be supported by geological interpretation. It is sometimes used for determining functional relationship of predictor variates with the criteria variates to generate dependent parameter over mapped region by interpolation at regular grid. The accuracy of map generation is centered on the midpoint of the 2D grid/grid cells. The computed trend value and actual values may show a strong divergence away from the midpoint. The margins are the most inaccurate zones. Extrapolation of trend equation beyond the control point will result in very erroneous values.

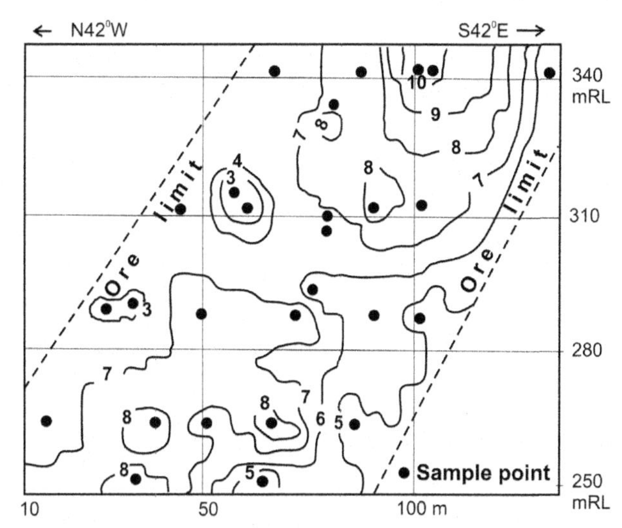

FIGURE 9.12 Fourth degree trend surface map of zinc content of orebody 7W, Balaria deposit, Rajasthan, India.

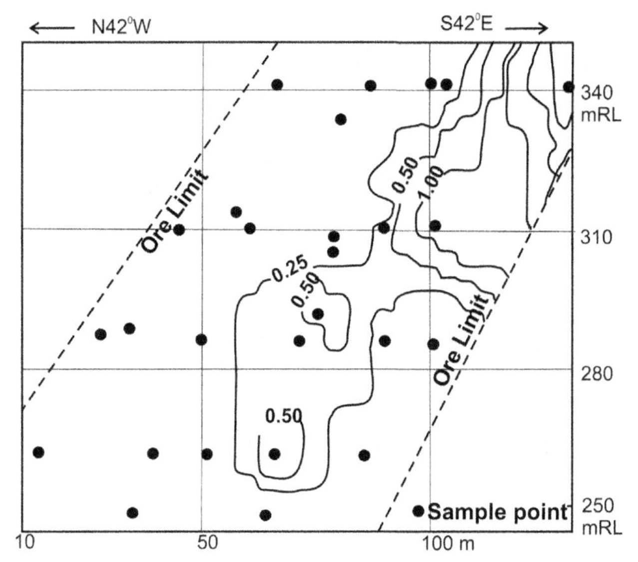

FIGURE 9.11 Fourth degree trend surface maps of lead content of orebody 7W, Balaria deposit, Rajasthan, India.

The trend surface map can indicate (a) possible continuity of mineralization in depth and strike direction, (b) inherent pattern of mineralization like metal zoning and association and (c) the grain of longer and shorter axis of variations. It may often generate false anomalies since the separation of noise into positive and negative set is closely linked and they together sum to zero. Smoothening of higher values and enhancement of lower values is an inbuilt part of the procedure. Subset trend maps show no relation to a composite trend maps. In spite of those constraints, these are utilized for the study of generalized pattern of global variation.

9.2.20. Moving Average

The moving average is a technique of smoothening of data of occasionally erratic high and low values. It is an estimator operating within a pattern of adjacent values. The operator governs the weighting factor. The pattern may be linear combinations, circular, rectangular, square window etc.

9.2.20.1. Moving Weighted Average of Block Mean

A sizable ore block is removed during mining operation. The minimum size depends on geological behavior and extraction method. It will vary from deposit to deposit and mine to mine. However, large blocks are less subject to extreme grade fluctuation due to volume-variance relationship.

TABLE 9.10 Computed Value of Zinc Content of Orebody 7W, Balaria Deposit, Rajasthan, India

Level	Latitude	% Zn (sampled)	% Zn (computed)	Level	Latitude	% Zn (sampled)	% Zn (computed)
343	68	6.11	6.05	288	102	4.85	4.27
343	86	7.17	7.46	263	15	6.44	6.67
343	102	11.18	10.24	263	38	8.28	8.26
343	105	10.01	10.62	263	54	6.81	7.92
343	134	4.41	4.49	263	67	8.54	7.04
314	46	6.63	6.75	263	84	4.51	4.81
314	60	6.37	6.21	250	5	7.30	7.15
314	82	7.70	7.55	250	35	8.22	8.23
314	91	8.73	7.88	250	64	4.13	4.27
314	101	7.44	7.17	317	59	6.34	6.17
288	28	6.20	6.19	286	34	6.29	6.41
288	49	7.58	6.98	330	81	6.88	7.18
288	74	8.74	7.92	309	80	6.11	7.56
288	90	5.34	6.96				

Data source: Unpublished PhD Thesis, Haldar, 1982.

The weight (W) of a block is to be determined to establish a moving average based on block mean value ($\tilde{Y}$). A block moving average design is shown in Fig. 9.13. $\tilde{Y}$ will be the estimated grade of block-"1" from the means of exploration sample in blocks 1-9. $\tilde{Y}$ is the estimated grade of ore to be produced from the center block and $\overline{Y}_1$ is the mean assay value of exploration core in the same block. Once the equation is obtained the sample design will move to unknown region. The equation is determined from the past records of the grade of ore removed from developed region of the same mine.

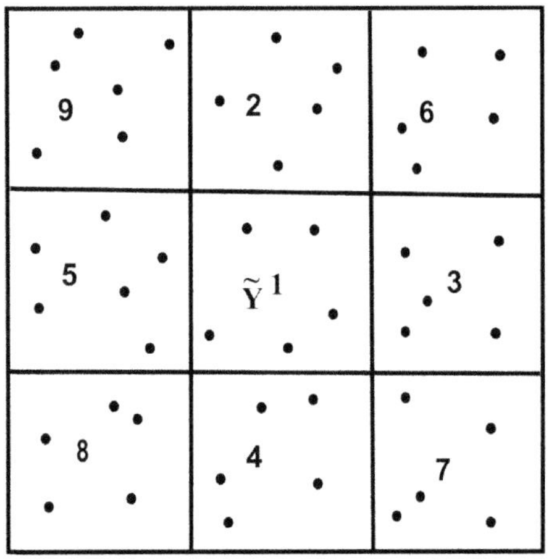

$$\tilde{Y}_{ij} = \textstyle\sum_{k=1}^{n} W_k Y_k$$

$$Y = b_o + \textstyle\sum_{i=1}^{n} b_i \overline{Y}_i$$

$$Y = \text{Actual from past record}$$

$$\overline{Y} = \text{Mean of exploration data}$$

$$b = \text{Coefficients}$$

FIGURE 9.13 Diagram showing concept of "Block" estimation by moving average method.

The estimated $\tilde{Y}$ is based on the weighted sum of adjacent observation Y.

$$\tilde{Y}ij = \textstyle\sum_{k=1}^{n} W_k Y_k$$

$$\tilde{Y} = \text{estimated value,}$$

$$W = \text{weight}$$

$$Y = \text{adjacent block value}$$

The sum of total weight must be equal to 1,

$$\textstyle\sum_{k=1}^{n} W_k = 1$$

9.2.20.2. *Moving Arithmetic Average*

The arithmetic average of adjacent samples on either side controlled by a window can give an estimation for

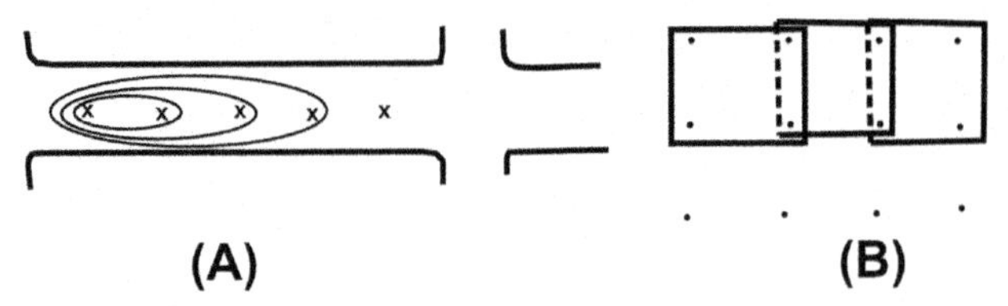

FIGURE 9.14 Concept of moving arithmetic average grade computation of (A) underground workings and (B) geochemical grid samples on surface.

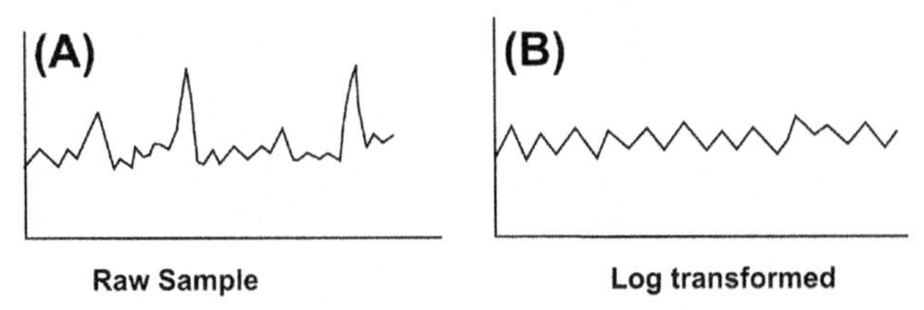

FIGURE 9.15 Transformation of (A) fluctuating raw samples to (B) smooth values.

non-heterogeneous ore distribution as in case of drive or grid samples (Fig. 9.14A and B).

9.2.20.3. Other Mathematical Technique

The occurrence erratic high values in the spatial distribution as in case of gold, silver, platinum and similar precious metals can be smoothened by log transformation (Fig. 9.15A and B).

9.3. MISCLASSIFIED TONNAGE

Krige (1951) [47] observed that blocks of ore estimated as high grade using the old traditional method of evaluation constantly indicate a lower true grade after the mill feed grade reconciliation. Similarly, estimated low-grade blocks signify a constant higher true grade. The true versus estimated grades of all blocks are plotted as in Fig. 9.16. An elliptical envelope is observed with a major axis (least square fit) not parallel to the 45° line from the origin.

All the estimated versus actual points would fall on or close to the 45° line if the blocks been estimated correctly. The estimation technique is expected to provide close proximity between computed and actual production sample grade. In practice, a number of blocks would have a similar estimated grade with wide variation from true grade. This indicates that any estimated value has a bias (error) when compared with the true block values except where the line of best fit intersects the 45° line. Although a bias exists, the average of all the block estimates would not necessarily be much different than the true average grade. If we apply a desired cutoff grade, four regions namely, I, II, III and IV will be formed on the estimated true grade distribution.

- Region-I is comprised of all those blocks that have both an estimated and true grade above the cutoff grade which would be and should be mined.
- Region-II includes all those blocks that have an estimated and true grade below the cutoff grade which would not and should not be mined.
- Region-III includes blocks estimated to be above cutoff grade but have true grade below the cutoff grade. These blocks would be mined when they should not and would thus reduce the mine head grade.
- Region-IV is comprised of blocks estimated to be below the cutoff grade, but which have a true grade above the cutoff. This block will not be mined when they should have been and thus will reduce the mine head grade.

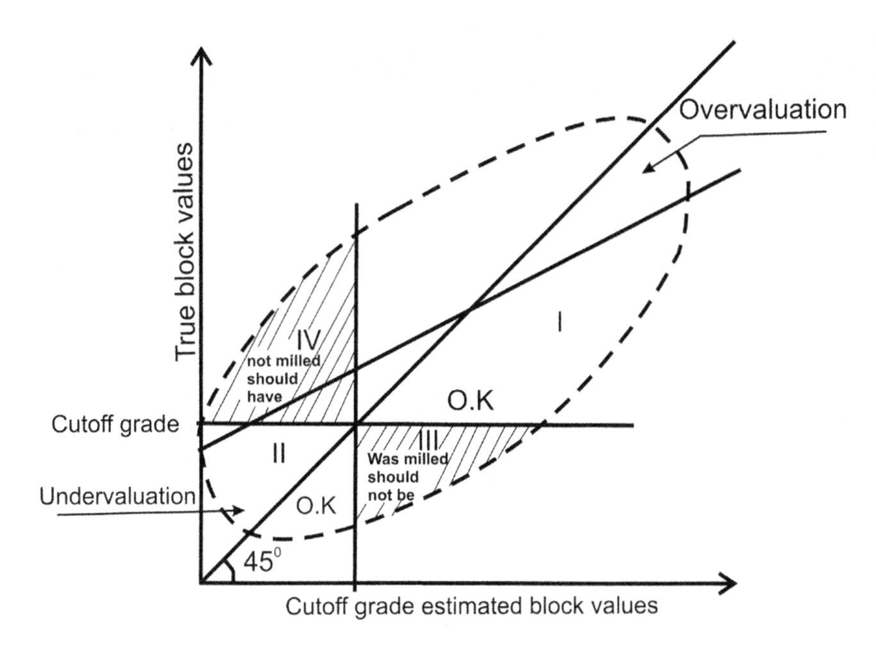

FIGURE 9.16 Misclassified tonnages as conceived by Krige (1951) [47] portray that the "estimated" and "true production" grades vary considerably due to inadequate sampling, estimation flaw and complexity of mineralization.

It is the envelope of the dispersion or variance of the block grades that causes after the application of cutoff grade, the difference between the expected or planned mine head grade and the true grade. It is therefore necessary to employ that technique which decreases the variance in the block estimates, i.e. decrease the spread of block estimates so that the regression line moves to 45°. Secondly, it is necessary to employ an estimating technique that will reduce the variance of the error of estimation or estimation variance so that the envelope narrows and ideally becomes the 45° line. These two requirements are compatible and essential.

9.4. GEOSTATISTICAL APPLICATIONS

These problems of grade tonnage mismatch and wider grade variances of estimated blocks have been resolved by developing the regionalized stationary and variability of metal distribution within the deposit.

9.4.1. Block Variance

Let us assume that we have large number of sample points distributed evenly spaced within the orebody. The body is divided into a number of blocks of equal size. An average grade can be computed for each block by taking the arithmetic or weighted average of all the sample points falling in that block. The simplest case would be equal number of blocks with one sample in each block. The other extreme case would be only one block representing the whole deposit that contains all sample points. If the variances of the block grades are plotted against the size of the block, the block variance relationship is obtained (Fig. 9.17).

Obviously, the variance of the block grades decreases with the increase of block size. As each block value is an average of all the sample points in the block and mean of the ore deposit is the mean of all the block values, the mean of the deposit remains constant. The variance of the errors between the true grade of a block and that one estimated from a center sample point generally decreases with increase in block size.

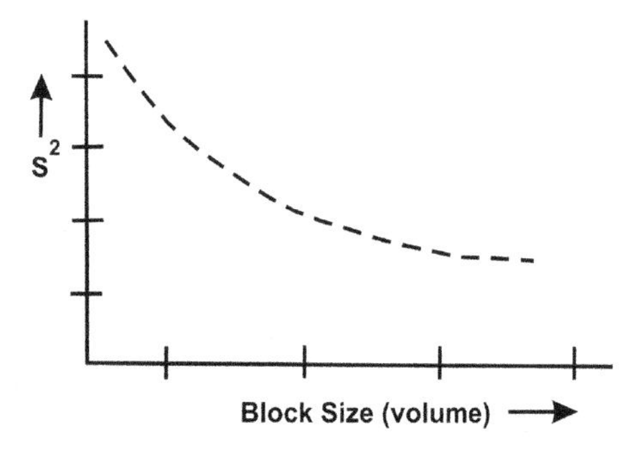

FIGURE 9.17 Concept of mutual relationship between sample volume size and variances.

The estimation of block size will depend on the mine planning and it is necessary to select an estimation technique which will take into account of all surrounding samples to reduce the variance of the estimation error. An orebody model has to be developed based on geological and statistical information. Some geological concepts that need to be considered are as follows:

(a) Better the orebody continuity smaller the error associated with extrapolating a given grade over larger area,
(b) Influence of a sample over an area may be related to a geological zone of influence that may vary in different direction (anisotropy),
(c) Some ore deposit, such as gold, silver and platinum may exhibit a nugget effect (large variation in grade over small distance) and therefore larger samples may be required since a sample is a volume (support) and not a point.
(d) Demarcate geologically distinct areas within a deposit so that different estimation procedure may be required (nonstationary).

9.4.2. Semi-variogram

The most natural way to compare two values is to consider their differences. Let us assume that two samples $Z(x)$ and $Z(x+h)$ are located at two points, x and $x+h$ with two different percent metal grades. The second sample is "h" meters away from first. This value expressing the dissimilarity of grade between two particular points is of very little implication. The average difference for all possible pairs of samples at "h" meters apart throughout the deposit will be geologically significant. The difference between any pair of sample will be either positive or negative and the sum of the all pairs will be misleading. It is therefore logical to square the differences, sum them and divide by the number of pairs to understand the average variance of paired samples at certain distance apart. This dissimilarity function is expressed by the model equation:

$$2\gamma \rightarrow (h) = \text{Average } \Sigma\{Z(x) - Z(x+h)\}^2$$

Where, $2\gamma \rightarrow (h)$ is the "semi-variogram" or simply "variogram" function. It is the function of a vector, in other words, a distance and the orientation of that distance. It articulates how the grades differ in average according to the distance in that direction (Fig. 9.18).

The semi-variogram is thus stated as:

$$\gamma(\boldsymbol{h}) = \mathbf{1/2N}\sum_{i=1}^{N}\{Z(x) - Z(x+h)\}^2$$

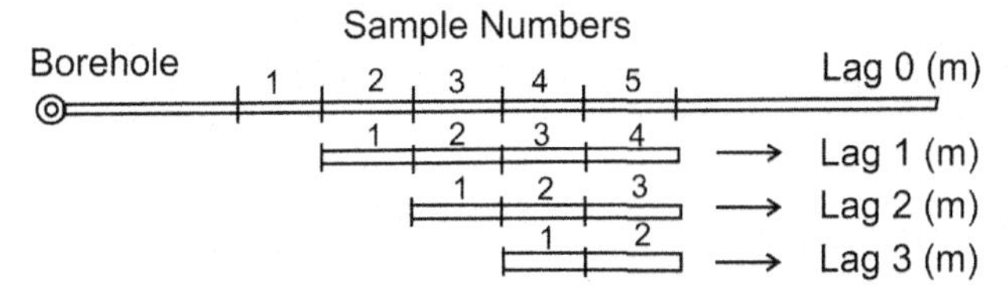

FIGURE 9.18 Concept of semi-variogram by sliding sample data string at certain space (Lag in m) interval.

where, $\gamma(\boldsymbol{h})$ = semi-variogram function value
N = number of sample pairs
$Z(x)$ = value at point (x), may be grade, width, metal accumulation etc.
$Z(x+h)$ = value at point ($x+h$) i.e. h distance away from point (x)

The simplest semi-variogram computation can be by sliding mineralization in one direction along a borehole (Fig. 9.18). Let there be a set of five samples of 1 m length each along a borehole as 2, 6, 3, 9 and 12% Zn. The sample string slides by one step (1 lag) i.e. at 1 m, 2 m, 3 m lag to compute the average variance and continue.

Sample (%Zn)	Lag	Difference	Difference^2	
2	1m			
6	2	4	16	$\sum$ Diffence^2 = 70
3	6	−3	9	No of pair = 4
9	3	6	36	Av. variance = 70/8
12	9	3	9	Semivariance = 8.75
2	2m			
6				
3	2	1	1	$\sum$ Diffence^2 = 91
9	6	3	9	No of pair = 3
12	3	9	81	Av. variance = 15.16
2	3m			
6				
3				$\sum$ Diffence^2 = 85
9	2	7	49	No of pair = 2
12	6	6	36	Av. variance = 21.25

This can be represented in semi-variogram plot of lag against $\gamma(h)$ function along x- and y-axis respectively (Fig. 9.19). The freehand or fitted curve is extended downward to intersect the variance axis. If it touches above the origin this part is known as: "nugget effect (C_0)". The curve then rises up to the maximum variance (population variance δ^2) at a particular lag equivalent distance and levels out or flattens. This distance is known as "range (a)". Each sample value has an influence up to about 2/3 of the range. The extension variance C_1 is the difference between population variance and nugget effect.

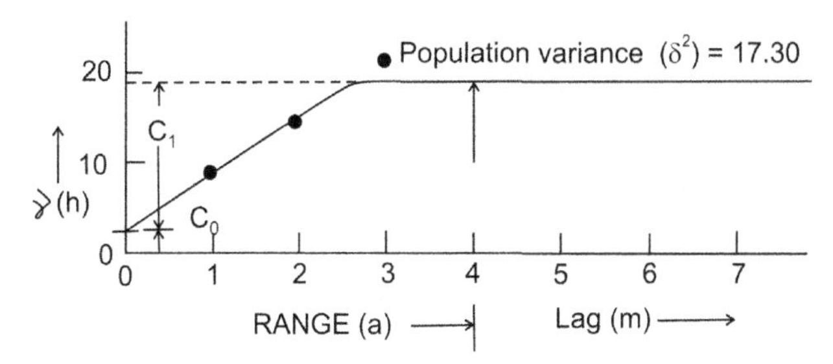

FIGURE 9.19 Drawing of standard semi-variogram along the drill hole samples.

In case of 2D grid sample data the variogram is computed at different directions. Sample points on either side of the variogram line, within acceptable limit, are projected on it. Variance between points is computed and grouped under similar lag range. This process is repeated on all samples and arranged in serial order as per lag. Number of participating pairs contributing to the sum of differences square is known and a variogram is computed. Similarly variogram in other directions within the study area are computed and compared.

In practice, theoretical semi-variogram is never realized. The gamma function $\gamma(h)$ is estimated from limited points and is called the experimental variogram (Fig. 9.20).

The Fig. 9.20 shows a number of sample points equidistant apart with their sample values, say thickness, next to each point. If we wish to compute $\gamma(\mathbf{1})$, that is, the semi-variogram value of all samples points at 1 unit distance (lag) apart, irrespective of direction then. There are 17 sample pairs at 1 unit apart.

$$\begin{aligned}\gamma(1) &= 1/2N(1-2)^2 + (2-2)^2 + (2-3)^2 \\ &\quad + (2-1)^2 + (1-4)^2 + (4-3)^2 + (2-1)^2 \\ &\quad + (1-1)^2 + (1-2)^2 + (1-2)^2 + (2-2)^2 \\ &\quad + (2-1)^2 + (1-1)^2 + (2-4)^2 \\ &\quad + (4-1)^2 + (3-3)^2 + (3-2)^2 \\ &= 31/34 = 0.91\end{aligned}$$

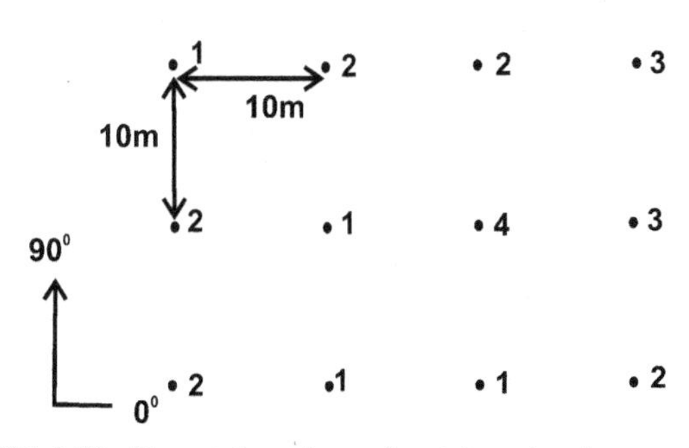

FIGURE 9.20 Computation of experimental semi-variogram of grid sample data along various directions such as 0° and 90°.

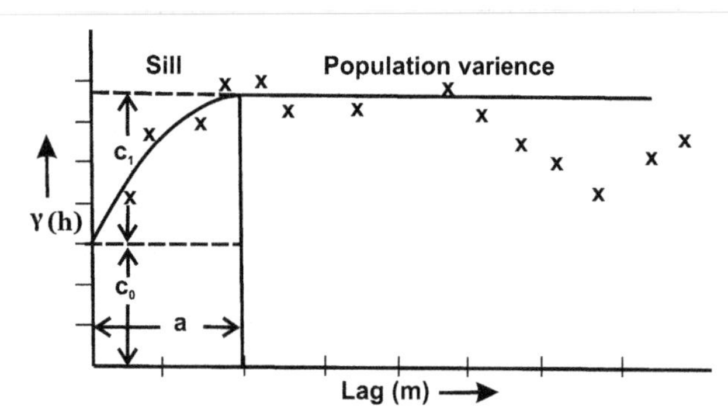

FIGURE 9.21 Typical semi-variogram plot with curve fitting and follow the population variance after reaching zone of influence.

The next shortest sample distance is $\sqrt{2}$ unit distance apart with total sample pairs of 12.

$$\therefore \gamma\left(\sqrt{2}\right) = 26/24 \text{ or } \gamma\left(\sqrt{1.4}\right) = 1.08$$

The gamma function for all the possible lags are computed and plotted as $\gamma(\boldsymbol{h})$ against respective lag. The semi-variogram will look like as given in Fig. 9.21.

9.4.2.1. Properties of Semi-variogram

9.4.2.1.1. The Continuity

The continuity is reflected by the rate of growth of $\gamma(h)$ for small value of "h". The growth curve exhibits the regionalized element of the sample. The steady and smooth increase is indicative of the high degree of continuity of mineralization till it plateaus off at some distance. This is known as the structured variance or EV (C or C_1) read on the $\gamma(h)$ scale. A typical semi-variogram of coal deposit showing good continuity is given in Fig. 9.22.

9.4.2.1.2. Nugget Effect (C_0)

The semi-variogram value at zero separation distance (lag = 0) is theoretically 0. However, the semi-variogram often exhibits a nugget effect (>0) at an infinitely small separation distance. The complex mineral deposits, like the base metals, may occur as nuggets, blobs and often concentrated as alternate veinlets resulting rapid changes over a very short distance. The gamma functions extrapolated back to intersect with y-axis (Fig. 9.23). The positive measure of $\gamma(\mathbf{0})$ is the magnitude of the random or unexplained elements of the samples and called the nugget effect (C_0). The noble and precious metals like gold, silver and platinum group of elements often exhibit very high nugget effect even up to total unexplained variance. This creates lot of uncertainty in continuity of mineralization and grade estimation leading to excessive sampling.

There are three possible reasons for nugget effect of various magnitudes.

(1) Sampling and assaying errors,
(2) Smaller microstructures or nested variogram at the shorter distance where no borehole data exist,
(3) Combination of both.

9.4.2.2. Semi-variogram Model

The next step is to fit a model subsequent to the determination of the experimental points of a semi-variogram. There exist numbers of semi-variogram models which fulfill certain mathematical constraints. The models can broadly be divided into two groups: (1) those with a Sill and (2) those without Sill. A Sill implies that once a certain distance "h" is reached the values of $\gamma(h)$ does not increase although it may oscillate. The models without a Sill have growing values of $\gamma(h)$ for increasing values of "h".

9.4.2.2.1. Features of Semi-variogram

The Sill ($C_1 + C_0$)

The growing semi-variogram curve normally reaches a plateau after certain lag and equals to the population variance. This is known as the Sill. In practice the gamma

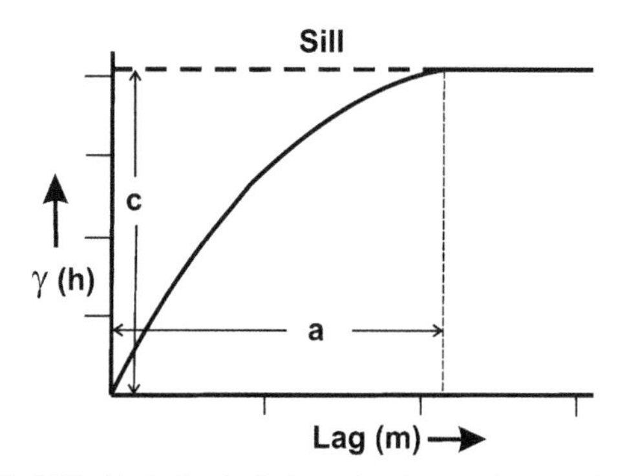

FIGURE 9.22 Typical spherical semi-variogram from coal deposits without any nugget effect.

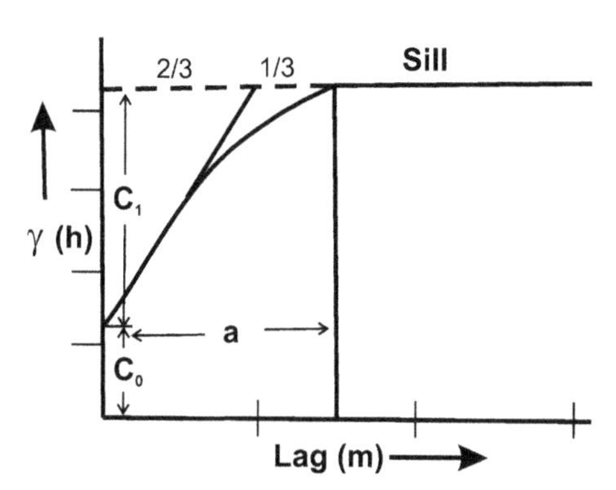

FIGURE 9.23 Typical semi-variogram from massive base metal deposits with moderate UV (C_0).

function may fluctuate on Sill line because of metal zoning, layering and hole effect.

$$\text{Sill } (C) = C_1 + C_0$$

9.4.2.2.2. The Zone of Influence (a)

The zone of influence is that neighborhood beyond which the influence of a sample disappears, or in other words, the samples become independent of each other. It is standard practice to characterize the zone of influence in a given direction by the distance at which the semi-variogram eventually reaches a plateau and levels out. Generally the first few points of $\gamma(\boldsymbol{h})$ are joined by hand or by polynomial regression and extended to the Sill. The intersection with Sill is 2/3 of zone of influence or range (a).

9.4.2.2.3. The Isotropic-Anisotropies

The anisotropies are easily depicted by computing the variograms in different directions. For example, if points in the east-west orientation of Fig. 9.20 are selected then the values of the semi-variogram would be:

$\gamma(1) = 15/18$

$\gamma(2) = 12/12$

$\gamma(3) = 5/6$

Similarly, values in north-south direction or in any other direction could be computed and the semi-variograms compared. If the semi-variograms exhibit the same features irrespective of the direction then the underlying structure as defined by the semi-variograms is said to be "isotropic". However, in many cases the semi-variograms display different features in various directions and the structure is said to be "anisotropic". In such condition an average semi-variogram can be considered.

The types of semi-variogram with and without Sill are given in Figs 9.24 and 9.25.

(a) Spherical model

$$\gamma(h) = C_0 + C_1(1.5h/a - 0.5(h^3/a^3)), \text{ for } h < a, \text{ and}$$

$$\gamma(h) = C_0 + C_1, \text{ for } h \geq a$$

Where, C_0 = nugget value

C_1 = explained variance

$C_0 + C_1$ = sill

h = distance between points

a = zone of influence or range

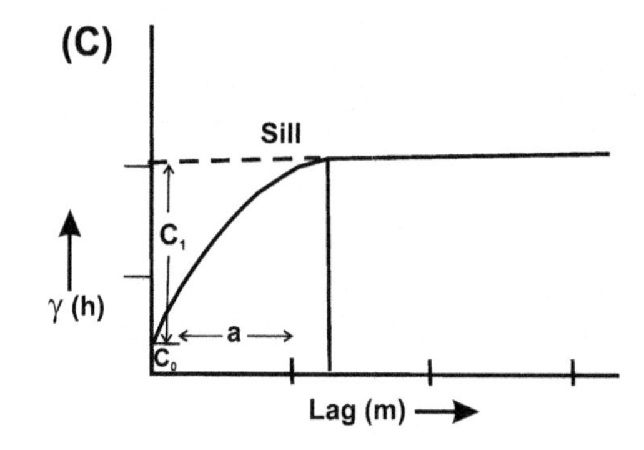

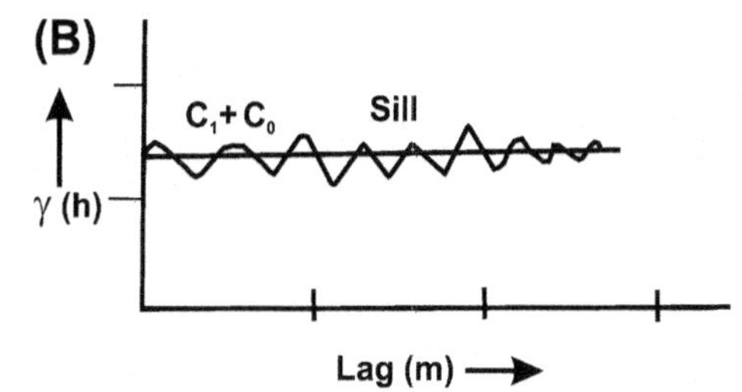

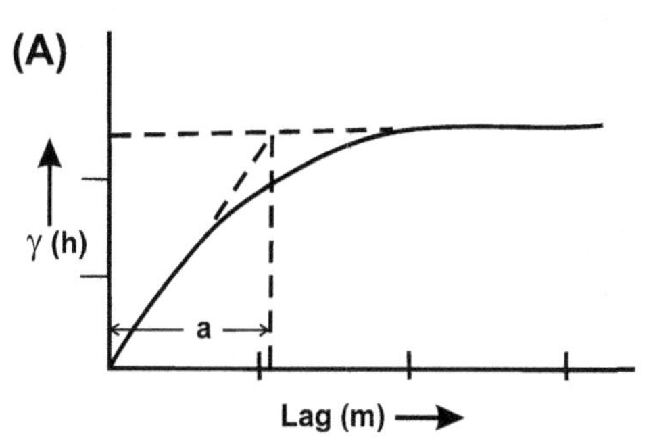

FIGURE 9.24 Semi-variogram models with Sill (A) spherical, (B) random and (C) exponential.

(b) Random model

$$\gamma(h) = C_0 + C_1, \text{ for all } h$$

(c) Exponential model

$$\gamma(h) = (C_0 + C_1)(1 - e^{-h/a}), \text{ for } h < a$$

The types of variogram without Sill are (Fig. 9.25):

(d) Linear model

$$\gamma(h) = (\alpha^2 h/2), \text{ where } \alpha^2 \text{ is a constant}$$

(e) Logarithmic model

$$\gamma(h) = 3\,\alpha \log^h$$

(f) Parabolic model

$$\gamma(h) = 1/_2 \alpha^2 h^2$$

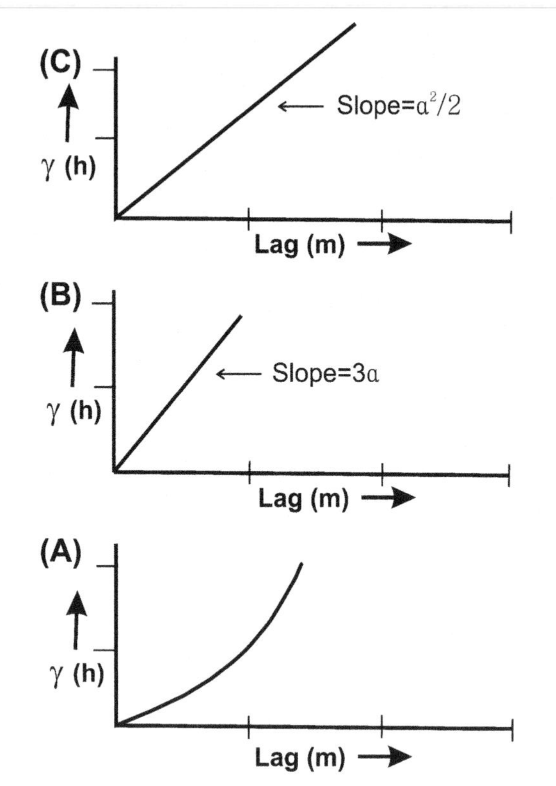

FIGURE 9.25 Semi-variogram model without Sill (A) parabolic, (B) logarithmic and (C) linear.

The following points are to be observed for construction of experimental semi-variogram:

(1) Most semi-variograms (>95%) used in geostatistical applications belong to spherical model. Sill is usually taken as being equal to the δ^2 to fit the variogram. A straight line through the first few points should intersect the line of the Sill at two third of the slope.

(2) In practice, a minimum of 50 pairs of samples should be used for computation of experimental points. Some 100 sample values may be required to achieve this, although adequate models can be fitted with less data.

(3) As the distance "*h*" increase, the number of sample pairs will decrease, thereby indicating less reliability of $\gamma(h)$ for larger distance. Only the initial 5–10 experimental points will be required to fit a model.

9.4.2.3. Smoothing Semi-variogram

The sample data in the previous example are equally spaced. However, in practice even with a regular drilling pattern the coordinates of the borehole collars will be irregularly placed to some odd meters away from the grid line. If a semi-variogram model is computed at a specific distance "*h*" it may so happen that no sample pair may be found at that "h". The $\gamma(h)$ could be computed for calculated values of "*h*" between the data but many of the distances "h" might only have one sample pair. This would result in a random pattern of experimental semi-variogram and no model could be fitted to the statistically valid experimental points.

The effect of data at irregular intervals can be thought of a "noise" and that the data need to be smoothed to find the underlying semi-variogram. The optimal smoothing can be done by (a) step interval and (b) angular tolerance.

(a) Step interval

The simple procedure is to select the experimental points for a range of distance Δh. For example, if the data are irregularly spaced at an average distance of 100 m, then a step interval of 110 m could be used with a point value computed for all sample pairs from 0 to 100, 100 to 200, 200 to 300 m, etc. Semi-variogram could be computed at different step intervals of +5, 10, and 20 with the most of variable results occurring for short intervals. As the step value increase the filtering effect reduces the "noise" and a smoother and more statistically viable model will be obtained. Over-smoothing will mislay the underlying structure and a straight line of experimental points at population variance will be obtained incorrectly suggesting a random model with no special correlation.

Referring sample data from Fig. 9.20 the $\gamma(h)$ in east-west direction:

$$\gamma(0-2) = 27/30$$

$$\gamma(2-4) = 5/6$$

The average of the distance is used for the single value of $\gamma(h)$

$$\gamma(1.4) = 27/30$$

$$\gamma(3) = 5/6$$

(b) Angular regularization

The other method of smoothing is based on the fact that irregular sample data may not fall along specific directions when anisotropic behavior is apprehended. For example, no pair of samples may lie along the east-west axis (this will be called 0°-180°). This method of smoothing is not to calculate the semi-variogram for a single discrete direction but for a direction and range $\Delta\alpha$ to either side of it (Fig. 9.26). Therefore, with direction of 0° angular regularization of ±45° and a step interval of 1 unit distance in our example, the values of $\gamma(h)$ are for the exclusive-inclusive range:

$$\gamma(0-1) = \gamma(1) = 15/18$$

$$\gamma(1-2) = \gamma(1.7) = 28/36$$

$$\gamma(2-3) = \gamma(2.3) = 33/30$$

$$\gamma(3-4) = \gamma(3.3) = 6/12$$

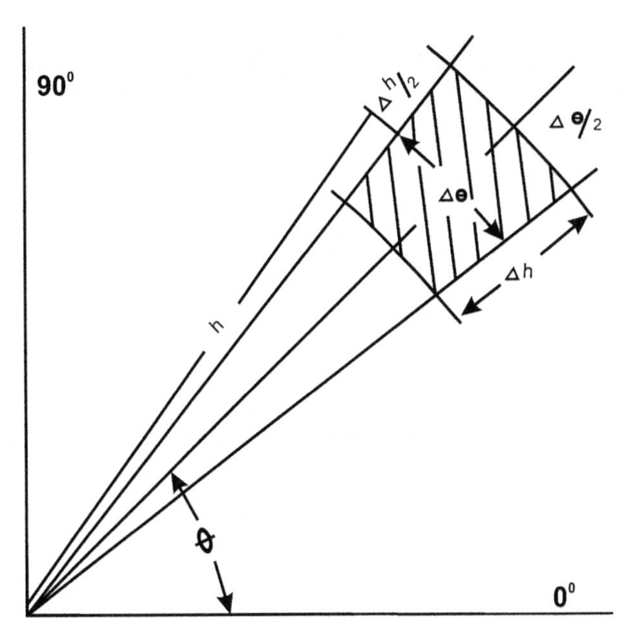

FIGURE 9.26 Angular regularization—sample falling in sketch area are accepted by tolerance in distance (Source: David, 1988) [17].

The greater the distance for each step and the wider the angle around a specific direction, the greater is the smoothing. Insufficient smoothing could be difficult to fit a model. Over-smoothing will destroy the actual character.

9.4.3. Estimation Variance

The estimation can be defined as the variance of the error made in estimating the grade of a panel or block of ore by assigning to it the values of the samples lying in and around it within the zone of influence. It is apparent from the semi-variogram that the relative location of the samples to the block will have an influence upon the weighting coefficients that can be assigned to each sample value.

The $Z(V_i)$ are true unknown grades of the blocks V_i and $Z^*(V_i)$ are the linear combinations, where $Z\times(V_i) = \sum_{j=1}^{n} a_j Z(X_j)$ of the sample grades at locations X_j ($j = 1, \ldots, n$). Then, $Z^*(V_i) - Z(V_i)$ will be the error made in assuming that the value Z^*_i extends over V when the true value is V_i and VAR $(Z_i - Z^*_i)$ is the variance of this error.

$$\text{VAR (Error)} = \text{VAR}\ (Z_i - Z_i^*)$$

If a block value is estimated by a sample or a number of samples each given an equal weight, then the variance of the error δ_e^2 associated with the estimated mean is:

$$\delta_e^2(W_s \text{ to } W) = +2\ \bar{y}\ (W_2;\ W) - \bar{y}\ (W_s;\ W_s) - (W;\ W)$$

where, W_s = sample support

W = block

Or, in presence of nugget effect C_0/n is to be added, where n = number of samples.

In the general case where the weights are different for each sample that is used to estimate the block value, the weights still add up to 1.0 so that the estimation is not biased, then the formula for the estimation variance:

$$\delta_e^2 = 2\Sigma_{i=1}^{n} b_i\ \bar{y}\ (w_i;\ W) - \Sigma_{i=1}^{n}\ \Sigma_{j=1}^{n} b_i b_j \bar{y}\ (w_i;\ w_j) - \bar{y}\ (W;\ W)$$

Where, b_i or b_j = the weights for each sample, $n = 1, I$ or j

Therefore, block value = block value $= \sum_{i=1}^{n} b_i Z(w_i)$

Where, $Z\ (w_i)$ = sample value and $\sum_{i=1}^{n} b_i = 1$.

It is possible to vary the two weights applied to w_1 and w_2 in an infinite number of combinations. For each combination a new δ_e^2 can be estimated, but only one combination will minimize δ_e^2. The method of determining b_1 and b_2 to minimize δ_e^2 is called "Kriging". The optimal weights are derived from Krige's equation dependent on the semi-variogram or spatial correlation of distance function and direction.

9.4.4. Kriging

Prof. D. G. Krige initiated and developed the application of geostatistics to the valuation and optimization of ore in South African gold mines in early 1970s. Prof. Matheron named the estimator as "Kriging". The principles of the estimation procedure are:

(1) It should be a linear function of the sample value x_i. Block value of $W = \sum_{i=1}^{n} b_i x_1 + b_2 x_2 + \cdots + b_n x_n$ where, b_i is the weight given to sample w_i.

(2) It should be unbiased. The expected value (μk) should be equal to the true block value (μW).

$$E\{(\mu\text{k} - \mu\text{W})\} = 0$$

(3) The mean squared error of estimation of μW should be a minimum.

$$E\{(\mu\text{k} - \mu\text{W})^2\} = \text{a minimum}$$

The kriging estimator (μk) satisfies these conditions of linear function, unbiased estimation and a minimum variance. The corresponding error of estimation from sample (w_s) to block (W) is the kriging error (δ_e^2). Kriging estimator is also known as the Best Linear Unbiased Estimator (BLUE).

A block W of 3 × 3 is estimated using two samples w_1 in the center and w_2 in the corner (Fig. 9.27). A set of simultaneous equations are used to resolve for the optimum set of weights assuming the population mean unknown:

$$\sum_{i=1}^{n}\sum_{j=1}^{n} b_i b_j\ \bar{y}\ (w_i;\ w_j) + \lambda = \sum_{i=1}^{n} b_i\ \bar{y}\ (w_i;\ W)$$

Where, λ = the lag range multiplier.

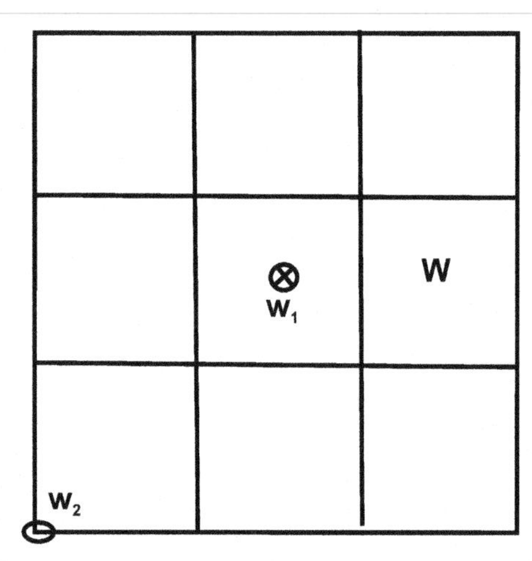

FIGURE 9.27 Block (W) estimation representing nine subblocks with one sample in center (w_1) and one sample at corner (w_2).

The error of variance δ_e^2 can be estimated from the kriged weights by:

$$\Sigma_e^2\ \Sigma_k^2 = -\ \bar{y}\ (w;\ W) + \Sigma_{i=1}^{n}\ b_i\ \bar{y}\ (w_i;\ W) + \lambda$$

For example, and ignoring λ for the moment,

$$b_1\ \bar{y}_{11} + b_2\ \bar{y}_{12} = \bar{y}_{1}W$$

$$b_1\ \bar{y}_{21} + b_2\ \bar{y}_{22} = \bar{y}_{2}W$$

The simultaneous equations when written in matrix form:

$$\begin{bmatrix} \bar{y}(w_1w_1) & \bar{y}(w_1w_2) \\ \bar{y}(w_2w_1) & \bar{y}(w_2w_2) \end{bmatrix} \times \begin{bmatrix} b_1 \\ b_2 \end{bmatrix} = \begin{bmatrix} \bar{y}(w_1;W) \\ \bar{y}\ (w_2;W) \end{bmatrix}$$

or

$$\begin{bmatrix} 0 & 1 \\ 1 & 0 \end{bmatrix} \times \begin{bmatrix} b_1 \\ b_2 \end{bmatrix} = \begin{bmatrix} 0.536 \\ 0.882 \end{bmatrix}$$

$$\text{Therefore, } b_2 = 0.536$$

$$b_1 = 0.882$$

However, $b_1 + b_2$ must be equal to 1, but in this case $(0.536 + 0.882) = 1.1418$. Therefore, the lag range multiplier is to be introduced to ensure $b_1 + b_2 = 1$. The new matrix is,

$$\begin{bmatrix} \bar{y}(w_1w_1) & \bar{y}(w_1w_2) & 1 \\ \bar{y}(w_2w_1) & \bar{y}(w_2w_2) & 11 \\ 1 & 1 & 0 \end{bmatrix} \times \begin{bmatrix} b_1 \\ b_2 \\ \lambda \end{bmatrix} = \begin{bmatrix} \bar{y}(w_1;W) \\ \bar{y}(w_2;W) \\ 1 \end{bmatrix}$$

or

$$\begin{bmatrix} 0 & 1 & 1 \\ 1 & 0 & 1 \\ 1 & 1 & 0 \end{bmatrix} \times \begin{bmatrix} b_1 \\ b_2 \\ \lambda \end{bmatrix} = \begin{bmatrix} 0.536 \\ 0.882 \\ 1 \end{bmatrix}$$

$$\therefore b_2 + \lambda = 0.536$$

$$b_1 + \lambda = 0.882$$

$$b_1 + b_2 = 1$$

Solving b_1, b_2 and λ

$$b_1 = 0.673$$

$$b_2 = 0.327$$

$$\lambda = 0.209$$

$$b_1 + b_2 = 1 \text{ as required}$$

The mean estimate for the block is:

$$0.673\ Z(w_1) + 0.327\ Z(w_2)$$

And the error of estimation is:

$$\begin{aligned}\delta_e^2 &= -0.683 + 0.673 \times 0.536 + 0.327 \times 0.882 \\ &\quad + 0.209 \\ &= 0.175\end{aligned}$$

$\delta_e = 0.418\%$, CL which is independent of grade.

In summary, we observe that if we estimated the block value by different sample configuration the following errors were made:

In block estimated from corner sample, $\delta_e^2 = 1.081$.

In block estimated from center sample, $\delta_e^2 = 0.389$.

In block estimated from corner and center samples with equal weights,

$$\delta_e^2 = 0.235$$

In block estimated from corner and center samples using kriged weights,

$$\delta_e^2 = 0.175$$

The kriging method is referred to as BLUE. It is best since minimum error variance is obtained, unbiased because the weights sum to 1.0, which means the expected block value is equal to the true block value, and a linear estimator in its application.

9.4.5. Kriging Estimation—An Example

Point and block kriging can be used to estimate the value at a point or block. Semi-variogram is a prerequisite of kriging estimate. The accuracy of kriging estimation is, to large extent, proportional to the precision factor of variogram. Point kriging has been suggested as a test for a semi-variogram model. After a model is selected each sample point is kriged in turn without its value being used in the matrix. The kriged sample values are compared with the actual sample values and the model adjusted till the errors are minimized. This is known as "Jack Knife" test.

The stepwise computation of kriging for a part of ore-body (sublevel stope) is demonstrated by seven sample points. The sample details, variogram for kriging at central point "Q" of block ABCD (25 × 25 m) is given in Table 9.11 and Figs 9.28 and 9.29.

The semi-variogram of the total population provides the $C_0 = 0.8$, $C_1 = 2.5$ and $a = 50$ m. Five samples namely 1, 2, 3, 5, and 6 are selected which fall within the radius of influence of 30 m from computation point Q. The distance d_1 (1 to Q), d_2 (2 to Q), d_3 (3 to Q), d_5 (5 to Q) and d_6 (6 to Q) are required for interpolation. In case of kriging technique, in addition to that, one has to compute distances and variances between 1 to 1, 1 to 2, 1 to 3, 1 to 5, 1 to 6, 2 to 1, 2 to 2, 2 to 3, 2 to 5, 2 to 6, 3 to 1, 3 to 2, 3 to 3, 3 to 5, 3 to 6, 5 to 1, 5 to 2, 5 to 3, 5 to 5, 5 to 6, 6 to 1, 6 to 2, 6 to 3, 6 to 5 and 6 to 6, i.e. with five samples additional 5 × 5 inter-sample distances are computed. Inter-sample variances eliminate the shadowing effect present in sample distribution within the zone of influence.

TABLE 9.11 Sample Coordinates and Grade from a Sublevel Stope

Sample no.	Coordinate (m) Easting	Level	% Zn
1	314	83	2.40
2	314	93	2.76
3	314	107	4.84
4	288	74	5.98
5	288	90	6.38
6	288	105	7.04
7	298	76	3.63

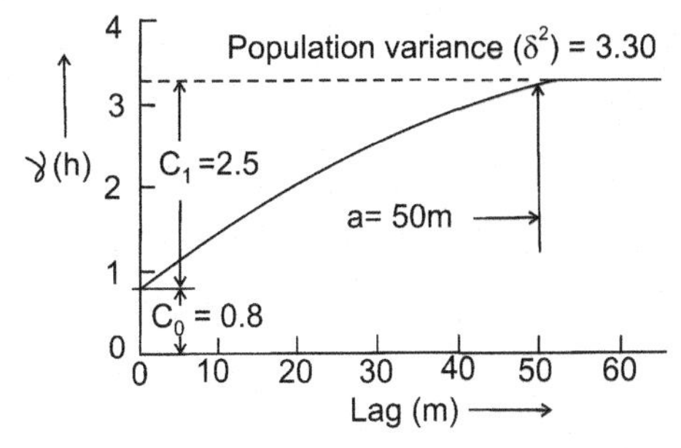

FIGURE 9.28 Semi-variogram parameters of samples from Table 9.9.

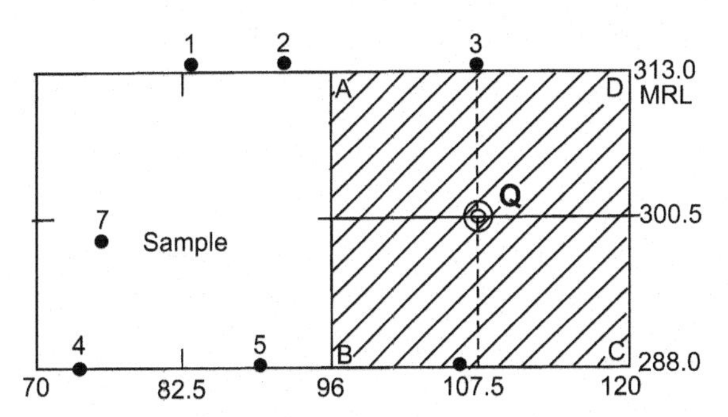

FIGURE 9.29 Conceptual block kriging from surrounding samples.

The variogram signifies relation between the distances and variances, and the equations can be formulated in the matrix form such as:

$$\begin{bmatrix} V_{11} & V_{12} & V_{13} & V_{15} & V_{16} \\ V_{21} & V_{22} & V_{23} & V_{25} & V_{26} \\ V_{31} & V_{32} & V_{33} & V_{35} & V_{36} \\ V_{51} & V_{52} & V_{53} & V_{55} & V_{56} \\ V_{61} & V_{62} & V_{63} & V_{65} & V_{66} \\ a_1 & a_2 & a_3 & a_4 & a_5 \end{bmatrix} \times \begin{bmatrix} a_1 \\ a_2 \\ a_3 \\ a_4 \\ a_5 \\ \lambda \end{bmatrix} = \begin{bmatrix} V_1Q \\ V_2Q \\ V_3Q \\ V_5Q \\ V_6Q \\ 1 \end{bmatrix}$$

Where, $a_1 + a_2 + a_3 + a_4 + a_5 = 1$ (weights for coefficients) and $V_{11}, \ldots, V_{nn}$ represent the variances instead of the distances which are measured earlier. In the present data set the following equation is obtained:

$$a_1\ 0.0 + a_2\ 1.8 + a_3\ 2.5 + a_4\ 2.7 + a_5\ 1.9 = 2.2$$

$$a_1\ 1.8 + a_2\ 0.0 + a_3\ 2.8 + a_4\ 2.5 + a_5\ 2.1 = 1.9$$

$$a_1\ 2.5 + a_2\ 2.8 + a_3\ 0.0 + a_4\ 1.9 + a_5\ 1.6 = 2.3$$

$$a_1\ 2.7 + a_2\ 2.5 + a_3\ 1.9 + a_4\ 0.0 + a_5\ 1.8 = 1.8$$

$$a_1\ 1.9 + a_2\ 2.1 + a_3\ 1.6 + a_4\ 1.8 + a_5\ 0.0 = 1.8$$

$$a_1 + a_2 + a_3 + a_4 + a_5 = 1.0$$

This relates the inter-assay variance to the variance contribution to point Q. Solution of the above equations will provide five weighting coefficients for five sample assay and another coefficient λ for the kriging estimator. The coefficients are computed as:

$$a_1 = 0.10,\ a_2 = 0.31,\ a_3 = 0.07,\ a_4 = 0.31,$$

$$a_5 = 0.21 \text{ and } \lambda = 0.27$$

Therefore, the estimated value at Q will be:

$$a_1 \times 2.4 + a_2 \times 2.76 + a_3 \times 4.84 + a_4 \times 6.38$$
$$+a_5 \times 7.04 = 0.1 \times 2.4 + 0.31 \times 2.76 + 0.07$$
$$\times 4.84 + 0.31 \times 6.38 + 0.21 \times 7.04 = 4.82\% \text{ Zn}$$

After obtaining the estimated value (4.82% Zn) one is interested to know the variance associated to it. If we look back to our process of computation we find that the extension variance from Q to 1 (block-point covariance) is 2.2, while we utilized only 0.10 (a_1) of this variance to get an estimate of Q. Similarly for the second sample 0.31 has been utilized out of 1.9 and so on. The summation of the variances utilized for estimation at Q will be:

$$a_1 \times 2.2 + a_2 \times 1.9 + a_3 \times 2.3 + a_4 \times 1.8 + a_5 \times 1.8$$
$$= 1.94$$

The total COV utilized for the estimation i.e. block-point COV (1.94) and point-point kriging COV λ (0.27) is 2.21. The variogram of the orebody can explain 1.72 in terms of variance for a block size of 25 × 25 m (an average distance of 12.5 m from any point to another point within the block). The residual variance of the estimate is $2.21 - 1.72 = 0.49$ which is not explained by the estimate at Q and will be equivalent to 0.7 in terms of standard deviation. The mean grade (4.82% Zn) has a possible variability of 0.7 over the mean. The confidence interval at 95% probability will be $(0.7/\sqrt{5} \times 1.96$ or $\pm 0.61)$. The error of the mean is 4.82% Zn ± 0.61. Keeping other conditions constant if we change the block size to 15 × 15 m the EV equivalent to 7.5 m is 1.5 and the error of mean by simple computation will be ±0.74. It can be concluded that as the block size decreases the CL value (range) will increase.

The kriging methodology rests largely on precision of the variogram which in reality is not smooth and indicative of disturbances inherent within the samples. Since the variogram used is a generalized average one, the estimate may differ from the true value. Local trend existed in the data set should be deleted before kriging. This will make the process complex for mine block estimation.

9.4.6. Benefits of Geostatistics

(1) Statistical techniques provide sequential analysis of exploration campaign with broad parameters at any stage (refer Chapter 15, Section 15.9.14).
(2) Variogram quantifies range of influence of sample in all directions indicating isotropic/anisotropic nature of the deposit.
(3) Variogram extracts the explained (C_1) and nugget (C_0) variance of orebody.
(4) Point, area and block estimation by (kriging) represent BLUE techniques which minimize error variance.
(5) Computes estimation error.
(6) Estimate average grade and tonnage of individual blocks for the purpose of planning selective mining and grade control.
(7) Optimize sampling design (refer Chapter 15, Section 15.9.15).
(8) Analysis of estimated and actual performances.

FURTHER READING

Kotch and Link (1986) [46], Davis (1973) [19] and Sahu (2005) [63] are excellent to start statistical applications in geology and program writing. Krige (1951) [47], 1962 [48], 1978 [49] concerned with lognormal mine grade valuation and uncertainties associated with production grade. Matheron (1971) [54] introduced concept of regionalized variability and the kriging interpolation. Isobel Clark (1982) [13] introduces textbook of practical aspects. David (1977 [16] and 1988 [17] described reserve estimation. Rendu (1978) [58] discussed semi-variogram with mathematical treatment. Journel and Huijbregts (1978) [43] and Issaks and Srivastava (1989) [40] are advance textbook in mining geostatistics. Henley (1981) [37] introduced the concept of Nonparametric Geostatistics.

Chapter 10

Exploration Modeling

Chapter Outline

"Boundless imagination may conceive an idea → Idea may give birth to a concept→ Concept may perceive a model and → finally a Geoscientist shall discover a mineral deposit."

—Author.

10.1. DEFINITION

Modeling is an investigative technique. It uses a physical or mathematical representation of a system or theory, which accounts for all or some of its parameters. The process building is conceptualization of a logical illustration of all activities (the model) related to the phenomenon under investigation. Exploration modeling covers interrelationship ranging between geology, geochemistry, geophysics, sedimentology, stratigraphy, structure, host rock affinity, genesis and exploration input in the context of water, petroleum, coal and mineral/ore deposits. Models are developed based on observations of the past, data input of the present and interpretation to make an experimental and empirical prototype for the future search. It visualizes the "Unknown" to reach the expected goal. Thus, the individualistic approach makes the model quite distinct and different for the same object (deposit). Poorly constructed model can be misleading, counterproductive and at times end up at confused state. A model can be modified and corrected with the incoming information. The model follows same principle anywhere under similar geological conditions with allowance for local effect.

The broad objectives and strategies are to be defined in clear terms so that the orientation can be streamlined accordingly. The purposes of modeling are target selection, future investment, resource augmentation, exploration optimization, economic evaluation and midterm corrections under multidisciplinary activities. The objectives depend on technological support, financial capacity and market strategy of the exploring/mining/ore-dressing agencies. Geoscientist has to follow a dynamic routine, a disciplined planning, integration and interpretation of multidimensional functions and make determined efforts to attain success of mineral discovery.

10.2. TYPES OF MODEL

The various geological/exploration models are classified as follows:

(1) Descriptive
(2) Conceptual
(3) Genetic
(4) Mineral Deposit/Belt
(5) Predictive
(6) Statistical and geostatistical
(7) Orebody
(8) Grade-Tonnage
(9) Empirical
(10) Exploration.

10.2.1. Descriptive Model

This is the first step in modeling approach by collection, study, analysis and reinterpretation of existing information

Mineral Exploration. http://dx.doi.org/10.1016/B978-0-12-416005-7.00010-6

available as published literature, maps on various scales, ground geochemical and airborne geophysical data. It includes description of all measurable physical and chemical properties of occurrences and deposits. Geological parameters include depositional features, tectonic setting, age of formation, etc.; micro-level features of deposit itself covering location, discovery history, surface signatures, rock types, mineralogy, structure, deformation, metamorphic grade, ore control, geochemical and geophysical response, presence of halos, evaporates, tonnage and metal content. This information is accessible either free or on cost depending on country's policy. This forms the base for applying RP/PL to the State Government authorities.

10.2.2. Conceptual Model

The conceptual model represents ideas and relationships between them. At this stage a mental image of the object or process, describing general functional relationships, present or predicted, between major components in a system is conceptualized. These mental images can be, and are, often translated into simplified schematic written descriptions and abstract visual representation. The purpose is to formulate activity plans and flow diagrams. A working model would be under continuous development and modification. It is based on experience to visualize similar geological process likely to exist elsewhere. It includes assumptions on stratigraphy, mineral properties, dimensionality and governing processes.

10.2.3. Genetic Model

It is the perception of ore genetic process based on direct and indirect evidences and knowledge of host environments. It includes overall insight of how the geological forces act to influence the formation of ore deposits. These descriptive and interpretative features classify the process of formation as igneous, sedimentary, hydrothermal, SEDEX, VMS or VHMS and layered mafic-ultramafic intrusive. Most deposits are deformed, metamorphosed and remobilized and that may obscure primary structure. Many corroborating evidences are collected for specific type of deposit before advocating particular genetic process. The genetic model plays a significant role such that a layered mafic-ultramafic intrusive is significantly indicative for hosting Platinum Group of Elements closely associated with chromium-nickel-copper-gold-silver mineralization such as Bushveld, Sudbury, Stillwater, Sukinda-Nausahi layered igneous complex. The distinguished examples of deposits known for genetic model include Broken Hill base metals, NSW, Australia (SEDEX), Kidd Creek copper-zinc, Timmins, Canada (VMS) and New Brunswick zinc-gold, Bathurst, Canada (VHMS).

10.2.4. Mineral Deposit/Belt Model

Mineral Deposit/Belt model is systematically arranged information describing some or all important characteristics, variations within a group and type of known deposits. It uses results of previous investigations to foresee geological nature of wider area or belt. Belt is divided into a number of blocks (cells/sub-cells) of equal size. The block sizes are based on the dimension of the existing deposits and complexity of data being processed (Fig. 10.1). The "Control" block must have at least one deposit in it.

The possibility of mineralization in the unknown cells can be predicted using interrelationship of key geological attributes of the control cells. Mathematical techniques used are cluster, multiple regression and characteristic analysis. The derived equation can be extended under similar geological conditions (Fig. 10.2) for anticipating new exploration targets.

10.2.5. Predictive Model

Predictive model is a mathematical technique used to forecast the prognostic mineral deposits, metal resources and expected areas likely to host mineralization. The model, thereby, assigns priorities for initial reconnaissance activity and look for prospective areas. There are many mathematical equations available. The popularly convincing one is Zipf's Law. This model has been effectively applied for exploration target appraisal for oil field resource of Western Canada, gold deposits of Western Cordilleras, copper deposits of Zambia, gold, uranium, copper-lead and tin deposits of Australia and zinc-lead, copper and iron ore deposits in India.

Professor G. P. Zipf (1949) [77] stated in his book "Human behavior and the principle of least efforts" that "the biggest is twice as big as number two, three times as big as number three, and so on". This concept applies where a free and natural competition mechanism exists e.g. free migration of people between lesser and better developed cities. The size of population between big cities shows similar distribution pattern when ranked. In case of ore deposits, though, such an assumption is difficult to make: perhaps the so-called free competition mechanism may be controlled by favorable geological conditions such as lithology and structures for migration of elements.

1,1	1,2	1,3	1,4	1,5	1,6	1,7	1,8	1,9	1,10
2,1	C.C							C.C	
3,1			C.C						
C.C								C.C	
					C.C				
7,1									

FIGURE 10.1 Creation of cells (4 × 4, 5 × 5, 10 × 10, km × km) for mineral deposit/belt model and C.C stands for "Control Cell" with geological characteristics including definite presence of mineralization.

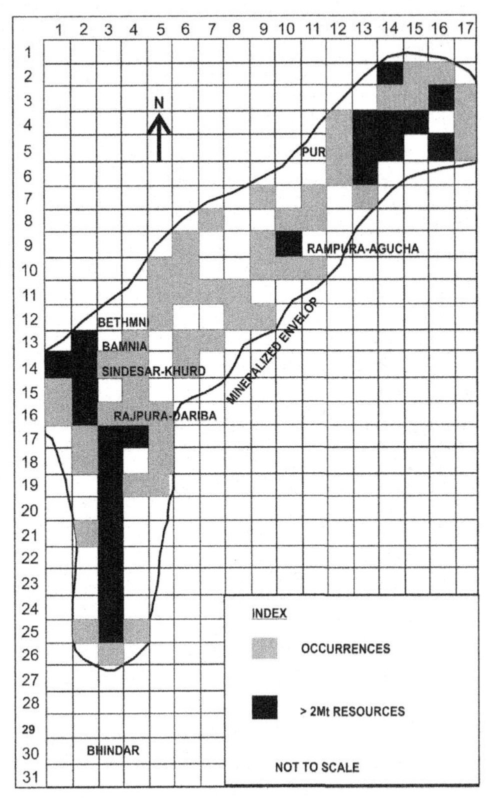

FIGURE 10.2 Expected exploration targets through deposit modeling showing established mineral resources (black), prospective target area (gray) and barren subblocks (blank) at Rajpura-Dariba-Bhendar Belt, Rajasthan, India.

The law is the limiting case of the Pareto distribution in statistics. The series is expressed as:

$$X, X/2, X/3, \ldots X/n$$

$$S_m/S_n = (n/m)^k$$

Where, S_m is the size of item of rank m, S_n is the size of item of rank n, and k is a proper fraction. Zipf gave many distributions with $0.5 \leq k \leq 2.0$.

10.2.5.1. Statement of the Model

The metal content (Zn + Pb) of 24 zinc-lead deposits in India are ranked and plotted on double logarithmic paper. The plot of raw data show irregular curve. The existing largest deposit is shifted to second position and fitted to the above equation of Zipf's Law. The plot of new series (Fig. 10.3) demonstrates ideal straight line. The new series is comprised of 61 deposits, 24 existing and 37 deposits of prognostic category.

The model indicates that about 75% of zinc metal was yet to be discovered in 1985 (Fig. 10.4). The prospective areas for new investigation are suggested. Continued exploration established additional +100 Mt reserves of high grade till 2011 in existing deposits (Rampura-Agucha) and new discoveries (Bamnia-Kalan and Sindesar-Khurd).

10.2.6. Statistical and Geostatistical Model

The Statistical-Geostatistical model is a part of resource and reserve appraisal system. The traditional methods of ore reserve estimation developed based on the characteristics and value of the surrounding ground. The procedures do not consider any valid way to measure the reliability or uncertainty of the tonnage and grade. When things go wrong, "experience" factors are commonly applied to make necessary corrections. This often results to write-off payable deposits and overvalue low-grade deposits. The conventional procedures occasionally deliver ore of estimated grades to the mill, particularly for short- and midterm period.

Techniques based on classical statistical theory provide solutions to the problem. The techniques include sample mean, range, standard deviation, variance, frequency distribution, histogram, correlation coefficient, analysis of variance, *t*-test, *f*-test and chi-square test for single variable and multivariate elements. Statistical methods provide deposit tonnage and grade with overall confidence limits. Classical statistical method is based on the assumption that sample values are randomly distributed and are independent of each other. It does not include the inherent geological variance within the deposit and ignores spatial relationship.

D. G. Krige (1962) [48] and subsequently Matheron (1971) [54] developed the theory of regionalized variables (RV). It is a random quantity that assumes different values based on its position in space within some region. This technique produces the "best linear unbiased estimate" (BLUE) of reserve and yields a direct quantitative measure of the reliability.

The statistical and geostatistical applications are capable to recognize the distribution pattern, identify characteristics of elements, establish correlation between variables and estimation of tonnage-grade with associated level of confidence. The techniques are competent to optimize sampling interval by sequential analysis during ongoing exploration and midterm corrections. Statistical and Geostatistical procedures can assist in optimization of exploration drilling for specific objectives (Fig. 15.23).

10.2.7. Orebody Model

The orebody model is conceptualizing the 3D perspective of the deposits. This can be done by creating series of drill hole sections with delineation of orebody. The geological interpretation is logically controlled by preparation of longitudinal sections and level plans so that a 3D view can be accomplished. One can experience an orebody model of

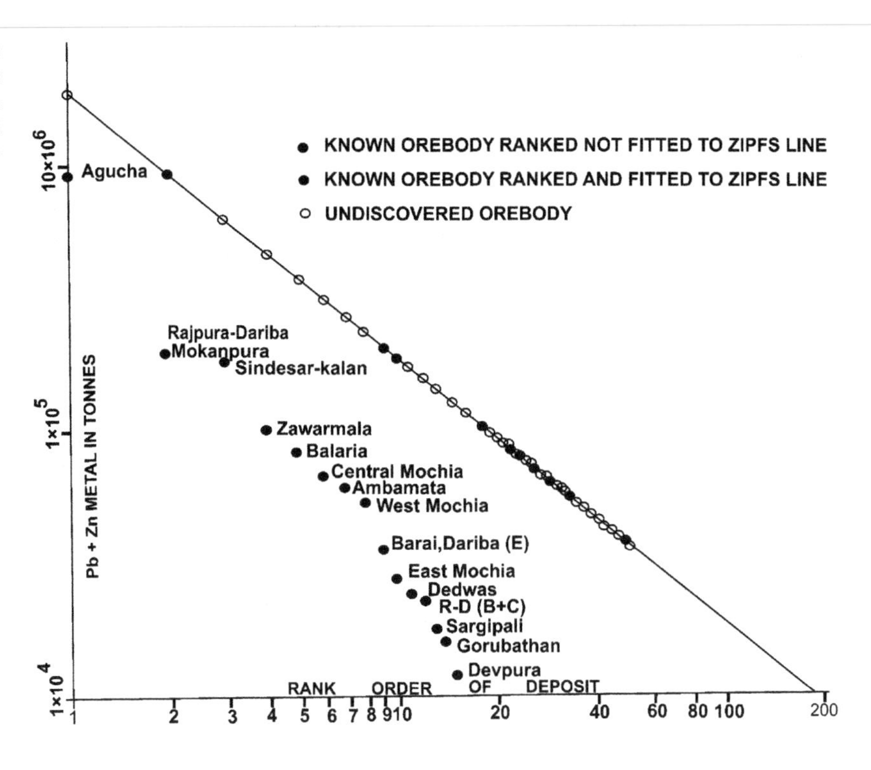

FIGURE 10.3 Twenty-four zinc-lead deposits in India, ranked and fitted to Zipf's Law in 1985. The irregular series fitted well at one step shifting in the Zipf's equation; +100 Mt reserves added including discovery of two hidden deposits by 2011.

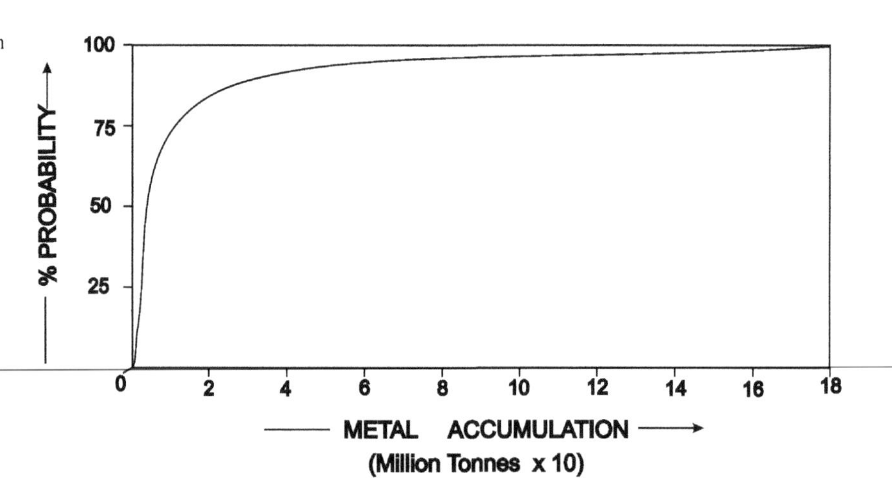

FIGURE 10.4 Probability of a deposit less than or equal to a given metal accumulation.

clay or plaster of paris for demonstration. The 3D orebody view with mine working layout can easily be achieved by using Geology-Mining software (Fig. 10.5) from all sample information. This is done by joining number of points from top to bottom of the footwall and hanging wall boundaries of the orebody between two adjacent sections on either side and known as wire framing.

The model can generate small subblocks for estimation of tones and grades. The block size is equivalent to daily/weekly/quarterly/annual production (Fig. 10.6). It is created at advance stage of exploration having sufficient physical and chemical properties of orebody with high level of confidence. The information enables to forecast grade and other parameters and used for mine design, planning, scheduling, blending and quality control. The bench and block plans are prepared with graphic facilities of user-friendly commercial software. The models enable planners to select effectively the most promising means of extracting ore both physically and economically.

10.2.8. Grade-Tonnage Model

Grade and tonnage models can organize a comparative status of the deposits in a region and help in delineation.

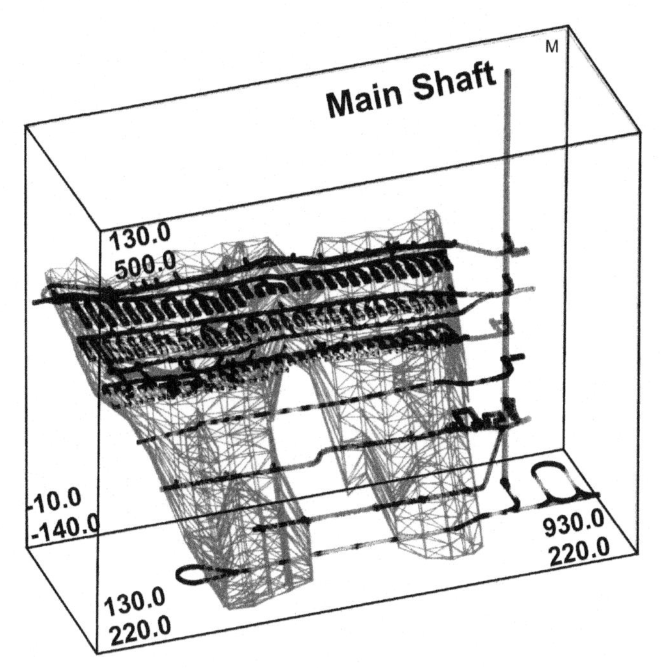

FIGURE 10.5 3D orebody wire-frame model based on 50 × 50 m drill interval of main lode (South) at Rajpura-Dariba mine, India, processed by DATAMINE software in 1991.

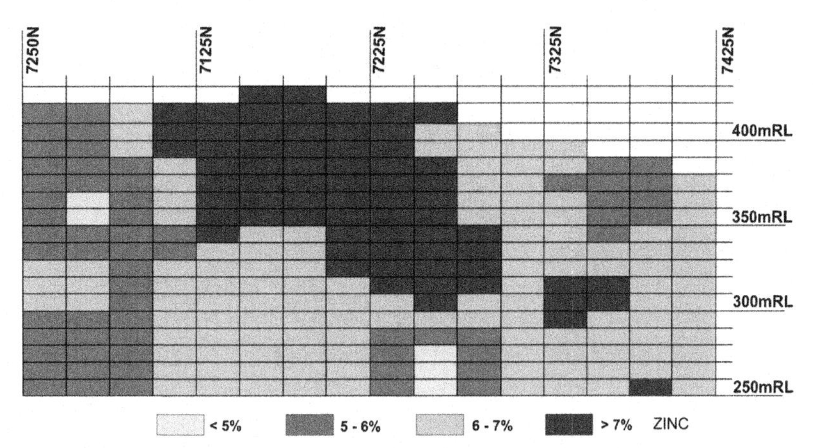

FIGURE 10.6 Zinc distribution on longitudinal section of a base metal deposit at 25 × 25 × 10 m block size showing the metal grade useful for forecasting, scheduling and monitoring.

Number of existing worldwide zinc-lead deposits are grouped with respect to its pre-mining size and grade (Table 10.1) to develop a grade-tonnage model. The tonnages versus total grade of the deposits are plotted in logarithmic scale (Fig. 10.7). The graphical presentation can be compared that provides a clear picture of deposit type and mined accordingly. The model provides information to speculate likely tonnage and grade of undiscovered deposits in the region. It also helps in assessment of economic analysis of these prognostic resources.

The present grade-tonnage model outlined four distinct types of clusters:

(a) The group-I is represented by world-class high grade-high tonnage zinc-lead-silver deposits, viz., Red Dog, Sullivan in North America, HYC, Broken Hill, Mt Isa, Century, Hilton in Australia and Rampura-Agucha in India. The reserves (+100 Mt), Zn + Pb grade (+11%) and silver content (+50 g/t) rank these deposits economically and exceptionally attractive for easy investment decision for quick and high return. These types of deposits qualify for large production capacity, quick payback period and long mine life.

(b) The group-II includes medium-tonnage and medium-grade deposits like Lady Loretta, Cannington, Lennard Shelf in Australia, Rajpura-Dariba, Sindesar-Khurd and Kayar in India. These deposits are economically significant for reserves (+10 Mt), grade (8-10% Zn + Pb) and rich silver content.

TABLE 10.1 Pre-mining Tonnage and Grade of Important Zinc-Lead-Silver Deposits in America, Australia and India (Source: Haldar, 2007) [33]

Deposit	Reserve (Mt)	% Zn	% Pb	% TMC	Metal (Mt)	g/t Ag	Age (~Ma)
A. North America							
Red Dog	150	16.2	4.4	20.6	30.9	110	300
Sullivan	170	5.5	5.8	11.3	19.21	59	1468
B. Australia							
Broken Hill	158	8.9	6.0	14.9	23.5	55	1650
HYC	227	9.2	4.1	13.3	30.2	60	1640
Century	118	10.2	1.5	11.7	13.8	36	1595
Lady Loretta	8.3	18.4	8.5	26.9	2.2	125	1647
Mount Isa	150	7.0	6.0	13.0	19.5	150	1652
Hilton	120	10.2	5.5	15.7	18.8	100	1652
Lennard Shelf	17	5.5	4.0	9.5	1.6	10-75	380
C. India							
Rampura-Agucha	107	13.5	2.0	15.5	16.6	54	1804
Rajpura-Dariba	25	8.0	2.2	10.2	2.6	100	1799
Sindesar-Khurd	37	6.7	3.2	9.8	3.6	154	1800
Bamnia	5	5.7	2.5	8.2	0.4	100	1800
Sindesar (E)	94	2.1	0.6	2.7	2.5	20	1800
Mokhanpura	63	2.2	0.7	2.9	1.8	10	1800
Kayar	11	10.4	1.6	12.0	1.3	10	~900
Mochia	17	4.3	1.8	6.1	1.0	40	1710
Balaria	16	5.9	1.2	7.1	1.1	36	1702
Zawarmala	7	5.0	2.2	7.2	0.5	40	1708
Baroi	6	0.0	4.6	4.6	0.3	50	1700
Deri	1	10.2	8.0	18.2	0.2	15	990
Ambaji	7	5.3	3.3	8.6	0.6	15	990

(c) The group-III deposits are comprised of medium tonnage (−10 Mt) and low grade (6-7% Zn + Pb) and low silver. The deposits like San Felipe in Mexico, Bou-Jabeur in Tunisia, Khanaiguiya in Saudi Arabia, Mochia, Balaria, Zawarmala, Baroi, Deri, Bamnia and Ambaji in India are marginally breakeven. The deposits continue mining because of existing infrastructure and periodical increase of metal price.

(d) The group-IV consists of Pering in South Africa, Gradir in Montenegro, Sindesar Kalan and Mokanpura in India with high resources (30-100 Mt), very low grade (<2% Zn + Pb) and negligible silver content. These big deposits look forward to large low-cost mine production, application of density media separation (DMS) for pre-concentrate generation, innovative technological breakthrough in metallurgy and rise in metal price.

The grade-tonnage relationship of a deposit can be visualized by plotting of reserve in the *x*-axis in logarithmic scale and total average metal grade in *y*-axis in normal scale (Fig. 10.8). The curve provides a quick evaluation of a deposit for its mining viability between tonnage and required grade. Variation in grade within a deposit helps production scheduling to maintain an optimum even grade to the metallurgical plant.

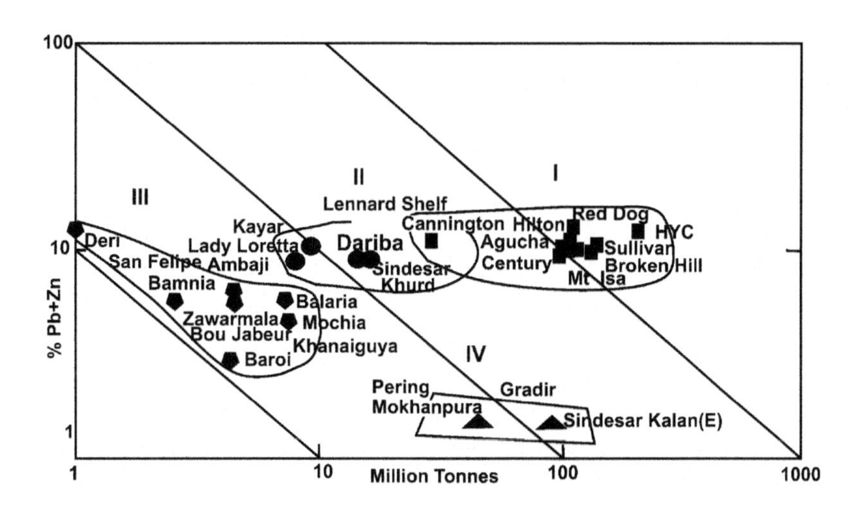

FIGURE 10.7 Grade-tonnage diagrams for explored zinc-lead-silver deposits in the World showing four clusters of very high and very low grade with various sizes of resources.

10.2.9. Empirical Model

Empirical model is a simplified representation of a system or phenomenon that is based on experience or experimentation. Modeling revolves around the need of future projection of possible availability of commodity under similar geological condition to satisfy ever-growing demand of society. The system can be developed by grouping or categorization of deposits on the basis of past experience in exploration. It can be used for future mineral search under similar favorable host environment. The key parameters are identified from the known deposits to conceptualize the module and targets. The target is to be defined in clear terms such as type of resources and reserves, type of mineral grades, structurally controlled high-grade deposits, localization of orebody (near surface or deep seated) and metal to metal ratio such as Zn to (Zn + Pb) ratio between 40 and +90. It can be demonstrated that the Palaeo-proterozoic zinc-lead deposits of Australia and India have many common features like surface signature, host rock assemblage, linear arrangement, grade of metamorphism, presence of geochemical halos, metal zoning and age of deposition. The key geological features between Australia and India are as follows:

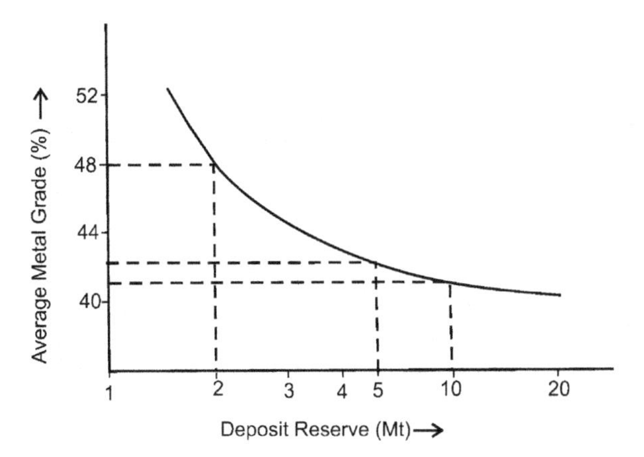

FIGURE 10.8 Conceptual presentation of grade/tonnage curve for broad valuation of chromite or bauxite properties aiming at investment for mine development.

(a) Surface signature: Surface oxidation as gossan is common key evidence of a sulfide deposit underneath. Size, shape and attitude of gossan can be an important tool for drawing exploration campaign after initial geochemical sampling. However, deep-seated body will be devoid of any such remnant. Most Proterozoic deposits like Broken Hill, Mt Isa in Australia, and Rajpura-Dariba, Rampura-Agucha, Saladipura, Khetri and Jagpura in India show excellent preservation of gossan.

(b) Mineralization: Essentially strata-bound and often stratiform, disseminated to massive, layered to fracture filled, sphalerite, galena, pyrite, pyrrhotite, chalcopyrite, sulfosalts with recoverable value-added metals like, Ag, Au, Cd etc.

(c) Host rock: Essentially single/multiple unit of volcano-sedimentary origin having tuffaceous component representing graywacke, black shale, siltstone, dolostone, calc-silicate, phyllite and schist with or without graphitic material. Primary sedimentary structures are obscured by subsequent various grades of deformation and metamorphism.

(d) Metamorphism/deformation/shearing: Deposits are metamorphosed, deformed and intensely sheared that vary between lower green schist and upper amphibole-granulite facies and strong penetrative deformation. HYC mineralization is un-metamorphosed.

(e) Carbonaceous matter: Most Proterozoic deposits are associated with carbonaceous matter with varying proportion.

(f) Evaporites (gypsum/barite): Presence of gypsum and barium indicates shallowing and deepening of fluctuating sedimentary basin evolution in which metal deposition events happen through exhalative metal-bearing fluids, addition of metal from country rocks to form a mineral occurrence. Many of the SEDEX deposits contain barite and gypsum.

(g) Halos: Presence of pyrite, siderite, Mn, carbonaceous material is common feature enveloping mineral body. If identified correctly halo can be good guide for regional exploration.

(h) Time of mineralization: All Australian zinc-lead deposits in Broken Hill, McArthur-Mount Isa basin belong to Palaeo-proterozoic age, ranging between +1600 and 1700 Ma. Major Indian zinc-lead deposits in Aravalli-Supergroup of NW Indian shield belong to same time group ranging between 1700 and 1800 Ma. Sullivan deposit, Canada, dated as Middle Proterozoic age of 1470 Ma.

10.2.10. Exploration Model

The procedures of mineral exploration have been discussed in greater detail in this book between Chapters 1 and 9. Exploration models promote the idea of generating prospects using a series of decision loops based on successive inputs from different sources. Each input contains new data to aid in decision making process. Model starts with study of existing literature and regional maps followed by various stages of Reconnaissance → Prospecting → Exploration → Phases-I, II, III → Estimation → Feasibility study. A dynamic model presented by logical flow diagram spells out objectives, time and cost along with decision making criteria to continue, modify and temporarily discard (stack in shelf) at any stage of activity. The exploration activity can be summarized in the following stages.

(A) Regional geological activity:

(Reconnaissance Permit)

The mineral exploration starts with study of demand-supply scenario, existing literature, satellite maps and imageries to support new investment, followed by application of mathematical/forecasting model. A regional field check is required for presence of surface signatures, suitable stratigraphic formation and favorable host rock, structure like shear, lineament, path finder elements and hallo from geochemical, response from airborne geophysics. The final goal would be identification of mineral province, belt.

(B) District geological criteria:

(Large Area Prospecting or Prospecting License)

The activities involve detail geological mapping, geochemical and ground geophysical survey, prioritization of targets and diamond and reverse circulation drilling. The result would be estimation of reserves and resources.

(C) Local geological activity:

(Mining Lease)

This is the final stage of exploration comprising of detail close space drilling to derive reserves and grades with higher confidence. At this stage the deposit is thoroughly understood covering mineralization features, stratigraphic control, host rock assemblage, structure, lineament and tectonic setting.

10.3. MODELING: A HOLISTIC DYNAMIC APPROACH

The mineral exploration flow diagram is sequentially synthesized step by step. It evaluates the property at the end of each stage for economic significance and opens two alternative paths suggesting either to "level pass" and continue successive exploration activity or to store provisionally in shelf for the future with changing environment (Fig. 10.9). The objectives of the search and preparation requirements are defined at the beginning. It is proposed to analyze the demand-supply scenario of the mineral or group of minerals at national and global level for prioritizing the long-term investment policy of the country. Once the mineral(s) is/are identified, the existing literature on occurrences, geological packages, exploration history, preferential host environment in space, characterization of the deposits in the country and elsewhere are compiled. The common key parameters are discussed with linkage if exists. The understanding of the stratigraphic horizon is essential to define a broad target area.

This guides to look for the favorable host environment. A preliminary field check along with few geochemical samples from probable rocks will indicate the significance of the area for submission of "RP". The work envisages are acquisition and interpretation of existing airborne data if any, stream sediment sampling and airborne electromagnetic survey with advance configuration system. The survey is carried out both in frequency and time domain with higher penetration capacity to identify deep-seated metallic bodies. Regional mapping, checking the existing maps, ground geophysics and test drilling will be able to identify anomalies and set the priority of targets with recommendations. The status of mineralization is expressed under UNFC code of Economic axis, Feasibility axis and Geology axis (E3, F3, G4) indicating the lowest level of confidence. The same tests can be performed through USGS or JORC Standard Codes.

If the result passes the first level, "Large Area Prospecting" or "Prospecting License" is applied to conduct general exploration covering detail mapping of the target

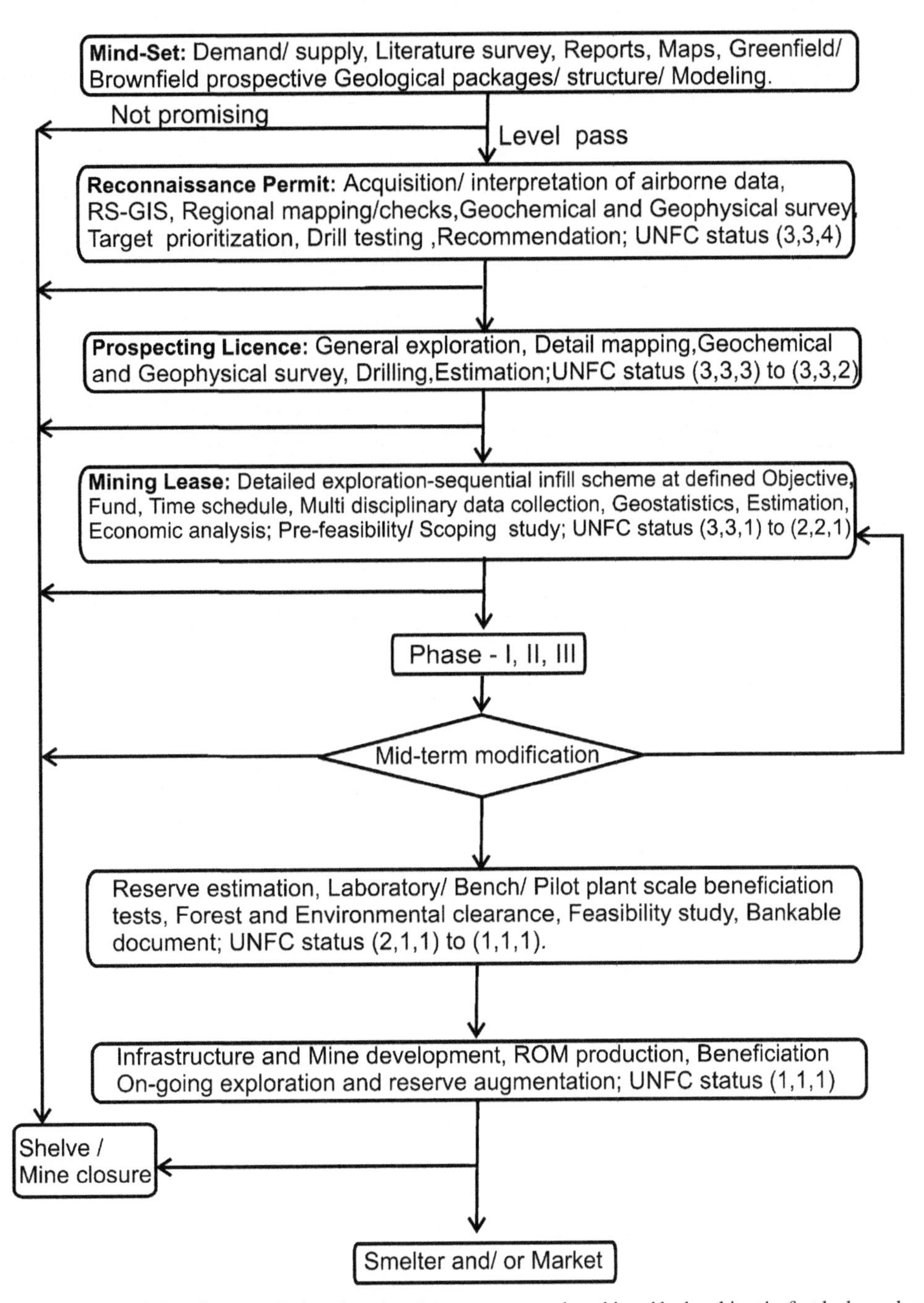

FIGURE 10.9 Holistic concept and flow diagram of mineral exploration system as can be achieved by breaking the first lucky rock exposed in the field to transport the metals to the market.

area along with geochemical and ground geophysical survey, broad base drilling and estimation of resources. The status of mineralization is marginally higher confidence of UNFC code (3, 3, 3) to (3, 3, 2).

Positive outcome allows submission of "Mining Lease". A scheme for sequential exploration with defined objectives, fund and time schedule for each stage is drawn. The activities include multidisciplinary data collection, statistical and geostatistical tests, reserve and resource estimation, economic viability, and Scoping study/Prefeasibility study. Midterm modification and corrective course of action is introduced at the end of each phase. The confidence status ranges between (3, 3, 1) and (2, 2, 1).

The estimation of ore reserves, preparation of geological reports, laboratory/bench/pilot plant scale of beneficiation test works, forest and environmental clearance, Prefeasibility and Scoping study, Feasibility and Bankable reports are completed at appropriate stage. The outcome will empower the final institutional investment decision. The confidence status is very high and varies between (2, 1, 1) and (1, 1, 1) of UNFC Code.

At confidence level of (1, 1, 1) the infrastructure and mine development starts followed by regular ROM production with matching beneficiation plant, if needed. Ongoing exploration continues aiming at "2 tonnes" of ore replacement for "1 tonne" of mine production. This is the World over mining philosophy.

Resource prediction model-Zipf's Law, concept of variable cutoff, and grade-tonnage relationship will strengthen the conceptualization of exploration strategy.

10.4. LIMITATIONS

Geological processes are events that occur on a geological timescale ranging between millions of centuries, hundreds of meters and thousands of kilometers. Compare to this the everyday models from physics and engineering operate at laboratory units and scale of human lifetime. Geological concepts represent an abstraction of nature and the numerical model represents a tremendous simplification of geological concept. Geological models are conceptualized with the physical, chemical and biological processes observed from stratigraphic sequences and syn-depositional and postdepositional phenomena introduced by external and internal tectonic forces. Uncertainties are present at various levels and these limitations are to be appreciated. Outcomes need to be compared with observations. A geological model can be considered predictive only when the applied parameters are within a reasonable range of values, supported either by measurements or by some type of geological reasoning, and only when the difference between the model outcome and observations is within an acceptable range of uncertainty. There is no guarantee with respect to the uniqueness of result in a particular situation and hence these should be expressed in a probabilistic rather than in a single-value manner.

FURTHER READING

The book "Exploration Modeling of Base Metal Deposits" by Haldar (2007) [33] will be adequate for this topic and many more related experiments.

Chapter 11

Elements of Mining

Chapter Outline

"Principle of mining is not maximizing production but to aims at zero waste generation with long-term sustainable development for non-renewable wasting assets."

—Author.

11.1. DEFINITION

Mining is the process of excavating ore minerals along with minimum quantity of waste rocks from the Earth's crust for the benefits of the mankind. The activities consist of handling loose ground, drilling and blasting of hard rocks, removal of broken materials from working place, and supporting the ground for safe operations. Various mining methods are available to exploit different types of deposits. The prime objectives in all cases are to mine the orebody in the safest and most economical way without sacrificing the interest of conservation of minerals—a nonrenewable wasting asset. The choice of mining techniques for an orebody depends on:

- Nearness to the surface,
- Nature of overburden,
- Shape, size, regularity and continuity,
- Strike, dip, thickness and rock strength,
- Nature of mineralization,
- Host and wall rock condition,
- Stripping/overburden and ore to waste removal ratio,
- Possibility to minimize internal and external dilution,
- Availability of infrastructures,
- Cost of mining and mineral dressing,
- Production target and resource-reserve status,
- Value of primary, associated commodities and value-added elements.

The first choice of hard rock mining is by adopting open-pit techniques, if the orebody is exposed or exists near to the surface. Underground mining methods are appropriate to that part of the orebody where open-pit operation is uneconomic due to high ore to overburden ratio. Deep-seated deposits are exclusively mined by underground methods. An open-pit mine continues and evantually converts to underground method at later period if the

Mineral Exploration. http://dx.doi.org/10.1016/B978-0-12-416005-7.00011-8

orebody persists beyond the economic limit of the open-pit option. The mining can be categorized as:

- Small/medium/large-scale production,
- Manual/semi-mechanized/fully mechanized operation,
- State/Private/Joint Venture ownership.

The underground soft rock mining of coal, lignite, rock salt, potash possesses different characteristics than hard rock metal mining like zinc-lead, copper, gold, platinum, uranium, iron, chromium and manganese as given in Table 11.1.

The various mining methods can broadly be classified on physical status as surface and underground. The underground stoping methods are further subdivided into soft-bedded type deposits and hard rock veins and massive type based on rock strength, mechanization, support system and production capacity (Table 11.2).

TABLE 11.1 Characteristic Features of Soft and Hard Rock Mining

Soft rock coal mines	Hard rock metal mines
Bedded, large horizontal extent, near uniform thickness, dip and quality.	Irregular in extent, thickness, shape, size, distribution, dip and metal content.
Sharp orebody contacts at footwall and hanging wall with waste rocks.	Orebody contacts are generally defined by cutoff principles.
RP, PL, ML at wide-space drilling, planning, 1-3 years development period.	RP, PL, ML at close-space drilling, planning, 3-5 years development period.
Limited sampling due to uniform nature of deposit.	Detail sampling from grass root to throughout the mine life.
Can accommodate minor surveying error.	High standard of surveying necessary to control undesired dilution.
Soft rock mining often uses electrically operated drills.	Hard rock mining requires compressed air power-driven drills.
Simple mine design, <5% development in waste rock, less waste handling.	Complex mining, +20% development in waste, more waste handling.
Produce inflammable and explosive gas.	Such problems are rare.
Limited mechanization.	High and sophisticated mechanization.
In general direct use ROM Ore. Upgrade process is simple and handling of rejects is limited.	Upgrade to concentrate by beneficiation and metal by smelting and refining. Tailing and slag disposal are critical.

11.2. SURFACE MINING

Surface mining is comparatively much cheaper than underground methods. It covers about 70% of global mineral production. The technique is most appropriate in case the orebody is either exposed or exists close to the surface. It can broadly be divided into several types depending on the nature of materials being handled. It can be unconsolidated placer (alluvial, colluvial and eluvial), gravel and mineral sand, bedded coal seam, blanket type bauxite, hard massive dipping base metals, iron ore, rock-phosphate, limestone, marble, kaolin and talc. The operation is either manual digging by picks and hammers to advance mechanized method. The surface mining process deals with removal of ore and large quantity of overburden in varying proportions. The ore to overburden ratio of bedded deposits is the proportion between vertical thickness of ore and the same of the overburden. The ratio between ore mined and overburden actually removed is called "stripping ratio". The thumb rule for planning ore and overburden ratio from past experiences is assumed for selecting mining method as:

(1) Manual quarrying 1:1.5
(2) Semi-mechanized quarrying 1:2
(3) Mechanized quarrying 1:4 to 10 and even occasionally >20 in case of high value noble metals
(4) Underground mining 1:<0.25

11.2.1. Placer Mining

Placer deposits are loose unconsolidated and semi-consolidated materials. It forms by surface weathering, erosion of the primary rocks, transportation and concentration of valuable minerals. Small deposits of gold, tin, diamond, monazite, zircon, rutile and ilmenite are common example. These minerals are recovered by small-scale miners working in the informal sector often using simple primitive artisan mining techniques. The process involves digging and sifting through mud, sand and gravel manually by bare hands or at best using simple tools and equipments like shovels, sieves, etc. This is similar to cottage industry of rural areas and characterized by low productivity, lack of safety measures, high environmental impact and low revenue earning for the Government.

The production and productivity can be increased by bucket dredging with or without centrifugal suction pump for large size properties of sand and gravels at the seacoast, deep offshore and deeper seabed poly-metallic nodules. The "bucket-dredge" can operate down to 20-30 m. The modern dredge can produce between 600 and 1500 tonnes per hour. Mineral concentration is done by jigs, cyclones, spirals and shaking tables. The bucket wheel excavator is a variation for dry sand mining. The machine excavates the mineral sand selectively and

TABLE 11.2 Classification of Stoping Methods by Physical Criteria, Rock Strength, Mechanization, Support System and Production Capacity

Deposit type/mining methods	Strength and support system	Mechanization and capacity
Surface mining—open pit		
Placer deposit	Loose sand, soft and unconsolidated, minimum support requires.	Total mechanization and high capacity—optional.
Shallow deposit	Medium hard, waste backfill.	-Do-
Massive deposit	Hard, massive, maximum waste backfill recommended.	Total mechanization with high production capacity.
Ocean bed mining	Soft and environment sensitive.	High mechanization.
Underground mining—bedded deposits		
Room and Pillar mining	Medium, roof collapses, sand fill.	Medium.
Longwall mining	Medium, roof collapse, support by stone, sand, timber and steel props.	Total mechanization and very high capacity.
Coal gas	Soft and medium, natural support.	Low capacity.
CBM	-Do-	Medium capacity.
Shale gas	-Do-	Low capacity.
Solution mining	-Do-	Low capacity.
Underground mining—hard rock		
Square set	Bad ground and pillar recovery, instant and total timber support.	Labor intensive, high cost, low production
Shrinkage	Narrow steep veins, occasionally backfilling required.	Less development, low mechanization, low OMS.
Cut and Fill	Bad ground, cable bolting, rib pillars and cement mix backfill essential.	Moderate mechanization and production.
Sublevel	Steep and wide body, crown and sill pillars, occasional fill support.	Advance mechanization and high production.
Sublevel top slicing	-Do-	-Do-
Sublevel long-hole drill	-Do-	Very high productivity
VRM	Massive orebody, mining in sequence of primary, secondary and fill panels, and backfill essential.	Advance mechanization, low cost, large capacity and high productivity.
Block caving	Large, massive, conditional support.	-Do-
Mass blasting	Pillars recovery, conditional support	-Do-

continuously feed the material to the hopper through a conveyor belt (Fig. 11.1).

11.2.2. Shallow Deposits

Strip mining process is most suitable for fairly flat shallow single-seam coal, lignite and other bedded deposits. The mineral layer is covered by an even thickness of overburden composed of soft topsoil and weathered rocks in succession. The soft and unconsolidated overburden can be stripped and removed either by dragline or shovel to expose a coal seam and certain metallic ores. The overburden might need drilling at grid spacing of 7.5×7.5 to 15×15 m depending on its hardness and thickness. The drill holes are charged with explosives and blasted. Drilling and blasting continues in advance with the movements of dragline and shovel.

Surface soil is often stripped separately and dumped as stockpile. The excavators either dispose of the overburden

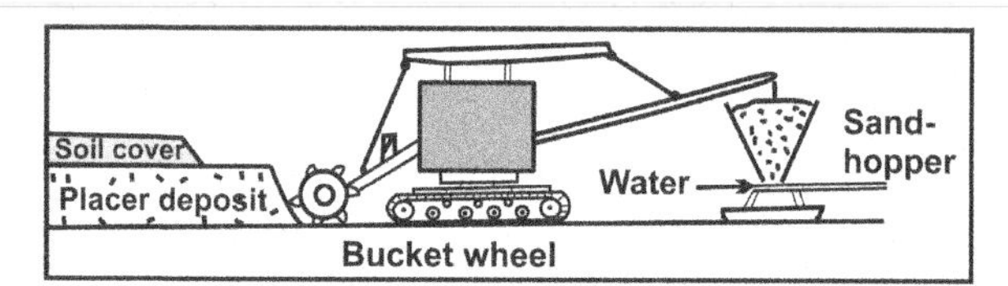

FIGURE 11.1 Schematic view of mechanized bucket wheel excavator in operation for beach sand mining to recover valuable minerals like gold, tin, diamond, monazite, zircon, rutile and ilmenite.

to a suitable location for land reclamation or store the waste material for future backfill after the coal/minerals are removed. The topsoil from the stockpile is spread back onto the reclaimed surface of the stripped mine. The new topsoil is often protected by seeding or planting grass or trees on the fertilized restored surface. The coal/metallic ore is usually removed by an exclusive separate operation. It uses smaller drill capable of drilling entire thickness of the seam or at suitable bench height if necessary. The blast hole spacing must be closer than that of the overburden rocks. The process involves charging with ANFO (ammonium nitrate mixed with diesel fuel oil) explosive and light blasting. This will avoid pulverization of coal. The broken coal or minerals are removed by shovel or front-end loader, crushed if required, screened to various size fractions and transported to beneficiation plant. The high wall of the mine opening is stable at 3 in 1 i.e. around 20° from vertical. The lumpy stockpile heap of overburden waste is stable at 30°-35° for shale and 35°-45° for limestones and sandstones. All measurements are with respect to the horizontal surface. The total cycle of ore and waste mining is given in Fig. 11.2.

In case of a deep-seated bedded deposit within permissible stripping ratio the overburden is removed by opening successive and progressive benches. It continues till sufficient area over the ore is exposed. The multiple seam mining is done by operating first pair of overburden and coal bed at a time and followed by second and third pairs in sequence. Finally, the total overburden rocks, stockpiled around the mine opening, are backfilled to the abandoned mine. The excavated land is reclaimed for future use.

11.2.3. Open-Pit Mining for Large Deposits

The open-pit or open cast mining method is the obvious choice for a property with wide area of mineralization exposed or exists close to the surface and continues to greater depth. The open pit is opening the orebody from the surface by separate removal of ore and associated waste rocks. It is the most economic option for a deposit up to that depth where the economic ratio of ore and waste can sustain. There are many advantages in open-pit mining method namely:

- Full visualization of exposed orebody and negligible ore loss,
- No ore is blocked, except crown pillar at the ultimate open-pit bottom to continue to underground mining,
- Greater concentration of operations, better grade control and blending,
- No need of artificial light in the day shift with natural ventilation round the clock,
- Greater safety, minimum mining hazards like gasification, roof and wall support,
- Easy draining/pumping of subsurface water,
- No restriction of working with heavy and bulky machineries,
- Lower capital and operating costs,
- Minimum mine development work and higher OMS leading to early production and quick return of capital invested (payback period).

There are few disadvantages like acquisition of surface right and rehabilitation of inhabitant people, loss of production due to extreme summer and winter, rain and snow, and handling of excessive waste rock.

A short, medium-term plan within the framework of a long-term design is prepared based on surface topography, 3D configuration of the orebody with respect to its shape, size, inclination, depth, grade distribution, hydrology etc. The first task in open-pit mining is to remove the topsoil, subsoil and overburden rocks in sequence. The overburden is dumped separately. It can revert back for replacement in reverse order. The next stage is to open the mineralized ground as first slot and known as "box-cut". The slot is then expanded to form bench system. The mining continues by advancing the benches horizontally within the broad framework of ultimate mine layout. The benches have two components i.e. floor for easy movement of man, machinery and materials and face (wall) to prevent collapsing. The pit maintains a critical slope angle both at footwall and hanging wall side, not exceeding 45° from the horizontal. The slope

FIGURE 11.2 A schematic view of complete operating cycle of mining for shallow-bedded deposits like coal and lignite seam.

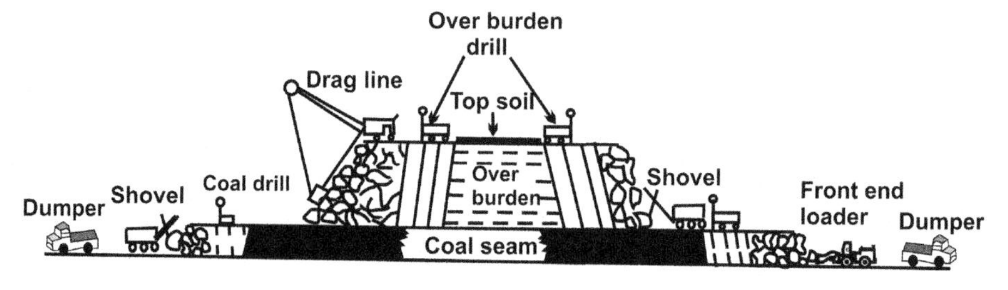

at footwall is in correspondence with the inclination of the orebody. The limit will be close to the orebody with minimum waste rock generation. The hanging wall slope is relatively shallow to reach to the deepest level of ultimate pit bottom (Fig. 11.3). The hanging wall benches generate the maximum waste rock from overburden.

The vertical height of benches varies between 5 and 10 m depending on width of the orebody, type of machineries deployed and to minimize footwall dilution. The minimum width of the benches is 18 m. The haul roads are 45-60 m wide at a gradient of 1 in <9 and permit easy movement of dumpers and other heavy machineries. The haulage road is connected to the main road for shifting ore to the surface stockpile and subsequently to the process plant. The waste rock is moved to the waste dump at appropriate location. The haul road runs along the periphery of the pit. It proceeds downward from upper to lower bench by developing ramps at suitable turning making a total haulage system (Fig. 11.4).

The total pit development and production activities include drilling, blasting, excavation, loading and transportation of broken ore (Fig. 11.5). The drill units are Jackhammer, Wagon drill and DTH hammers. The excavation and loading equipments are scrapers, bulldozer-ripper combination, front-end loaders, power shovels, draglines, bucket wheel excavators-bucket chain excavators

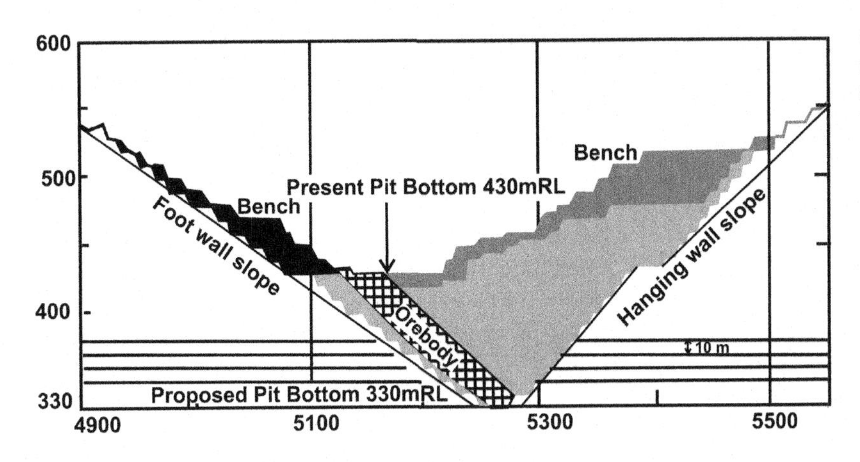

FIGURE 11.3 Schematic design of benches, footwall and hanging wall slope angle within stability and safety of miners and machineries and ultimate pit bottom in open-pit mine planning.

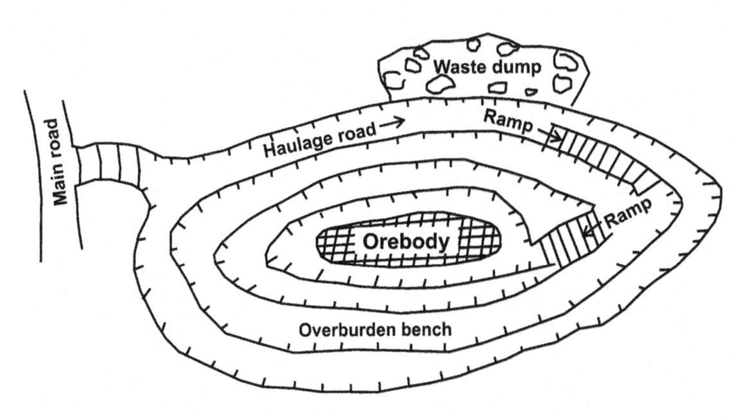

FIGURE 11.4 Sketch map to visualize the pit development plan showing the orebody at lower bench, selection of possible waste dump area, haulage road and ramp.

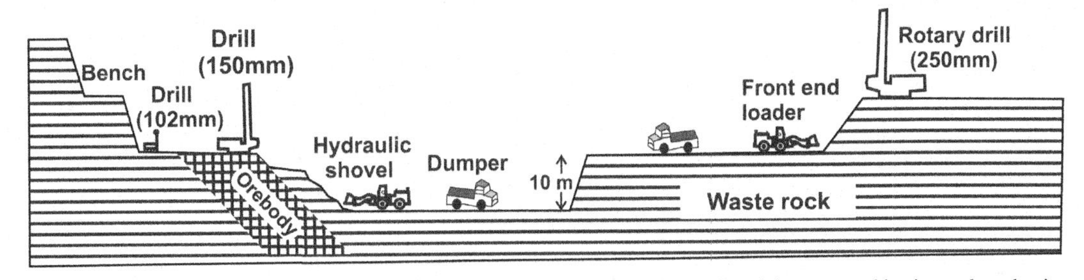

FIGURE 11.5 Schematic overview of mine activity schedule for ore and overburden drilling, shovel-dumper combination and production movements.

and graders. Transportation is done by various capacity trucks, dumpers, trains and belt conveyors.

The overburden waste rock is dumped as heap at a suitable location beyond the ultimate pit limit. The heap spreads both horizontally and vertically. The preferred location would be at the shortest distance over nonagricultural, non-forest land and non-drainage slope at the footwall side of the deposit. The dump material can be backfilled to the abandoned pit as the pit progresses or at the closure of the mine. The mine backfill process is the environmental reclamation of the worked out area. A view of a large working rock-phosphate mine is given in Fig. 11.6.

11.2.4. Ocean Bed Mining

Ocean water and ocean floor cover 70% of Earth's surface and host a vast variety of mineral resources namely salt (NaCl), potassium (K), magnesium-calcium [($MgCO_3$) and $CaMg(CO_3)_2$] sand, gravel, gypsum, poly-metallic manganese nodules (iron-manganese-copper-cobalt and nickel), phosphorites, seafloor volcanogenic massive and rich copper and zinc sulfides (VMS), with associated lead, silver and gold. Ocean water contributes formation of placer gold, tin, titanium, ilmenite, zircon, diamond and fresh water. The direct extraction of resources is limited to salt and potassium by evaporation, magnesium by electrolysis and fresh drinking water by commercial desalinization system. The fresh water is also produced by reverse osmosis process. The ocean bed gold, tin, titanium, diamond and manganese nodules (copper-nickel-manganese) are recovered by dredging technology from nearshore and deep ocean sediments. Dredging is an excavation operation carried out underwater, in shallow or deep seas with the purpose to collect bottom sediments and dispose at a different location. A dredger is a ship or boat equipped with a dredge. The unit can be used for mineral exploration underwater and seabed mining. Please refer Chapter 4, Section 4.5.14 and Fig. 7.31 for poly-metallic poly-nodule survey recovery by dredging.

The increasing population and the exhaustion of readily accessible terrestrial deposits undoubtedly will lead to broader exploitation of deep-seated land deposits and increasing extraction directly from ocean water and ocean basins.

11.3. UNDERGROUND MINING

The underground mining method is appropriate to that part of the orebody where open-pit operation is uneconomic due to higher ore to overburden ratio. The deposit is either deep seated or has significant vertical dimension. The purpose of underground mining is to extract the ore existing below the surface safely and economically. The aim is to generate as

FIGURE 11.6 View of Jhamarkotra rock-phosphate mine with haulage road and series of overburden benches. The mine is planned for 7 km long, 700 m wide and 280 m ultimate open-pit limit at 2 Mt ore and 16 Mt overburden waste per annum capacity (December 2008).

little waste as possible. The excavation of steeply inclined thin (vein type) and massive orebody proceeds either upward by "overhand" or downward by "underhand" stoping method. The underground mining techniques are broadly grouped under two categories:

(A) Soft rock mining: Coal, lignite, potash, rock salt, and
(B) Hard rock mining: Copper, lead-zinc, gold, chromite, uranium, platinum-palladium.

11.3.1. Mine Access

The orebody is accessed from the surface to the underground by various ways depending on the topography, configuration of the mineralization and stages of operation i.e. detail exploration, mine development and regular mine production. The access to the orebody is planned at minimum distance to reduce initial cost of development and the cost of movement of the man and material throughout the mine life. The selection of entry type should satisfy the most economic ore-waste haulage scheme. A mine may have single and in general multiple types of entry system. Care should be taken not to block the valuable ore in pillar around the any access tunnel or shafts. There must be at least two entry systems in combination for safe return route of the miners trapped inside due to any accident. The multiple entries render air intake-outlet for improved underground ventilation. Roof and walls of all the mine entries passing through weak zones must be properly supported by timber, steel, rock and cable bolting, cement grouting, rock creating and concrete walls.

11.3.1.1. Adit

Adit is a typical doorway to reach the orebody located inside the hill. The entry is by excavating horizontal tunnels preferably from the footwall side of the hill slope or above the valley level (Fig. 11.7). The dimension in the initial stage is 2 × 2 m for exploration and delineation of the orebody. The dimension is widened up to 4 × 3 m in the event of viable mining proposition. The adit serves the purpose of development and production above the valley level. The

FIGURE 11.7 An adit—horizontal entry above the valley level to reach the orebody at optimum distance for underground mining at Zawarmala, India.

broken ore from above the adit level can be brought to the portal in trains, conveyor belts, and rubber-tired trucks. Any water seepage above the adit level can be drained without pumping. In certain favorable situation the adit can serve the way for development of underground shaft as at 424 m level, Kolihan Copper Mine, Rajasthan. The adit can function for improving the underground ventilation system. The walls and roof of the adit are supported by concrete, if required, and painted white for better illumination. The cost of Adit development is relatively much cheap compared to any other entry system to the underground.

11.3.1.2. Incline

An orebody existing at shallow depth can be accessed by a sloping road, called "incline", driven from or above the valley level. An incline is moderately dipping access at an angle ~30° from horizontal or 1 in 4 to 1 in 5 so that the miners can negotiate the slope less strenuously. Rail tracks are laid in the incline for haulage of machine, ore and waste through bucket fitted to mechanized head gear-pulley system (Fig. 11.8). Inclined length beyond 150-200 m is not advisable. The size of the incline may vary between 2 × 2 m for exploration and initial mine development. It is expanded to 4 m wide × 2 m high for regular production. The roof and walls are supported for safety reason, if required. The cost of incline is moderately higher than adit development. It is suitable for near-surface small to medium size deposits at initial low production target. The incline can serve as air intake or outlet for improving ventilation system of underground mines.

11.3.1.3. Decline

The access to the underground mine can be achieved fast by developing a road called "decline" or "ramp" driven from the valley level or inside the mine, at 1 in 9. The slope enables driving of rubber-tired equipments like jeep, heavy-duty dumpers and earthmoving machineries. The decline is designed to develop as a spiral tunnel moving downward which circles either the flank or around the deposit. It is connected to each mine level. The decline begins with a box-cut at the surface fully protected by iron structure, bricks or concrete and acts as the portal (Fig. 11.9). It may also start from the wall of an open-pit mine or underground workings at desired level. The dimension can start from 4 to 5 m wide × 3 m high and expanded to size suitable for the safe movement of mine cars, dumpers and other machineries. The cost of development is moderately higher than adit or incline but much cheaper than shaft. Moreover, this trackless mining can reach the orebody and mine levels at much faster time and provides much higher rate of early production by using heavy-duty low-profile dumpers. Decline is the most accepted entry system to reach the levels at the shortest time at highest speed of development. Ramp is globally adopted mine entry system today including Kolihan Copper Mine and Sindesar-Khurd zinc-lead-silver mine in India.

11.3.1.4. Shaft

Shaft mining is the earliest form of underground method by excavating a vertical or near-vertical tunnel from the top down. The deposit located or continues in-depth is accessed

FIGURE 11.8 Mine entry systems by incline at a moderately steep angle ~30° to reach the orebody at shortest distance, primarily preferred during exploration to confirm ore characteristics and sample for beneficiation test works at Sindesar Khurd Project, India.

FIGURE 11.9 Mine entry systems through decline or ramp at a slope of up to 1 in 9 with fastest development to the bottom most levels and suitable for earthmoving machineries and heavy-duty dumpers at Sindesar Khurd Mine, India.

by sinking a well on stable ground adjacent to the footwall side of dipping orebody. It is known as "shaft" in mining terminology (Fig. 11.10). The location in the deposit or in the hanging wall side will require protective pillars to maintain stability of the shaft as mining progresses. The ore blocked in shaft pillar may be lost forever. The shafts are sunk as circular, square and rectangular openings. The shaft may be inclined following the dip of the deposit to avoid increasingly longer crosscuts to the ore at greater depth. But vertical shafts are techno-commercially preferred for better efficiency of ore hoisting. The shafts are permanently lined with concrete or steel. The finished diameter of shaft varies between 4.2 m (capable of accommodating a single cage and tub) and 6.7 m (a pair of tandem cages capable of accommodating two tubs). A mine shaft is frequently split into multiple compartments. The largest compartment is typically used for the cage, a conveyance used for moving workers and supplies to the underground. It functions in a similar manner to an elevator or lift in high-rise building. The second compartment is used for one or more skips, used to hoist ore and waste to the surface. The large mines have separate shafts for the cage and skips. A skip is a large open-topped container designed for automatic loading on to an ore bin. The shaft system is equipped with multi-rope friction winders, one for cage and the other for a skip with a counterbalancing weight. The cage has payload of 3.5 tonnes or 50 men. The skip bears a payload of 5 tonnes. The cages run on rigid guides. The skip along with its balancing counterweights run on guided ropes to prevent undue swing. The third compartment is used for an emergency exit. It is equipped with an auxiliary cage or a system of ladders. Additional compartment houses mine services such as high-voltage cables and pipes for transfer of water, compressed air or diesel fuel and air intake and exhaust for ventilation.

The shaft starts from the valley level and under certain circumstances from the ultimate open-pit bottom and from underground working. The ultimate shaft depth should be about 5 m below the bottom of the minable orebody. The deepest shaft in the world is reported from Driefontein (4000 m) and Tau Tona (3900 m) gold mines, Johannesburg, Merensky Reef platinum-palladium mine (2200 m), South Africa, Timmins copper-zinc mine (2682 m), Ontario, Canada, and Mount Isa copper-zinc-lead mine (1800 m), Queensland, Australia. The depth of shafts in India are at Kolar gold mine, Champion Reef (2010 m), Mosabani copper mine (685 m), Jaduguda uranium mine (640 m), Rajpura-Dariba zinc-lead mine (600 m), Zawar group (452 m, Fig. 11.10) and Khetri copper mine (475 m). The shafts can broadly be classified into two types depending upon its purpose. The main shafts are designed for hoisting ore and waste to the surface and lowering heavy machineries.

FIGURE 11.10 Main shaft with steel headframe and other structure in the center, and conveyor belt at right hand corner for ore transfer to mineral dressing plant at Mochia Mine, Rajasthan.

The auxiliary shafts are generally designated for man winding, material transport, stowing and backfilling material for open stopes and ventilation. The main shaft is located at the center of gravity of the orebody so that the distance of ore transportation is optimum from any point. The shaft must be protected from any damage by keeping sufficient solid stable block of rock and even ore on all sides as shaft pillar.

11.3.1.5. Raise and Winze

"Raise" and "winze" are miniature form of shaft. A vertical internal connection between two levels and sublevels of a mine is called a "winze" if it is made by driving downward. The "raise" is made by driving upward (Fig. 11.11).

The downward excavation for winze is carried out by hanging chain ladders or fixing iron ladder from the levels and sublevels. The upward driving is done with the help of fixing iron ladder upward and making temporary platform for overhand drilling and blasting. The development speed can be expedited by employing mechanized raise climber fitted with guide rails and fitted with temporary platform for drilling upward (Fig. 11.12).

11.3.1.6. Level

A typical underground mine has a number of near-horizontal "levels" at various depths below the surface and spread out from the main mine access (adit, incline, ramp and shaft) to the orebody. The levels are designated by prefix as main, sub, haulage and exploration depending on the purpose. The main levels are in vertical separations at an average height of 30, 60 and 120 m based on method of mining for development and production. The sublevels are intermediate levels for stope drilling at 10, 15 and 30 m. The haulage levels are used for transportation of ore to the crushing chamber and ore bin for hoisting to the surface (Fig. 11.11). The exploration levels are used for diamond drilling to enhance the tonnage and grade with higher precisions at lower cost and accuracy. The exploration levels are planned to utilize for mining purposes later.

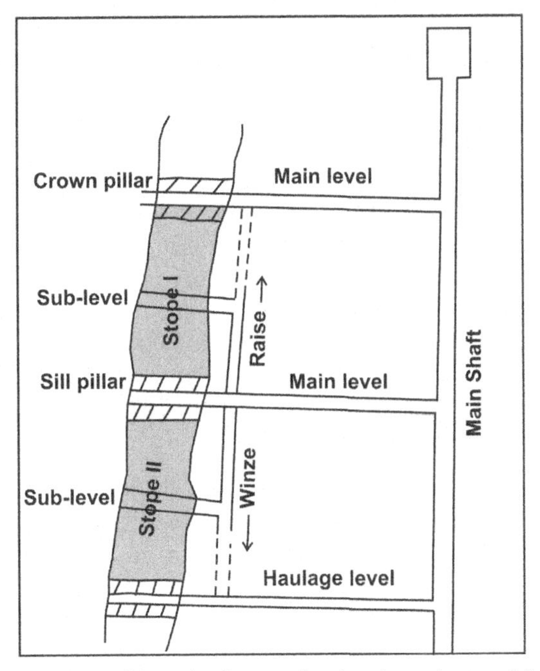

FIGURE 11.11 Schematic diagram showing the underground levels, sublevel, raise and winze used for mine development and production.

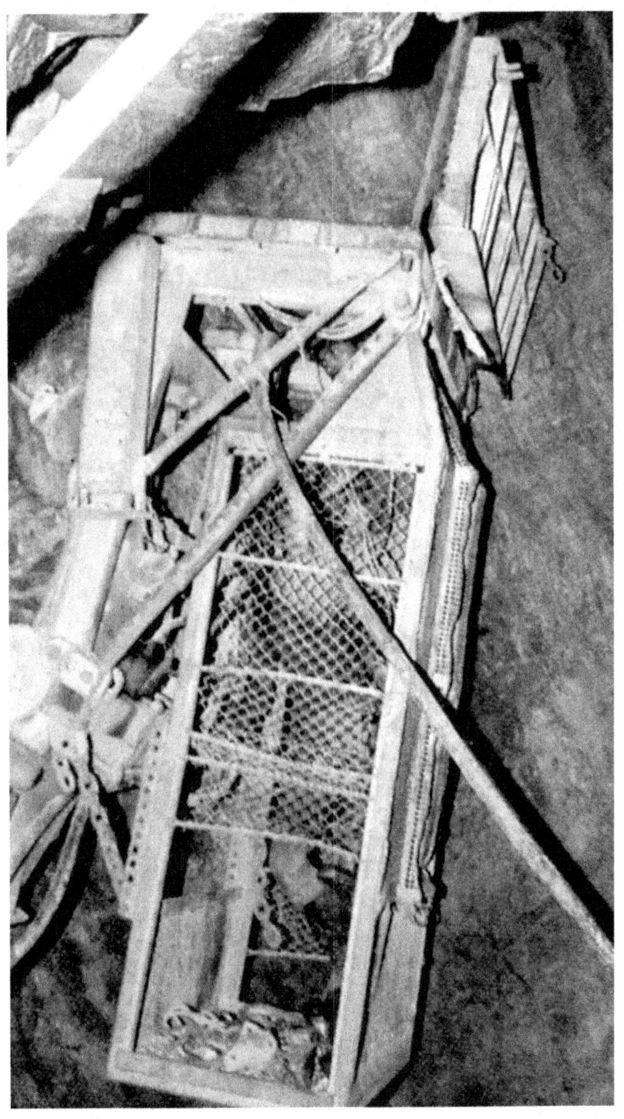

FIGURE 11.12 Alimak raise climbers are often in use for vertical or inclined excavation to join between levels and sublevels.

11.3.1.7. Drive and Crosscut

Drives and crosscuts are horizontal tunnels developed in a level with a cross section between 2 × 2 and 4 × 3 m. When the tunnels run nearly parallel to the orebody is called "drive". The drives are named as sill, drill, haulage drive depending on location and function (Fig. 11.13). The drives are very useful for exploring the orebody over the strike length from one end to other and serve the transport route of ore and development waste from the stopes.

The workings running at right or acute angle across the elongation of the orebody are called "crosscut" with further

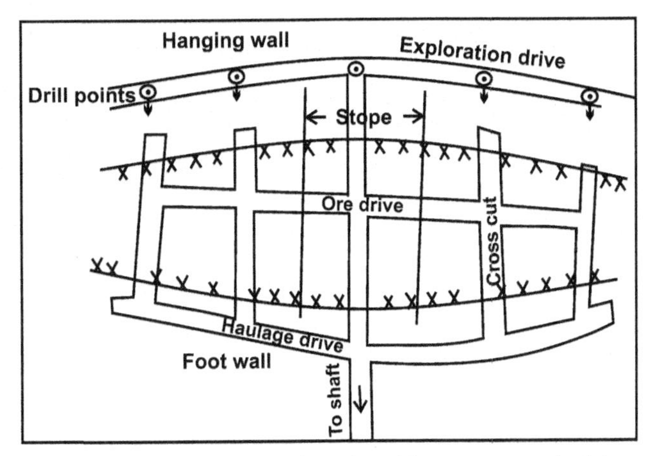

FIGURE 11.13 Schematic level diagram showing the various drives, crosscuts and mining stopes and purposes.

classification as main (Fig 11.14), drill, ore and exploration crosscut. The crosscuts are immensely useful for exposing the orebody across for true width, mine development, production and ore transfer passage.

11.3.1.8. Stoping and Stope

"Stoping" is the removal of the broken ore from an underground mine leaving behind an open space known as a "stope". Stope is a 3D configuration of in situ ore material designed for mining as an independent subblock in underground mining. Stopes (Fig. 11.13) are excavated near perpendicular to the level into the orebody. The excavated stopes are often backfilled with tailings, development waste, sand and rocks from nearby area. The fill material is mixed with cement at various proportions to increase strength of solidification. There are various stoping methods.

11.3.1.9. Mine Pillar

Ore left in situ to support the ceiling on various technical grounds is called "pillar". Pillars are of three main types viz. crown, sill and rib. The ore which is blocked and left at the top of the stope to prevent collapsing of upper level is called "crown pillar" (Fig. 11.11). Similarly the ore which is left below the stope to prevent collapsing of the active stoping is called the "sill pillar" (Fig. 11.11). If some part of the ore is left between two adjacent stopes as in vertical retreat mining (VRM) method or around some permanent structure like mine shaft, it is called the "rib pillar". Majority of the pillars, except around permanent structure, can be recovered later after the mining is completed between the two main levels and void stopes are filled. The reserves blocked in pillars are of "proved category" and grouped as "Other ore". The category changes to "developed" status at the time of recovery.

11.3.2. Underground Mining of Bedded Deposits

Underground mining of bedded deposits deals with extraction of soft ore. It is largely done for coal and

FIGURE 11.14 View of a main crosscut starting from the central main shaft to the orebody at Boula-Nausahi chromite deposit and operates as main production haulage (December 2009).

sometime for rock salts like halite and sylvite. The method in common practice is driving through tunnels, passages, and openings. All are connected to the surface for the purpose of removal of broken ore. Excavating equipments cut, break and load the soft coal to a size suitable for haulage. Continuous loadings of coal, having low density and high bulk nature, on a conveyor belt are the best choice. Alternatively, the coal is drilled, and the resultant holes are loaded with explosives and blasted in order to break the coal to the desired size. There are various methods of mining bedded deposits depending on the thickness and depth of the seam.

In order to protect the miners and equipment in an underground coal mine, much attention is paid to maintain and support a safe roof or overhead ceiling for the extraction openings. A long face or working section of coal, some 200 m in length, is operated at one time. The miners and machinery at the working face are usually protected by hydraulic jacks or mechanical/timber props which are advanced as the coal is extracted.

11.3.2.1. Room and Pillar Mining

Pillar mining, also known as "Room and Pillar" or "Bord and Pillar" mining, is the most common type in underground coal mining because of its flexibility at low capital expenditure. The method is suitable for narrow (2-4 m thick) coal seam free from stone bands. The seams are located at moderate depth having strong roof and floor which can stand for long period after development. The name indicates that "rooms" are the entryways within the seam and series of equidistant pillars of coal are left standing to support the roof of the mine (Fig. 11.15). A typical design would have the entryways (rooms) with a width between 3 and 4.8 m employing coal-cutting machines. The pillars vary between 12 and 49 m from center to center depending on the size of the gallery, depth of mining and rock condition. Pillars can be square and rectangular. The excavation height varies between 2 and 3 m. The nature of immediate roof is the deciding factor. About 30% of planned production is obtained during the development stage once the seam is touched.

The mine development starts after the main connecting road reaches the mining horizon from shaft or incline or decline. The excavation carries in a set of "Panels" confined in a district. Each district is separated from adjacent district either by solid coal or by brick and stone barrier. Essential roads/galleries are developed between two panels for passage of men, machineries, drainage, ventilation and stowing. A road driven along the dip of the seam is called "Dip gallery" or "Dip". One of the heading is called the "Main Dip" or "Main Rise". A road driven along the strike of the seam is called "Level Gallery" or "Level". A roadway in barren rocks connecting two or more coal seams is called a "Drift" or "Stone Gallery". A gallery in the process of driven is called a "Heading". The moving form of any working place is "Face". The panels are sealed off and isolated in case of any emergency arising out of spontaneous heating or fire in the panel.

The coal pillars are recovered in sequence after completion of development work. The pillars are removed in combination of coal cutter, loaders, shuttle cars and train haulage. Continuous mining is performed by integration of continuous-cutting machine and conveyor haulage grouping. Continuous mining utilizes equipment with a large robust rotating steel drum fitted with tungsten carbide teeth to scrape coal from the seam. After removing coal the roof is allowed to break and collapse by caving into the voids known as "Goaf". Pillars are split into small "Stooks" and extracted. The number of splitting depends on

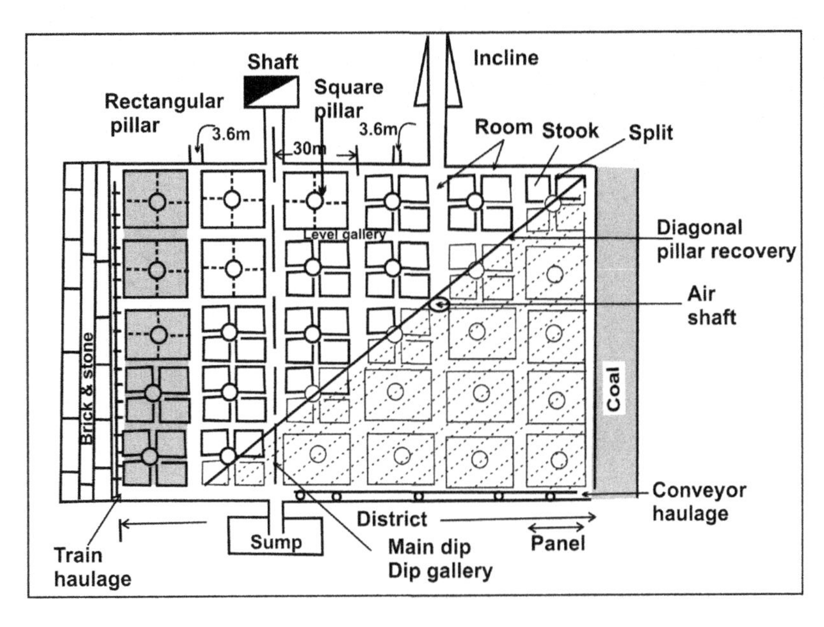

FIGURE 11.15 Schematic diagram of "Room and Pillar" mining method suitable for narrow coal and lignite deposits.

size of the pillar. The pillars are recovered in a sequence of diagonally, or otherwise based on strata condition. The surface may develop cracks and subside. The goaf is completely packed with incombustible material by "stowing" method. The filling is critical where water bodies or infrastructure exists on the surface or in case of multi-seam mining. The complete pillar mining activities and sequence of pillar extraction are depicted in Fig. 11.15. The Room and Pillar system can be employed in mining any blanket deposit of iron ore, bauxite, base metals, stone and aggregates, talc, soda ash and potash.

11.3.2.2. Longwall Mining

Longwall underground mining is a technique where a wall of 250-400 m long of coal is mined in a single slice. The method is suitable for thin seam of 1-2 m thick extended over 3-4 km even at a greater depth. It requires high capital investment toward large capacity continuous coal cutter and self-advancing hydraulic roof support system. Mine development is bare minimum to attain production at full capacity at the earliest. The method provides maximum yield with highest extraction of coal recovery.

Longwall mining works on laying one long face extending the entire working section of the coal seam. A double-ended ranging drum shearer continuous coal-cutting machine removes the coal and transfers it to a series conveyor. The line of action is always in one direction leaving the void (goaf) behind. The roof over the goaf is allowed to collapse or partially or completely supported by stone and sand. A void strip of 3-6 m wide immediate to the advancing face is supported by timber, steel props, bards and chocks in a systematic manner. In modern large-scale mining the roof of the advancing face is propped by self-advancing hydraulic support. The "longwall advancing" starts extraction of coal from the vicinity of the main entry and proceeds outward to the boundary of the panel. The roadways are made in the mined-out area behind the face. The "longwall-retreating" blocks the panels completely on all four sides and face retreats along the roadways toward the main entry (Fig. 11.16). One of the two roadways on either end of the longwall face is called "Gate road" or "Haulage gate" used for ore transport and air intake. The other one "Tail gate" serves as material supply and air outlet.

"Short-wall" mining method is a variation to longwall and involves the use of a continuous mining machine with moveable roof supports. The coal panel dimensions are 50-60 m wide and about 500 m long due to factors like geological strata. The continuous cutter operates to remove the coal slices. The mined coal is then dumped onto a face conveyor for haulage. The roof is supported by specially designed shields which operate similar to longwall mining.

11.3.2.3. Coal Gas Mining

The coal seam with better continuity located at a greater depth of 500-800 m with low calorific value are prima

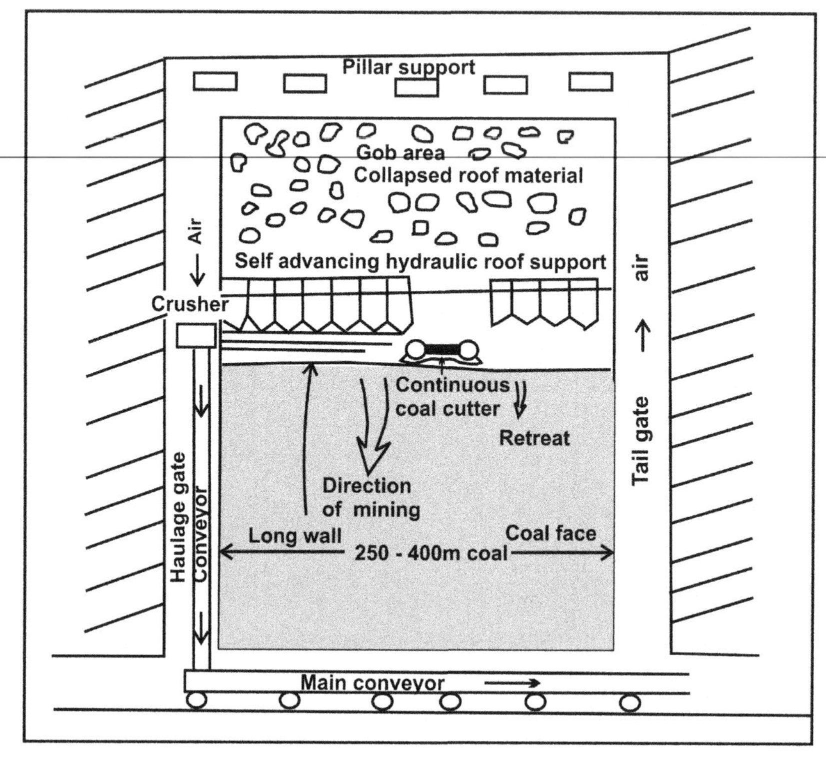

FIGURE 11.16 A general layout of longwall advancing with highly mechanized large production and simultaneous backfilling support system for coal and lignite deposits.

facie uneconomic for mining. This coal resource can be converted into a combustible gas by in situ underground gasification process. The large quantities can be viable fuel for power generation. The natural methane gas also forms during compression and heating of organic matters over geological time period and entrapped within the coal seam.

The underground coal gasification process includes drilling of one well into the non-mined coal seam for injection of the oxidants (water/air or water/oxygen mix). The production well is drilled some distance away to bring the gas to surface (Fig. 11.17). The coal seam is ignited through the first well and burns at high temperatures, generating carbon monoxide (CO), carbon dioxide (CO_2), hydrogen, oxygen and large quantities of methane (CH_4) at high pressure. The coal faces carry on burning and the oxidants are injected for spreading the flaming along the seam. The process continues laterally through the entire seam between the injection and the production well. The injected oxidants react with coal to form a combustible gas which is brought to the surface in a production well, cleaned and used as a fuel. A cavity is formed as the coal burns and the roof collapses. The operation is expected to be more cost effective with greater the lateral growth and longer the life of gasification. Once the coal gasification is exhausted a new set of wells are developed in the nearby area.

11.3.2.4. Coal Bed Methane

CBM is a primary clean energy source of natural gas. Development and utilization of CBM is of great social and economic benefits. It is a clean burning fuel (compressed natural gas—CNG) for domestic and industrial uses. The extraction of CNG reduces explosion hazards in underground coal mines. Large amount of methane (CH_4) is often associated with some of the coal seam. It is generated either by microbiological or by thermal process as a result of increasing heat at greater depth during coal formation. Coal seams are often saturated with ground water at high pressure. CBM recovery is a method of extraction of methane from coal by drilling a well into the deposit (Fig. 11.18). Methane will readily separate by reducing water pressure with partial pumping of water. The gas moves to the well and pipe out to surface. It is compressed and sold to the market. The extraction process involves drilling of 100s of wells with extensive infrastructural support facilities. United States currently produces 7% of the natural gas (methane) from CBM. India is endowed with huge reserves of bituminous coal of Paleozoic and Tertiary ages within the CBM window at depths of nearly 250-1200 m.

11.3.2.5. Shale Gas Mining

Shale gas and oil shale are significant enrichment of natural organic matter in fine-grained shale rock. The favorable location of shale gas is in various sedimentary basins pertaining from Carboniferous to Cretaceous age. Shale gas has been on the global energy map since the 1950s. The product has been technologically and economically accessible into important source of natural fuel in USA since late 1990s. The interest for shale gas is increasingly gaining prominence throughout the entire world as low-cost energy resources. Shale is in general has low permeability to allow significant fluid or gas flow both horizontally and vertically. Therefore, gas production of commercial quantity requires advance in hydraulic fracturing within the shale rock for additional permeability. Vertical holes are initially drilled to reach the target shale horizon. Then horizontal directional drilling technique is employed for number of horizontal holes from one parent vertical hole. Hydro-fracturing is done by introducing fluid (mainly water) under high pressure to allow the gas to move to the production well (Fig. 11.19). The horizontal drilling has

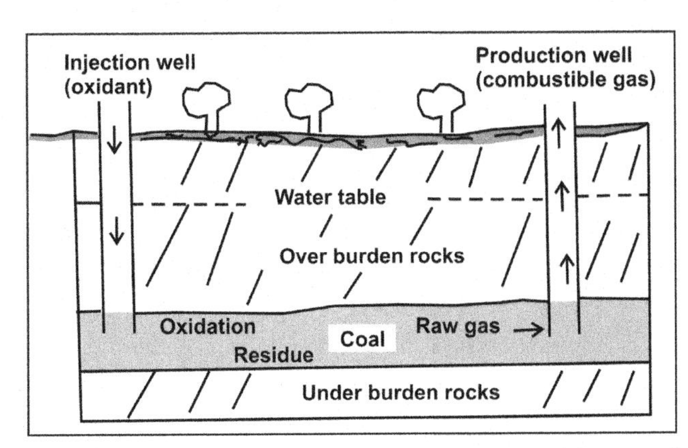

FIGURE 11.17 Schematic view of gasification process of deep-seated low-calorific coal seam through series of injection and production wells.

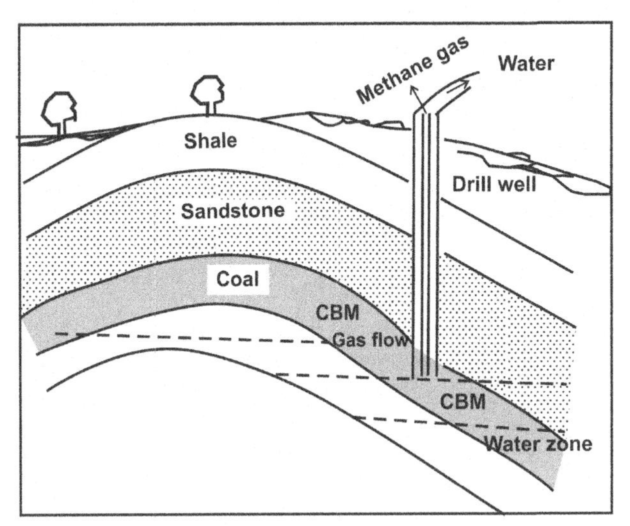

FIGURE 11.18 Schematic view of CBM mining as popular source of clean CNG used for domestic and industry purposes.

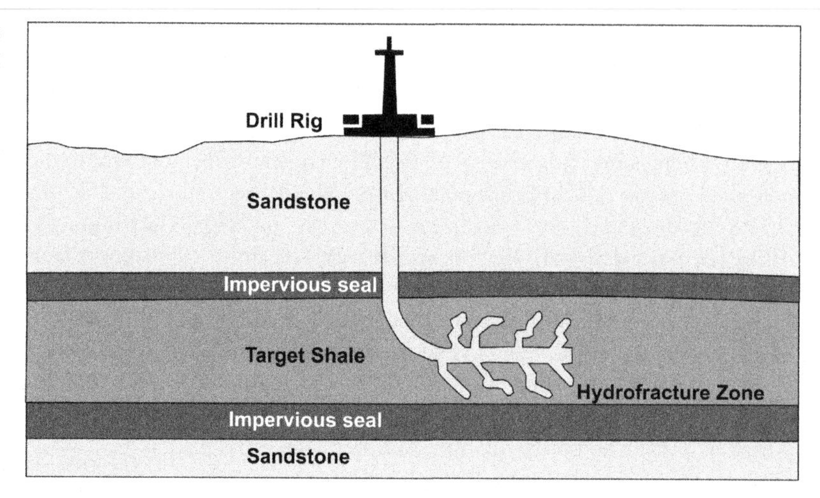

FIGURE 11.19 Schematic view of shale gas mining by hydro- or gas-fracturing. The technique is becoming a popular source of energy today everywhere.

made shale gas profitable in the past decade making shale gas an important source to have in an energy company's asset portfolio.

11.3.2.6. Solution Mining

Deep-seated rock-salt deposits can be mined by introducing fresh water by a powerful pump with large-diameter double tube pipe into the orebody. The water flows into the ground and salt is dissolved into solution (brine). The system has a smaller diameter inner pipe that forced the brine to the surface. The return salt solution is stored in tank. The brine is finally pumped to the chemical plant for ultimate recovery of clean salt (Fig. 11.20).

11.3.3. Underground Hard Rock Mining

The underground metalliferous mining deals mainly with vein type to massive orebodies like gold, platinum, chromium, zinc-lead, copper, uranium and manganese generally hosted by hard rocks. Stopes are developed with 3D perspectives. Drives and crosscuts are developed at successive levels which are connected vertically. The ultimate mining precedes either upward by overhand or downward by underhand stoping. The common underground mining methods are discussed hereafter.

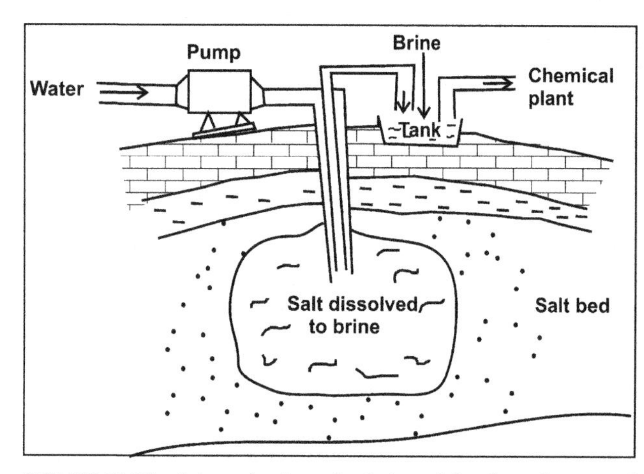

FIGURE 11.20 Schematic view of solution mining from deep-seated rock salt at low cost.

11.3.3.1. Square Set Stoping

The square set stoping is applicable for high-grade thin to medium orebody with weak walls and back needing instant support system. The technique is suitable for recovery of fractured ore remnants and pillars with extremely bad ground condition. The method is costly and labor intensive but suitable regardless of size, shape and depth. The ore recovery is better with efficient grade control. The waste can selectively be left in the stope as filling. Timbering is dominant feature in square set stoping causing fire hazards and high accident rate.

The concept is very similar to "Room and Pillar" mining method. The term square set indicates regular framed timber support of the void area immediately after the ore is removed. The ore is excavated generally by overhand drilling in small rectangular blocks large enough to be supported by standard square set of timbers (Fig. 11.21). Ore can be mined in horizontal, inclined and vertical panels. The underground operation utilized a square set mining method at very high-grade Phoenix copper mine, British Columbia, Canada, Broken Hill zinc-lead mine, Australia, Sullivan zinc-lead mine, Canada, and Namtu-Bawdwin lead-silver-zinc mine, Myanmar.

11.3.3.2. Shrinkage Stoping

The conventional shrinkage method of stoping is practiced in steeply dipping (70°-90°) narrow veins with regular ore boundaries and thickness ranging between 3 and 12 m. The length of shrinkage stope varies between 50 and 100 m

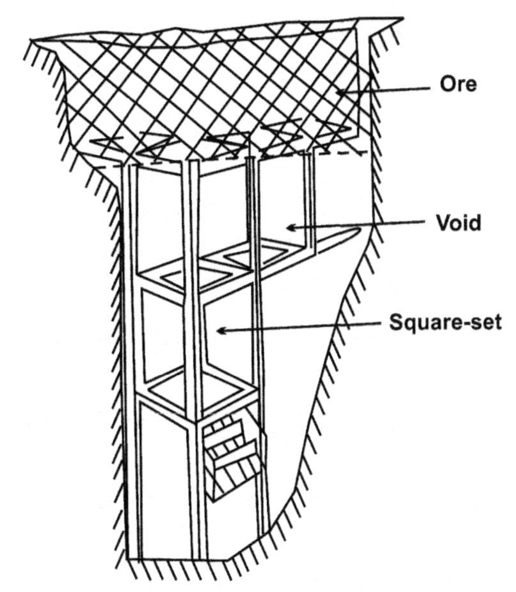

FIGURE 11.21 Square set mining design is suitable to recover high-grade ore blocked in weak and fractured pillars needing instant support system.

with height between 40 and 60 m. The development work involves by connecting the two main levels by raises at both ends of the stope. The raises provide through ventilation for working and man-ways at regular vertical interval. The next activity is excavation of a preparatory level 5-10 m above the haulage level. The two levels are connected by a number of conical box-hole raises at 8-10 m center. Preparatory level is then stripped up to side-walls exposing the whole width of the orebody. The cycle of overhand drilling by 33 mm diameter Jackhammer drills and blasting with ANFO explosive continues end to end in the strike direction. The average lift in the stope is between 2 and 2.5 m/blast. The blasted ore makes a greater volume than in situ ore due to swell factor. Therefore, about 35-40% of the broken ore is withdrawn (shrink) from the advancing stope. It provides working space between muck pile and the new overhead face to drill for the next ore slice (Fig. 11.22). The drilling, blasting and partial ore removal continues up to the upper pillar drive. Once the top of the stope is reached all the ore is removed from the stope. The stope may be backfilled or left empty depending on the rock conditions. Power factor achieved in the stope is around 250 g/t of ore broken. The method yields a stope productivity of 30-50 tonnes per day. This method was practiced at Zawar zinc-lead and Mosaboni copper mines earlier. The method is abandoned due to low OMS.

11.3.3.3. Cut and Fill Stoping

Horizontal flat back Cut and Fill mining method is applicable under wide range of conditions from small to large deposits of irregular outline, flat to steep dipping deposits. The cost of preparation and development is much lower than other methods. The production starts quickly requiring less manpower. The cost of filling operation is high. The development work involves preparation of a haulage drive along the orebody at the lower main level and an undercut at 5-10 m above. There will be number of separate raises for the purpose of man-ways, ore-pass and filling material. The stopes are 90 m long and divided into three panels of 30 m each, separated by 5-m-thick rib pillars. The producing panel is excavated by overhand drilling using Jackhammer drills of 33 mm diameter and wagon-mounted COP-89 drifters. Excessive damage of the roof is experienced by vertical overhand drilling in case of weak rock condition. The pattern can be changed to 3-m-long horizontal holes by air-leg-mounted Jackhammers in two rows on a 1.8-m-high vertical face. The roof is pre-supported by 12-m-long cable bolts at 2 × 2 m grid, repeated every 7-10 m lift. The blasting is done in horizontal slices not exceeding 3 m in the roof leaving 5 × 5 m post pillars at 15 m centers along strike and 12 m across if the ground condition dictates.

The mucking is carried out by 1.3 m^3 electric "Load Haul and Dumps" (LHDs) operating on consolidated fill.

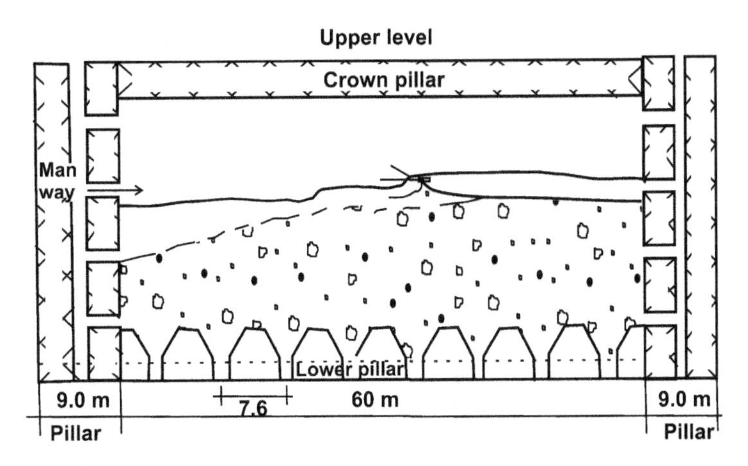

FIGURE 11.22 Schematic diagram of shrinkage stope in operation at Mochia mine, Zawar, India, to selectively recover high-grade vein-type sulfide ore.

It loads and hauls the broken ore up to the nearest ore-pass which opens at a track-facilitated haulage drive. Diesel or battery-powered locomotive pulls a train of 5-tonne granby cars. The locomotive carries the ore up to the main ore-pass over an underground jaw crusher with output of −150 mm size. The primary crushed ore is hoisted to surface in 6-tonne-skip driven by 697 KW koepe winders. The stopes are backfilled with waste rock, sand and +32 micron size classified mill tailings mixed with 5-15% cement. Cyclic drilling, blasting, loading and filling in three panels of stopes constitute the Cut and Fill mining operation (Fig. 11.23). Each stope can produce 200-250 t/d. The rib, crown and sill pillars are recovered later. The post pillars are left inside the fill material. Cut and Fill method is widely deployed at copper-zinc deposit at Cobar, New South Wales, zinc-lead deposit of New Broken Hill Consolidated, and zinc-lead deposit at Rajpura-Dariba, India.

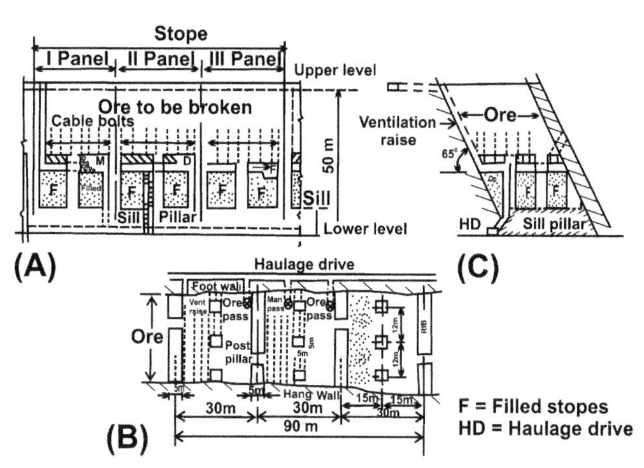

FIGURE 11.23 Schematic diagram of flat back Cut and Fill mining with post pillar stopping at Rajpura-Dariba mine, (A) longitudinal section, (B) plan and (C) cross section. The operation is in a cycle of drilling, mucking and filling.

11.3.3.4. Sublevel Stoping

11.3.3.4.1. Conventional Sublevel Stoping

The conventional sublevel stoping (Fig. 11.24) is applicable for steeply dipping thick mineralized lenses with regular boundaries. The orebody is strong enough to separate easily from stable hanging wall and footwall. The stope is designed between two main levels at a vertical separation of 67 m. Haulage drive is at 20 m from the ore contact. The draw point crosscuts are at 9 m center to center at bottom main level. The crosscuts end with development of undercut or trough or box-hole for drawing ore. There are two raises on either end of the stope limit for development of three sublevels which act as drill drive within the orebody. Stope development involves driving of the first sublevel (preparatory level) from the long pillar raise 10-12 m above the ore draw point level. The second and third sublevels are developed at about 10 m vertical intervals leaving a 4- to 5-m crown pillar above the topmost sublevel. One of the raises serves as service for man-way and the other for opening the slot up to a height of about 4 m above the widened preparatory level. This follows stripping of slot-raise between the preparatory level and second sublevel up to footwall and hanging wall. It creates a free face for parallel down-hole drilling of 7.6 m long and 36-mm diameter on a sublevel bench. Blast holes are drilled with 1 m spacing and 1.20 m burden. Care is taken to puncture the hole into the roof to ensure blasting of humps, if any. The powder factor is comparatively low at 125 g/t of ore broken, achieved in primary stoping necessitated secondary blasting of large size boulders. The stope productivity is 100 t/day.

11.3.3.4.2. Sublevel Top Slicing

Sublevel top slicing method (Fig. 11.25) is scientific development of conventional sublevel stoping for increased level of production. Stope development starts with the excavation of a set of 67-m-long ore and man-pass raises, located on

FIGURE 11.24 Conventional sublevel stopping is applicable for steeply dipping thick orebody under competent ground condition.

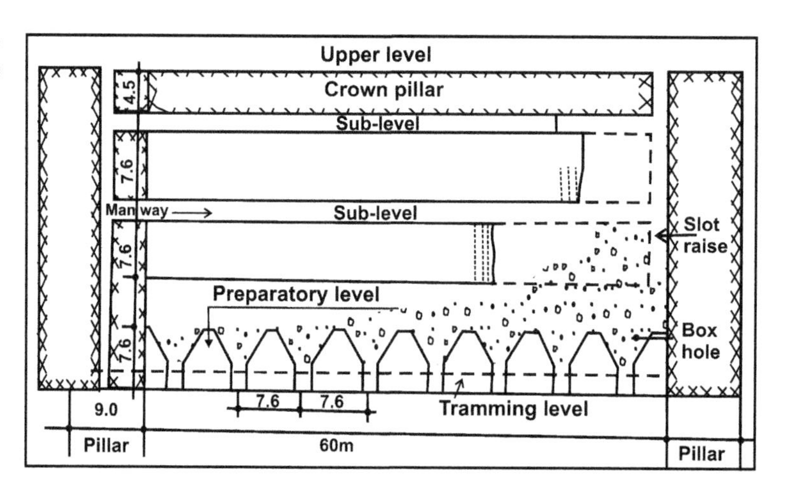

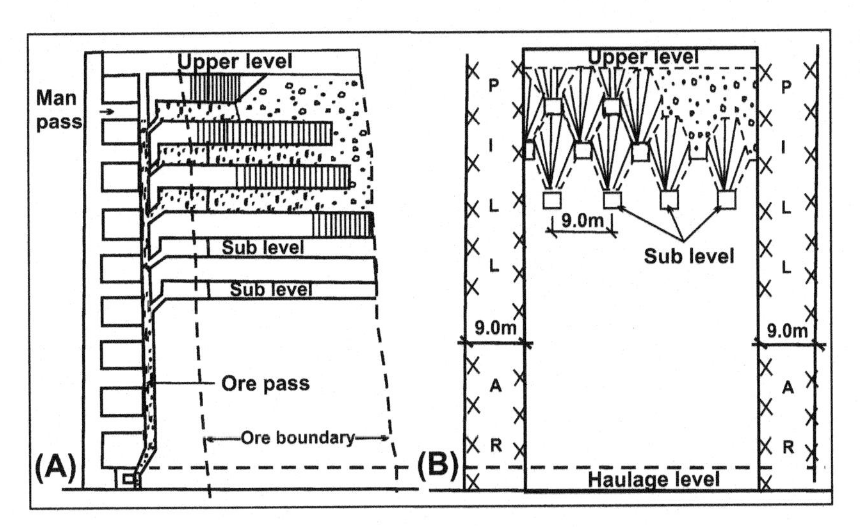

FIGURE 11.25 Conventional sublevel top slicing mining method yields high productivity.

either side of the haulage drive. The high-speed raising with safety is done by using Alimak raise climber. Sublevels are driven from the man-pass raise toward and up to within 5 m of the ore-pass raise. A finger-raise is then excavated from the ore-pass raise for the sublevel to puncture in it. Each sublevel is developed by drives and crosscuts at 10 m interval between the rib pillars. Sublevel top slicing starts by stripping of a slot-raise between two sublevels near the hanging wall. Blast-holes, drilled in an upper fan pattern, covering the ore boundary between three sublevels formed a trough for withdrawal of ore. The ore is blasted against broken material, retreating from hanging wall toward the footwall. The sublevel interval in due course of time would increase to 9.6 m between sixth and seventh levels. This method could yield an overall productivity of 1.66 tonnes per man-shift against 0.85 tonne from 6.7 m level interval stope. The ore recovery is high.

Stope drilling can be improved by 57 mm diameter blast-holes in fan pattern, drilled by Simba Junior rigs in sublevel top slicing. Explosives are in a combination of ANFO and Anodets to achieve faster and more efficient blasting. A productivity improvement of 70 tonnes per shift can be achieved by utilizing 1.5 m^3 capacity diesel-operated LHD thereby increasing the productivity from each loader to 200 t/shift. Ore, dumped from the levels, into the nearby ore-passes, is drawn at the bottom through pneumatic chutes into 5-tonne capacity grandby cars to be hauled by a 5/8-tonne trolley wire locomotive. Grandby cars dumped the ore into the bunker over an underground double toggle Jaw crusher.

11.3.3.4.3. Sublevel with Long-Hole Drilling

Sublevel with long-hole drilling (Fig. 11.26) is further innovative modification of sublevel stoping. The stope starts from an extraction level, driven 13 m above the haulage level. An orebody or trough drive and an extraction drive, parallel to each other with interconnecting crosscuts at 10 m center, is developed. Sublevels are driven at 25 m vertical interval above the extraction level. A dog-legged raise, funneled at the top, between haulage and extraction level, serves as the ore-pass. Inclined raises along the orebody are driven from extraction level to the top level at both ends of the stope. The stope is generally 100 m long. Hanging wall raise is used as slot-raise while the footwall raise is used as the man/material-pass for sublevels. The position of slot-raise is usually located at the center of the stope, which retreats on both sides. The slot is extended sideways by blasting against the slot-raise to create a free face for sublevel ring blasting. The blast-holes are drilled at 360° rings, 1.5 m spacing and with a 1.5-m toe-burden for stope blasting. DTH hammer drills with 150-200 mm diameter and 60-70 m long for level and Simba Junior (57 mm diameter) for trough section are used.

The diesel-, battery- or electric-operated LHDs collect and transport broken ore from the stope and dump into the nearby ore-pass. A ramp at a convenient location between haulage and extraction level can be developed for movement of LHD and other material. The broken ore is loaded at the main haulage level through pneumatic chutes into a train of 5-tonne grandby mine-cars and hauled by a trolley-wire locomotive. Overall productivity of 2.2 t/man-shift is achieved from these stopes. It reduces stope development from 2 to 0.7 m for every 100 tonnes of production. The stope productivity increases from 3 to 16 t/m due to increased spacing and burden. The large-diameter DTH holes allow amenability to explosives as base charge to ANFO.

11.3.3.5. Vertical Retreat Mining

VRM, also known as vertical crater mining (VCR), is applicable for large massive wide orebody with moderate to steep deep at reduced cost. The orebody size and shape are relatively regular with extensive depth continuity.

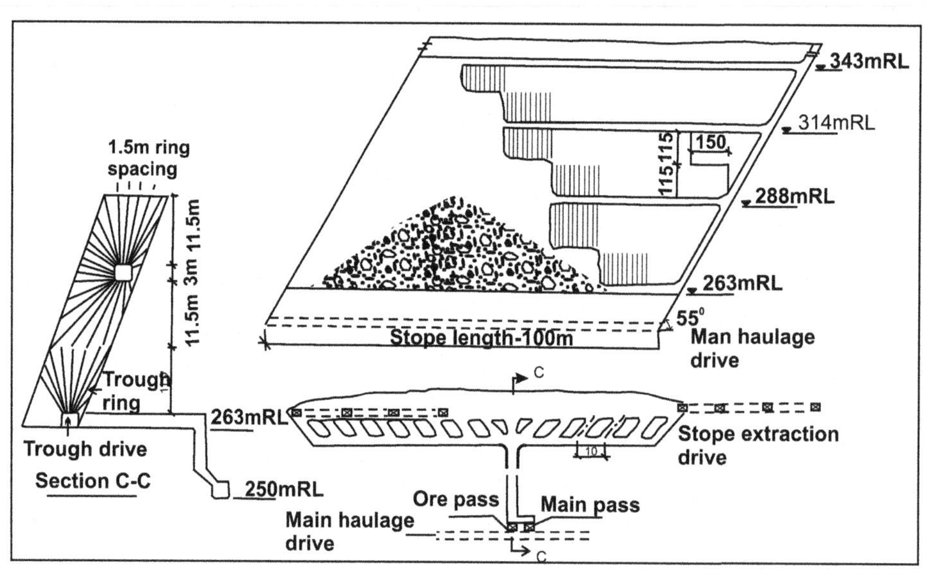

FIGURE 11.26 Sublevel stoping with long-hole drilling is innovative modification and mechanization for further higher productivity with or without backfill.

The technique is well accepted with the development of drills capable of drilling large-diameter holes up to a length of about 125 m. The DTH and In-The-Hole (ITH) drill units with 150-165 mm diameter work both in vertical and inclined planes as determined by the dip of the orebody. The stope portrays uniform shape and designs relatively smaller stoping panel of 15-30 m along the strike of the orebody (Fig. 11.27). The vertical height can be 100 m including crown and sill pillars of 12 m each. Stope development starts with excavation of a haulage drive at main level, an extraction drive 12 m above and two drill drives at 40-45 m vertically above. Ramp, 1 in 7, connects the uppermost drill drive and extraction drive for movement of men, heavy equipments like DTH, ITH, radio remote-controlled LHDs and material. An access to each panel is achieved at extraction and drill drives through 3.5×3 m crosscuts developed at 15 m interval.

VRM is an application of spherical charge blasting. The ore crosscuts at extraction level serves as initial free face for spherical blasting while those at drill drives accommodate the drill unit. Blast-hole drilling of 15-16.5 cm diameter holes is drilled by CD 360 ITH drilling machine. The drilling is done downward in a predetermined pattern from drill crosscuts. Few holes are designed up to the lower crosscut, followed by flank-holes to cover a 15-m width of the stope. The holes drilled in the section up to the lower crosscut are utilized for spherical charging and blasting at the beginning. The explosive is loaded only at the bottom of each hole for each successive blast. This allows the ore to be broken into the bottom sublevel in ascending horizontal slices using the same blast-hole. No slot-raise is required. This breaks off a slice of ore from the bottom of the ore block and drops into the draw point at extraction level. Broken muck is partially drawn by removing just enough ore to create a sufficient void for the following blast. The ore left in the panel acts as a support to the walls and control wall dilution. After the last blast in a panel the ore is quickly drawn. The last chunk of ore during completion of

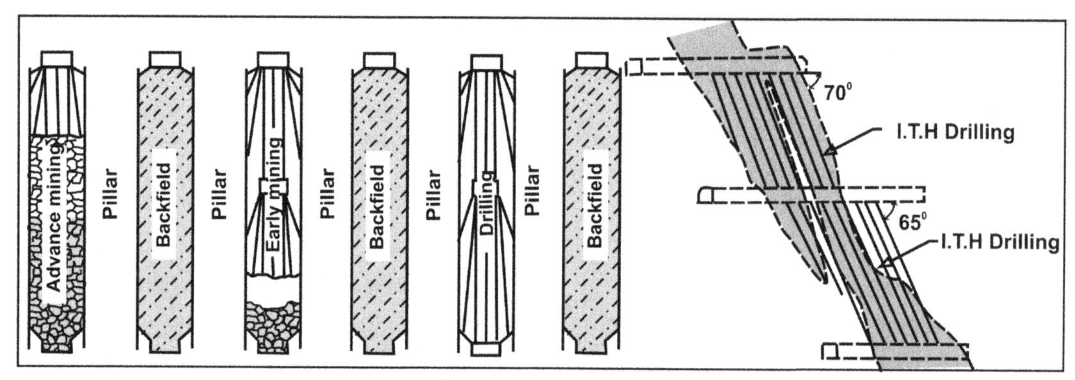

FIGURE 11.27 Schematic view of VRM method with cyclic sequence of long-hole drilling, mucking and filling of alternate stopes and pillars for a vertical height of 60-100 m in large-volume regular orebody.

ore drawl is done by radio remote-control LHD without exposing the operator into the empty stope. The void is backfilled with classified tailings mixed with 5-15% cement for consolidation. The sequence of mining would be excavation of 60 m center primary blocks, leaving strong pillar of 45 m wide on either side, followed by secondary and lastly the rib pillar blocks. The large planned waste rock can be isolated and left during blasting.

The method is extensively used for recovery of ore blocks at lower levels of Rajpura-Dariba mine, Rajasthan, yielding a tonnage factor of 20 t/m of drilling along with better fragmentation, higher ore recovery and productivity. This mining method is particularly safe to men and machineries because the miner does not enter the area where the ore is blasted and drawn using radio remote-controlled LHDs. Drilling and loading are done from the top levels. The method also eliminates the need to support the ground in the running stope after each blast.

11.3.3.6. Block Caving

Block cave mining is applicable to sufficiently large, amply massive and often uniformly low-grade orebody usually with steep to vertical dip. The ore mass should be naturally fractured and weak enough to cave under gravity. It is a mass-mining method at low cost that allows for the bulk mining and extraction of large quantity at a time. The method is suitable for existing large open-pit mines beyond pit bottom limit and to extend the operations underground by block caving. The proposed underground mine layout and development is complicated. The development, drilling and blasting requirements for ore production are minimal. The developmental work includes excavation of haulage drive, draw points with set of finger-raises and an undercut. The ore above the undercut is fractured motivated by initial long-hole blasting. Thereby large block of ore and overburden move downward under gravity, crush each other to grind the ore to smaller sizes. In-stope secondary blasting or hydraulic breaking may be required to treat the boulders. The ROM quantity and grade are continuously monitored. The drawl is discontinued in the event of excessive dilution by wall collapse or mixing of overburden waste rocks. Block caving is an economical and efficient mass-mining method, where rock conditions are favorable.

11.3.3.7. Mass Blasting

Underground mining practices witness a large quantity ore to the tune of 0.5-1.0 Mt being blocked in various pillars. Pillars provide mine stability, support and safety for future. This valuable ore blocked in pillars is extracted by mass blasting suitable for open stopping methods to increase mine life and safe condition at reduced cost. The stability of the mine, in the light of pillar blasts, can be evaluated using geotectonic software to decide the sequence of extraction of pillars. The study can predict mine condition likely to prevail after the mass blast.

One of the large underground pillar blast was carried out in 1994 at Mochia mine of Zawar group consuming 145 tonnes of explosive in a single relay bast yielding 0.55 million tonnes of ore. The pillars composed of 48-m-thick crown and complimentary 43-m-thick rib. The crown pillar was split into 24 horizontal slices by drilling fan rings of 115 mm diameter. The rib pillar was split into 10 vertical slices by drilling fan rings of 165 mm diameter. The crown pillar was fired downward and the rib pillar laterally into the preexisting stoping voids of 2W and 3W stopes (Fig. 11.28). The powder factor was 3.79 t/kg and the cost per tonne of ore was ₹ 66 (~1 to 1.5 US$) only.

11.4. MINE MACHINERY

The contemporary open-pit and underground mines perform on sophisticated high-end advance mechanization. Multiple large-diameter DTH and ITH drill units, diesel/compressed air/electric-powered radio remote control large capacity loaders, +85 t rubber-tired dumpers and trucks, conveyor, rail and aerial ropeways for ore transportation are common practices. The mine machineries can broadly be classified based on the type of operations such as

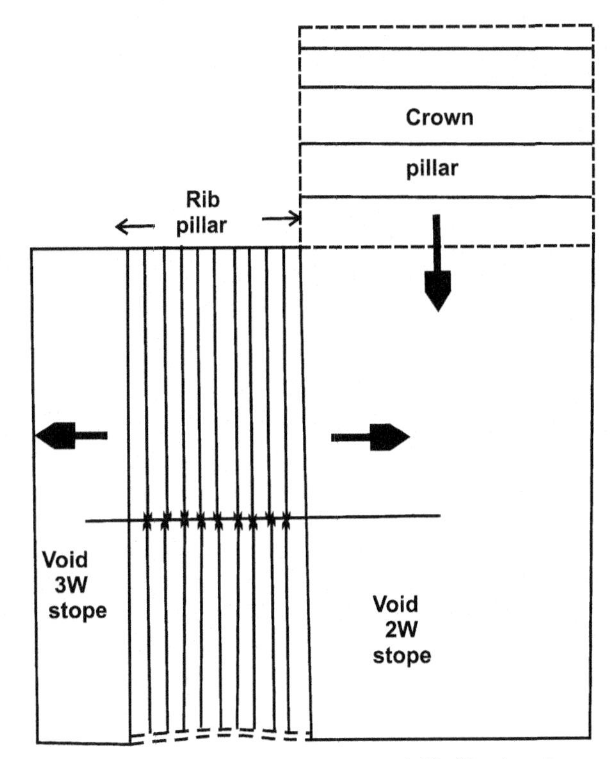

FIGURE 11.28 Schematic firing of crown and sill pillars by relay mass blasting.

development, production drilling, collection, loading and transportation.

11.4.1. Drilling

Mining of moderate to hard compact rocks requires drilling, either in dry or wet condition and blasting for generating broken material. Various types of drill machines are available based on hole diameter and capacity to drill. Drill units for mining purposes are essentially noncore type with tungsten carbide bit. Drilling is carried out by combined percussive and rotary action. Machines are run generally by compressed air or electric power and rarely by diesel. Common drill units are Jackhammer, Wagon drill, DTH hammer and ITH.

Jackhammer is a compressed air-operated drill. The unit is lightweight (15-25 kg) and operated by single person. The hole diameter varies between 30 and 38 mm and drills depth between 3 and 6 m. Common applications are development of benches in small open-pit mine. The machine is in regular use in underground mine face development (refer Chapter 7, Section 7.2.2.2 and Fig. 7.3) and overhand drilling in shrinkage stope.

Wagon rock drills operate usually by compressed air and are mounted on a portable frame fitted with rubber tire-three wheels or robust crawler chain or tractor for easy movement to the hole site during active mine operation (Fig. 11.29). The hole diameter varies between 50 and 115 mm for a depth of 3-40 m. The mast of the drifter is usually 3 m long which can be changed as per requirement. The mast is capable of rotating from vertical to horizontal and drilling is done at any angle. The conventional Wagon drills have been modified to heavy duty using hydraulic pressure. Wagon drills are used for surface and underground mine production.

Simba Junior, manufactured by M/s Atlas Copco, is pneumatic screw feed unit, designed specially for underground mining to drill in 360° rotation. The hole diameter varies between 51 and 57 mm with maximum drilling length of 25 m. The drill unit is attached with two columns and easy to move from one place to another. It is suitable

FIGURE 11.29 Robust crawler chain-mounted Wagon drill in operation at Jhamarkotra rock-phosphate mine to drill 5-10 m bench height and easy movement to hole sites.

for tunneling, drifting and long-hole stope drilling (Fig. 11.30). Simba Junior series of drill machines are suitable for mines of low to medium production.

The challenges of extremely high-speed mine development and large production are met by extra heavy-duty innovative designed mechanized drill jumbo rocket boomer rigs (Face Master by GHH (Gute HoffnungsHütte) and Mine Master, Atlas Copco) with single (Fig. 11.31), twin and three boom for drilling blast-holes of diameters between 41 and 76 mm and of net length up to 3.8 m. The unit can move 12 km per hour by hydrostatic or mechanical power shift tramming systems up to 15° gradeability. The unit can be used in underground workings both for coal and metal mining. Design allows effective drilling of blast-holes in underground workings of height up to 3.6 m. Similarly, automatic rock bolting machines are available.

FIGURE 11.30 Column-mounted pneumatic screw feed Simba Junior drills suitable in underground mines to drill in 360° rotation at Rajpura-Dariba mine, India.

The drill jumbo boomers move easily to the sites and set the direction and angle of boom fast and accurately, superfast rate of penetration and simple operation by single person with high productivity (Fig. 11.32).

"Down the Hole" or "DTH Hammer" drill is meant for drilling borehole in hard compact rock formation (Fig. 11.33). It combines both percussion and rotary action and powered by compressed air. If required, DTH units are provided with mud pump for negotiating the loose overburden. The hole diameter varies between 115 and 165 mm for a depth of 125 m. The drilling tool is a cross bit or a button bit. In DTH hammer drill the assembly of the drill bit and its short-length pipe is called the DTH hammer. The lowering and hoisting of the drill rods are operated by a chain. The drill is leveled with three hydraulic jacks. The drilling is usually vertical with a possible swing of 20° on either side from vertical in some model. DTH hammer rock drills have been leading the way in performance, reliability and longevity since their introduction 15 years ago.

"In the Hole" or "ITH" drill, similar to DTH, is especially useful for VRM methods, drills at compound angles from the vertical and horizontal directions. Low headroom

FIGURE 11.31 Single boom drill jumbo with basket diesel-powered articulated chassis and permanent four-wheel drive.

FIGURE 11.32 Close view of single-hand operator face drilling by Tamrock Single Boom electric-powered jumbo drill with straight 14 ft steel feed operation.

is achieved by the employment of a double-acting hydraulic cylinder. The ITH drills are extensively used in large massive mineralization for quick evaluation.

Crawler-mounted (continuous tracks) "Rock breaker" is not a drilling machine, but works on percussive motion. But it is an important equipment of mining machinery family. It is essentially used to break oversize boulders before feeding to standard crusher. It is powered by diesel oil. The machine is mounted on crawler chains and can move to any desired location (Fig. 11.34).

FIGURE 11.33 Tire-mounted DTH hammer drill rigs in operation for drilling 60 m vertical height of VRM stopes at Rajpura-Dariba mine during 1990s.

11.4.2. Mucking

"Muck" is loose rock or ore or clay material that has been fragmented as a result of blasting in a working face or stope in open-pit or underground mining excavation. "Mucking" is the process involved in loading, hauling and transporting the muck away from the stockpile, dump or worksite. Mucking of broken ore or waste rock from a small or big mine is carried by various types of machineries. The size of the mucking units changes with respect to the bulk of the material to be moved from open-pit and underground operation.

"Bulldozer" is diesel power-driven units mounted on either rubber-tired wheels or crawler chains. It has a pusher blade in the front fitted with a bucket. It is very often used in the initial development stage of open-pit mine to level the ground by dozing for setting infrastructure. It is also used for digging the soft earth or weathered rocks.

FIGURE 11.34 Crawler-mounted (continuous tracks) hydraulic hammer type Rock breaker in operation to break large size stones of ROM ore at mine outside.

"Scrapers" are diesel-operated tired wheeled units with a cutting blade at the bottom. The blade cut and pushed a thin slice of soft material in open-pit mining. The "ripper" is an attachment in place of a blade and cuts deep furrows in the ground as it moves. Scraper and ripper work in combination with bulldozer.

"Power shovels" and "draglines" (Fig. 11.35) are diesel or electrically powered heavy-duty machines mounted on revolving deck or crawler. It has a long light boom, crawler chain and bucket controlled by cables. The bucket is lowered down in the soft earth or blasted loose rock, coal or other ore. The unit works from upper bench or level and successively moves to lower benches. The bucket digs and drags by controlling the cable to load broken materials. The bucket is hoisted by cable and unloads the contents to suitable location by swinging movement of the boom.

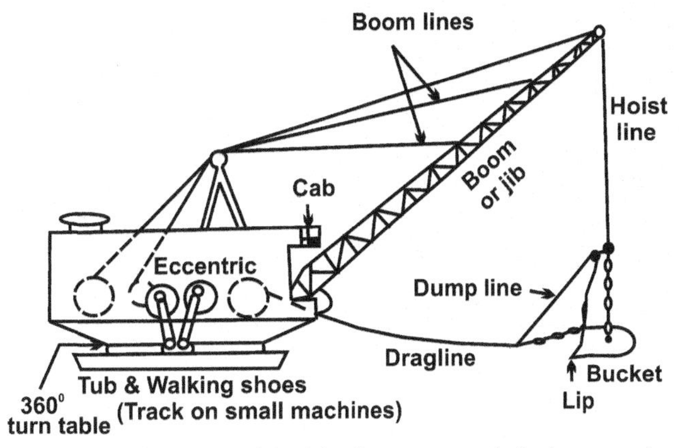

FIGURE 11.35 Schematic diagram showing the function of dragline to remove the broken ore and waste at mine operation site.

Load Haul and Dump units (LHD) represent a total in-site solution of ore transportation, particularly, for underground mining. It is essentially a tire-mounted tractor with a bucket as the front attachment. The bucket can be lowered to the ground, pushed to load broken ore and raised to dump hydraulically or mechanically. It is also known as "front-end loader" or "pay loader". These units run on compressed air or electric power and rarely by diesel for environmental restriction. The capacity of each unit varies widely depending on quantity of ore to be handled in a day. The LHD loads the ore from the face or draws pint of the stopes, hauls it to few 10s of meters and dumps it to low-profile dumper, rail or ore-pass. LHD is driven by an operator or remote control for drawing last part of ore from open stopes. The capacity and technology varies from compressed air-driven 1 m^3 CAVO-310 Hopper Loader (Fig. 11.36), diesel or compressed air-driven 1.5 m^3 LHD (Fig. 11.37) and finally 2.7 m^3 electric-powered radio remote control operations (Fig. 11.38) to be the mainstay of ore transport from VRM or bulk mining. The other common mucking machineries are "bucket wheel excavator" and "bucket chain excavators".

FIGURE 11.36 Compressed air-driven four-wheel drive "CAVO loaders" on four extra rubber molded tires with 1 m^3 capacity are useful in small-scale underground mine transportation.

FIGURE 11.37 Diesel/compressed air-driven 1.5 m^3 LHD transporting the broken ore from working stopes to loading points.

11.4.3. Transporting

The main transportation from open-pit mine to ore stockpile is carried out by trucks and dumpers. The capacities range between 25 and 85 t (Fig. 11.39). Low-profile dumpers are used in underground mines having access by decline. Conveyer belts are commonly used to transport ore in underground mines, with access to shaft or incline. Occasionally environment friendly battery-operated locomotives are utilized to move the ore to the shaft and hoisted to surface in buckets or skips. The ore is emptied mechanically into bins beneath the surface head frame for transport to the mill.

11.5. MINE EXPLOSIVE

An "explosive" is a solid or liquid chemical substance which can instantly produce a sudden expansion of the material accompanied by development of high temperature and pressure. The explosive explodes with high vibration and loud noise when ignited by a flame, heat or sudden shock. "Detonator" gives violent shock to the explosives to create an explosion. The process is called "detonation". The explosion breaks and shatters a large volume of rock mass depending on the type, power, quantity and design of the blast. The commonly used commercial explosives are:

(1) Gun (black) powder: Potassium nitrate (75%), charcoal (15%) and sulfur (10%).
(2) Flash powder: Fine metal powder (aluminum or magnesium) and a strong oxidizer (potassium chlorate or per chlorate).
(3) Ammonium nitrate (AN): Ammonia gas with nitric acid.
(4) Nitroglycerine: Gelatinous, semi-gelatinous and powdery.
(5) Armstrong's mixturc: Potassium chlorate and red phosphorus.
(6) Trinitrotoluene (TNT): Yellow insensitive crystals that can be melted and cast without detonation.
(7) ANFO: AN-Fuel Oil explosive is ammonium nitrate mixed with diesel oil.
(8) Slurry explosives: Mixture of TNT and ANFO in water.
(9) Emulsion explosive: An oxidant, a fuel, and an emulsifier.
(10) Gelatin: Nitrocellulose in nitroglycerine and mixed with wood pulp and sodium nitrate or potassium nitrate.

11.6. ROCK MECHANICS AND SUPPORT SYSTEM

Rock mechanics is the study of geotechnical responses of geological phenomena. It involves measurement of the stability of rock strata and soil deposited within the

FIGURE 11.38 Electric-powered radio remote control operation of 2.7 m^3 capacity.

mined-out artificial voids. The knowledge and indication of the geotechnical characteristics including rock strength parameters is evidently the first step in rock mechanic technique. This is further supplemented by study of weathering, joints, shears, faults and other rock deformation. The critical parameters are determination of the rock quality designation (RQD) and "rock mass rating" (RMR) from drill core. The measurements include laboratory testing of the strength of various rocks likely to encounter during mining. The observations are processed by numerical modeling to conceptualize the support system in order to protect mine, miners, equipments and underground infrastructure. Roof support is accomplished with timber, concrete, steel column, rail, bar and most commonly with roof bolts. "Roof bolts" are long steel rods used to bind the exposed roof surface to the rock behind it.

Local ground support is used to prevent minor rock falling from the roof and walls. All excavations may not

FIGURE 11.39 Hydraulic shovel loading and 85 t Haulpac dumper at Jhamarkotra rock-phosphate open-pit mine, Rajasthan (December 2008).

require local ground support. Standard metal wire mesh is a screen with approximately 10 × 10 cm openings. The wire mesh is attached to the roof or walls using pointed anchor bolts grouted with cement. The local fall of fragmented rocks can be prevented by "Shotcrete". The process of shotcrete requires spray or injects concrete or mortar with high pneumatic velocity through the nozzle of a hose onto the loose fractured surface. The concrete is a mixture of cement with sand, gravel, stone chips, fly ash, chemical and water. Shotcrete undergoes placement and compaction due to the projected force. It can impact onto any type or shape of surface, including vertical or overhead areas. The mixture coats 50- to 100-mm-thick layers on the roof and walls preventing smaller rocks from falling.

The subsidence of the roof can be supported by timbers. The timbers are cheap and easily available. Seasoned timbers are generally used. The timber bars are arranged as a prop between the floor and the roof of the mine opening (Fig. 11.40). However, use of timbers is discouraged because it generates deforestation in the long run.

Rock bolting is common practice as ground support to prevent major ground failure. Holes are drilled into the roof and walls. Long metal bars are inserted to hold the ground together. Point anchor or expansion shell bolt is a metal bar of 20-25 mm in diameter and 1-4 m in length. As the bolt is tightened the expansion shell located at the top end expands and the bolt tightens holding the rock together. Cable bolts are used to bind large masses of rock in the hanging wall and around large excavations. Cable bolts are much larger than standard rock bolts, usually between 10 and 25 m long. Cable bolts are grouted with cement (Fig. 11.41).

The large open stopes are backfilled by sand, mixture of cement and sand, cement and rock or cement and classified tailings where large volume of orebodies are to be mined at depth or where leaving pillars of ore is uneconomic. This method is as the refilled stopes provide support for the adjacent stopes, allowing total extraction of economic

FIGURE 11.40 Timber chock support system to prevent collapse of bad ground.

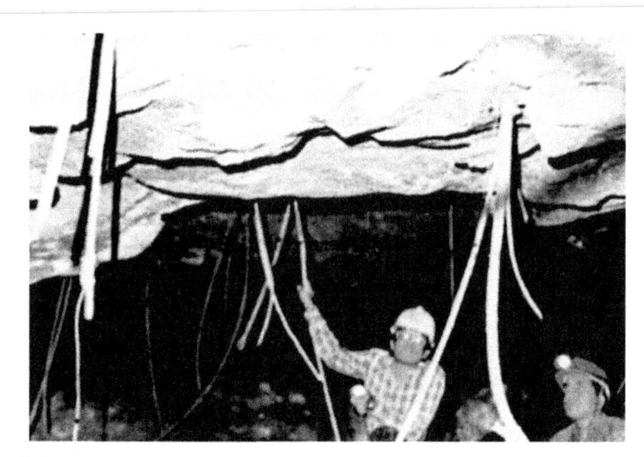

FIGURE 11.41 Cable bolting is suitable to bind large rock masses for extended period and adopted as appropriate support system in bad ground.

resources (refer Section 11.3.3.3 Cut and Fill and Section 11.3.3.5 VRM).

11.7. MINE VENTILATION

Mine ventilation is one of the most important aspects of underground mining. Ventilation is required to clear toxic methane gases as in coal mine, remove fumes from blasting and occasional exhaust from diesel equipments. Ventilation provides fresh air to underground and at the same time removes noxious gases as well as dusts that might cause lung disease like silicosis. In deeper mines ventilation is also required for cooling the workplace for miners. The primary sources of heat in underground hard rock mines are depth factor from surface, blasting, rock temperature, machinery, auto compression and fissure water. Underground mine ventilation is regulated by air intake through one of the incline or main shaft and air outlet through other incline or ventilation shaft fitted with exhaust fan.

11.8. MINE CLOSURE

All mines get closed one day (Fig. 11.42), permanently or provisionally, due to complete removal of ore, short of demand, fall of metal price, uneconomic operations, natural calamity and legislative order. The closure of a mine has significant social and economic implications for the surrounding regions and the people, employees, their dependents and others. In the event of closure of an existing mine, some procedures have to be followed beforehand. "Final Mines Closure Plan" should be submitted 1 year prior to the proposed closure. Also required is "Progressive Mine Closure Plan", within 180 days from date of notification in the format and guidelines amended in MCR, 1960 and MCDR, 1988, issued by the IBM. It involves effective planning of the landscape covering land, mine, river, water, waste dump, and tailing disposal after mining. The program encompasses ongoing rehabilitation process to restore to

FIGURE 11.42 The Sullivan mine in British Columbia closes after 105 years. It is the Canada's longest lived continuous mining operation and produced 16.00 million tonnes of lead and zinc metals, as well as 9000 t of silver. (Credit: Kimberly Daily Bulletin, 21 December 2001.)

a level acceptable to the society as self-sustained ecosystem and not become a burden.

The mine closure plan includes an introduction discussing the lease areas affected by mining, the impacted communities and stakeholders, decommissioning and infrastructure plan, the mine area rehabilitation and finally the social and economic impacts of mine closure. The closure plan must highlight a brief description of the deposit, available mineral reserve and resources, method of mining and mineral processing and reason for closure. It should include reclamation and rehabilitation of mined-out land, water and air quality, waste, topsoil and tailing dam management, existing infrastructure, disposal of mining machineries, compensation and alternative rehabilitation, time schedule and costs.

11.9. MINING SOFTWARE

Computer applications in geological data processing and mine design have become a boon in mineral industry. In-house software development is time consuming and requires availability of expertise. Even then it can satisfy only to certain limited extent. Vendor software is becoming more and more user friendly day by day. The continuous development over the last three decades enabled the software to handle the simplest to most complicated situations. Vendor software takes into account all possible needs for all possible users. The software is continuously on update to satisfy the global customers/users.

In-depth knowledge of computer and intricacy of the software is not essential to use the software. An adequate training and practice, and specially practice, can enhance the skill of the user. However, it is important that the user is well versed about the subject, data input and the possible result expected at the end of the processing. One should never blindly depend on the black box. No single software can provide total solution to the mining industry. A combination of in-house and vendor software can take care of all possible alternatives including effect of local area. There are number of geology and mine planning software readily available in the global market and names of some commonly used packages are mentioned here without giving any prejudice or degree of qualitative position. Each one has its strength and deficiency. One has to pick the software which suits the requirement best. No hardware and software will work unless trained human resource is stationed who has a zeal to work. The staff training by vendor should be an integral component while selecting the software. In any case format of data collection should be standardized during the exploration. There are five files, four of them are raw data and fifth one is processed information as mentioned below. The file format is in ASCII/EXCEL/Dbase with standard decimal specified. Many exploration companies capture data online while collection and electronically upload to the process computer. The data collection files are:

(1) COLLAR file
The COLLAR file contains the data related to details of starting point of the borehole (BH) and chain of sample. The Latitude/Longitude coordinates can be national or local grid after grid rotation as convenient following UTM system. *Z* coordinates represent level

with respect to Mean Sea Level (MSL). The borehole number must be unique all through the database.

Borehole number	*X* or Longitude/ easting (m)	*Y* or Latitude/ northing (m)	*Z* or level (m)	BH depth (m)
RA080	55.00	−700.00	384.42	110
RA065	115.00	−150.00	387.50	210

(2) SURVEY file
The SURVEY file contains the survey or deviation data along the course of the borehole.

Borehole no.	Depth at (m)	Angle (°)	Bearing (°)
RA080	0	−50	320
RA080	50	−51	321
RA080	90	−51.5	321

(3) ROCK file
The Rock file contains the alphanumeric codified rock type intersected in the borehole.

Borehole no.	From (m)	To (m)	Rock type
RA080	0.00	11.50	SOIL
RA080	11.50	23.25	GBSG
RA080	23.25	60.00	GBMS
RA080	60.00	110.00	GBSG

(4) ASSAY file
The ASSAY file contains the details of chemical analysis of the samples collected. Sampling is generally carried out with unequal length. The original samples can go through an intermediate process to convert the samples to equal length and composite to larger length to reduce variance. Equal length samples are most suitable for statistical analysis.

BH no.	From (m)	To (m)	% Pb	% Zn	g/t Ag
RA080	28.50	29.50	0.50	1.25	10
RA080	29.50	31.00	1.90	15.25	45
RA080	31.00	32.25	2.10	12.00	55
RA080	32.25	34.00	1.55	9.50	40

(5) SAMPLE file (De-surveyed)
The SAMPLE file is the processed file of each sample, unequal, equal or composite length with central coordinates within the deposit and its corresponding chemical values.

BH no.	*X*	*Y*	*Z*	% Pb	% Zn	g/t Ag
RA080	40.00	−700	370.00	0.50	1.25	10
RA080	39.35	−700	369.50	1.90	15.25	45
RA080	39.10	−699	368.10	2.10	12.00	55
RA080	38.90	−699	367.80	1.55	9.50	40

The list of software vendors alphabetically arranged is:

(1) DATAMINE International, UK, is the world's leading developer, trainer and maintainer of integrated technology for total solution to the mining industry encompassing geological data capture, geostatistical evaluation, mine planning, production scheduling, environmental issues with full virtual reality system.
(2) GDM BRGM, France, provides 3D GeoModeler for processing geological, geochemical and geophysical data integrated with GIS applications.
(3) GEMCOM-SURPAC International Inc., Australia, develops software designed to automate and integrate key operations for exploration, resource evaluation, mine planning, mine design, mine operations.
(4) GEOVARIANCES is a French Independent Software Vendor for development and maintenance of geostatistical data modeling related to mining and environmental issues.
(5) LYNX, Vancouver, Canada, provides site setup and configuration functions along with real-time monitoring, control, logging, automatic report printing, and on-screen history of customizable user interface of geological information. It has ability to create multiple screen displays with critical information in plain view with option to view details. Drill-down displays provide an intuitive means to monitor many sites.
(6) MEDSYSTEM, Tucson, Arizona, USA, develops and supports software for mine design, evaluation, geologic modeling, surveying, geostatistics and scheduling on open-pit, underground, coal, metal and nonmetal mines.
(7) MINEX 3D, Australia, an associate of GEMCOM, is a comprehensive software package that synthesizes the geological data to arrive at the 3D modeling.
(8) PC MINE, Vancouver, Canada.
(9) SURPAC, Australia, is a total system covering geological assessment, survey and mine planning.

FURTHER READING

Deshmukh (1995) [21 and 22] discussed the elements of mining technology in two volumes in a very simplified way. An introduction to mining by Thomas (1973) [71] is a classical work and worth reading. Introduction to Mining Engineering by Howard (1987) [35] and Underground mining methods—engineering fundamentals and international case studies, edited by Hustrulid, et al. (2001) [38] are valuable additions in mining technology.

Chapter 12

Mineral Processing

Chapter Outline

Minerals are wanted in the highest state of purity for their end uses.

—Author.

12.1. DEFINITION

The definition of minerals, rocks, ores, prime and associated commodities, trace elements and gangues are defined in Chapter 1. All these, by definition, are naturally occurring substances possessing definite chemical composition, atomic structure and physical properties. The minerals in general occur in certain heterogeneous association with complex interlocking boundaries. The properties of common metallic and nonmetallic minerals have been summarized in Table 2.2. The metal content and other chemistry varies widely between minerals. Minerals as well as metals are wanted in the highest state of purity (say, 99.99%) for their end uses.

"Mineral processing", also known as, "mineral beneficiation", "mineral engineering", "mineral dressing" or "ore dressing," has been defined as the science and art of separating valuable metallic and nonmetallic minerals from insignificant gangues. Post-mining activity involves mineral beneficiation and extractive metallurgy. The mode of operations passes through the process of liberating and separating ore minerals into valuable products and unworthy waste as rejects. This is done by exploiting the characteristic differences in physical and chemical properties of different minerals and applying variety of techniques. It is described as the value-added processing of raw materials (ROM ore) to yield marketable intermediate products (copper concentrate) or finished products (silica sand) containing more than one valuable minerals and separation of gangue (tailing). The ROM components consist of the following:

(a) Building and decorative stones like granite, marble, limestone,
(b) Industrial minerals like calcite, fluorite, apatite, diamonds and gemstones, barite, wollastonite, bauxite,
(c) Metalliferous minerals like chalcopyrite, sphalerite, galena, bauxite, hematite, and
(d) Precious metals like gold, silver, platinum, palladium in native form.

The process should not, under any circumstance and even at any intermediate stage, alter the physical and chemical identity of the parent minerals for the subsequent treatment (smelting). The metalliferous concentrate is further treated

Mineral Exploration. http://dx.doi.org/10.1016/B978-0-12-416005-7.00012-X

by extractive metallurgy, either Hydrometallurgy or Pyrometallurgy and Electrometallurgy for extraction of metals in the purest form.

Mineral beneficiation involves four prime types of activities:

(1) "Comminution", "liberation" and/or "particle size reduction";
(2) "Sizing and separation" of particles by screening or classification;
(3) "Concentration" by taking advantage of physical and surface chemical properties;
(4) "Dewatering" or "solid-liquid separation".

"Liberation" is the release of valuable minerals between themselves and from the associated gangues at the coarsest possible particle size. The optimum particle size for best liberation is seldom achieved due to complexity of intermixing natural characteristics. When valuable minerals and gangues are interlocked in a particle, it is known as "middling". "Sizing" is the separation of particles according to their size. "Concentration" is the separation of minerals into two or more products such as valuable minerals in concentrates, gangues in tailing and locked particles in the middling. Middling fractions are misplaced particles and often associate with either concentrate or tailings. "Dewatering", in stages, produces relatively dry concentrate at desired moisture content. The moisture reduction reduces cost of long-distance shipment and prevents hazards during transportation.

Mineral beneficiation technique is oriented to conservation of mass (material balance) and accepts best possible grade of concentrate at highest possible recovery efficiency of each mineral (mineral balance).

12.2. ORE HANDLING

Ore handling at the mine site, open pit or underground is one of the important activities prior to mineral beneficiation. The several tasks are generation of proper size of broken ROM material, cleaning, removal of harmful substances, transportation to plant site, creation of stockpiles with even size and grade and analysis of mill feed. The ores from various sources of mines, stopes and draw points are successively reduced to smaller size. The intermixing of material by blending of ores from different sources maximizes homogeneity and consistency before feeding to the process operation.

12.2.1. Cleaning

ROM ore contains pieces of iron and steel broken from mine machinery, drill rod and bit, part of woods from mine support arrangements along with clays and slimes out of nonmetallic gangue minerals. Each of these materials is harmful to the process system as it can jam and damage the crusher-grinder-pulverizer, chock the screens and clog the flotation cells and filtration media. It must be removed at an appropriate stage of operation. The removal is done by hand sorting, electromagnetic separation over conveyor belt and washing of clay and slimes.

12.2.2. Transportation

ROM ore is transported to the mineral processing plant by various means. It depends on the size of operation say 1000-10,000 tpd and distance to be covered between the mine and the plant which can vary from few 100 m to 10s of kilometers. In case of long-distance transportation, it is done by aerial ropeway and heavy-duty trucks and dumpers. If the distance is short, the standard rubber belt conveyor with support rollers at the bottom of belt (Fig. 12.1) is the most popular and widely used method of handling loose bulk material. A set of conveyors negotiates the distance, sharp turns and gradient. The basic principle is to operate continuous transportation system in the direction of the gravitational movement at the shortest possible distance between supply and delivery points.

12.2.3. Stockpile

Directly salable ROM ores are stockpiled (Fig. 12.2) at mine head with various size fraction and grade. The ore can directly be loaded to the transport vehicle for movement to the customer destination.

FIGURE 12.1 Standard open surface rubber belt conveyor system with support rollers at the bottom of belt for ore transportation over kilometers.

FIGURE 12.2 Hand or screen sorted chromite ore stock piles of different specifications for direct sale and transportation to ferro-chrome plant.

Storage of raw material as buffer stockpile is also necessary for uninterrupted operation of a continuous process plant. Ore is stocked as reserve at every stage of output/input of stope draw points, mine dispatch sites, and crusher-grinder feeder for uninterrupted feeding in the eventualities of any breakdown at any unit of operation. The quantity of storage depends on the size of operation, frequency of unexpected breakdown and shutdown of individual units for routine maintenance. As a general practice 3-6 days of requirements are stocked at every site. The stockpiles can be flat, conical and elongated stack with chutes at the bottom for auto-feeding to conveyor belt or loading equipments (Fig. 12.3). Feeder mechanism regulates a smooth flow of ore movement from storage bins to the delivery point in an automatic conveyor network. Many types of feeders in use are chain, apron, roller, rotary, revolving disc and vibratory.

12.2.4. Weighing, Sampling and In-stream Analyzer

Physical weighing of ore and moisture, sampling and chemical analysis are necessary at every stage of operation namely mill feed, concentrate and tailing. The information provides proper specification of end products and metallurgical accounting.

(a) Weighing

Weighing of moist ore is conducted at every stage of ore movement from mine draw point, surface transport system

FIGURE 12.3 Close conveyor system (top) transporting ROM ore automatically unloaded from shaft cage to bin and interim stockpiles in the right with chutes at the bottom for auto-feeding to mineral process plant.

and conveyor belts at the process plant. It is done by multiplying number of ore-filled mine cars, trucks and dumpers, and aerial ropeway buckets with the tonnage factor of broken ore. The weights are compared by physical surveying the volume/tonnage of the stockpiles and results of continuous automatic weighing devices fitted with the conveyor belt. The moisture content is determined by randomly taken grab samples from different parts of stockpiles, dumpers, mine cars and conveyors. The samples are heated on open oven or by thermal dryer to evaporate the moisture. Weights of moist and subsequent dry ore are obtained for the same grab sample. The moisture content is determined by:

$$\% \text{ Moisture } = \{(\text{Wet weight} - \text{Dry weight})/(\text{Wet weight}) \times 100\}$$

The dry weight of the concentrate is determined by deducting the moisture factor. The weight of the tailing is the difference between weight of feed and concentrates. This is compared by estimation of tailing flow over the running hours of the flotation plant.

(b) Sampling

Various sampling methods are discussed in Chapter 7. The grab sampling of ROM ore is regularly done to assign the mine production grade. However, the grade of assorted broken ore and fine fragments cannot be précised due to wide range of particle size and heterogeneity of the material. Better sample grade is expected from mill feed material at −12 mm size. The best possible grade representation is at mill discharge point of −75 μm size when it becomes extremely homogeneous. Sampling of fine fragments can manually be collected at regular intervals of 15, 30 or 60 min by two ways. It is either by taking grab of mill feed (−12 mm) from the conveyor belt or by collecting mug full of slurry (−75 μm) at ball mill discharge of feed, concentrates and tailing. Advance mineral process plants are equipped with automatic samplers at the incoming and outgoing product flow at preferred preset time interval. The device (Fig. 12.4) is a simple collector or cutter which moves mechanically at constant speed and interval across the whole stream of material either in dry or slurry form. The sample container is large enough to hold material for an operating shift. The samples are analyzed through facilities available at the operating unit.

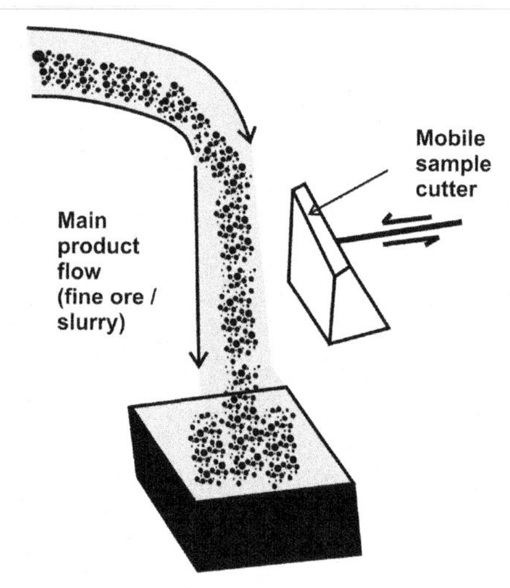

FIGURE 12.4 Sketch diagram of automatic unbiased mill sampling arrangement of pulverized slurry ball mill product flow (top) by to and fro moving cutter (middle) at certain time interval.

(c) In-stream analyzer and process control

Mineral beneficiation, particularly base and noble metals, is very sensitive to optimum use of reagents, recovery of metals and clean concentrate. It is due to high fluctuation of feed grade to the flotation cells yielding loss of metals to tailing. Off-line analytical procedures that have been discussed at Chapter 4 are not suitable in case of changing quality of feed. The process is not capable of continuous in-stream detection and spontaneous corrective measures. This is surmounted by complete unbiased concentrator automation. The circuit is comprised of three major integrated units: probe or sensor, in-stream analyzer and digital process control system.

In-stream X-ray analyzer (Fig. 12.5) employs sensors acting as a source of radiation which is absorbed by the sample causing fluorescent response of each element. The analyzer probes are installed in feed, concentrates and tailing streams. Metal content (Pb, Zn, Cu, Fe, Cd, Ag, Au etc.) and pulp density, in the form of electrical signals, from the probes (sensors) are conveyed in electronic circuits (detector generating a quantitative output signal) to

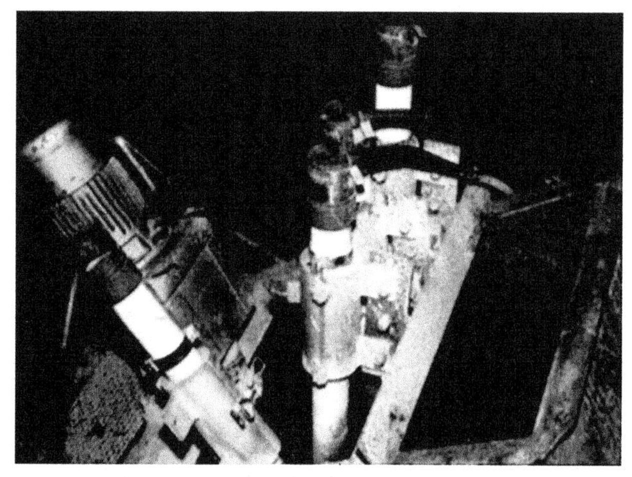

FIGURE 12.5 Mill sampling arrangement by in-stream analyzer. The probe installed in slurry stream of feed (conditioner) and reject (tailing) for continuously sensing the metal grades and simultaneous digital process control of reagents.

a digital computer in the control room. A continuous screen display or/and printout showing the elemental dispersion at every minute or minutes is available for manual or automatic control of reagents in the flotation process. The field instruments for the flotation circuits comprise of pH and metal probes, magnetic flow meters with control valves for reagent dosing pumps. The system improves the recovery of each metal as well as concentrate grade. The regulated feed-reagents, apart from improved metallurgy, result in significant savings of reagent cost.

12.2.5. Particle Size Analysis

Mineral processing techniques depend a lot on the particle behavior which in turn varies with its size. Therefore, size analysis is of great significance to determine the quality of grind and establish the degree of liberation of valuable minerals between them as well as from the gangue at various particle sizes. Sizing is the general expression for separation of particles. The simplest of sizing processes is screening i.e. passing the particles to be sized through a sieve or series of sieves. Sieves consist of nets made of iron rails, rods and wire having specific aperture so that preferred size fraction can pass through it. Sieves are designated by standard mesh number with aperture size expressed in micron (μm) as given in Table 12.1. (micron or μm = one millionth of a meter = 10^{-6}). The bold figures in Table signify the optimum mesh number and aperture size for flotation of zinc-lead ore.

TABLE 12.1 Standard Wire Mesh Sieve Size (BS 1796) Uses in Various Industries

Mesh number	Nominal aperture size (μm)	Mesh number	Nominal aperture size (μm)
3	5600	36	425
3.5	4760	44	355
4	4000	52	300
5	3350	60	250
6	2800	72	212
7	2360	85	180
8	2000	100	150
10	1700	120	125
12	1400	150	106
14	1180	170	90
16	1000	**200**	**75**
18	850	240	63
22	710	300	53
25	600	350	45
30	500	400	38

12.3. COMMINUTION

Mineral deposits are composed of ore and gangue minerals intimately associated in an intricate mosaic. Minerals are assembled in various proportions with forms varying between very fine to extremely coarse grain size depicting layered, veins, stringers, to complex structure (Fig. 12.6).

The individual mineral may occur as inclusion within another type and often with interlocking boundaries (Fig. 12.7). Until and unless the individual minerals are unlocked and completely liberated from each other the concentration process of the minerals rich in value cannot be perceived. This is achieved by comminution (particle size reduction).

The particle size of ore is progressively reduced to optimum fraction for separation by a method suitable to the physicochemical properties of the minerals. The initial comminution process begins during the mining operation through blasting of in situ orebody, draw by excavators or scrapers and move out to ore stockyard as ROM. The fragment size at this stage is heterogeneous varying anywhere between 1.50 m and fines. The large boulders are reduced by rock breaker at the ore dump yard (refer Fig. 11.34) or on the grizzly and by onsite crusher. The ultimate size reduction in a successive sequence of

FIGURE 12.6 Camera image of polished surface of rich ROM ore showing yellow-brown sphalerite, gray-black galena and gangue minerals (white calcite, dolomite and quartz crystals).

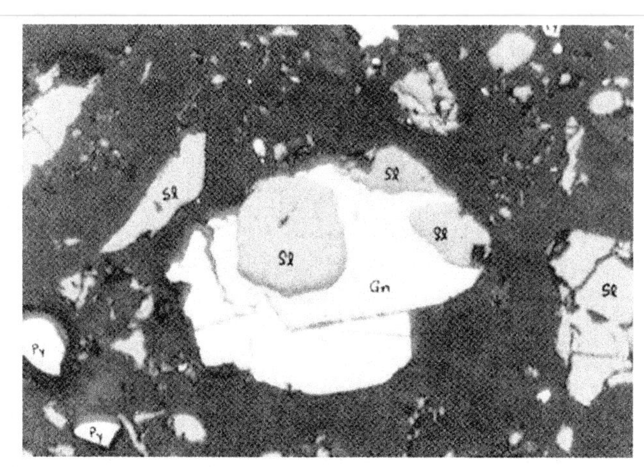

FIGURE 12.7 Polished thin section under microscope showing coarse inclusion and interlocking gray sphalerite (Sp) and white galena (Ga) in the matrix of host rock.

crushing, grinding and pulverizing is an integral part at the process plant. The individual mineral grains are liberated (Fig. 12.8) to highest extent and ready for froth flotation or any suitable beneficiation techniques.

12.3.1. Crushing

Crushing is accomplished by compression of the ore against rigid surface or by the impact against surface in a rigidly constrained motion path. Crushing is usually a dry process and carried out on ROM ore in succession of two or three stages namely by (a) primary, (b) secondary and (c) tertiary crusher.

(a) Primary crusher

Primary crushers are heavy-duty rugged machines used to crush ROM ore of (−)1.5 m size across. These large size ores are reduced in the primary crushing stage for an output product dimension of 10-20 cm. The common primary crushers are of "Jaw" and "Gyratory" type.

"Jaw crusher" reduces the size of large rocks by dropping them into "V"-shaped mouth at the top of the crusher chamber. It is created between one fixed rigid jaw and a pivoting swing jaw set at an acute angle to each other. The compression is created by forcing the rock against the stationary plate in the crushing chamber as shown in Fig. 12.9. The opening at the bottom of the jaw plates is adjustable to desired aperture for product size. The rocks remain in between the jaws until it is small enough to set free through this gap for further size reduction by feeding to the secondary crusher.

The type of Jaw crusher depends on input feed and output product size, rock/ore strength, volume of operation, cost and other related parameters. Heavy-duty primary Jaw crushers are installed in underground for uniform size reduction before transferring the ore to the main centralized hoisting system. Medium duty Jaw crushers are useful in underground mine with low production (Fig. 12.10) and in process plant. Small Jaw crushers (refer Fig. 7.32) are installed in laboratory for preparation of representative samples for chemical analysis.

"Gyratory crushers" essentially consist of a long conical hard steel crushing element suspended from the top. It rotates and sweeps out at a conical path within the round hard fixed crushing chamber (Fig. 12.11). The maximum crushing action is created by closing the gap between the hard crushing surface attached to the spindle and the concave fixed liners mounted on the main frame of the crusher. The gap opens and closes by an eccentric drive on

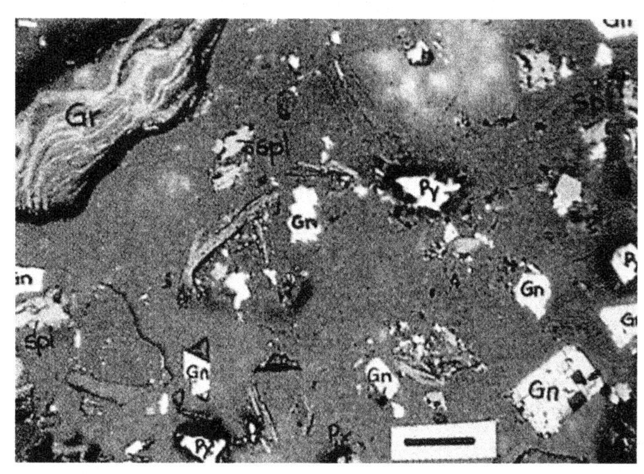

FIGURE 12.8 Final milling completes the liberation of free sphalerite, galena, pyrite, graphite and gangues and ready for froth flotation or any suitable process.

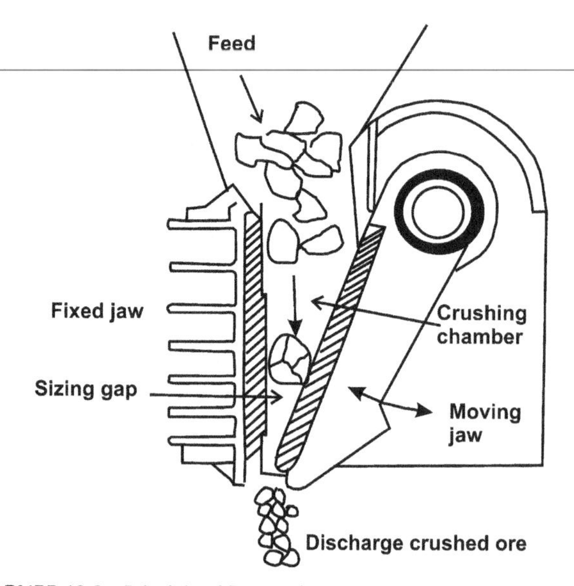

FIGURE 12.9 Principle of Jaw crusher showing the path of lumpy feed ore to fragmented product crushed under high pressure of fixed and moving jaws. (Source: modified after Ish Grewal, Met-Solve, Canada [29]).

FIGURE 12.10 Medium size Jaw crusher in operation in underground mine for crushing ROM ore before transferring to surface at Zawarmala, India.

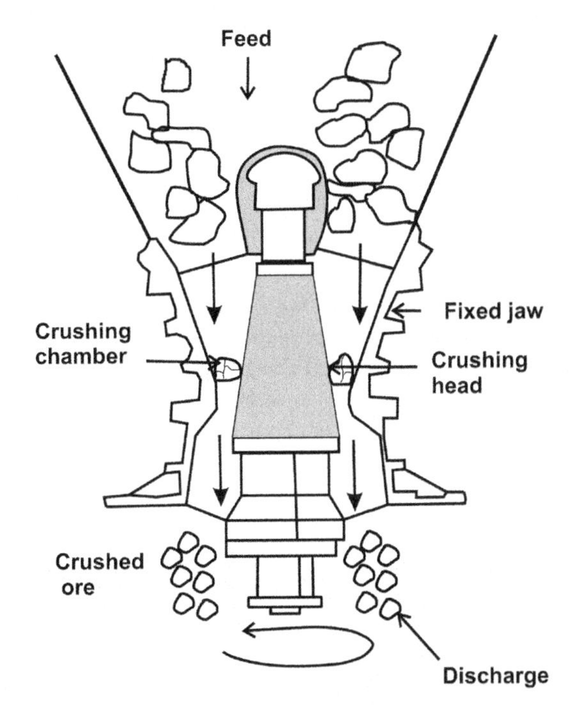

FIGURE 12.11 Working principle of Gyratory crusher for breaking lumpy ore pressed between fixed jaw and rotating conical head (Source: modified after Ish Grewal, Met-Solve, Canada [29]).

the bottom of the spindle that causes central vertical spindle to gyrate.

(b) Secondary crusher

Secondary crusher is mainly used to reclaim the primary crusher product. The crushed material which is around 15 cm in diameter, obtained from the ore storage, is disposed as the final crusher product. The size is usually between 0.5 and 2 cm in diameter so that it is suitable for grinding.

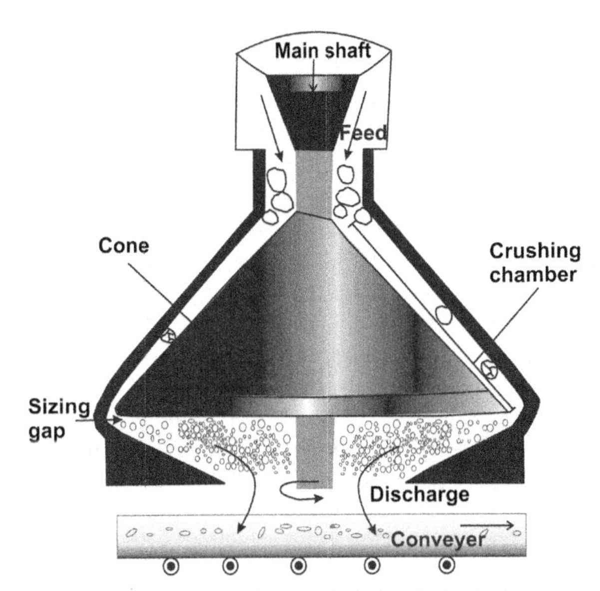

FIGURE 12.12 Schematic diagram depicting the basic elements and function of Cone crusher.

FIGURE 12.13 A working Cone crusher in operation and performing both secondary and tertiary crushing function.

Secondary crushers are comparatively lighter in weight and smaller in size. It generally operates with dry clean feed devoid of harmful elements like metal splinters, wood, clay etc. separated during primary crushing. The common secondary crushers are Cone, Roll and Impact type.

"Cone crusher" (Fig. 12.12) is very similar to the gyratory type, except it has a much shorter spindle with a larger diameter crushing surface relative to its vertical dimension. The spindle is not suspended as in Gyratory crusher. The eccentric motion of the inner crushing cone is similar to that of the Gyratory crusher.

A working Cone crusher is shown in Fig. 12.13. It can perform as tertiary crusher when installed in close circuit between the secondary crusher and the ball mill to crush any overflow material of vibratory screening.

"Roll crushers" consist of a pair of horizontal cylindrical manganese steel spring rolls (Fig. 12.14 and Fig. 7.33 for laboratory scale) which rotate in opposite directions. The falling feed material gets squeezed and crushed between the rollers. The final product passes through the discharge point. This type of crusher is used in secondary or tertiary crushing application. Advance Roll crushers are designed with one rotating cylinder that rotates toward a fix plate or rollers with differing diameter and speed. It improves liberation of minerals in the crushed product. Roll crushers are very often used in limestone, coal, phosphate, chalk and other friable soft ore.

"Impact crushers" (Fig. 12.15) employ high-speed impact or sharp blows to the free falling feed rather than compression or abrasion. It utilizes hinged or fixed heavy metal hammers (hammer mill) or bars attached to the edges of horizontal rotating disks. The hammers, bars and disks are made of manganese steel or cast iron containing chromium carbide. The hammers repeatedly strike the material to be crushed against a rugged solid surface of crushing chamber breaking the particle to uniform size. The final fine products drop down through the discharge gate while the oversize particles are swept around for another crushing cycle until it is fine enough to fall through the discharge gate. Impact crushers are widely used in the quarrying industry for road and building making chips. These crushers are normally employed for secondary or tertiary crushing.

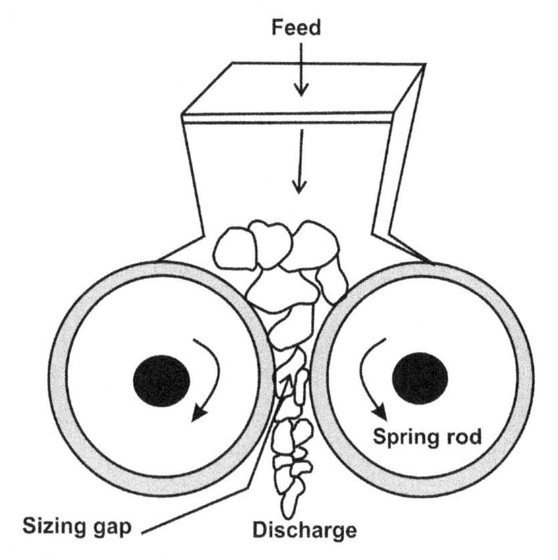

FIGURE 12.14 Conceptual diagram depicting the basic elements of Roll crusher.

(c) Tertiary crusher

If the size reduction is not complete after secondary crushing because of extra hard ore or in special cases where it is important to minimize the production of fines, tertiary re-crushing is recommended using secondary crushers in a close circuit. The screen overflow of the secondary crusher is collected in a bin (Fig. 12.16) and transferred to the tertiary crusher through conveyer belt in close circuit.

12.3.2. Grinding Mill

Grinding is the final stage used in the process of comminution. It is usually performed in rotating cylindrical heavy-duty steel vessels either dry or as suspension in water. The loose crusher products freely tumble inside the rotating mill in the presence of agitated grinding medium. Grinding takes place by several mechanisms such as combination of impact or compression due to forces applied almost normally to particle surface, chipping due to oblique forces and abrasion due to forces acting parallel to

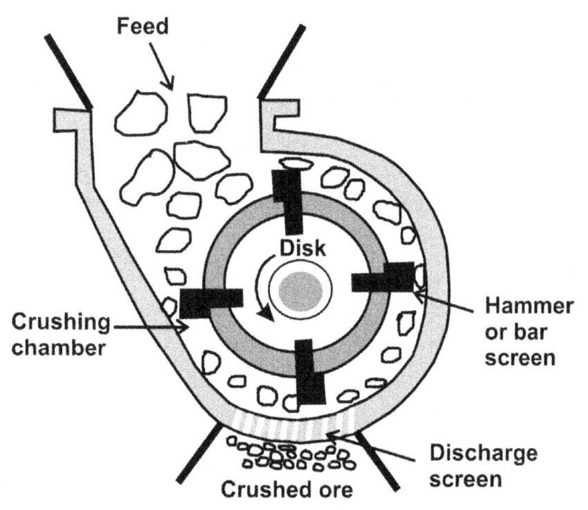

FIGURE 12.15 Schematic diagram showing the basic elements and function of Impact crusher (modified after Ish Grewal, Met-Solve, Canada [29]).

FIGURE 12.16 Close circuit transfer of oversize material from secondary crusher collected in ore bin (top) and auto-transfer to tertiary crusher by conveyor belt (bottom).

the surfaces. Grinding inside a mill is influenced by the size, quantity, type of motion and space between individual pieces of medium within the mill. The primary differences between these mills are in the ratio of diameter to the length of the cylinder and the type of grinding media employed. Grinding media can be steel balls, steel rods, hard rock pebbles or the ore itself and the mill is classified accordingly. The grinding mill reduces feed particles of 5-20 mm to optimum liberated size between 40 and 300 μm as required for beneficiation.

(a) Ball mill

Ball mills (Fig. 12.17) are short cylindrical vessels with a shell to diameter ratio of 1.5 to 1 and less. When the length to diameter ratio varies between 3 and 5 it is called tube mill. The grinding medium is high-carbon or cast alloy steel balls. The particle size of the feed usually does not exceed 20-25 mm. The grinding is caused by balls being moved up the side of the mill in such a way that they release and fall to the point where they impact the ore particles in trailing bottom region of the slurry. Ball mills are operated at higher speed so that the balls can be thrown up and strike back to the other wall with increased speed of hitting the ore particles. Ball mills are suited for finer grinding of hard and coarse feed. It is better suited for grinding base metals, phosphate and operated in close circuit with the flotation cells.

(b) Rod mill

Rod mills are long cylindrical vessels with length of the shell 1.5-2.5 times longer than its diameter. The breaking medium is steel rods. The rotating drum causes friction and attrition between steel rods and ore particles. As the mill rotates, the rods cascade over each other in relatively parallel mode and that prevents over-grinding of softer particles. The product discharge is either through central or end peripheral or overflow type. Rod mills can take feed particles as coarse as 50 mm to produce as fine as 300 μm. Rod mills are suitable for preparation of feed to gravity and magnetic concentration.

FIGURE 12.17 Standard grinding ball mill plays significant role to complete the liberation of ore and gangue minerals. It is in operation at Zawar Mine, India.

(c) Pebble mill

Pebble mills are similar to ball mill except that the grinding media is closely sized suitably selected rocks or pebbles. The rotating drum causes friction and attrition between rock pebbles (say, quartz or quartzite pebbles) and ore particles. Pebble mill operates at low cost with respect to grinding media, power consumption and maintenance. It has wide application in gold mines.

(d) Autogenous mill

Autogenous mills (AG) pulverize due to self-grinding of the ore without any additional breaking media. The drum is typically of large diameter with respect to its length, generally, in the ratio of 2 or 2.5 to 1. The rotating drum throws larger ore particles in a cascading motion which causes impact breakage of larger and compressive grinding of finer size. It operates at a lower cost. The AG are often integrated in large mineral processing operations. However, if the hardness and abrasiveness of the ore varies widely then it may result in inconsistent grinding performance.

(e) Semi-autogenous mill

Semi-autogenous mills (SAG) are essentially variation of AG with addition of steel balls along with the natural grinding media rectifies the problem of inconsistency in grinding. The total amount of balls in these mills ranges between 5 and 15% of the volume. Many of the present-day plants install SAG mills as primary or first stage grinding in combination with ball mills. It reduces the cost of media and replacement of rods. The maintenance cost in general is low. SAG mills are primarily used in the gold, copper, platinum, lead, zinc, silver, alumina and nickel industries.

12.4. SCREENING AND CLASSIFICATION

Particle size plays a critical role in mineral processing for beneficiation of valuable minerals suited during any particular downstream operation. Screening and classification are two distinct techniques of particle separation based on size. The relatively coarser particles are separated by screening. The screens are attached to all type of crushing units at feed and discharge stages. The oversize materials are diverted to re-crush and regrind devices. The undersize materials pass to the next finer stage for crushing or grinding. The particles that are considered too fine to be sorted efficiently by screening are separated by classification. The classifiers are attached to the grinding units in close circuit for treating over- and undersize particles accordingly.

12.4.1. Screening

The crushed particles are separated using a hard metallic screen having perforated surface with dimension of fixed and uniform aperture. Materials are dropped to the screen surface. The particles, finer than the openings, pass through the screen. The oversize particles are conveyed to the discharge end for re-crushing. Screening is generally difficult for very fine material and operated under dry condition or with less moisture content. The efficiency of screen performance is judged by the recovery of desired size and misplaced material in each product. The factors affecting screen performance depend on particle size, shape, orientation, feed rate, angle of discharge, percent open area, types of vibration, moisture content and the feed material. Different types of industrial screen are available each suited to handle a particular type of material. The screen may be non-vibrating or vibrating type. The later is more frequently used. Grizzly is an example of the non-vibrating type. The trommel screens, vibratory and gyratory, are examples of the later.

"Grizzly" is used for primary screening of very coarse materials. It is generally installed for sizing the feed to the primary crusher. A grizzly is fundamentally a horizontal or inclined set of heavy wear-resistant manganese steel rails or bars set in a parallel manner at a fixed distance apart (Fig. 12.18). Finer materials fall through the spacing of the bars. Oversize materials slide on the surface of the bars and are reduced on site by manual hammering, power pack rock breaker (Fig. 11.34) or local explosive pop blasting. Screens are static for very coarse materials. It can incorporate mechanism to shake or vibrate the screen to improve the performance.

"Trommel", revolving screen or drum separator is a horizontal or slightly inclined rotating cylindrical screen (Fig. 12.19). The material is fed at one end of the cylinder. The undersize material falls through the screening surface while the oversize is conveyed by the rotating motion down the incline to the discharge end. Trommel can separate several sized product by using a series of screen with coarsest to finest apertures. It can handle both dry and wet feed material. Trommel separators are cheap and most suitable for soil washing of coal and iron ore industry at higher end applications. Screening of aggregates and road materials is at the lower end.

"Vibratory" screen is the most common screening device found in mineral processing applications for the various types of material and particle sizes encountered. It works on the shaking motion of the surface and the resulting action imparted on the material being screened (Fig. 12.20). It can be arranged as multiple decks so that different particle size products can be obtained from a single feed.

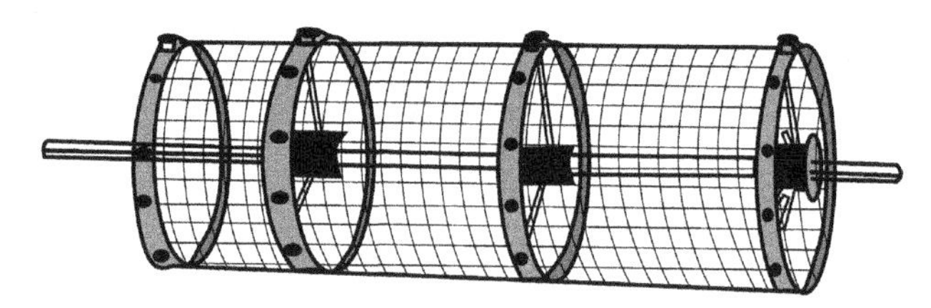

FIGURE 12.19 Trommel drum separator revolves in horizontal or low-angle axis to separate the size fractions (Source: after Wills, 2007 [76]).

FIGURE 12.18 ROM ore is transported to beneficiation plant and uploaded on Grizzly separator for preventive screening before feeding to primary crusher.

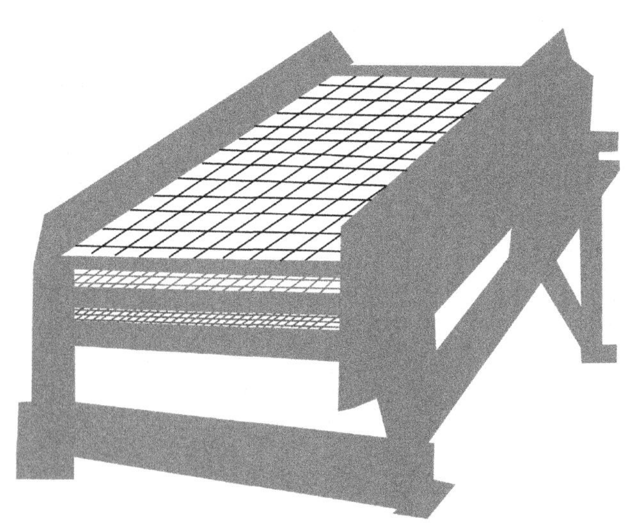

FIGURE 12.20 Schematic view of vibratory screen separator with multiple screen decks at reducing size fraction.

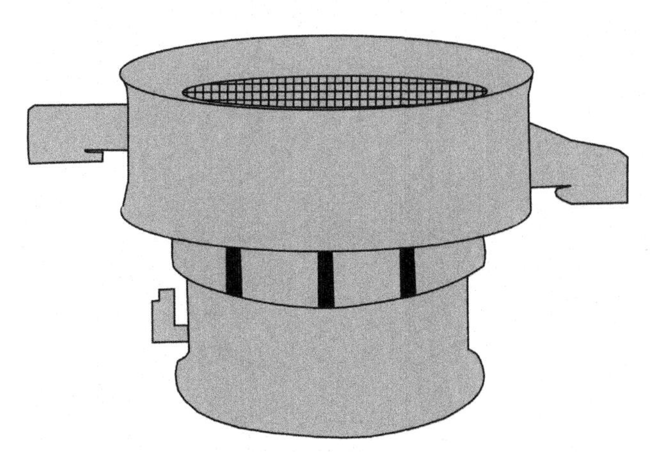

FIGURE 12.21 Schematic diagram of Gyratory screen separator with various available models.

"Gyratory" screen separators (Fig. 12.21) are removable and replaceable circular multiple decks or trays for different-sized products. It works both a gyratory and a slight vertical motion to the screen deck. Gyratory screen separators are ideal for large and fine solid particles and solid-liquid applications. It operates by specially designed motor mounted vertically at the center of the base plate of the screen.

12.4.2. Classification

Separation of particles by screening is not effective for exceptionally fine materials composed of different mineral mixture. Sorting into two or more mineral product of similar size is possible on the basis of the velocity with which the grains fall through a fluid media. This method of separation and concentration by difference in the settling rates due to variable particle size, shape and density in a fluid media is known as "classification". The fluid medium in general is water under modified condition such as rising at uniform rate, changing density, addition of suitable reagents and passing air bubbles. Classifiers consist of a sorting column in which a fluid rises at uniform rate. Particles introduced into the sorting column sink and report as underflow if their terminal velocities are greater than the upward velocity of the fluid. On the other hand, if their terminal velocity is less than the upward velocity of the fluid, it rises and reports as overflow. Classification equipments include hydraulic classifiers, horizontal current classifiers, spiral and rake classifiers and hydro-cyclones.

(a) Hydraulic classifier

Hydraulic classifier works on differences in the settling rates of particles of feed pulp against the rising water currents. The units are simple to design and consist of a series of conical sorting columns. The columns in series are successively larger in size with relatively lower current velocity. The relative rate of settling against the varying up-flow currents in each of the conical pockets accumulates coarser particles in the first to finest in the last container (Fig. 12.22). The sediments are removed from the bottom of the settling zone and treated accordingly. The very fine slimes overflow at the last column.

(b) Spiral classifier

Spiral classifiers are typically mechanical-driven devices. The unit drags coarse sandy sediment from the settled feed pulp by a continuously revolving spiral along the bottom of an inclined surface to a higher discharge point on one end of the settling tank (Fig. 12.23). The fines overflow at the other end. "Rake" classifier is a variation in mechanism of shifting the coarser component. The rakes dip into the feed pulp, move in an eccentric motion along an inclined plane for a short distance, then lift it up and go back to the starting point to repeat the operation. Spiral classifiers are usually preferred than rake type as material does not slide backward. The backward sliding of material happens when the rakes are lifted between strokes.

(c) Hydro-cyclone

Hydro-cyclone is the most important and widely used classifier in the mineral processing industry, particularly for

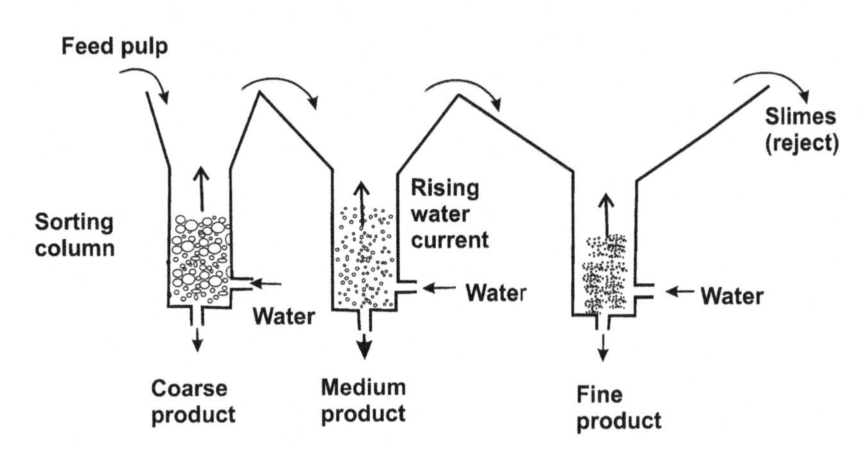

FIGURE 12.22 Schematic diagram showing the principle of hydraulic classifier in series of successively larger cone size with relatively lower current velocity (Source: modified after Wills, 2007 [76]).

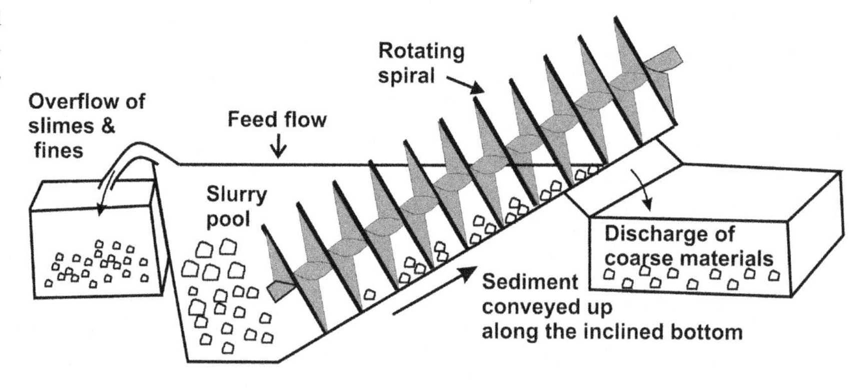

FIGURE 12.23 Principle of mechanical spiral classifier separating coarse materials from fines. (Source: modified after Ish Grewal, Met-Solve, Canada [29]).

base metals. It is normally installed in close circuit between grinding and conditioning path for flotation of complex base metal ore. It consists of a cylindrical section at the top connected to a feed chamber for continuous inflow of pulp and an overflow pipe. The unit continues downward as a conical shape vessel and opens at its apex to underflow of the coarse material (Fig. 12.24). The feed is pumped under pressure through the tangential entry which imparts a spinning motion to the pulp. The separation mechanism works on this centrifugal force to accelerate the settling of particles. The velocity of the slurry increases as it follows a downward centrifugal path from the inlet area to the narrow apex end. The larger and denser particles migrate nearest to the wall of the cone. The finer or lighter particles migrate toward the center axis of the cone and reverse its axial direction. It follows a smaller diameter rotating path back toward the top. The oversize discharge fractions return back to the mill for further grinding, while the undersize overflow move to the conditioning tank for flotation. The hydro-cyclones perform at higher capacities relative to their size and can separate at finer sizes than most other screening and classification equipments.

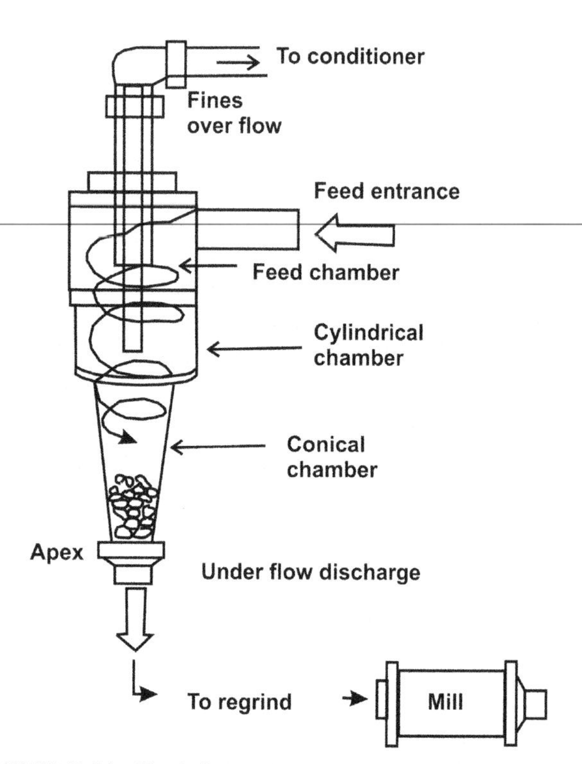

FIGURE 12.24 Sketch diagram showing the working principle of hydro-cyclone in close circuit classification.

12.5. CONCENTRATION

Minerals rarely occur as purest form and seldom used directly. Most of the nonmetallic and all metallic orebodies contain the valuable elements in widely varying state between parts per million and percentage. Ore as produced from the mine head needs to be beneficiated or upgraded to intermediate stage or final form for industrial uses. The process of upgradation is called "concentration" and upgradation product is the "concentrate". The concentration is performed by various methods of exploiting the physical and chemical behavior of the materials. Beneficiation processes are many, such as leaching, sorting, gravity, magnetic, electrical, dense media and flotation.

12.5.1. Leaching

Leaching is a process that extracts metals directly from the low-grade ores (LGOs), often in the oxide form and rejects accumulated in old tailing pads. This is a slow process and can take several months for metal extraction. Many a time leaching is economically effective to convert unviable property to profitable venture. The main leaching reagents are diluted hydrochloric, sulfuric and nitric acids. Acid leaching of low-grade Cu, Au, Ag, Pt deposits, and tailing pad is done without incurring much cost in mining, crushing and milling. Cleaning of iron ore or limonitic stained quartz sand by diluted sulfuric acid is a common industrial practice.

LGO, generally from open-pit overburden, is blasted, loaded and transported to the primary crushers. The coarse

FIGURE 12.25 Copper heap-leaching at Whim Creek, Philbara, Australia (Credit: Dr Tim Sugden, Venturex Resources Limited, Australia).

ore shifts to the heap leach pad. The leach pad is covered by a series of pipes and hoses which sprinkle shower of diluted acid solution to the low grade, often oxidized, ore. Metal is dissolved and flows to a pond at the bottom of the pad (Fig. 12.25). Leached solution containing the dissolved metal is pumped to the solvent extraction (SX) circuit. It looks like a series of agitation tanks or cells. The SX process concentrates and purifies the metal leach solution. Metal is recovered by the electro-winning cells at high electrical current efficiency. Adding of special chemical reagents to the SX tanks binds the metal selectively. Metal is stripped and separated easily from the reagent for reuse.

12.5.2. Ore Sorting

The process of mineral concentration and cleaning was conceptualized centuries earlier by selective hand sorting of desired or undesired particles of lumpy size by mere appearance, color, texture, heaviness etc. Hand sorting was common practice to separate rich ore as concentrate and wood or iron pieces as cleaning process from the ROM ore. Hand sorting is still a popular method in small-scale mining operation for separation of waste rock and specific sizing of ore for commercial purposes (Fig. 12.26).

The sorting techniques are changed to mechanical mode by adopting optical, electronic and radioactive properties for large-scale industrial applications. This could be possible due to distinct contract between the valuable ore and the waste gangue minerals with respect to their physical properties. The critical attributes are light reflectance (base metals and gold ore, limestone, magnesite, barite, talc, coal), ultraviolet ray (wolframite, sheelite), gamma radiation (uranium, thorium), magnetism (magnetite, pyrrhotite), conductivity (sulfide ores) and X-ray luminescence (diamond). The main objective of mechanical sorting is to reduce the bulk of the raw ROM ore by rejecting large volumes of waste material at an early stage. The process utilizes two-stage separations. The first stage involves

FIGURE 12.26 Hand sorting and commercial sizing by manual works at low cost in small-scale mining operation at Nausahi chromite mine, India.

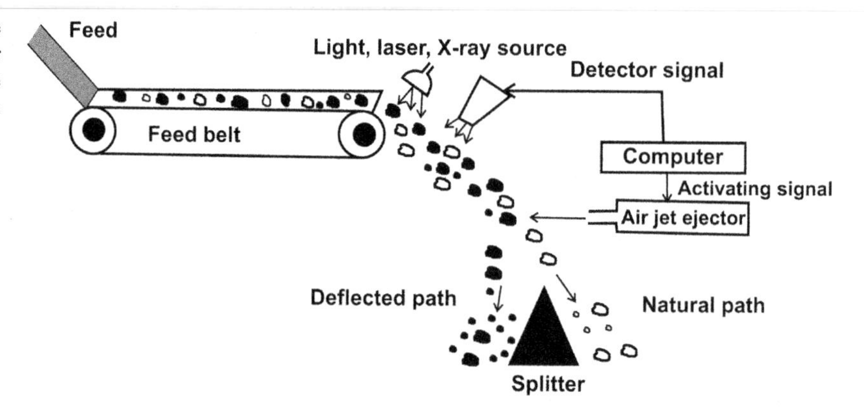

FIGURE 12.27 Schematic diagrams showing the principle of an automatic electronic ore sorter for separation of ore and gangue at lumps and coarse form at reduced cost. (Source: modified after Ish Grewal, Met-Solve, Canada [29]).

primary crushing of feed that liberates pre-concentrate and barren rejects. The second stage performs re-crushing, grinding and processing to produce final concentrate and tailings. This two-stage operations will substantially lower the cost of large volume of crushing, grinding and subsequent process of upgradation to produce marketable final concentrates.

A fully automatic electronic sorting device is comprised of an integrated circuit of an energy source, a process computer, a detector and an ejector (Fig. 12.27). The ROM ore at desired fragment size, preferably washed, moves on a conveyor belt or vibrating feeders at uniform speed and released maintaining a natural flow of the stream of ore particles. The energy elements like light rays, laser beam and X-ray converge from the source and reflect from the surface of the rocks passing through the sorting zone. The nature of reflectance is sensed by the detector system and sends signals to the computer. The amplified signal activates an air jet at the right instant and intensity to eject the particle from the stream. The accepted and rejected particles are dropped in separate stack around a conical splitter.

12.5.3. Gravity Concentration

Gravity concentration is a proven process for mineral beneficiation. The gravity concentration techniques are often considered where flotation practice is less efficient and operational costs are high due to extremely complicated physical, chemical and mechanical considerations. The gravity separations are simple and separate mineral particles of different specific gravity. This is carried out by their relative movements in response to gravity along with one or more forces adding resistance to motion offered by viscous media such as air or water. Particle motion in a fluid depends on specific gravity, size and shape of the moving material. The efficiency increases with coarser size to move sufficiently but becomes sensitive in presence of slimes. There are many types of gravity separators suitable for different situations. There are many devices for gravity concentration. The common methods are manual pans, jigs, pinched sluice and cones, spiral concentrator and shaking table to name a few.

(a) Panning

Panning as a mineral or metal recovery technique was known to ancients since centuries past. Gold panning was popular and extensively practiced in California, Argentina, Australia, Brazil, Canada, South Africa and India during the nineteenth century. Panning is done by manual shaking of tray containing riverbed sand and gravels, alluvial deposits containing precious metals like gold, silver, tin, tungsten etc. The shaking of the tray separates sand, stones and fine-grained metals into different layers by differential gravity concentration (Fig. 12.28). The undesired materials are removed. This is primitive practice at low cost and generally in practice at small scale by the local tribal people.

(b) Jig

Jigs are continuous pulsating gravity concentration devices. Jigging for concentrating minerals is based exclusively on

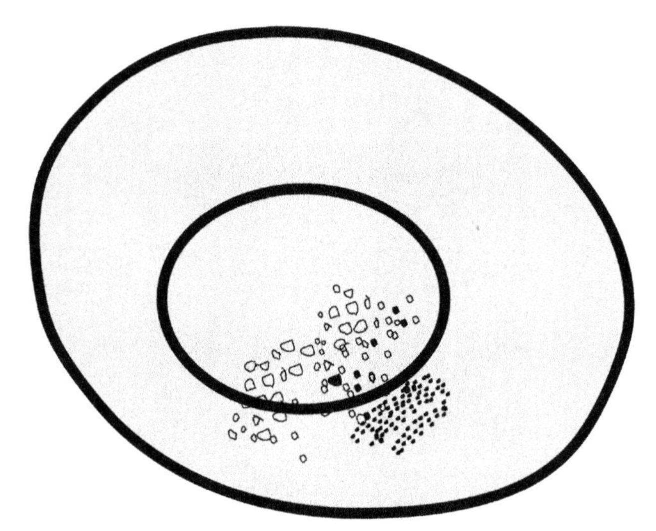

FIGURE 12.28 Recovery of gold, platinum and other heavy valuable minerals by panning is extensively practiced today in remote tribal areas.

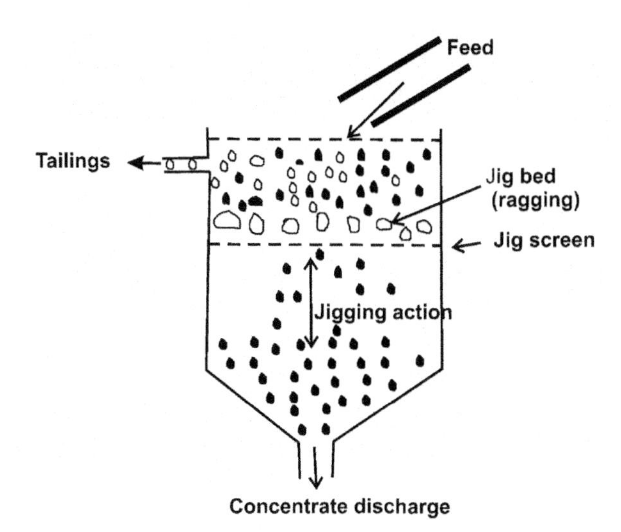

FIGURE 12.29 Conceptual diagram illustrates the basic principles of Jig concentrator.

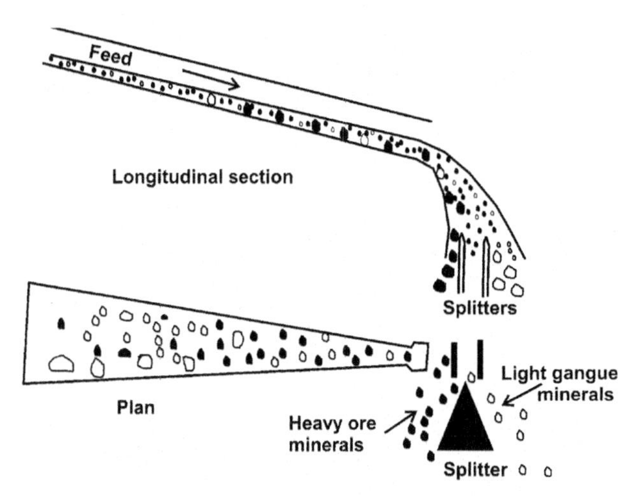

FIGURE 12.30 Conceptual diagram portrays the basic principles of pinched sluice concentrator.

differences in the density of the particles. The elementary jig (Fig. 12.29) is an open tank filled with water. A thick bed of coarse heavy particles (ragging) is placed on a perforated horizontal jig screen. The feed material is poured from the top. Water is pulsated up and down (the jigging action) by pneumatic or mechanical plunger. The feed moves across the jig bed. The heavier particles penetrate through the ragging and screen to settle down faster as concentrate. The concentrate is removed from the bottom of the device. Jigging action causes the lighter particles to be carried away by the cross flow supplemented by a large amount of water continuously supplied to the concentrate chamber. Jig efficiency improves with relatively coarse feed material having wide variation in specific gravity. Jigs are widely used as efficient and economic coal cleaning device.

(c) Pinched sluice and cones

Pinched sluice and cones is an inclined trough made of wood, aluminum, steel and fiberglass, 60-90 cm long. The channel tapers from about 25 cm in width at the feed end to 3 cm at the discharge end. Feed consisting of 50-65% solids enters the sluice and stratifies as the particles flow through the sluice. The materials squeeze into the narrow discharge area. The piling causes the bed to dilate and allows heavy minerals to migrate and move along the bottom. The lighter particles are forced to the top. The resulting mineral strata are separated by a splitter at the discharge end (Fig. 12.30). Pinched sluices are simple and inexpensive device. It is mainly used for separation of heavy mineral sands. A large number of basic units and recirculation pumps are required for an industrial application. The system is improved by development and adoption of the Reichert cone. The complete device is comprised of several cones stacked vertically in circular frames and integrated.

(d) Spiral concentrator

Spiral concentrator is a modern high-capacity and a low-cost device. It is developed for concentration of LGOs and industrial minerals in slurry form. It works on a combination of the solid particle density and its hydrodynamic dragging properties. Spirals consist of a single or double helical conduit or sluice wrapped around a central collection column. It has a wash water channel and a series of concentrate removal ports placed at regular intervals along the spiral. Separation is achieved by stratification of material caused by a complex combined effect of centrifugal force, differential settling and heavy particle migration through the bed to the inner part of the conduit (Fig. 12.31). The most extensive application is treatment of heavy mineral beach sand consisting of monazite, ilmenite, rutile, zircon, garnet etc. It is widely used to upgrade chromite concentrate. Two or more spirals are constructed around one central column to increase the amount of material that can be processed by a single integrated unit.

(e) Shaking table

Shaking table consists of a sloping deck with a rifled surface. A motor drives a small arm that shakes the table along its length, parallel to the rifle pattern. This longitudinal shaking motion drives at a slow forward stroke followed by rapid return strike. The rifles are arranged in such a manner that heavy material is trapped and conveyed parallel to the direction of the oscillation (Fig. 12.32). Water is added to the top of the table and perpendicular to the table motion. The heaviest and coarsest particles move to one end of the table. The lightest and finest particles tend to wash over the rifles and to the bottom edge. Intermediate points between these extremes provide recovery of the middling (intermediate size and density) particles.

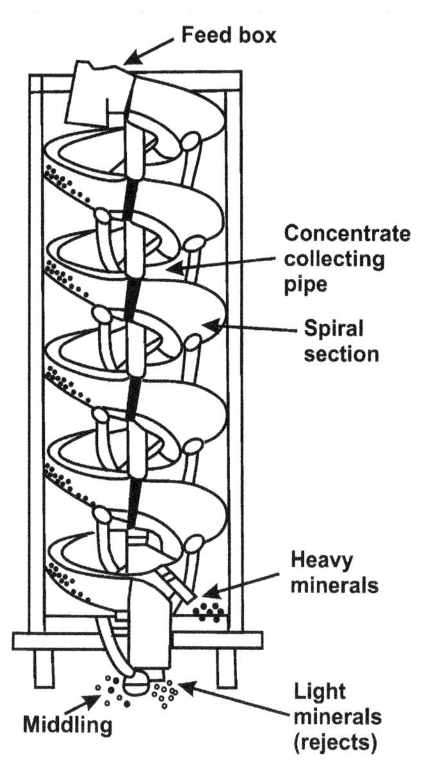

FIGURE 12.31 A typical sketch diagram of spiral concentrator working at chromite process plants, Sukinda layered igneous complex, India.

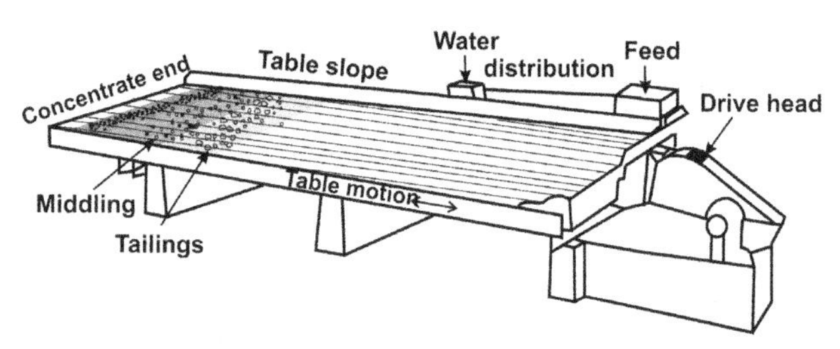

FIGURE 12.32 A typical model diagram of shaking table in recovery of gold, platinum etc.

Shaking tables find extensive use in concentrating gold. It is also used in the recovery of tin and tungsten minerals. These devices are often used downstream of other gravity concentration equipments such as spirals, Reicherts cone, jigs and centrifugal gravity concentrators for final cleaning prior to refining or sale of product.

(f) Multi-gravity separator

Multi-gravity separator (MGS) is a new development in flowing film concentration expertise which utilizes combined effect of centrifugal force and shaking (Fig. 12.33). Centrifugal force enhances the gravitational force and obtains better metallurgical performance by recovering particles down to 1 μm in diameter. It would otherwise escape into tailing stream if other conventional

FIGURE 12.33 MGSs at Rajpura-Dariba mineral process plant, India, during early 1990s.

wet gravity separators like jigs, spiral, table etc. are used. The principle of the system consists essentially in wrapping the horizontal concentrating surface of a conventional shaking table into a cylindrical drum and then rotates. A force, many times greater than the normal gravitational pull, is exerted by this means on particles in the film flowing across the surface. This enhances the separation process to a great extent. MGS in close circuit with lead rougher cells of graphite schist-hosted sulfide ore improves the lead concentrate metallurgy from 20 to +40% Pb. Graphitic carbon content reduces simultaneously from >10 to less than 3%. Presence of graphitic carbon interferes with the flotation of sulfide ore resulting in low metal recovery and unclean concentrate. MGS improves the metallurgical recovery and quality of concentrate for graphite carbon-bearing sulfide ore and high alumina-bearing fine iron ore. MGS technique is working successfully at Rajpura-Dariba zinc-lead plant and all iron ore plant in India by decreasing graphitic carbon and alumina respectively. MGS improves 42.9% Cr_2O_3 with 73.5% recovery from the magnetic tailings of Guleman-Sori beneficiation plant in Turkey.

12.5.4. Magnetic Separation

Magnetic separations take advantages of natural magnetic properties between minerals in feed. The separation is between economic ore constituents, noneconomic contaminants and gangue. Magnetite and ilmenite can be separated from its nonmagnetic RFM of host rock as valuable product or as contaminants. The technique is widely used in beneficiation of beach sand. All minerals will have one of the three magnetic properties. It is ferromagnetic (magnetite, pyrrhotite etc.), paramagnetic (monazite, ilmenite, rutile, chromite, wolframite, hematite, etc.) or diamagnetic (plagioclase, calcite, zircon and apatite etc.). Commercial magnetic separation units follow continuous separation process on a moving stream of dry or wet particles passing through low or high magnetic field.

The various magnetic separators are drum, cross-belt, roll, high-gradient magnetic separation (HGMS), high-intensity magnetic separation (HIMS) and low-intensity magnetic separation (LIMS) types.

(a) Drum separator

Drum separator consists of a nonmagnetic drum fitted with three to six permanent magnets. It is composed of ceramic or rare earth magnetic alloys in the inner periphery (Fig. 12.34). The drum rotates at uniform motion over a moving stream of preferably wet feed. The ferromagnetic and paramagnetic minerals are picked up by the rotating magnets and pinned to the outer surface of the drum. As the drum moves up the concentrate is compressed, dewatered and discharged leaving the gangue in the tailing compartment. The drum rotation can be clockwise or counterclockwise and the collection of concentrate is designed accordingly. Drum separator produces extremely clean magnetic concentrate.

(b) Cross-belt separator

Cross-belt separator consists of a magnet fixed over the moving belt carrying magnetic feed (Fig. 12.35). The magnet lifts the magnetic minerals and puts across the field leaving the gangue to tailing. The system is widely used in mineral beach sand industry for separation of ilmenite and rutile. However, it is replaced by rare earth role magnetic and rare earth drum magnetic separators.

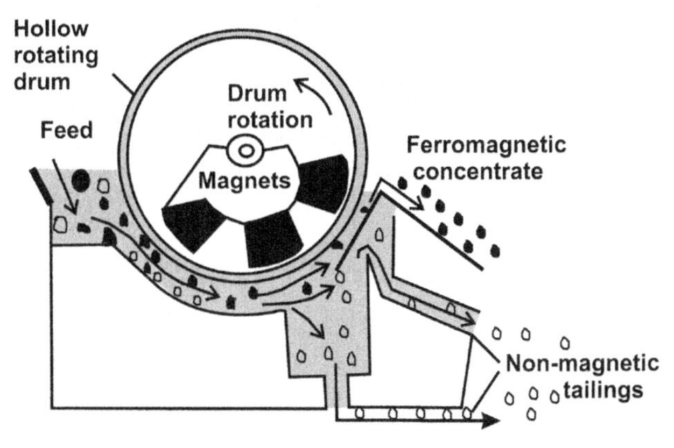

FIGURE 12.34 Sketch diagram showing the principle of drum magnetic separator. (Source: modified after Ish Grewal, Met-Solve, Canada [29]).

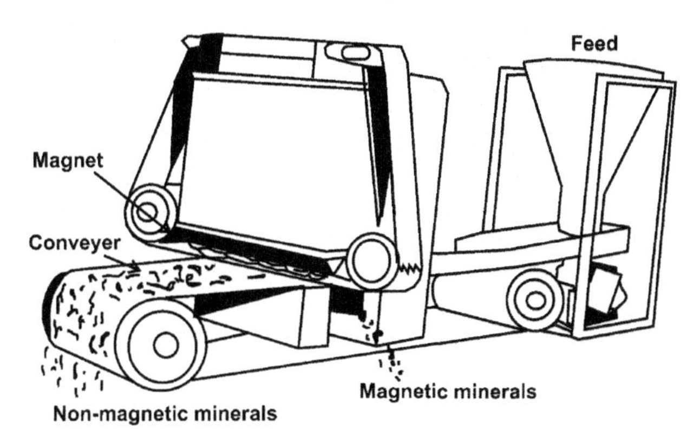

FIGURE 12.35 Schematic diagram showing cross-belt magnetic separator. (Source: modified after Wills, 2007 [76]).

12.5.5. Electrostatic Separation

Electrostatic separation works on natural conductivity properties between minerals in feed. The separation is between economic ore constituents, noneconomic contaminants and gangue minerals. The common units are high-tension plate and screen electrostatic separator. Electrostatic plate separators work by passing a stream of particles over a charged anode. The electrostatic minerals lose electrons to the plate and are pulled away from the other particles due to the induced attraction to the anode. The stream of moving particles is preferred between 75 and 250 μm, dry, close size distribution and uniform in shape for efficient separation to occur. The technique is used in separating monazite, spinel, sillimanite, tourmaline, garnet, zircon, rutile and ilmenite from heavy beach or stream placer sand. The electrostatic technique with local modification is extensively used in Australia, Indonesia, Malaysia and India bordering Indian Ocean for separation of mineral sands.

12.5.6. Dense Medium Separation

DMS or heavy medium separation (HMS) works on sink-and-float principle of minerals with variable specific gravity as in the case of coal and shale, sulfide ore in carbonaceous host rock. DMS in industries uses organic liquids, aqueous solutions and thick suspensions in water or pulp of heavy solid in water. This is the simplest method following wet gravity separation. The minerals lighter than the liquid medium will float and those denser than it will sink (Fig. 12.36). It helps very well as pre-comminution rejection of bulk of gangue material and pre-concentration of LGO. DMS technology is being very well adopted in investment of mining by pre-concentration of minerals from as low as 1-2% zinc grade. The technique helps in rejection of bulk of the gangue minerals prior to grinding for final liberation. A good example out of many can be cited from Pering zinc open-pit mine, South Africa, with 50 Mt reserves at 1.4% in situ Zn + Pb grade. Simplified DMS process reduces the ROM volume by pre-concentrate mass pull of 22% at 5.2% Zn + Pb.

12.5.7. Flotation

Since beginning of the twentieth century, flotation is the most flexible and adaptable mineral beneficiation technique. Mineral separation by flotation works on

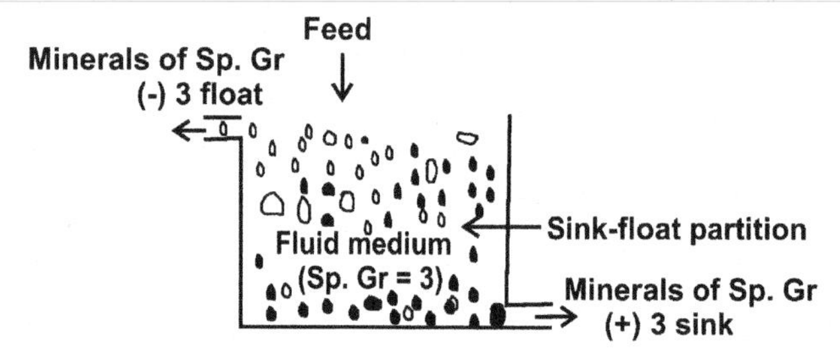

FIGURE 12.36 Conceptual diagram illustrating the principles of DMS process to separate coal and shale and promote pre-concentrates from low-grade sulfide ore and carbonate gangue minerals.

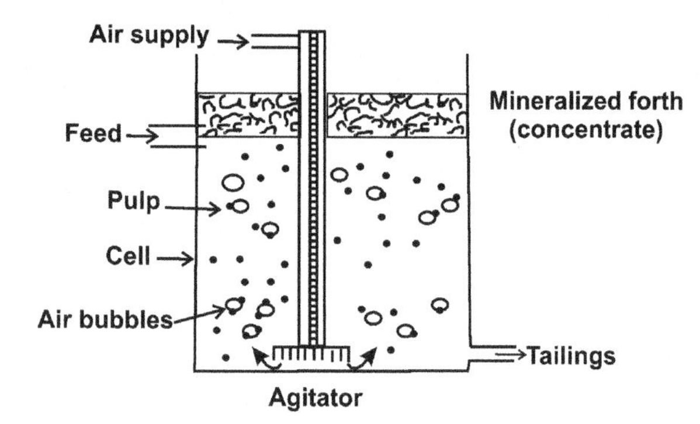

FIGURE 12.37 Schematic diagrams showing the principle of froth flotation and total function inside a flotation cell.

physicochemical surface properties of valuable and unwanted gangue minerals. It is being continuously modified for low-grade complex sulfide ores like lead-zinc, lead-zinc-copper, nickel-platinum-gold, tin, fluorite, phosphate, fine coal and iron ore at a lower cost with better recovery. The processes are known as froth and column flotation.

(a) Froth flotation

Froth flotation process produces froth of selective mineral agglomerates and separates them from other associated metallic components and gangue minerals. The physicochemical surface properties of optimum fine size fraction make some specific minerals hydrophobic. The particles turn into water repellent by coming in contact with moving air bubbles in the presence of certain reagents. The forth portion moves up leaving other metallic minerals and gangue (tailing) below, stabilizes for a while and collects as concentrate for further cleaning (Fig. 12.37).

The mineralized froth (concentrate) stabilizes for a while at the top of the cell (Fig. 12.38), overflows and moves to cleaner, filter and dryer in sequence to form mature salable product. The concentrate is the raw material for extracting metals by smelting and electro and chemical refining.

The continuous process of separation in a commercial plant happens in a series of containers called "cells"

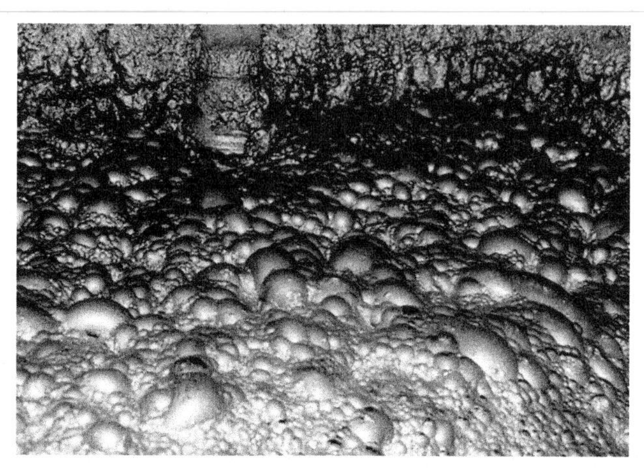

FIGURE 12.38 Formation of mineralized froth (lead concentrate) from active flotation cell and moves to lead cleaner.

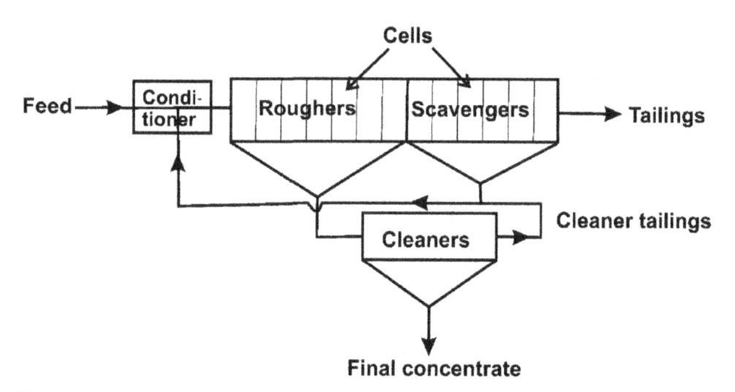

FIGURE 12.39 Schematic diagram of process flow sheet illustrating conditioner-rougher-scavenger-cleaner cells including formation of concentrate and reject.

forming a "bank" (Fig. 12.39). The products of final grinding mill pass through a conditioner tank where the pulp is conditioned in a few seconds to couple of minutes in the presence of Xanthate and Methyl Isobutyle Carbinol (MIBC). The pulp then enters the first few cells of the bank charged with reagents, known as "rougher cells". It moves up some of its valuable hydrophobic minerals as rich froth. Rougher concentrate gets diluted and refloated in the downstream "cleaner cells" to produce highest grade concentrate. The last few cells in the bank, known as "scavengers", contain relatively low-grade froth. This low--grade concentrate along with the tailings of the cleaner cells usually recirculates in the rougher cells. New feed enters the conditioner rougher cells → scavenger cells → cleaner cells until the barren tailing overflows the last cell in the bank.

Bank of flotation cells are installed in rows in the floor of the ore dressing plant in close circuit for generation of multiple concentrates of respective minerals (Fig. 12.40).

There are three main types of reagents used in flotation, namely, collectors, frothers and regulators. Each individual set of reagents plays a specific role in mineral processing.

FIGURE 12.40 A typical bank of flotation cells in operating circuit.

(a) "Collectors", also known as "promoters", are organic compounds which make the surface of certain selected minerals water repellent. These reagents are added to the pulp in the conditioner tank and ball mill. The mineral surface absorbs the collectors during conditioning period making them hydrophobic. Froth is formed when they come in contact with flowing air bubbles and float to the surface. Sodium Isopropyl Xanthate and Potassium Amyl Xanthate are the commonly used collectors.

(b) "Frothers" are surface-active chemicals that concentrate at the air-water interface. It prevents air bubbles from coalescing or bursting by lowering the surface tension of the slurry. Frothing properties can be persistent or nonpersistent depending on the desired stability of the froth. Pine oil and alcohols such as MIBC are most commonly used frothers.

(c) "Regulators" or "modifiers" are used to modify the action of the collector by intensifying or reducing water-repellent effect of mineral surface conditions. This is done to assist in the selective flotation of minerals. Regulators can be classed as activators, depressants, or pH modifiers. Regulators may activate poorly floating minerals such as sphalerite by adding readily soluble copper sulfate. Similarly regulators can depress certain minerals rendering it hydrophilic and preventing their flotation. Minerals like pyrite and arsenopyrite can be depressed by adding sodium cyanide or lime, so that a differential flotation can be performed on a complex ore. Nigrosene reagent is used for maximum depression or elimination of graphitic carbon from zinc-lead sulfides in graphitic host rock. Separation of graphitic carbon aids in producing high-quality clean zinc and lead concentrates. Chemicals that change the pH of the slurry are also used as modifiers. An alkaline condition of medium is preferable in the flotation process where most of the collectors are stable. Alkaline environment minimizes the damage done by corrosion of cells and pipelines. pH modifiers include lime, soda ash and sulfuric acid. It can act as activators and/or depressants by controlling the alkalinity and acidity of the slurry. Modifiers can also counteract interfering effects from the detrimental slimes, colloids, and soluble salts that can absorb and thereby reduce the effectiveness of flotation reagents.

The design of mineral processing circuit requires integration and assembly of various unit operations. It starts from crushing-grinding-flotation with generation of valuable concentrates and rejection of tailing in continuous process. It is desirable to conduct laboratory, bench and pilot plant scale test works at the appropriate stages of exploration and mine development activities. It is done before adopting a commercial plant flow diagram of complex mineral assemblages. The representative sample is obtained by compositing duplicate mineralized core covering the characteristics of entire deposit and bulk sample collected from initial mine development. The grind size, concentrate grades and other quality, recovery of valuable minerals, type of reagents and cost parameters are optimized.

(b) Flotation of zinc-lead ore

Majority of the zinc-lead deposits belongs to Palaeoproterozoic age and are hosted by dolomite, calc-silicate and mica-schist with or without graphite. Feed grades are typically 3-15% Zn, 1.0-2.5% Pb, and 40-150 g/t Ag with varying amount of pyrite and pyrrhotite. The typical grades of lead concentrate, zinc concentrate and tailings represent <65% Pb, <55% Zn, and ~0.20% Pb and ~0.50 to 1% Zn respectively. Silver and cadmium are recovered in lead and zinc concentrates in that order.

The ROM ore is crushed by primary Jaw crusher to yield a product size of −150 mm. The fraction between 150 and 50 mm is fed to secondary Cone crusher. The −50 mm size fraction of secondary crusher discharge and the screen undersize is fed to the tertiary Cone crusher. The final crusher product of −12 mm fraction size is fed to the ball mill for wet grinding. The ball mills run in close circuits with hydro-cyclones and yield a product containing 68% of −74 μm size particles. The specific particle size is essential for optimum liberation of valuable minerals from gangue material (Fig. 12.41). Zinc sulfate and sodium cyanide are added in the ball mill for the depression of sphalerite and pyrite respectively.

The cyclone overflow from the ball mill moves to the lead flotation circuit where Sodium Isopropyl Xanthate and MIBC acids are added as collector and frother respectively for lifting galena fragments. The circuit consists of rougher and scavenger flotation cell banks. The lead concentrate

FIGURE 12.41 Zinc-lead ore grinded to 68% of -74 μm particle size to yield highest liberation of ore and gangue minerals.

FIGURE 12.43 Zinc concentrate with grade up to 55% Zn is achieved based on feed quality.

FIGURE 12.42 Lead concentrate with grade up to 65% Pb is achieved based on feed quality.

FIGURE 12.44 Tailing from Zn-Pb process plant with grade around 0.20% Pb and 0.50-1.00% Zn is lost based on feed quality.

(Fig. 12.42) from lead rougher and scavenger is transferred to lead cleaner cells for washing till specific quality is achieved. Sodium cyanide is added to the lead cleaner cells for further depression of pyrite. The scavenger concentrates and the cleaner tailing are recycled to the lead rougher cells for additional recovery.

The overflow from the lead scavenger cells is fed to the zinc conditioner. Various reagents are added for specific purpose. These are sodium cyanide for pyrite depression, copper sulfate as sphalerite activator, Sodium Isopropyl Xanthate as collector and MIBC as frother. Zinc flotation is achieved in two stages of rougher and scavenger cells. The zinc rougher concentrate is transferred to the cleaning circuit for further upgrade. The scavenger concentrate and tailing of cleaner cells are recycled to the zinc rougher cells. The total zinc concentrate moves to zinc cleaner cell to achieve clean concentrate with grade up to 55% Zn (Fig. 12.43).

The final overflow of zinc bank cells moves out of the cells as rejects (Fig. 12.44) and transferred to tailing ponds or void filling in underground mines.

A typical process flowcharts of zinc-lead beneficiation plant and products are illustrated in Fig. 12.45.

Concentrates of lead and zinc are thickened in respective thickeners and pumped to individual filtration plant. The filtration is done by disc vacuum, drum or belt filters. The concentrates are dried using rotary thermal dryer or exposed to natural drying. The final lead and zinc concentrates are transported to respective smelters by road, rail and sea route. The eventual reject of the flotation plant (tailing) is pumped either to a nearby tailing dam and/or to the underground stopped out voids for filling as support system and prevents from collapse. Five to 10% cement is added for strengthening the underground fill. The tailing is allowed to settle in the dam. Water is reclaimed through a network of percolation wells and pumped back to water treatment plant for industrial reuse. A typical analysis of

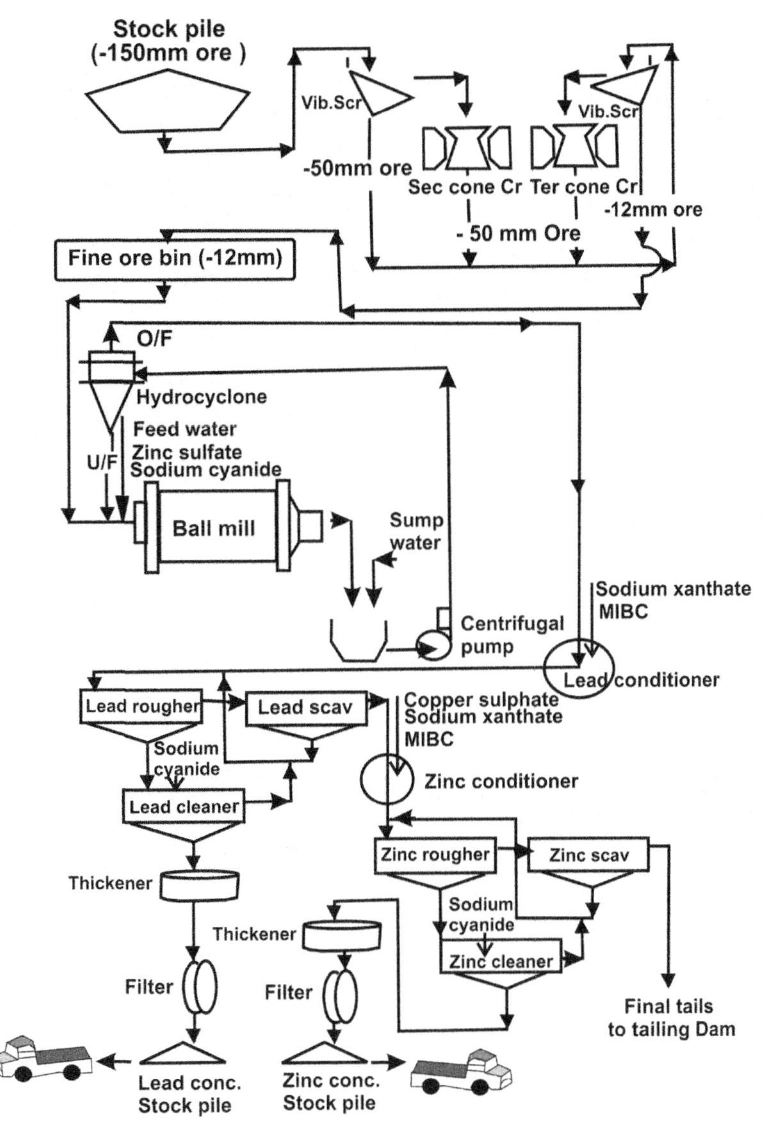

FIGURE 12.45 A typical process flowchart of zinc-lead beneficiation plant.

LGO and process products is given in Table 12.2. Zinc and lead metals are extracted by "pyrometallurgical" or "hydrometallurgical" process and electrochemical refining to the level of 99.99% metals. The precious elements recovered by electric refining are silver, cadmium, cobalt etc. The other important by-product recovery during smelting is sulfur to produce sulfuric acid.

(c) Flotation of copper ore

Copper ore containing about 0.5-2% copper as disseminated and stringers are reduced from 150 to 25 mm size in Jaw and Cone crushers. The crushed ore is grinded in ball mills to -74 μm size. The ore is processed by froth flotation comprising of rougher and scavenger cells. The commonly used reagents for flotation of copper ore are Xanthates as collector, pine oil as frother, sodium cyanide as depressant and lime to maintain the slurry alkalinity. Concentrate in the form of froth and slurry contains +20% copper. It is thickened in rake-type unit from a level of 30% solid content to 60%. The concentrate grade can further be elevated to 30% Cu. Recovery of associated molybdenum is achieved by integrated column flotation using cleaner concentrate. The thickened slurry is treated in vacuum disc filters to remove water. The powder concentrate contains about 10-12% moisture. The moisture content in the concentrate is further reduced to a level of 0.2% before charging to flash and anode furnace. The final treatment is electrochemical refining to obtain copper cathode of 99.99% purity. Precious metals recovered during electrorefining are gold, silver, tellurium, and selenium. The other

TABLE 12.2 Typical Analysis of LGO, Lead, Zinc Concentrate and Tailing

Parameters	Ore	Lead concentrate	Zinc Concentrate	Tailings
% Pb	2.0	65	1.5	0.13
% Zn	4.0	3.5	55	0.36
% Fe	5.50	5.5	6.5	4.5
Ag g/t	40	800	90	9
Cd g/t	200	160	2300	24
% Ca	12			15
% Mg	6			7
% Acid insoluble	28	1.30	1.20	28

Ore to concentrate ratio:1:15
Lead recovery is 85% and zinc recovery is 91.60%

important recovery during smelting is sulfur gas to produce sulfuric acid. A typical flowchart is given in Fig. 12.46.

The multi-metal copper-zinc-lead ore is treated by integration of three complete banks in series, each one for copper, lead and zinc to produce respective concentrates as discussed above. In case of complex ore metallurgy bulk concentrate is produced containing all the three primary metals and value-added elements like gold, silver, cadmium, cobalt, molybdenum, selenium and tellurium.

(d) Flotation of iron ore

Iron ore minerals, particularly hematite and goethite, are beneficiated by combination of size fraction, pre-concentration and flotation in stages (Fig. 12.47). Iron ores require removal of silicate impurities of a finer size by flotation for higher grade products of +60% Fe. The ROM ore at 400-600 mm is fed to the primary crusher with product set at −40 mm. The crushed product is screened at two stages. The overflow of the first screen (+40 mm) is re-crushed. The underflow of the first and overflow of the second screen i.e. −40 and +10 mm size are directly sent for loading as blast furnace grade. The underflow (−10 mm) is passed through classifier. The undersize of −1 mm is sent to the tailing pond. The overflow of −10 and +1 mm is grinded in ball mill to produce 200 μm product size. The pulp is subjected to hydro-cyclones for separation of slimes and removal of silica. The collectors such as amines, oleates, sulfonate or sulfates are used for the flotation of silica. Magnetic or gravity separation is introduced at any suitable stage for pre-concentration. The final iron ore fines are converted to pellets.

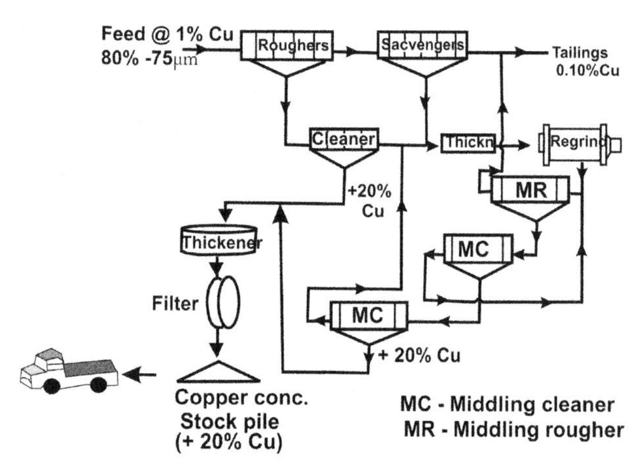

FIGURE 12.46 A typical flow diagram of copper ore beneficiation.

(e) Flotation of rock-phosphate ore

Rock-phosphate, the main raw material for fertilizer and phosphoric acid, occurs as high-grade ore (HGO at +30% P_2O_5), medium-grade ore (MGO at 20-30% P_2O_5) and LGO (at 15-20% P_2O_5). The HGO acts as a critical lifeline for the economic success of the project. It is mined, crushed to a desired size and sold straightway to fertilizer plant. The very HGO acts as sweetener for blending with MGO to increase the volume of direct salable product. The LGO is beneficiated by reverse/inverse flotation techniques.

The LGO is grinded to −74 μm size for liberation of main mineral constituents such as phosphate, carbonate and silica. The fine product is subjected to flotation after conditioning with fatty acid salt as collector and reduction of silica content in the alkaline circuit at the first stage. Addition of phosphoric and sulfuric acids in the second stage acts as pH controller for removal of carbonate in the froths. The phosphate component is depressed in the acid circuit and sent to thickener, filters and dryer to form final salable concentrate. The P_2O_5 content in the concentrate is +30% with concentration ratio around 2.5.

(f) Flotation of coal

The present-day highly mechanized coal mining (longwall mining) generates up to 25% fines of around 250 μm to 20 mm size. These fine fractions require upgradation by separation of high ash content. Flotation and DMS is the effective way to recover the fine coal. Petrochemical products such as diesel oil, kerosene and liquid paraffin are the most commonly used collectors. The two types of inputs for coal processing are high-value coking coal for pyrometallurgical industries and low-value thermal coal for power generation.

(g) Column flotation

Concentrates are often contaminated with the presence of excess undesired elements in zinc (silica), lead

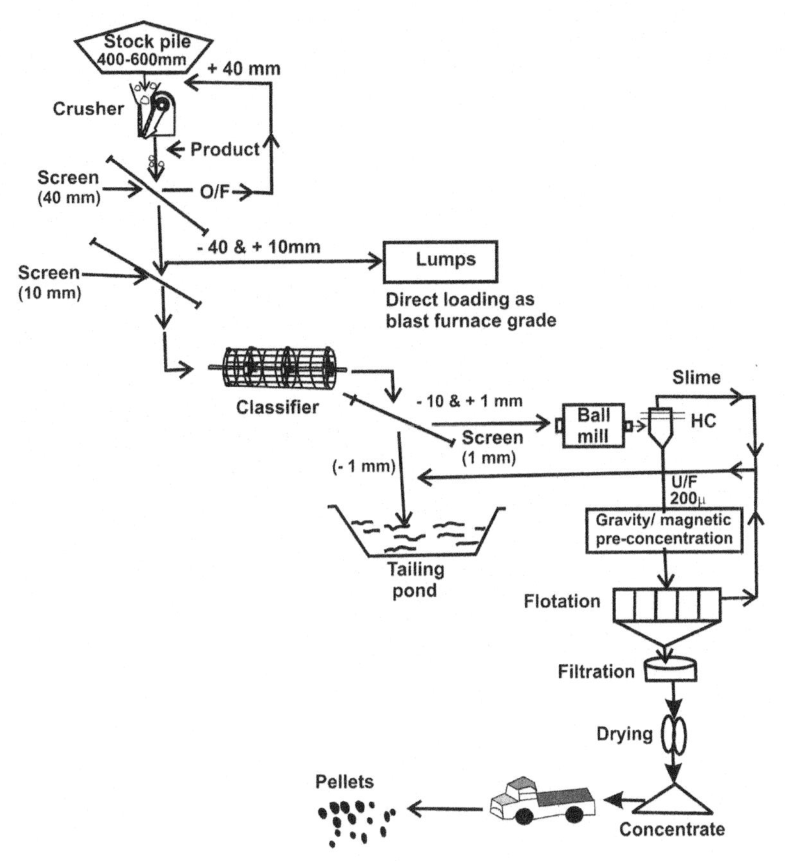

FIGURE 12.47 A complete flowchart of iron ore beneficiation.

(graphitic carbon), copper (silica), molybdenum, chromium, fluorite, manganese, platinum, palladium and titanium. Flotation column is a significant recent development in mineral processing industry on account of efficient single step cleaning action to upgrade fine-sized concentrate. It offers improved metallurgy at low cost, simplified circuit and easy control compared to conventional cells.

Column flotation works on countercurrent principle. The pulp moves down. The rising swarms of fine air bubbles generated by a gas sparger installed at the bottom carry valuable mineral particles to the froth at the top of the column. The three distinct design features that distinguish from the conventional flotation are (1) addition of wash water at the top of the froth, (2) absence of mechanical agitator and (3) bubble generation system. A schematic flotation column is shown in Fig. 12.48. Column flotation consists of two distinct zones, namely, collection and cleaning. In the collection zone, falling particles from the feed slurry are contacted by countercurrent with rising bubble swarm. Hydrophobic particles collide with and attach to air bubbles and are transported to the cleaning zone. Hydrophilic and feeble hydrophobic particles are removed from the bottom of the column. In the cleaning zone, water sprinkles over the top of the froth, thus providing a clean wash of the concentrate and further liberation of fine gangue particles from the valuable minerals. The liberated fine gangue particles move to the tailing zone and removed.

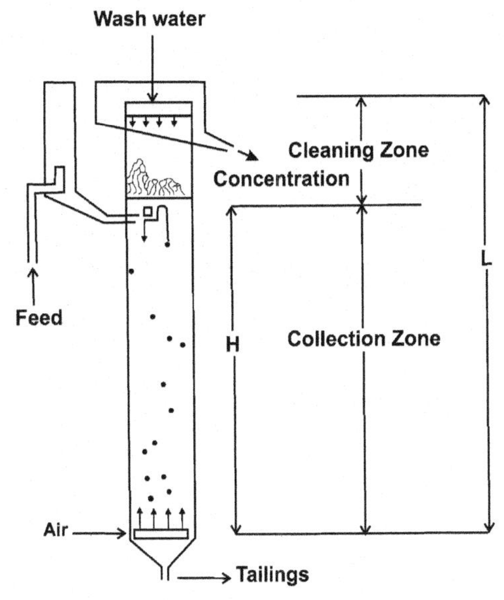

FIGURE 12.48 Basic features of flotation column collection zone (H) cleaning zone. Rejects move to the tailings.

Columns are installed and integrated between the concentrate cleaner tanks and the thickener. Columns have proved effective for cleaning applications and achieved upgrading in a single stage with improved metallurgy as seen in Zambian copper, Canadian molybdenum-copper, chromium and coal in United States, zinc-lead in Australian and India. Industrial flotation columns are round (Fig. 12.49) or square in cross section, typically 5-8 m in diameter and 10-15 m height. Column flotation improved the zinc concentrate metallurgy from 48% Zn with 5-7% SiO_2 at 80% recovery to +50% Zn with <3% SiO_2 at 85% recovery.

12.5.8. Dewatering

Most ore beneficiation methods require large volume of water. It is necessary in the process of separation of various valuable and gangue minerals. The final concentrates as produced contain high proportion of moisture. Smelters, captive or custom base, are generally located at long distances from mining- beneficiation sites due to inadequate infrastructure. Shipment of concentrate in pulp form to long distances is not advisable. Pulp transport by road, rail or sea route is unsafe even at exorbitant high cost. Therefore, dewatering or solid-liquid separation is performed to generate dry concentrate. However, partial presence of water is desirable, say between 5 and 10% moisture content, for easy handling and safe transport. Metal losses are expected if the moisture content is totally dry or too low. It often becomes serious environmental issue on account of spreading air-driven concentrate in dry and dust form. Dewatering is done at successive stages of sedimentation or thickening, filtration and thermal drying.

"Sedimentation" is natural gravity settling of the solid portion of the concentrate pulp. It takes place in a cylindrical thickening tank in the form of layers (Fig. 12.50). Pulp is fed continuously from the top of the tank through pipe. The clear liquid overflows out of the tank. The thickened pulp settled at the bottom is taken out through a central outlet. The deposition process can be accelerated and the settled solids can be pushed toward the central outlet by rotating suspended radial arms performing as automatic rake mechanism. Sedimentation process would produce thickened pulp of 55-65% solids by weight.

"Filtration" is the second stage of solid-liquid separation, normally after thickening, by means of a porous medium. The most common filter media is cotton fabrics but can be extended to any one of jute, wool, linen, nylon, silk and rayon. The filter pads allow liquid to percolate and retain the solid on the outer surface. The filter media is washed and cleaned at regular interval for better performance and longevity. Several types of filter mechanisms are in use. The most widely used filters in mineral processing applications are disc, drum and horizontal type. Filtration produces moist filter cake of 80-90% solids.

"Disk filters" are used with vacuum filtration equipment. It is made of several large discs (Fig. 12.51). Each disk consists of sectors that are clamped together. The ribs between the sectors are designed in a radial fusion narrowing at the center. The semidry feed enters from the

FIGURE 12.49 Industrial flotation column in operation at Rajpura-Dariba plant, India, with advantages like better recovery and higher concentrate grade at lower operating costs.

FIGURE 12.50 A view of thickener with rake mechanism in the first phase of dewatering of concentrate by sedimentation or gravity settling of solids.

FIGURE 12.51 A view of disc filter in the second phase of dewatering of concentrate.

FIGURE 12.52 A view of drum filter, an alternative process of dewatering concentrate.

side. The disc rotates slowly so that cake forms on the face of the disc and semidry cakes are lifted above the slurry. The cake is suction dried. It is removed by scraper blades fitted on the side of each disc and pushed to discharge chutes. Generally, disk filters are used for heavy-duty applications such as dewatering of lead-zinc-copper concentrate, low-grade iron ore-taconite, coal, and aluminum hydrate.

"Horizontal belt filter" consists of a highly perforated horizontal rubber drainage conveyor deck fitted with filter media. Slurry is fed at the starting point of the deck and moves to the other end. Filtration starts partly by gravity and partly by vacuum mechanism attached to the bottom of the moving drainage deck. The cake is discharged as the belt reverses over a roller.

"Drum" or rotary drum filter works on the same principle as that of a disc filter. The drum is mounted horizontally and rotates in slow motion (Fig. 12.52). The surface of the drum is tightly wrapped with filter media and divided into several compartments, each one attached with drain lines. The filter is partially submerged in slurry feed. The drum rotates slowly through the slurry and produces filtered cakes while moving out of the submergence level. Partially dry cakes are removed by a combination of reversed air blast and automatic scraper knife.

"Drying" of concentrate is done prior to shipment. Rotary thermal dryer is widely used for production of final salable concentrate. It consists of a long cylindrical shell mounted on a roller at little slope to rotate the unit in uniform speed. Hot air at about 980 °C is passed inside the cylinder through which the wet feed moves from feeding point to discharge end by gravity. Dry concentrate at 5–10% moisture moves on conveyer to the stockyard before being loaded onto trucks or rail wagons as required for shipment.

12.5.9. Tailing Management

Ore to concentrate ratio ranges between 2:1 and 10:1 and even more depending on richness of the deposit. The quantity of fine rejects or "tailing" is between 1 and 15 tonnes to produce 1 tonne of concentrate. A mine with annual production of 1 Mt copper/zinc-lead ore will generate fine tailing anywhere between 0.80 and 0.90 Mt. This huge amount of fine tailing is to be handled and disposed in careful manner without disturbing ecological balance of the surroundings. Tailing from the scavenger bank is pumped to tailing thickener. The underflow of tailing thickener, at maximum attainable density, is pumped to the tailing dam (Fig. 12.53) located at a nearby suitable distance from the plant. The modern tailing dams are constructed with deep-rooted walls and cement spread on floor to prevent any leakage of tailing water through cracks and fissures. The dam on the downstream is erected on alternate layers of gravels and sand so that water percolates through the bottom and settles in a tank.

Tailing thickener overflow, excess water from tailing pond through water collection wells and the underflow of tailing dam is reclaimed and recycled on continuous basis following the principle of "0" discharge. The reagent-mixed water is pumped to the plant for industrial reuse after lime treatment. A thin layer of gravel is placed directly over the tailing surface for dust mitigation. The top surface of tailing pond is reclaimed by direct vegetative stabilization after the dam is full and dried up. It is done by growing grasses, shrubs and trees to arrest the blowing of dry sand, a source of air pollution. The vegetated cover aimed at shedding rainfall runoff during humid season.

The alternative management of tailing disposal is to pump the dense slurry to the open underground voids of cut and fill, sublevel and vertical crater retreat stopes. The tailing in the initial fill is mixed with 10% cement to make a strong barricade at the stope mouth and subsequently by 3-5% cement for strong stabilized support system.

FIGURE 12.53 View of tailing dam in a hilly terrain that would eventually be filled over time and rehabilitated by vegetative stabilization.

12.6. METALLURGICAL ACCOUNTING

Plant performance is judged by the cost of operation, quantity, purity and recovery of the valuable products. Metallurgical accounting is important to control the operation at every stage. These key parameters are identified and computed by material balance using simple equations of two products (Eqns 12.1 and 12.2) for single metal. The single metal may be zinc and two products are zinc concentrate and tailing. The equation can be extended for calculation of multiproduct of multi-metal deposit including recovery of value-added by-products. The multi-metals may be from zinc, lead and copper deposit producing multi-product of three concentrates zinc, lead and copper. The value-added by-products of cadmium, silver and gold are recovered with zinc, lead and copper concentrate respectively. The procedure utilizes few basic inputs of the process plant. The model is demonstrated by plant parameters given in Table 12.2. The steps are illustrated considering only zinc metal for simplicity. The zinc-lead deposit in clean dolomite host rock is designed for 3000 tpd mine and matching milling capacity. The average feed, concentrate and tailing grades are 4.00% Zn, 55% Zn and 0.36% Zn respectively.

The material balance by equalizing input and output:

$$T_f = T_c + T_t \qquad (12.1)$$

The metal or mineral balance by equalizing total metal or mineral input and output:

$$T_f \times G_f = T_c \times G_c + T_t \times G_t \qquad (12.2)$$

where,

T_f = Weight of feed in tonnes; G_f = % Grade of metal or minerals in feed,

T_c = Weight of concentrate(s) tonnes; G_c = % Grade of metal or minerals in concentrate,

T_t = Weight of tailing in tonnes; G_t = % Grade of metal or minerals in tailing.

12.6.1. Plant Recovery

The plant recovery is the percentage of total metal recovered in the concentrates from the feed ore. A recovery of 86% means that 86% of metal has been recovered in the concentrate and 14% metal is lost in the tailing. If the ore contains more than one metal (say Pb, Zn and Cu), then the weighted cumulative recovery in concentrates and loss in tailing is to be mentioned.

$$\text{The plant recoery} = (T_c \times G_c/T_f \times G_f) \times 100\% \qquad (12.3)$$

$$\text{Or } [\{(G_c \times (G_f - G_t)\}/\{G_f(G_c - G_t)\}] \times 100\%$$

$$= 91.60\ \%$$

12.6.2. Ore to Concentrate Ratio

It is the ratio of the weight of the feed to the weight of the concentrates produced. Higher the grade of the deposit lower will be the ratio and vice versa. The concentrate grade and recovery are the metallurgical efficiency factors. From Eqn 12.2,

$$F/C = (G_c - G_t)/(G_f - G_t) \qquad (12.4)$$

where, F/C represents the ratio between concentrate and feed

= (55 − 0.36)/(4.00 − 0.36)

= 15 i.e. 15 tonnes of feed will yield 1 tonne of concentrate.

TABLE 12.3 Plant Performance and Metal Balance of a Zinc-Lead Mine

Item	Weight (tonnes per day)	% Zn metal	Weight of zinc metal (tonnes)	Distribution % metal
Feed	3000	4.00	120	100
Concentrate	200	55.00	110	91.60
Tailing	2800	0.36	10	8.40

12.6.3. Enrichment Ratio

The enrichment ratio is the ratio between concentrate grade and feed grade and expressed as G_c/G_f i.e. $55/4.0 = 13.75$

12.6.4. Metal Balancing

The metallurgical accounting or metal balancing is computed by shift/day/week basis with the help of automatic sampler at feed, concentrates and tailing points. However, the mineral beneficiation is a continuous process linking flow of material at the crushing, grinding, flotation, concentration, tailings, thickener, filters, dryers and stockpiles. The weight of final concentrate is measured in the stockpile at regular interval. The metallurgical balance at short time interval, say shift, day or week is an approximation to take preventive measures in the plant. The results of cumulative long-term tenure, such as monthly/quarterly/annual are more authentic. The computation is done with Excel spreadsheets using two-product formula. It can be programmed for cumulative results and performances. Commercial software is available. Table 12.3 enumerates the plant performance and metal balance of single metal and concentrate for a day's production. In case of a multi-metal deposit, say copper-lead-zinc, the metallurgical balance will be complicated and to be expanded on the same principle.

12.6.5. Milling Cost

Milling cost including all individual subdivision is crucial in identifying high cost functions where improvements in performances would be beneficial to the project profitability. The cost cannot be generalized due to local variable costs on labor, energy and water. It will also be affected by the type and complexity of the ore being processed. A summary of the standard fixed and variable process plant operating costs is summarized in the Table 12.4.

12.6.6. Concentrate Valuation

The main objective of mineral beneficiation is to optimize the financial return of the ore at various combination of recovery efficiency and grade of the concentrates. This will depend on the average value of the metals at international market i.e. London Metal Exchange (LME), transport

TABLE 12.4 Operating Cost Summary (%) per Metric Tonne of Ore Beneficiated by Froth Flotation

Item	% Cost
Fixed costs	
Manpower	30
Overheads	1
Subtotal	31
Variable costs	
Power	20
Water	2
Reagents	20
Grinding media (crushing and grinding)	3
Crusher liners	2
Mill liners	2
Thickening and filtration	1
Tailing disposal	2
Laboratory costs	5
Maintenance costs	7
Concentrate transport	5
Subtotal	69
Total	100

charges to smelter and costs of metal extraction. The harmful impurities like Hg, As, U and precious trace and rare earth elements such as Au, Ag, Cd and Co will affect the value of the concentrate either as penalty or bonus respectively. The net smelter return (NSR) is calculated for any combination of recovery efficiency and grade as:

$$\begin{aligned} \text{NSR} = {} & \text{Value of contained metal} \\ & - (\text{Smelter/Refinery charges} + \text{Transport costs} \\ & + \text{Bonus} - \text{Penalty}). \end{aligned}$$

The cost of producing zinc includes cost of concentrate and conversion to metal (smelter + refinery). These costs are affected by local factors such as energy and labor where the mine, beneficiation plant and smelter are located. The average cash cost of producing zinc during 2010 was US$ 1800/tonne. The raw material i.e. ore and concentrate account for US$ 1200/tonne and conversion cost of US$ 600/tonne. Most of the smelters made profit against average LME price of US$ 2150/tonne. The cost for other metals and minerals can be estimated in the same way.

12.7. ORE TO CONCENTRATE AND METAL

Mineral processing or mineral beneficiation or upgradation involves handling of three primary types of ROM ore material which has been blasted, fragmented and brought out from in situ position. These materials can be used directly or by simple or complex processing and even applying extractive metallurgy like hydrometallurgical or pyrometallurgical methods. The categories are as follows:

(1) Rocks such as granites, marble, limestone, building stones, sand, coal and clays,
(2) Industrial minerals like quartz, diamond, gemstones, fluorite, apatite, zircon, garnet, vermiculite, barite, wollastonite, and
(3) Metalliferous deposits having varied metal content like in gold, platinum, chromite, chalcopyrite, sphalerite, galena, bauxite, hematite, magnetite, etc.

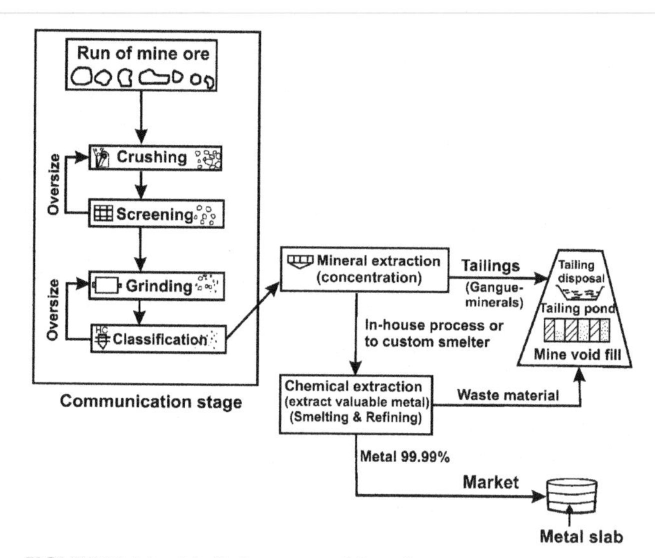

FIGURE 12.54 A holistic conceptual flow-diagram showing the journey from ROM ore to concentrate and ultimately to metal production (Source: modified after Ish Grewal, Met-Solve, Canada [29]).

The journey from ROM ore to concentrate and ultimately to metal has been conceptualized. The various unit operations used for liberation, separation, concentration and extraction have been discussed in the previous pages of this chapter. The activities and the typical sequence of operations in the process plant have been diagrammatically summarized in Fig. 12.54.

FURTHER READING

Gaudin, way back in 1939 [26], published an unparallel text book on principles of mineral dressing. Since then researches in the subject have observed sea changes in development of new concepts, mechanization for more efficient recovery at lower cost. The outcome of these progressive improvements and development of modern concept on ore beneficiation has systematically been dealt through "An introduction to the practical aspects of ore treatment and mineral recovery" edited by Wills (2006) [76]. The book will provide enormous support to students of higher studies in Mineral Engineering. "Introduction to Mineral Processing" by Grewal Ish (2010) [29] is an interesting reading. It includes many diagrams that make it easy to understand the subject.

Chapter 13

Mineral Economics

Chapter Outline

"Mines are not found, but made, usually at great cost and often with significant risk."

—Author.

13.1. DEFINITION

Mineral industry forms the backbone of a nation's economy and growth. It is an opportunity-based investment venture with high degree of risk associated at each stage of activities. The risks revolve around geological uncertainties, technical competency, needs of the society, commercial feasibility, economic viability, political stability and will of the Federal and Regional Governments. The risks can be minimized by generating adequate information during the various phases of exploration and critical economic analysis to safeguard the investment. The activities are largely scientific and technical in nature. The procedures have been discussed in previous chapters on various exploration techniques, estimation with appropriate level of 'Due Diligence' through QC and QA, mining and beneficiation. The political stability, will of the Governments and regional attitude are socioeconomic attributes. It is dealt by the involvement of Federal-State Governments and Private Entrepreneurs for the overall economic and social sustainable development of the area in particular and country as a whole.

The commercial and economic aspects are not in the hands of the investor and mainly rely on global market scenario. The feasibility analysis can indicate conversion of mineral resource to marketable commodity with adequate return on investment. The mineral resources are made ready for the end users by four well-defined processes, namely, exploration, development, production and extraction. The investment decision for each stage spins around interrelated components of "Resource", "Risk" and "Revenue".

13.2. INVESTMENT PHILOSOPHY

Mineral exploration program and subsequent development to a producing mine need investment of different magnitude without earning any revenue during the initial exploration stages. The return, revenue or benefits of different magnitude on investment realize in the later phase of mine production, mineral processing and smelting. Investment in the early phase of exploration i.e. "Reconnaissance" may fail to discover any mineral deposit and total expenses end in loss. The second phase i.e. "Large Area Prospecting" or "Prospecting" may delineate a mineral body with certain resource and metal content. The deposit may or may not be viable at this stage without further investment. The right answer is likely to arise at the end of "Detail exploration" and provide an economic deposit with reasonable assurance. The project may still fail to deliver the benefits due to some unknown factors. These uncertain clouds indicate that the mineral industry is prone to be of high-risk venture. But success in one deposit out of 100 attempts will compensate the earlier losses and make the investment profitable.

Mineral Exploration. http://dx.doi.org/10.1016/B978-0-12-416005-7.00013-1

The demand of minerals in national economy promotes the basic stimulus for investment. But who are the prospective investors under such speculative investment in mineral industry? The investors are from either Federal/ State Government and Government Enterprises for socio-economic developments or Private Sectors for business opportunities. The investors may operate as individual entity or as Joint Venture partners, big and small, organized and unorganized structure. Each one works with its own philosophy on investment.

Small unorganized owners cannot absorb the shocks of failure. Thereby they do not invest big sum on exploration and development. Their targets are small mineral property like silica sand, garnet, soap-stone, marble, limestone etc. The mines are in general open pit. The area of operation is local with particular type of mineral(s). Both the investment and revenue are of low magnitude.

Private organized sectors invest on exploration and subsequent programs mainly on financial criteria, strive to maximize profits and urge to attain return on investment at the shortest time. The primary and long-term objectives of maximizing profit are achieved by increasing production size, cost reduction and control in market share. This is done by increasing OMS, mechanization, dilution control and higher recovery efficiency in mine, ore dressing and smelting plant. These houses are specialized for particular type of minerals like iron ore, bauxite, base metal, chromite, nickel, platinum-palladium, coal, etc.

Government won companies work on optimum profit with long-term objectives of economic growth and self-sufficiency of the country as a whole and socioeconomic development of the region in particular. The Government conducts regional exploration program covering all minerals and outline mineral base regions and generates mineral inventory of the country. The objectives of the Government sectors are many folds, such as, overall socioeconomic growth of undeveloped area, generation of employment, reducing demand-supply goals and enhance foreign exchange reserve.

13.3. STAGES OF INVESTMENT

Three well-defined stages of investment cum activities are exploration, development and production. The interrelated practices are capable to convert once unknown mineral resource into a profitable commodity. The sequential approach of reconnaissance, prospecting and detailed exploration can establish an economic deposit. The development phase provides creation of infrastructure facilities, establishes mine entry system and designs mineral processing roots. There will be net outflow of cash or "Negative Cash flow" during exploration and development. The expenditures are capitalized. In the production stage the operating cost is met through the revenue generated ending with "Positive Cash flow" or cash inflow of different magnitude into the project.

"Cash flow" is the difference between all cash inflows (Revenue) less all its cash outflows (Costs). There are two types of cash flow connected with any investment.

(a) Cash flow before tax

The cash flow before tax (CFBT) of investment is the actual flow of money into and out of the business without consideration of **tax** deductibility. The method acts as an incentive by the Government to promote sustainable development of remote areas for the welfare of the local inhabitants.

(b) Cash flow after tax

The cash flow after tax (CFAT) does account for tax liability. It is essentially the CFBT less tax liability.

Net cash flow of one iron ore deposit with 8 years of mine life is computed at Table 13.1. The Net Present Value (NPV) is positive before tax and negative after tax over the

TABLE 13.1 **Simplified Presentation of Net Cash Flow Computation before and after Tax for the Iron Ore Deposit under Investment Showing (+) and (−) Loss/Revenue Respectively**

Year	Net CFBT (million $)	Net cash flow after 20% tax (million $)	Activity
0	−2	−2	Prospecting
1	−2	−2	Exploration
2	−50	−50	Development
3	5	4	Initial production
4	10	8	Production buildup
5	20	16	At full production
6	20	16	At full production
7	20	16	At full production
8	20	16	At full production
9	10	8	Reduce production
10	5	4	Mine closure starts
NPV at 12%	10.14	−1.10	Mine closed

mine life. The deposit can only be viable if the tax is exempted by the authorities as an incentive to promote the local development. The NPV computation is discussed at Section 13.4.2.1.

13.4. INVESTMENT ANALYSIS

The investment opportunity of mineral project is evaluated and compared with the cost at different stages of exploration, development and production vis-a-vis the expected revenue to be earned during the first 10-15 years of mine production. If the benefits are higher than the associated cost then the opportunity is lucrative and worth considering. The method of investment analysis begins with estimation of the resource to be spent on exploration, development, production, royalty, taxes and other activities versus the revenue expected to be received from the sale of end products. The phasing of activities, investment, cash flow and chances of failures over years is conceptualized in Fig. 13.1A, B and C. Fig. 13.1B clearly indicates that the cash flow during exploration and development is negative and becomes positive with the commencement of production.

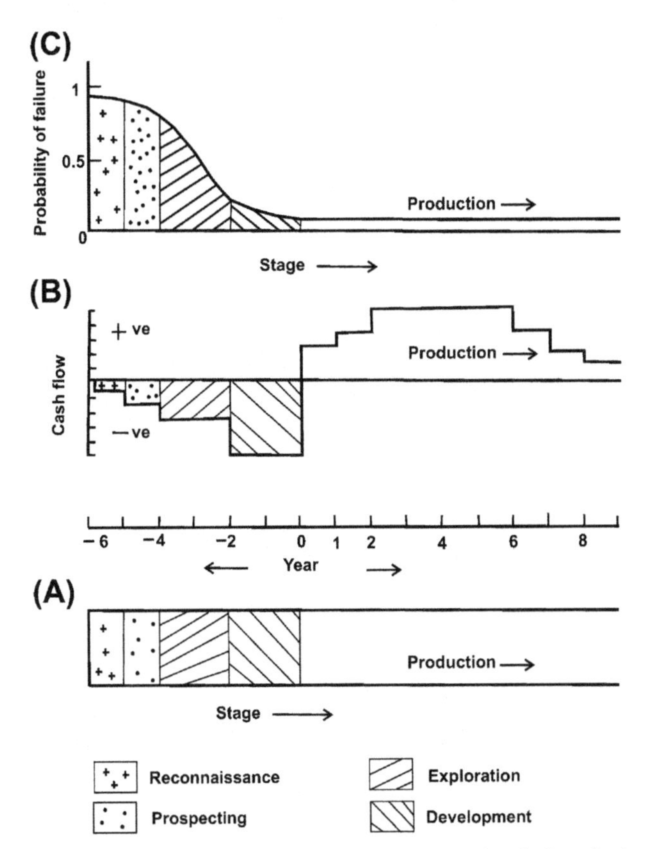

FIGURE 13.1 Schematic phasing of investment and cash flow distribution in mineral deposit: (A) stages of exploration and development, (B) cash flow and (C) probability of failure.

Various financial analyses are available for taking investment decision of a project. The two main types of financial techniques include undiscounted and discounted methods. The investors work on either of the two principles depending on his business viewpoint and values.

(1) "The bigger-the-better" (bigger benefits are preferred than smaller).
(2) "The bird-in-the-hand" (early benefits are preferred to late benefits), Johnson (1988) [41].

13.4.1. Undiscounted Method

Undiscounted method is simple and quick to apply in getting an essence of investment climate to get back the original investment rapidly. The easily understandable computation is the "Payback period" of the original investment. It is the number of years required for the cash income from the project equal to the initial investment incurred (Table 13.2).

On the face value, Project-A with 3 years Payback period would be preferred over Project-B with 4.25 years. But, Project-B will be more profitable at higher Internal Rate of Return (IRR at Section 13.4.2.2) of 20% on long-term bigger benefits

"Average Accounting Rate of Return (AARR)" is another simple and quick way of computing the return on investment by dividing the average annual net profit by the initial investment (Table 13.3).

$$\text{Average net profit} = 70/4 = 17.5$$

$$\text{Investment} = 100$$

$$\text{AARR} = 17.5/100 = 0.175 \text{ or } 17.5\%$$

TABLE 13.2 Payback Period Method (Cash Flow in Million US$)

Year	Project-A	Project-B	Activity
0	(−) 60	(−) 70	Exploration and development
1	20	10	Mine production
2	20	10	Mine production
3	20	20	Mine production
4	10	20	Mine production
5	10	40	Mine production
6	10	50	Mine production
Payback period	3 Years	4.25 Years	
IRR	15.5%	20.0%	

TABLE 13.3 AARR (Cash Flow in Million US$)

Year	1	2	3	4	Total
Net profit	10	15	20	25	70

13.4.2. Discounted Method

If the Fig. 13.1 is critically reviewed then the two amounts i.e. investment made during the early life of the project (−ve Cash flow) and the revenue received later (+ve Cash flow) cannot be compared directly as discussed at undiscounted method. It must be adjusted for the time value of money. Time value concept of money expresses as "a dollar received today is worth more than a dollar received tomorrow". An investor can simply deposit his capital fund preferably to the nationalized banks or other reputed financial institutions and earn a risk-free income without sweating his toil. Then why shall one invest his resources, mind and time in risky mineral base proposal without being compensated by enough benefits for receiving payment tomorrow rather than ensured regular earnings? Therefore, the value of future cash inflow (revenue − cost) must be discounted before comparing with the capital to be invested today to develop a project tomorrow.

Time value of money concept is expressed by the formula and elaborated at Table 13.4.

$$A = P(1+i)^n$$

where A = accumulated value of an investment (P) annually computed over (n) years at a variable interest rate (i).

The investor can invest US$ 100 today and will receive US$ 152 at the end of year three. This is the key concept in computing the present value (PV) of money to be received at different future years. Money received annually at the end of different years is not equal in value of today and therefore must be discounted to its PV of today. The PV is estimated by the formula below and elaborated at Table 13.5. The PV of US$ 100 after 10 years at discounted rate of 12% is worth US$ 32.20.

$$\text{PV} = 1/(1+i)^n$$

TABLE 13.4 Simplified Presentation of "Time Value of Money" Concept—Compound Interest Rate Computation at 15% over 3 Years

No. of years	Original investment ($)	15% Compounded interest	Accumulated value (US$)
1	100	* (1.15)	115
2		* (1.15) (1.15)	132
3		* (1.15) (1.15) (1.15)	152

* = Multiplication with principal investment.

13.4.2.1. NPV Method

The Net Present Value (NPV) is the sum of differences between a stream of discounted revenues and cost. It is computed by the following formula:

$$\text{NPV} = (R_0 - C_0) + \frac{(R_1 - C_1)}{(1+r)} + \frac{(R_2 - C_2)}{(1+r)^2} + \frac{(R_n - C_n)}{(1+r)^n}$$

where NPV in the year 0 is the sum of revenues (R) minus cost (C) in year 0 through n, adjusted back to the present using a discount rate, 'r'. A positive NPV indicates that the total expected revenues exceed total expected costs and suggests investment in the project. Similarly, a negative NPV will reject the project investment proposal. NPV = 0 is the break-even point of "no loss no gain" situation. A hypothetical Cash flow data are computed in Table 13.6.

TABLE 13.5 Simple Way to Compute PV up to 10 Years

No. of years from present	Original amount received (US$)	Discount factor at 12%	PV (US$)
0	100	* 100/1	100
1		* 100/(1.12)	89.28
2		* 100/(1.12) (1.12)	79.72
3		* 100/(1.12) (1.12) (1.12)	71.18
4		* 100/(1.12) (1.12) (1.12) (1.12)	63.55
5			56.74
6			50.66
7			45.23
8			40.39
9			36.06
10			32.20

* = Multiplication with principal investment.

TABLE 13.6 Cash Flow Data for a Hypothetical Mineral Project

Year	Revenues	Expenditures	Net cash flow*	Activity
0	0	5	−5	Reconnaissance
1	0	10	−10	Prospecting
2	0	20	−20	Detailed exploration
3	0	20	−20	Development
4	10	5	+5	Initial production
5	20	10	+10	Production at 50% capacity
6	40	20	+20	Production at full capacity
7	40	20	+20	Production at full capacity
8	60	25	+35	Increase production, OMS and cost cutting

*Net cash flow = [revenues (earnings) − cost (expenditures)] Million US$.

$$\begin{aligned}\text{NPV} &= (-5) + (-10/1.1) + [(-20)/(1.1)^2] \\ &\quad + [(-20)/(1.1)^3] + [(5)/(1.1)^4] \\ &\quad + [(10)/(1.1)^5] + [(20)/(1.1)^6] \\ &\quad + [(20)/(1.1)^7] + [(35)/(1.1)^8] \\ &= -5 - 9.09 - 16.53 - 15.03 + 3.42 + 6.21 \\ &\quad + 11.29 + 10.26 + 16.33 \\ &= 1.86 \text{ million \$ } (\textit{marginal case})\end{aligned}$$

13.4.2.2. IRR Method

"IRR", "Discounted Cash Flow Rate of Return (DCF ROR)" or "ROR" is an alternative method used to evaluate investment opportunities. It is applied in capital budgeting to measure and compare the project profitability. The principle and computation is similar to NPV. IRR is defined as the discount rate that equates the total discounted income with the total discounted costs of a project over the life period. It is simply the ROR at which NPV equals to zero.

$$\text{NPV} = \sum_{n=0}^{n} (Cn/(1+R))^n = 0$$

where Cn is the cash flow related to particular period (n) usually in year (0, 1, 2, ... n)

or

$$\text{NPV} = 0(R_0 - C_0) + \frac{(R_1 - C_1)}{(1+\text{IRR})} + \frac{(R_2 - C_2)}{(1+\text{IRR})^2} + \frac{(R_n - C_n)}{(1+\text{IRR})^n}$$

where R and C are revenues and costs respectively over the life of project in year 0 through n. The investor accepts the project with IRR greater than its minimum acceptable ROR. IRR for project data at Table 13.6 has been computed as 10.93% greater than discount rate of 10% assumed for accepting the proposal. IRR value obtained is zero if and only if the NPV is zero. NPV and IRR yield same conceptual decision of "accept or reject" for investing in new project proposal. NPV and IRR are computed by numerical or graphical method.

13.5. SOURCES OF INVESTMENT RISK

The activities and associated risks in mineral sector are looked in three broad ways and be addressed for Project Evaluation and Due Diligence.

(1) Scientific-technical aspect
  (a) Geological uncertainties: ore reserves and grade.
    (i) Reconnaissance (Blind investment—success rate is 1 in 100 and even 1000).
    (ii) Prospecting (Ray of hope—success rate is 1 in 10-100).
    (iii) Detail exploration (Better—success rate is 1 in 1-10).
    (iv) Adequacy of drilling and core recovery.
    (v) Accuracy of sampling, assaying and interpretation.
    (vi) Check studies through QC and QA.
    (vii) Mining: selection of mining method and rate of production.
    (viii) Mineralogical control on metallurgical responses of ore.

(ix) Equipments: capital and operating costs, reliability, spares and supplies.
(x) Processing and extraction planned along with time component.
(xi) Environmental due diligence.
(xii) Magnitude of investment progressively increases and necessitates to safeguard.

(b) Staff competency
(i) Competency at all levels of techno-managerial staff at highest standard.

(c) Economic viability
(i) Decision assumed to occur with certainty at the time of investment.
(ii) Revenue judged under ideal condition without future variations of diversified activities responsible for discovery of deposit to marketing of end product.
(iii) Adverse change of even one variable upsets entire decision of "Win" or "Loss".
(iv) Inherent risk to be addressed properly.

(2) Market aspect
(a) Variation in demand-supply scenario.
(b) Lower prices than expected.
(c) Foreign exchange rate on import and export of machineries and commodities.
(d) Future inflation.

(3) Sociopolitical aspect
(a) Future vision and commercial necessity for development of the country.
(b) Property transaction.
(c) Legal aspects.
(d) Export-import viewpoint of the country.
(e) Political stability/instability of the Government.
(f) Will/attitude of the Government
(g) Bureaucracy in licensing.
(h) Work culture and labor unrest.
(i) Sensitive environmental and forest issues.
(j) Excessive royalty, taxes and other regulatory policies.

13.6. INVESTMENT RISK AND SENSITIVITY ANALYSIS

The undiscounted (Payback period and AARR) and discounted (NPV and IRR) methods of evaluating mineral property for investment decision were assumed to occur under certainty at the time of investment. The cash flow is judged under ideal condition without considering future variations for a series of diverse activities responsible for discovery of a deposit to marketing of end products. Adverse change of even one variable will upset entire decision for profitable (Win-Win) or loss-making (Loss-Loss) venture. Therefore, the inherent risks associated with mineral industry have to be addressed appropriately. Adequate safeguard is to be taken well in advance to protect the investment and divert it to alternative opportunities.

There are technical risks during exploration regime. These are inadequacy of drilling, inaccuracy of sampling and analysis, unreality in interpretation, error in estimation of ore reserve and grades, selection of incorrect mining method, rate of production, process route and extraction planned along with time component. Market, economic or business risks depend on variation in demand-supply scenario, lower prices than expected, change in foreign exchange rate on import and export of machineries and commodities. Political risks are instability of Government, bureaucracy in licensing, labor unrest, sensitive environmental and forest issues, excessive royalty and taxes and other regulatory policies. These factors will influence the costs and revenues adversely.

The degree of uncertainty will vary between safe, low- and high-risk investments. Investors of the high-risk category have to be compensated with additional premium in profitability. Commonly used methods for minimizing the risk in mineral opportunity are in combination of gradually increasing or decreasing discount rates of individual variables while valuating future costs and revenues. If the discount rate is increased in steps, a set of cash flows will be generated that are negative in the beginning and generally positive at the end of mine life. The resulting NPV will also gradually reduce to negative passing through zero (Table 13.6). The NPV is substantially and marginally positive at 10 and 13% discount rate. It changes to low and high negative at 16 and 20% discount rate. The investor has to accept or reject the project proposal based on his perception/capacity of risk absorption.

The second way of risk analysis is to adjust cash flow by reducing the revenue by risk-free 10% over the base case with discount rate unchanged at 10% (Table 13.6). This will provide risk-adjusted NPV of (−) 4.94 with more certainty to reject the investment opportunity.

The third and most accepted method is "Sensitivity Analysis". In addition to computing the NPV using the most likely future cash flows, the Sensitivity Analysis method calculates a series of possible outcome taking into consideration all possible variation of each variable and in combination. The outcome forms a corridor of cash flow analysis and NPV reflecting the best and worse combinations. The analysis identifies the critical variables, such as increase in capital and operating costs, capacity utilization, and decrease in grade, tonnage and metal price that are sensitive to NPV and influence it most. Those critical variables are reviewed carefully to improve reliability for less risky decision. Sensitivity Analysis has been exemplified from Table 13.7 by three alternative scenarios, the base case (standard and simple), upside or best case (most optimistic) and down side or worse case (most pessimistic).

TABLE 13.7 Risk Adjustments in NPV Computation (after Gocht et al. 1988) [27]

Adjust the discount rate

Discount rate	*NPV*
10% (Risk free)	US$ 3.68 million
13%	US$ 0.56 million
16%	US$ −2.03 million
20%	US$ −4.74 million

Adjust the cash flows

	Revenues		Expenditure		Net cash flow	
Year	Unadjusted	Certainty equivalent	Unadjusted	Certainty equivalent	Unadjusted	Certainty equivalent
0	0	0	5	5	−5	−5
1	0	0	10	10	−10	−10
2	0	0	20	20	−20	−20
3	10	9	5	5	+5	+4
4	20	18	10	10	+10	+8
5	40	35	20	20	+20	+15
6	40	34	20	20	+20	+14

NPV at 10%	
Unadjusted	US$ 3.68 million
Certainty equivalent	US$ −4.74
Sensitivity Analysis	NPV (discount at 10%)
Upside case (increase revenue by 10%; reduce costs by 10%) US	$ 17.02 million
Base case	US$ 3.68 million
Downside case (reduce revenue by 10%; Increase costs by 10%)	US$ −8.93

Table format of Gocht et al. (1988) and computation by author.

The NPV in three options varies widely with expected high profit and high loss. The investor has to decide his game plan. It may so happen that all the three scenarios would have been either positive or negative NPV, and then the decision would be straightforward.

13.7. ECONOMIC EVALUATION OF MINERAL DEPOSITS

Economic evaluation of mineral deposit at any stage of exploration and development is assessed based on technical (geology, mining, processing and extraction), economic (cash flow, NPV, IRR, risk and sensitivity) and sociopolitical needs. The degree of precision of evaluation depends on the adequacy of information gathered and Due Diligence at that point of operation. The following information is collected during Geological (G), Feasibility (F) and Economic (E) studies:

(1) Geological stage (G)
 (a) Location and access to the deposit.
 (b) Terrain.
 (c) Climate.

- (d) Regional and deposit geology.
- (e) Regional and deposit structure.
- (f) Country rocks and mineralogy.
- (g) Host rock and mineralogy.
- (h) Rock quality and strength.
- (i) Adequacy of drilling and core recovery.
- (j) Sampling, assaying and interpretation.
- (k) Geo-technical characteristics of host rock.
- (l) Geo-hydrological characteristics of host rock.
- (m) Shape and size of the deposit.
- (n) Global tonnage and grade estimates with categorization.

(2) Feasibility stage (F)
- (a) Infrastructural facilities.
- (b) Mining method planned.
- (c) Mining recovery and ore loss.
- (d) Mine entries—adits, inclines, ramps, shafts.
- (e) Mine capacity.
- (f) Mine dilution.
- (g) Annual ore-waste ratio.
- (h) Production schedule and mine life.
- (i) Metallurgical test results.
- (j) Metallurgical recoveries.

(3) Economic stage (E)
- (a) Market factors:
  - (i) General economics
  - (ii) Market competition.
  - (iii) Short- and long-term demand-supply trends.
  - (iv) Short- and long-term price fluctuation at London Metal Exchange (LME).
  - (v) Short- and long-term price fluctuation of Foreign Exchange.
  - (vi) Equipment availability.
  - (vii) Raw material availability.
- (b) Cost factors:
  - (i) Total investment required.
  - (ii) Capital Expenditure (CAPEX).
  - (iii) Operating Expenditure (OPEX)—mine-mill-smelter-refinery.
  - (iv) Transportation costs.
  - (v) Tax and royalty.
  - (vi) Interest rate.
  - (vii) Depreciation criteria.
  - (viii) Cash flow.
  - (ix) Profit and loss.
- (c) Government factors:
  - (i) Foreign Direct Investment (FDI).
  - (ii) Mineral leasing policy.
  - (iii) Legal aspects.
  - (iv) Labor policies.
  - (v) Environmental and Forest policies.
  - (vi) Infrastructural support.
- (d) Social factors:
  - (i) Sustainability.
  - (ii) Infrastructure.
  - (iii) HRD all category and level.
  - (iv) Safety regulations.
  - (v) Education and vocational training.
  - (vi) Health care.
  - (vii) Industrial relation.
  - (viii) Social obligations.

13.7.1. Evaluation Process

The primary evaluation processes, such as, Payback period, AARR, NPV, IRR, undiscounted and DCF, Risk and Sensitivity Analysis have been discussed in the previous sections. The other economic factors that need to be defined for understanding and evaluating a mineral/mine project are as follows:

(a) Gross in situ value

In situ valuation is a fairly simple and straightforward method of valuating mineral deposits. The in situ metal content is the product of geological reserves (Measured + Indicated + Inferred) and percent metal grade of all commodities including value-added elements like Au, Ag, PGE etc. present in the deposit. Gross in situ value is the sum of each in situ metal content and respective metal prices.

(b) Mining and milling losses

The mining and milling losses are the metal unrecovered on account of part of geological reserves being left behind as pillars, dilution of geological grade by mining of waste or low-grade material and incomplete recovery of metals during the beneficiation. The value of the reserves blocked in pillars is added to revenue as and when recovered later. The recoveries decrease with the complexity of orebody and mineralogical control on metallurgical responses of ore. In general the mining and beneficiation recoveries vary between 75 and 95%.

(c) Smelting and refining charges

The value of the concentrate produced at mine head is determined after subtracting the metallurgical losses during smelting and refining, treatment charge per tonne of concentrate, refining charges per tonne of contained metal, credit for precious metals and penalty for deleterious contaminants. The metal loss, treatment charge, credit and penalty will vary between metals.

(d) Transportation cost

The cost of concentrate transportation from mine head to smelting point is deducted while valuing the concentrate. The transportation cost is a function of distance to be transported, number of transfer points and mode of shipment such as road, rail or ocean.

(e) Capital Expenditure

The capital costs or CAPEX are fixed assets or to add to the value of an existing fixed asset like buildings, equipments and other permanent facilities with a useful life extending beyond the taxable year. CAPEX includes all costs incurred and likely to be incurred up to production from the mine project. The capital investment for mine project includes Exploration and Delineation, Mine Development, Utility Plants, Co-Generation Plants, Concentrate extraction/ Tailing management, Infrastructure, Upgrading Facilities, Environmental Monitoring, Turnaround, Research and miscellaneous not associated with specific category.

Pre-production capital components are development of physical facilities such as mine access system (Incline, shaft, ramp), mine and process equipments, workshops, stores, offices and townships, schools, human resource development, vocational training center, industrial safety, community health care and recreations. There are other essential infrastructures such as roads, rails, power, potable and industrial water, etc.

Working capital component is a revolving fund required to operate the mine and process plant. The amount is returned as positive cash flow at the end of mine life.

Sustaining capital is required for timely replacement of equipments and any other anticipated major modification in the mine and mineral processing system.

(f) Operating Expenditure

The operating cost or OPEX is the ongoing cost for running the system. OPEX is the cost of production per tonne of ROM ore, waste, concentrates and overheads. The operating cost relates to two components: fixed and variable. The fixed costs are the salary and amenities paid to the employees, property taxes, license fees and overheads. The variable costs are directly associated with mining and milling activities for producing the ore, waste and concentrates. The variable costs include all raw materials and consumables, power, water, chemicals, maintenance, royalties and taxes. The annual operating costs will change with the volume of production. The operating cost will have a positive or negative effect under excess or low production on installed capacity.

(g) Revenue and Income

Revenue, Turnover, Sale, or Gross Sale refers to all the money the project receives for the value of the products or services to the customer during the life of active operation. Revenue at mine head is the tip line of cash flow computation associated with the anticipated development of the mineral deposit.

Net income refers to profit and represents the amount of money the company has leftover, if any, after paying all costs of production payroll, raw materials, taxes, interest on loans, etc.

(h) Inflation

Inflation is a rise in the general level of prices of capital and consumable goods and services over a period of time. It has direct effect on economy and profits. The capital and infrastructure may be replaced at a later time with much higher coat. The inflation of consumables and services are regular phenomena in all sphere of activities at rate between 5 and 10% (Fig. 13.2). Abrupt high rise in inflation due to whatever reasons renders adverse effect to the market economy with falling purchasing capacity for common man in the society. Many of the mining companies struggle with sharp increase in labor, energy and raw material costs all over the globe. The possible effect of inflation must be incorporated while evaluating the profit and loss of the project proposal.

(i) Depreciation

Depreciation is the spread of the CAPEX or assets over the span of the project life and generally at straight-line method of 10% or equal sum. The assets include machineries and permanent infrastructures including entry system like shafts. It is the method of reduction in the value of the asset across its useful life. The depreciation amount is deducted from the taxable income. Some mining companies follow accelerated depreciation of 20-30% in the initial years for tax benefits. The depreciated sum is added back to the CFAT deduction as it does not account for an actual flow of funds into or out of the project.

(j) Depletion

The depletion is the value of the ore (capital asset) mined each year and deducted from the taxable income. The depleted fund is similarly added back to the after-tax cash flow on the similar ground of depreciation. The deduction of depreciation and depletion allowances from the taxable

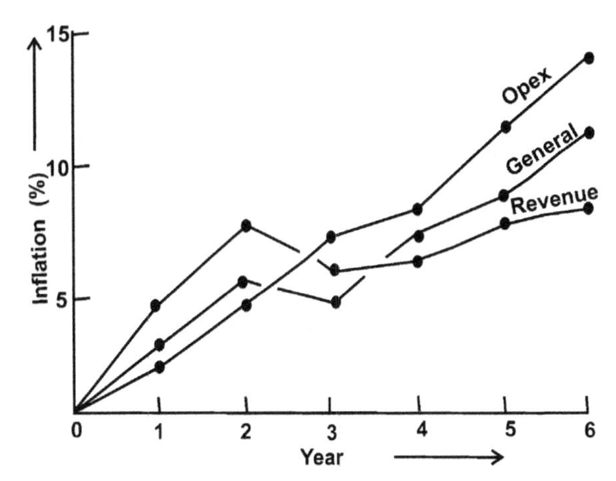

FIGURE 13.2 Differential inflation pattern of revenue, operating costs and general services.

income reduces the effective tax rate and acts as relief or incentive to the investors.

(k) Amortization

Amortization is the loan repayment installment model consists of both principal and interest. The simplest computation is the division of total amount by number of installment for the duration of the loan period. The amount of installment gradually reduces over time due to lower proportion of interest on unpaid principal. Negative amortization or deferred interest is a complex process of repayment that does not cover the interest due. The remaining interest owed is added to the outstanding loan balance.

(l) Mineral Royalties

Mineral royalties are paid to the State Government by the mine owners once the commercial production starts. Mineral production royalties usually take one of four basic forms: (1) A Flat Rate Unit of Production Royalty, (2) A Gross or Net Smelter Return Royalty, (3) A Net Revenue Royalty or (4) A Net Profits Royalty. The rate of Royalty varies widely from country to country and mineral to mineral. "Ad valorem royalties" are levied as a percentage of the total value of minerals recovered or the ex-mine value.

(m) Taxation payments

Tax payments are deducted as per corporate income tax to get an estimated time distribution of after-tax cash flow. The taxation rates vary widely between countries.

(n) Cash flow

Cash flow is the movement of the money into or out of a project. The Cash flow statement or Funds flow statement is the financial accounting in balance sheet showing the flow of cash in and cash out of the project. A schematic cash flow statement is given in Table 13.8.

TABLE 13.8 **Annual Cash Flow Statement in $**

Item head	Action
Gross sales revenue	(−) Transportation, smelting/refining and downstream ore processing charges. (−) Royalties (−) Less operating costs
Net operating revenue	(−) Non-cash items (−) Depreciation (−) Depletion (−) Amortization
Net taxable income	(−) Taxes (+) Credit
Net income after tax	(+) Non-cash items
Net operating cash flow	(−) Capital costs (initial and sustaining) (−) Working capital (−) Exploration costs (−) Acquisition costs (−) Land payments
Net cash flow	Investment decision, financial statement

The evaluation of mining project under consideration is necessary to determine the possibility of mining economically explored mineral resources. The outcome of the evaluation process will indicate (a) either proceed to next stage of detail activities along with investments or (b) withdraw from the venture. There are three types of mining investment evaluation models available in mineral and mining projects based on stages of exploration, resources/reserves adequacy and status. The studies include (1) Order of Magnitude Feasibility or Scoping study, (2) Pre-feasibility and (3) Feasibility studies.

13.7.2. Order of Magnitude Feasibility or Scoping Studies

The "Order of Magnitude Feasibility" or "Scoping studies" are initial financial appraisal of indicated mineral resources. The Scoping study is suitable for exploration projects under RP and more often under PL. It is more of conceptual type based on assumptions to decide on further detail exploration. One has to be optimistic regarding reserves and grades, mining and milling recoveries, costs and revenues, rather than very precise large volume of actual test data. Information on detail engineering design, method of mining and beneficiation, operating and capital costs are borrowed from experience, reports, case studies and published literature on similar type of deposits. This type of economic review being conducted during exploration tenure forms a groundwork and acts as an excellent guide to improve the area of information base. The main purpose is to generate the capability of the investor for "go or no-go decision". The Scoping study is reliable within 40-50%. An Order of Magnitude study of a base metal deposit during exploration stages is presented at Section 13.8.

13.7.3. Pre-feasibility Study

Pre-feasibility study is a detail approach on definite and more factual information with well-defined ore geometry, sampling and assaying with due diligence, reserve and grade with higher confidence up to ~80% accuracy, availability of infrastructure, proposed mining plans with scale of production, operating cost and equipments (not detailed engineering), bench scale mineral process route and recovery, and economic analysis including sensitivity tests, environmental impact and legal aspects. Experimental mining by opening small pilot pit or developing a crosscut underground for collecting representative samples, ore dressing pilot plant tests using bulk samples and other relevant detailed information may be required as a follow-up. Pre-feasibility study gives a more reliable picture of project viability within 70-80% reliability. The project is either under ML or ready to apply.

13.7.4. Feasibility Study

Feasibility study is the final phase of target evaluation based upon sound basic data with much greater detail analysis of the property toward development of mine and plant leading to regular long-term production. All previous estimates are modified and finalized with the availability of every detail on geology, engineering and economics. Majority of the ore reserves and grade is in the Partly Developed, Proved and Probable category. Detail engineering on mining and beneficiation plant completed. Capital and operating costs are set. Cash flow analysis with NPV, IRR and sensitivity to different assumptions regarding revenues, costs, discount rates, and inflation are realistic and more authentic. Environment impact assessment, permission from the Ministry of Environment and Forests, and Government formalities are expected to be cleared. Economic viability of the project is assured within 85-90%. In fact, the Feasibility report acts as a "Bankable" document for sources of finance from potential financial institutions, equity and Joint Venture including Foreign Direct Investments. The projects often listed in the standard Stock Exchanges such as American Stock Exchange (AMEX), Australian Securities Exchange (ASX), Bombay Stock Exchange (BSE), Canadian National Stock Exchange (CNSX) and New Zealand Exchange (NZX).

13.8. CASE STUDY OF ECONOMIC EVALUATION

13.8.1. A Zinc-Lead Project

Integrated geological, geophysical and geochemical survey followed by drilling during 90s established a high-grade steeply dipping zinc deposit in India with 9 million tonnes of reserve at 15% Zn + Pb + Ag. The mineralization occurs mainly within graphite-bearing mica-schist and partly in calc-silicate and quartzite, extensively intruded by pegmatite. Mineralization is controlled by lithology and structure. Pyrrhotite is the predominant sulfide mineral followed by sphalerite, galena, pyrite, and chalcopyrite in decreasing abundance. Mining is only feasible by underground method. Order of Magnitude study (Fig. 13.3A and Table 13.9) conducted during 1995, at 1100 US$ per tonne metals, ended with high NPV of 107 and 34 million US$ before and after tax.

Subsequent close space drilling during 2000 splits the orebody geometry and outlined three deep-seated en-echelon lenses separated by large waste partings. While the metal grades remain unchanged the reserve has been estimated at 3 Mt at 13% Zn, 2% Pb and silver. Mine life is reduced from 24 to 8 years. NPV, at 1200 US$ per metric tonne metal, changed to 14 and 1 million US$ before and after tax due to major reduction of ore reserve (Fig. 13.3B

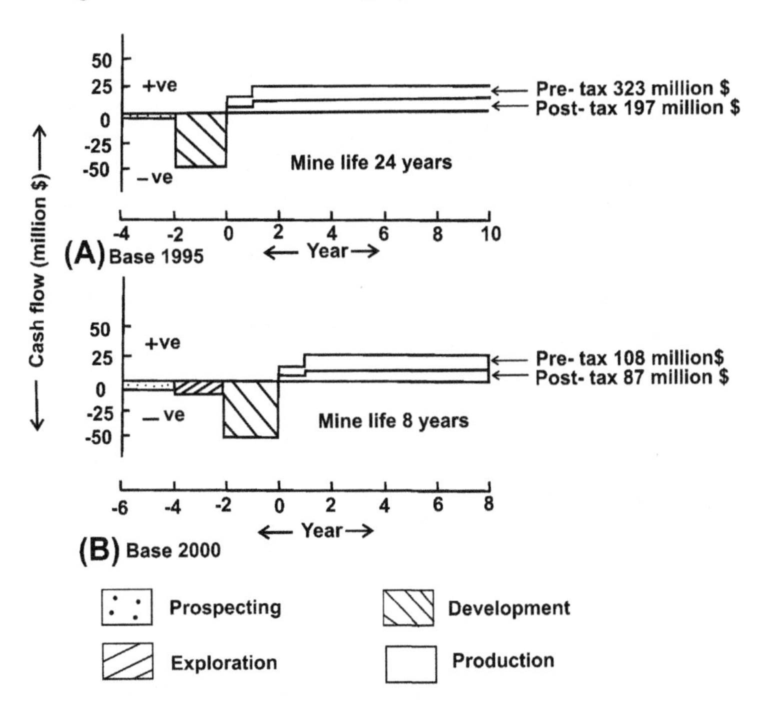

FIGURE 13.3 Phasing of investment and cash flow distribution of an exploration project during (A) 1995 and (B) 2000.

TABLE 13.9 Order of Magnitude Economic Study of a Zinc Deposit on Phased Exploration

	Parameters	Unit	(A) 1995 Base	(B) 2000 Base
	Ore reserves	(Mt)	9	3
	Grade (Zn + Pb) + Ag	%	13.28	15
	Mine capacity	(tpa)	300,000	300,000
	Mining recovery	(%)	80	80
	Mine life	Year	24	8
	Mine dilution	%	15	15
	Concentrate grade	%	52	
	Operating cost/tonne	$	40	
	Capital cost (CAPEX in million)	$	50	
	Treatment charges/t concentrate	$	180	
	Metal price	$	1100	1200
	Particulars		Million $	Million $
1	Gross in situ value			
	Zinc equivalent metal		1341.01	447
2	(a) Mine and milling loss		377.92	126
	(b) Smelting and refining charges		399.38	133
	(c) Concentrate handling and transport cost		36.66	36.66
	(d) (a + b + c)		813.96	271
3	Revenue at mine head (1-2)		527.05	176
4	(a) Operating cost		189.85	63
	(b) Capital sustaining cost		14	5
	(c) Total (a + b)		203.85	68
5	Gross income (3-4)		323.20	108
6	Depreciation allowance		50.00	50
7	Taxable income (5-6)		273.20	58
8	Tax at 46%		125.67	27
9	Net income (7-8)		147.53	37
10	Cash flow			
	(a) Before tax (5)		323.20	108
	(b) After tax (6 + 9)		197.53	87
11	Capital costs		50.00	50
12	Exploration cost		1.30	2.30
13	NPV			
	(a) Before tax		107.17	13.69
	After tax		33.64	1.06
14	Remarks on investment		Viable	Marginal

and Table 13.9). The economic valuation at that time (2000) suggested that the deposit is sensitive to ore reserve and other technical parameters. The project is marginally viable and needs more critical approach in decision making.

Exploration continued in the deposit due to high metal content and higher metal price during 2009-2011. The estimated resources increased to 10.6 Mt at 10.4% Zn and 1.6% Pb by the end of 2011. The metallurgical tests show that the ore is amenable to conventional flotation and two stages cleaning to produce zinc concentrate grade of 53.5% Zn at 88% recovery and lead concentrate grade of 63.8% Pb at 72.5% recovery. The current metal prices for both zinc and lead are about US$ 1850/tonne zinc, US$ 2000/tonne lead (January 2012). The investment strategy has competently changed in 2012 to mine the deposit with additional premium of silver recovery.

The lesson one learns from the above comparative economic evaluation study of the same deposit at different stages of exploration and understanding that investment decision is significantly sensitive to size and grade of the deposit, mining method, process recovery and metal price. The explored mineral deposits are normally not rejected but shelved for next opportune time.

13.8.2. A Zinc-Lead-Copper Silver Project

A massive poly-metallic deposit is exposed over 700 m in NW-SE direction and dips 70° → SW, width ranging between 5 and 30 m averaging 10 m. A large area of mafic volcanic rocks and associated fine sediments lie in contact with an irregular quartz-feldspar-porphyry intrusive. The main ore horizon is accessed and developed through adit and underground shaft. The mine production has been set at 2000 tpd or 730,000 tonnes per annum for 3.4 year mine life based on total ore reserves of 2.4 million tonnes at 6.76% Zn, 2.95% Pb, 0.35% Cu and 65.6 g/t Ag. Metallurgical recoveries of 88.7% Zn, 86.3% Pb, 60.4% Cu and 68.5% Ag are predicted. The total metal production has been estimated at 145,300 tonnes of zinc, 61,600 tonnes of lead, 5100 tonnes of copper and 3.5 million oz of silver. The Order of Magnitude or Scoping study is given in Table 13.10 for investment decision.

A comparative statement of cash flow on different alternatives is presented in Table 13.11 to understand the project viability.

The project is sensitive to low ore reserves and short mine life. Increasing the ore reserves from and nearby deposits with further exploration will open the opportunity for longer mine life which can take advantage of lower unit capital costs. The investment can be attractive by adding back the depreciation allowances to the after-tax fund.

TABLE 13.10 Scoping Study and Financial Model of Poly-Metallic Zinc-Lead-Copper-Silver Deposit

Parameters	Unit	Total
Mining and Processing		
Ore reserves	Mt	2.42
Ore mined	Mt	2.42
Ore milled	Mt	2.42
Waste	Mt	0.56
Zinc grade	%	6.76
Lead grade	%	2.95
Copper grade	%	0.35
Silver grade	g/t	65.6
Contained zinc	Million lbs	361.07
Contained lead	Million lbs	157.38
Contained copper	Million lbs	18.50
Contained silver	Million oz	5.11
Mine production and milling	At 2000 tpd/730,000 tpa/365 days	
Mine life	3.4 Years	
Zinc concentrate		
Zinc recovery	%	88.7
Zinc concentrate production	tonnes	270,023
Zinc concentrate grade	%	53.8
Zinc contained in concentrate	Million lbs	320.271
Lead concentrate		
Lead recovery	%	86.3
Silver recovery	%	68.5
Lead concentrate production	tonnes	123,212
Lead concentrate grade	%	50
Lead contained in concentrate	Million lbs	135.818
Silver contained in concentrate	Million oz	3.499
Copper concentrate		
Copper recovery	%	60.4
Copper concentrate production	tonnes	22,043
Copper concentrate grade	%	23
Copper contained in concentrate	Million lbs	11.18
Payable metals in concentrate		
Zinc	Million lbs	272.23
Lead	Million lbs	129.03

(*Continued*)

TABLE 13.10 Scoping Study and Financial Model of Poly-Metallic Zinc-Lead-Copper-Silver Deposit—cont'd

Parameters	Unit	Total
Silver	Million oz	3.32
Copper	Million lbs	10.79
Income statement		
Metal prices		
Zinc	$/lb	1.13
Lead	$/lb	0.85
Copper	$/lb	2.78
Silver	$/oz	12.61
Revenues		
Zinc	Million $	307.62
Lead	Million $	109.67
Copper	Million $	29.99
Silver	Million $	41.91
Total revenues	Million $	489.19
Total capital costs (CAPEX)	Million $	219.07
Cash operating costs (OPEX)		
Mining	Million $	74.39
Process plant	Million $	28.13
General administration	Million $	11.21
Zinc concentrate treatment charge	Million $	55.36
Shipping	Million $	20.41
Lead concentrate treatment charge	Million $	19.71
Shipping	Million $	5.99
Silver refining	Million $	1.40
Copper concentrate treatment charge	Million $	1.32
Shipping	Million $	1.55
Total cash operating costs (OPEX)	Million $	219.47
Depreciation for four years (CAPEX)	Million $	219.07
Total production costs	Million $	438.54
Income from operations	Million $	50.65
Taxes at 28%	Million $	35.23
Net income after taxes	Million $	15.42

TABLE 13.11 A Comparative Statement of Net Cash Flows at Different Alternatives

Year	Status	Revenue (Million $)	Net cash flow before depreciation and tax (million $)	Net CFBT (million $)	Net cash flow after 28% tax (million $)
0	CAPEX	(−) 219.07	0	(−) 219.07	(−) 219.07
1	Production	54.78	28.61	(−) 26.15	(−) 26.15
2	Production	196.93	105.60	50.84	36.60
3	Production	227.83	129.74	74.98	53.98
4	Production	9.65	3.75	(−) 49.01	(−) 49.01
Total		489.19	267.70	50.66	15.42
Payback period					2.9 years
IRR %			122.2	23.11	7.04
NPV at 10%			213.3	41.1	13.57

TABLE 13.12 Checklist for Investment in a New Mineral Deposit

(1) Deposit identity

(a) Name of the mineral deposit:
(b) Country/State/Province/District:
(c) Location:
(d) Nearest airport/seaport/railhead:
(d) Approach road condition:
(d) Climate:
(e) Average rain fall:
(f) Ground water table:
(g) Lease-hold/free-hold:
(h) Licensing Policy (RP/PL/ML)—transparent?

(2) Surface type

(a) Topography—plain land/undulation/hill:
(b) Barren/vegetation/forest cover:
(c) Mostly soil or rock cover:
(e) Weathered/fresh surface:
(f) Government or Private land:
(g) Any surface indication of mineralization:

(3) Brief geology

(a) Availability of geological map:
(b) Regional geological setting:
(c) Deposit geology:
(d) Country rocks and host rock(s):
(e) Regional/deposit structure:
(f) Deformation:

(4) Deposit/mineralization

(a) Plan area covered by deposit:
(b) Deposit boundary well defined?
(c) Strike length:
(d) Average width and range:
(e) Vertical extension:
(f) Exposed to surface/concealed in depth:
(g) Dip of the orebody:
(h) RFM:
(i) OFM:
(j) Gangue minerals:
(k) Principal economic mineral(s):
(l) Associated by-product minerals (s):
(m) Economic trace elements:

(5) Exploration inputs

(a) Details of soil/rock samples:
(b) Pits and trenches:
(c) Chips and channel samples:
(d) No. of boreholes drilled and interval:

TABLE 13.12 Checklist for Investment in a New Mineral Deposit—cont'd

(e) Core recovery:	(c) Modern workshop facility:
(f) Borehole survey:	(d) Housing:
(g) Average hole depth and drill interval:	(e) Health care:
(h) Standard of laboratories where analysis undertaken:	(f) Educational and recreation facilities:
(i) List of elements analyzed:	(g) Shopping:
(j) Methods and instruments used for analysis:	**(10) Miscellaneous**
(k) % Duplicate samples cross-checked with actual record of deviations between original and duplicate values:	(a) Political stability of the host country:
(6) Ore reserve/resources	(b) Relation between Federal and State Governments
(a) Reserve estimation methods followed:	(c) Will of each Government:
(b) Estimation cross-checked by other method:	(d) Tax and royalty:
(c) Total reserve estimated:	(e) Skilled labor, policy and work culture:
(d) Categorization of reserves (Proved/Probable/Possible) and proportion:	(f) Environment and Forest Acts and Regulations:
(e) Average grade of the deposit with range (all commodities such as % Pb, % Zn, % Cu, Ag (g/t), Au (g/t), etc.:	(g) Land protection for tribal inhabitants:
(f) Quantity of "float-ore and average grade":	(h) Environmental baseline and mining impacts:
(g) Plan area covered by float-ore:	(i) Ecotourism:
(7) Beneficiation test work	**(11) Special note**
(a) Laboratory scale*	(a) Proposal of investment based on "in situ ore reserve" or "exclusively on float-ore"?
(b) Pilot plant scale*	(b) Estimation actually on measured data or imaginary model?
(8) Economic analysis	(c) Visit to any near-by mining activities possible?
(a) Broad Order of Magnitude study conducted:	(d) Address/phone/e-mail of GS of the country.
(b) Pre-feasibility study conducted:	
(c) Feasibility study conducted:	
(9) Infrastructure	
(a) Power grid:	
(b) Water sources:	

**Result of test work if conducted (ore to concentrate ratio, concentrate grade, recovery etc.).*

13.9. SUMMARY

Investment in mining project is a challenging opportunity with associated risk at every phase of the mineral supply process. The degree of risk varies with highest in the reconnaissance and prospecting stage due to inadequate information about the deposit. The investor has to accept the total risk of project failure during this phase as the probability of success of discovery of an economic deposit is very low. So the investor has to be judicious and may look for alternatives before stepping in the risk-prone venture. However, once the deposit is properly delineated with confidence and investment decision is taken the risk shifts from geological aspects to technical, engineering and managerial factors. It includes selection of technology, mine design, production capacity, time and cost overrun. The project is exposed to various risks like optimization of productivity, scheduling, price cycles and other market factors during operating tenure. These risks can be conceptualized and minimized through adequate initial investment in sequential manner toward generation and economic evaluation of information considering all expected sensitive issues.

A checklist of information collection norms for investment in new mineral projects is given at Table 13.12.

FURTHER READING

Banerjee et al. (1997) [2] introduced the concept in the book Elements of Prospecting under valuation of a mine. Detailed discussions on the subject can be found in *International Mineral Economics* by Gocht et al. (1988) [27] *Part-II, Mineral Economics*, Chapters 5, 6 and 7 covering The Economic, Institutional, and Legal Framework for Mineral Development, Economic Evaluation of Mineral Deposits and Mineral Markets respectively with examples and statistics from different countries. Johnson (1988) [41] covered in detail with simplified computation for entire field of financial evaluation techniques and applications to mineral projects. Wellmer (1989) [75] and Evans (1999) [24] discussed the mine valuation techniques under Feasibility studies.

Chapter 14

Environmental System Management of Mineral Resources and Sustainable Development

Chapter Outline

A miner shall leave the mining area in better ecological shape than he found it.

—Author.

14.1. DEFINITION

The mineral wealth is finite and is a nonrenewable asset. Prospecting for minerals is costly and risky proposition. Mining in general is a land-base activity and has to be carried out only where the minerals occur. Ocean bed mining is limited to specific applications. The accelerated industrial and economic growth of a country necessitates rapid development of the mineral sector. Most of the basic industries in the manufacturing sector depend on this growth. Two basic issues confront this total process of exploration, mine planning, development, mining, processing, extracting and refining today. On one hand the developed and developing countries are in a race to achieve the highest level of economic and social progress. On the other side safeguarding the environment and maintaining quality and eco-friendly living conditions on a continued and sustainable basis is the basic human right. These are two sides of the same coin and complementary to each other. An optimum balance needs to be maintained between sustainable development and eco-friendly environment. Therefore, the policy makers and organizations carrying out the mineral development must promote the social well-being of the people living in and around the mining areas. This includes the local inhabitants who may or may not be concerned directly with the mining industry.

An Environmental System Management (ESM) is a powerful tool for managing the adverse impacts of activities of an organization on the environment aspects. ESM provides a structured approach to plan and implement environment protection measures. Organizations adopt ESM to improve environmental performance and in turn

Mineral Exploration. http://dx.doi.org/10.1016/B978-0-12-416005-7.00014-3

enhance business efficiency. ESM integrates environmental management into daily operations as well as long-term planning and other quality management systems.

The adverse impact of exploration, mining and related activities on the environment and the disruption of the ecological balance worldwide came into focus only few decades ago. India and other developing countries were equally responsible for this largely due to ignorance and consequences of long-term neglects. Remedial measures resulted in stringent enactments and rules to negotiate bumpy roads though some degradation effects are inevitable and unavoidable. The mineral sector has to be addressed in a way that causes least damage to the natural resources such as air, water, soil, biomass, and also to human communities and life forms. Sustainable mining with an integrative approach is critical and significant for the growth of the mineral sector.

14.2. ESM IN MINERAL INDUSTRY

The environmental pollution due to mineral industry results in the degradation of land, quality of the soil, vegetation, forest, air, water, human health and habitation and ecosystem. Any deterioration in the physical, chemical and biological quality due to human interference is a matter of serious concern. The magnitude and significance of the impact varies from mineral to mineral, geographical position, neighboring environment, and size and type of operations. The mineral sector works within the framework of NMP, MCR, and MMDR of the country (refer Chapter 1). At every phase of activity prior permission from State and Central Departments of Mines and Geology, BM, Ministry of Finance, Forests and Environment is mandatory.

The functional areas in the mineral sector can broadly be grouped based on purpose, type of operations and end products. Each area will be discussed separately with respect to activities causing expected sources of environmental degradation, identify the hazards and suggest the remedies. The causes, impacts and remedies are, in general, interrelated and overlapping. The functional areas are as follows:

(1) Exploration,
(2) Mining and Beneficiation,
(3) Smelting and Refining.

14.2.1. Exploration

The environmental effect is minimal during mineral exploration. The main features of mineral exploration program include surface mapping, airborne and ground geophysical survey, geochemical study by collection of soil, rock and water samples, excavations (pits, sumps and trenches) and drilling to various extent and magnitude. These activities should be carried out in such a way as to minimize its impact on the immediate environment. An appropriate compensation of land, agriculture and rehabilitation should be undertaken to satisfy the local inhabitants. A focus on community engagement process by facilitating employment opportunities to local community is important. The exploration program should include support, services, training and welfare to the community as a whole and youth in particular. This relation development model during exploration stages will pay dividends for future mining and related operations. This is the ideal time for development of fellow feeling and confidence easily with the local administration and community. The existing natural conditions should be maintained. Compilation and evaluation of existing and new data on satellite images, topography, geological maps, sample locations, geochemistry (presence of mercury and other toxic metals), mineral occurrences, and quality of air, water, vegetation and forests will be of great value for creating the environmental baseline. It will help to guide future course of environmental management program during and after the project life.

14.2.2. Mining and Beneficiation

The elements of mining and mineral processing are discussed at Chapters 11 and 12 in that order. The environmental impacts and consequential damages are generally high during these operations. The mining operations must be focused on the safety, environment, economy, efficiency and the community for a successful mining venture. The possible impact areas and their management can be grouped as follows:

(1) Baseline monitoring
(2) Land environment management
(3) Waste management
(4) Mine subsidence and management
(5) Mine fire and management
(6) Airborne contaminations and management
(7) Noise pollution and management
(8) Vibration and management
(9) Water management
(10) Hazardous process chemicals and management
(11) Biodiversity management
(12) Social impact assessment and management
(13) Economic environment
(14) Environmental impact assessment (EIA)
(15) Environmental management plan (EMP)
(16) Mine closure plan and management
(17) Mine rehabilitation and management.

14.2.2.1. Baseline Monitoring

Baseline monitoring is a significant component of monitoring programs for successful mining venture. Baseline monitoring commences at the Reconnaissance phase and continues to incorporate in Feasibility studies. It includes all relevant environmental, economic and social issues. The environmental baseline study will investigate the air quality and noise level, terrestrial soils, rainfall, hydrology, river flow and oceanographic condition, marine and freshwater quality and aquatic biota, forest fauna and flora, and finally community concerns about education, health care and economy. The baseline information identifies the possible impact areas for taking attention during the operating stages and their management. The system is continuously updated with periodical assessments to evaluate the extent of mining-related impacts and recovery following control of the impact or rehabilitation.

14.2.2.2. Land Environment Management

Land, in various forms, is a finite natural resource. The necessity of land is ever increasing due to rapid population growth in the developing countries and per capita enhanced industrial growth. One must analyze and understand that the quantum of land requirement during actual mining and beneficiation (Table 14.1) is small in comparison to other industries and urbanization. The minerals are mined at the sites where they exist. In general, mining activity occurs in remote places far away from cities. The possibility of land and soil degradation is expected at these remote locations only.

The types of impact on land, topography and soils and suggested environment management are as follows:

(A) Loss of agricultural and forestland

In case of open cast mining there will be complete loss of agricultural land and deforestation in and around the pit. Underground mining uses limited surface land for the entry system and infrastructure development. In either situation, adequate compensation is provided to the landowners by cash, employment and rehabilitation. New agricultural land is developed and aforestation is done under overall land-use planning. Mining in the unreserved forestland is replaced by enough plantations in nearby areas. Under normal circumstances no mining is permitted in reserve forest area.

(B) Topsoil and subsoil degradation and management

Various mining activities, particularly open pit, affect the topsoil and subsoil to a great extent by changing the natural soil characteristics e.g. texture, grain size, moisture, pH, organic matter, nutrients etc. In an ideal open-pit situation it is desired that the soil horizons within the selected mining limits are clearly defined. Topsoil and subsoil are removed separately, preferably by scraping, and stockpiled at an easily accessible stable land. These soils can selectively be relaid simultaneously to reclaim degraded land for agriculture or can be reused in future at the time of mine closure. As far as practicable the removed vegetation from the mining zone should be replanted at suitable areas.

(C) Changes of drainage pattern by blocking water and flash flood

The effect of unplanned mining and mine waste dumping will change the surface topography and thereby the local drainage pattern. The damage of natural drains and waste dumps may act as a barrier to the natural flow of rainwater resulting in water logging and flash floods which in turn will cause damage to agriculture and to local properties downstream. It will also affect the seasonal filling of nearby reservoirs and recharging of the groundwater around the area. The changes in the drainage pattern can be anticipated from the expected post-mining surface contours. Action plan for the surface drainage pattern can be designed accordingly. This planning is required particularly from the view of total water management and erosion control.

TABLE 14.1 Norms of Land Area Permissible for Exploration and ML in India

Stage	Land area (km^2)	Duration (years)	Anticipated activities
RP	<10,000	3 (nonrenewable)	Limited excavation for pitting, trenching and diamond drill site for sampling with no impact. No land requirement requires.
Large Prospecting Lease	500-5000	3-6	-same-
PL	1-500	3 + 2	- same-
ML	0.10-100 (Actual mining and processing in less than 1 km^2)	>20	1-10% of ML area acquires with clearance from MOM, MOF* and MOE. Creation of lumpy and fine waste. No reserve forestland permitted for mining and waste dumping.
Beneficiation/smelting	Limited area for process plant	Life of mine	Generation of fine waste, fluid, solid impurities and gases

** Ministry of Mines (MOM), Ministry of Finance/Forests (MOF) and Ministry of Environment (MOE).*

(D) Landslide

Opencast mining on hill slopes, particularly in areas of heavy rainfall, is vulnerable to landslides causing loss of human life, property and deforestation. This can be controlled by geo-technically designed slope of the mine and adequate support system.

(E) Unaesthetic landscape

Mining activity changes the land-use pattern and alters the surface topography by increased surface erosion and excavations. If proper reclamation is not done, this can result in unaesthetic landscape. Open-pit mines must be filled with mine waste rock as reclaimed land. It can be filled with rain or floodwater for fisheries, water sports etc.

(F) Land-use planning and management

The methods and procedures of land use are planned in detail before the actual mining starts. The status is periodically compared during active mining to maximize the benefits of better land use and to incorporate remedial measures in case of deviation. The mine area should be reclaimed to the best possible scenario at the time of mine closure. It is the responsibility of the mining company to take into account the cost of reclamation in the project cost. The reclaimed land should preferably be reverted back to the erstwhile landowners under a mutual agreement. If it does not work, the land can be developed for the local society based on the overall planning of the region. The modus operandi can be decided by representatives from the mining company, local inhabitants, local authorities and state planning department.

14.2.2.3. Waste Management

The type of waste and quantity likely to generate during mining, beneficiation, smelting and refining can be visualized form the Table 14.2.

FIGURE 14.1 Reclaimed land used for building zinc smelter at Vishakhapatnam, India.

The management of waste handling can be organized as given below:

(A) Lumpy waste

The coarse lumpy waste generated due to open-pit or underground mining can be used for reclamation of unused shallow/deep land in and around the mining area. This reclaimed land can be made into offices, industry (Fig. 14.1), community buildings, etc.

The low-lying uneven lands around the industrial area can be converted to amusement parks and playground (Fig. 14.2) using lumpy mine waste. It can also be used as solid waste fill of open-pit mined out voids.

(B) Fine waste

During mineral beneficiation, bulk of the fine waste generated is in the slurry form or tailings. The tailing is transferred through pipelines to the tailing ponds for settling. The top of the tailing pond can be developed as green grassy park, playground, picnic spot or for other alternative uses. The tailing mixed with 5-10% cement can be directly diverted to the underground as void filling for ground support.

TABLE 14.2 Likely Waste Generation, Coarse and Fines, during Mining and Beneficiation

Parameter	Open-pit mine	Underground mine
Ore to waste ratio	1:4, 5, ..., 10	<25%
Beneficiation (converted) fine waste (tailing in slurry form)	<90%	<90%
Computation at 1000 tpd mine or 300,000 tpa ore production		
Coarse lumpy waste	1,200,000 t	<75,000 t
Fine waste (tailing)	270,000 t at 90% of ore treated	
Smelting waste (reject)	Limited slag generation with value-added recoverable metals	
Refining waste (reject)	Limited fluid impurities with value-added trace elements	

FIGURE 14.2 Reclaimed land used for community and industrial sports and recreation at remote tribal hamlet of Zawar Group of Mine, India.

14.2.2.4. Mine Subsidence and Management

Subsidence is movement of ground, block or slope. It is caused by readjustment of overburden due to collapse and failure of underground operating mine excavation (Fig. 14.3), unfilled and unsupported abandoned mine and excessive water withdrawal. It can be natural or man made. Surface subsidence is common over shallow underground mines. Sudden subsidence of ground causes damage to man, material, topography, infrastructure and even mine inundation and development of mine fire.

The mine subsidence movement can be predicted by instrumentation, monitoring and analysis of possible impacts. The modification of underground extraction planning may help in minimum possible subsidence impact. Subsidence can be prevented by adequate support system (rib and sill pillars, steel and timber), cable and rock bolting, plugging of cracks, and backfilling by sand, cement-mixed tailing and waste rock.

14.2.2.5. Mine Fire and Management

Mine fire (Fig. 14.4) is a common phenomenon all over the mining world. This is especially true in many of the coal seams and sometimes in high sulfide (pyrite) rich deposits. Coal mine fire occurs due to the presence of high methane gas, instantaneous oxidation property of coal when exposed to open spaces and generation of excessive heat. The intensity of fire depends on the exposed area, moisture content, rate of airflow and availability of oxygen in the surrounding area. The nature of fire may be confined to surface outcrop, mine dump, open-pit benches and exclusively underground or even spread to surface. Fire in sulfide ore and concentrate is due to high pyrite-bearing dry stockpile exposed to open environment for long time under sun and heat.

Mine fire poses problems and is a matter of serious concern. The fire causes enormous impact with loss of economic, social and ecological nature. The losses are burning and locking of valuable coal reserves, polluting the air filled with excessive carbon monoxide, carbon dioxide

FIGURE 14.3 Surface subsidence over an operating underground zinc-lead open stopes without any loss of man or material.

FIGURE 14.4 An example of coal mine fire and recording of temperature by IR gun (left bottom) at Jharia coalfields, India. (Courtesy: Dr. A. Bhattacharya.)

and nitrogen, raising the surface temperature causing inconvenience to the people leaving nearby, damage of land, surface properties and vegetation, lowering the ground water table etc. The common diseases that affect the local inhabitants are tuberculosis, asthma and related lung disorders.

The nature of fire can be delineated precisely showing fire location, boundary (Fig. 14.5), intensity and direction of movement. The change of temperature and gas can be recorded and measured by surface instrumentation or by airborne thermal scanner. Surface thermal IR measurements are more commonly used. The temperature anomaly is measured by a handheld IR gun at the affected area on the surface or underground from various spots. The measurements are done in the predawn hours to minimize the effect of solar radiation. The prediction is done by simple contouring of temperature gradient or by applying different mathematical models. The depth and extent of fire can be determined by lowering probes into the fissures or along boreholes drilled in the affected areas. The temperature gradient is recorded by a digital recording unit connected by a long data transmission cable. This technique is less preferred due to the expensive drilling involved and frequent damage of transmission cable. The drill holes also act as catalysts for additional air supply to the fire activity. The third technique is by airborne IR survey. The region is mapped by low-flying aircraft or helicopter fitted with an IR scanner. The airborne interpretation is refined by simultaneous collection of ground information on weather, soil moisture and vegetation. The low-flying survey is now replaced by availability of precession data from satellite imageries and high-end GIS processing software.

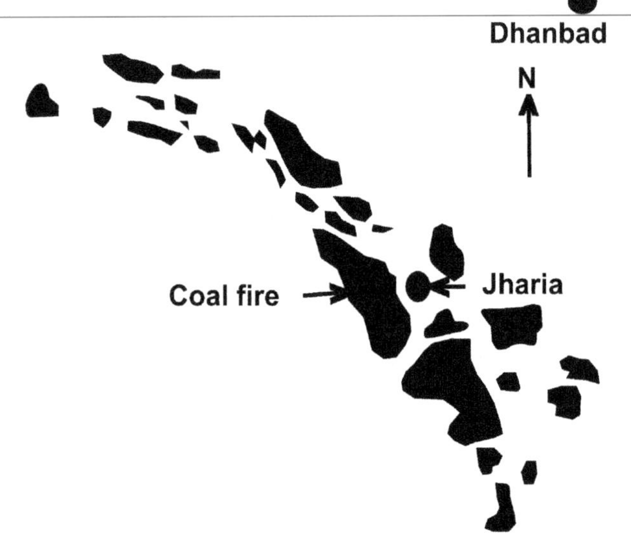

FIGURE 14.5 Schematic map of mine fire depicting the nature and movement of fire and planning for the remedies at Jharia coalfields, India.

Once the mine fire is properly delineated, it can either be stopped or checked from further spreading. The possible remedies are as follows:

(a) Stripping or digging the fire out physically.
(b) Injecting filling material like fly ash, water, mud, cement and sand to nonworking mines and voids through fissures, boreholes and other openings.
(c) Isolate by large-scale trenching, fireproof foam blanketing, impermeable layer of sand and debris, inert gas infusion, dry chemicals, foams.
(d) Plantation as much as possible to cool down temperature.
(e) Fast action at the earliest to prevent spreading and change of fire position.

14.2.2.6. Airborne Contaminations and Management

Airborne contaminants related to emissions of particulates (dust, diesel and silica) during exploration and mine production drilling, mining and beneficiation can produce significant sustainability impacts. Mines and beneficiation plants effect air quality by increasing concentration of any substance that is injurious to men, animals and plants. The pollutants can occur in solid, liquid and gaseous forms. The Suspended Particulate Matter (SPM) are small discrete masses of solids and liquids such as fine dust, smoke, fly ash, asbestos, lead, mercury, arsenic and other toxic metals. The gaseous pollutants are molecules of CO, SO_2, metal fumes, hydrocarbon vapor, acid mist, etc. The sources of air pollution are drilling, blasting, crushing and grinding, ore and waste handling, workshops, vehicles, etc. Air pollution causes injury to eye, throat, breathing passage and lungs of the workers and the local inhabitants. Chemical pollutants are responsible for serious diseases like birth defects, brain and nerve damage, pneumoconiosis, tuberculosis and cancer.

The air pollution in the mining complexes can be controlled by:

(a) Wet drilling applications in mine and grinding.
(b) Dust suppression through mobile sprinkler along the haulage road and fixed sprinklers in the waste dumps and stockpiles.
(c) Chemical treatment at haul roads.
(d) Selection of super quality mine explosives.
(e) Use of face mask.
(f) Installation of dust/gas extraction system at crushers.
(g) Ventilation fan and bag filters for cleaning of exhaust gases from refinery.
(h) Tall chimney to disperse residual gases after scrubbing, conversion and or cleaning.
(i) Systematic stacking of waste and vegetation over inactive benches.

FIGURE 14.6 Air pollution control by dust extraction system installed at ore dressing plant, India.

(j) Aforestations/green belt development around the mine periphery.
(k) Routine medical tests, monitoring and treatment of affected people.

Dust extraction and de-dusting facilities are installed in high dust creation area such as crushing, grinding and pulverizing areas inside the mine and more often in the beneficiation plant (Fig. 14.6). Gas cleaning system and Double Conversion Double Absorption (DCDA) sulfuric plant is set up in smelter to minimize emission and prevent sulfur dioxide and other intoxicated gases to the environment. Mercury removal plant aids to keep away from ingress of mercury in sulfuric acid and/or its entry into bio-cycle.

Notable way of controlling dust and gas can be affected by using tall chimneys and green belt development (Fig. 14.7).

TABLE 14.3 Ambient Noise Standards at Various Places

	Limits in dB(A)	
Area category	**Day (06-21 h)**	**Night (21-06 h)**
Industrial area (mine, plant)	75	70
Commercial area (office, market)	65	55
Residential area	55	45
Silence zone (hospitals, schools, churches)	50	40

14.2.2.7. Noise Pollution and Management

Every worker and resident in and around mining complexes deserves noise level within acceptable standards in his workplace and residential area. There is usually a necessity to understand and measure the existing ambient noise level as part of the environmental assessment process at the early stage of project formulation. Measurements are conducted by using an automatic noise logger over an adequate time period ideally during exploration. The recording is repeated prior to the mine being operational and while the mine is not operating to reflect the natural conditions. Excessive noise created by industrial machineries, transport vehicles and other associated sources increase beyond the acceptable level. Noise legislative framework is developed by most of the countries to combat problems caused by noise. In practice, routine monitoring is conducted periodically throughout the year at workplaces, schools and hospitals, places of worship, surrounding residences or other noise-sensitive receivers for taking corrective measures. Optimum noise level with variation around the standards is given in Table 14.3. The warning and danger limits are specified as 85 and 90 dB(A) respectively. No worker should be allowed to enter a workplace with noise level of 140 dB(A). All these standards are based on exposure of

FIGURE 14.7 Air pollution control by tall chimney and green belt development, October 2009, Paris.

8 h work-shift. The standards will vary between countries. Continuous exposure to high noise level cause deafness, nervousness, irritability and sleep interference. It also disturbs the wild life and ecosystem.

The source of noise must be identified for corrective measures at manufacturer's level. Systematic routine monitoring of noise level is done by Modular Precision Sound Level Meter which has a wide range of measurement capabilities. The remedies are as follows:

(a) Community liaison and involvement in the decision-making process.
(b) Periodical measurements and monitoring.
(c) Control measure at manufacture level.
(d) Change of blasting design and explosive control.
(e) Evacuation of people from the blasting area.
(f) Regulation of vehicular movements including night air traffic.
(g) Acoustic barrier and green belt development.
(h) Use of ear protection devices (earplugs, ear muffs) at workplace beyond 115 dB(A) to reduce noise level exposure.
(i) Location of residential and resettlement colonies away from noise-generating sources.

14.2.2.8. Vibration and Management

The major sources of vibration in the mining sector are drilling and blasting in open-pit and underground operations, heavy machineries deployed for breaking and transporting ore and high-capacity crushing and grinding units at beneficiation plant. The increasing size and depth of open-pit and large diameter long-hole blast in underground mines further aggravate the vibration. The other sources are movement of heavy vehicles around the workplace, workshop etc. Environment and mine safety authorities of several countries have laid down standards of acceptable vibration level to protect damages of existing structures and health hazards of workers based on their researches. The average ground particle velocity may not exceed 50 mm/s for soil, weathered and soft rocks. The limit for hard rocks is 70 mm/s. Any deviations in vibration level than standards may cause nervousness, irritability, sleep interference etc. Routine ground monitoring equipment can identify sources and nature of vibration.

The remedies are as follows:

(a) Modification measure at manufacture level.
(b) Change in blasting design by hole spacing, diameter and angle.
(c) Avoid overcharging, use of delays and improved blasting techniques.
(d) Use of superior quality explosive, explosive weight per delay and delay interval.
(e) Control of fly rocks.
(f) Green belt development.

14.2.2.9. Water Management

Water is known as life indicator and essential to sustain life. An adequate, safe and accessible supply must be available to all. The water can be classified into various groups depending on its source, use and quality. The primary sources of water are mostly from surface: oceans, rivers, streams, reservoirs and natural or man-made lakes. The other source is from subsurface aquifers that come to surface as springs. It can also be tapped by tube wells. In addition to the survival of human beings, animals and plants, it is also necessary for agriculture, industry and developmental activities. Water is rarely available in the purest form. It is usually polluted by various sources mainly through microbial, chemical and radiological aspects. Pollutants in the form of physical, chemical and biological waste make it unsuitable for uses. The physical pollutants are color, odor, taste, temperature, suspended solids and turbidity. The chemical pollutants are primarily related with geology and mining. The chemical contaminants are hardness, acidity/alkalinity, dissolved solids, metals (Fe, Pb, Cd, As and Hg), and nonmetals (fluorides, nitrates, phosphate, organic carbon, calcium, magnesium). The microbial hazards cause infectious diseases by pathological bacteria, viruses, parasites (protozoa and helminthes), microorganism and coli form. The chances of dissolved pollutants like Fl, Pb, Cd, As and Hg are high in water bodies in the vicinity of mining and beneficiation industries. This is due to the presence of pollutant elements in the ore-bearing host rocks and discharge of industrial effluents to the surface. Radiological hazards may derive from ionizing radiation emitted by radioactive chemicals in drinking water. Such hazards are rare and insignificant to public health. However, radiological exposure from other sources cannot be ruled out. Mining operations frequently cause lowering of ground water table due to pumping of water to make mining safe.

The guideline values for drinking water quality standards have been defined by World Health Organization ("WHO") 2008 with variations in acceptable limits from country to country. The recovered mine, plant and smelter water quality as reported from average Indian base metal mine is given in Table 14.4. Presence of pollutants causes poisonous and toxic effect to all living beings when the concentration is more than the permissible limits. It may risk the survival of the aquatic flora and fauna too. It needs treatment before use and preferably for industry purposes.

Water, being a scarce and essential material, needs to be used with long-term water management plan. The program must satisfy the industrial requirements as well as take care of domestic and agricultural needs of the surrounding villagers. The water balance exercise should cover study on requirements and availability of both quantity and quality

TABLE 14.4 Provisional Guideline Values for Drinking Water Quality by WHO versus Analysis of Typical Water Sample from average base metal Mines and Smelters, India

Parameters	Provisional guideline values for drinking water by WHO 2008	Recover water analysis from average base metal mines and smelters, India
Color	Colorless	Yellowish
Odor	Odorless	Pungent sulfur smell
Taste	Decent	Bad and rough
pH value	6.5-8.5	7.2-9.0
Total hardness (mg/l)	300	170-1000
Suspended solids (mg/l)	100	30-130
Total Dissolved Solids (TDS) (mg/l)	500-2000	400-2000
Arsenic as As (mg/l)	0.01	Nil
Barium (mg/l)	0.7	Not available
Boron (mg/l)	0.5	Not available
Cadmium as Cd (mg/l)	0.003	0.002-0.90
Chromium (mg/l)	0.05	Nil
Chlorides (mg/l)	250-1000	35-450
Copper as Cu (mg/l)	0.05-2.0	0.01-0.5
Cyanide (mg/l)	0.07	Not available
Fluoride as F (mg/l)	1.0-1.5	0.01-1.5
Iron as Fe (mg/l)	0.3-3.0	0.2-2.0
Lead as Pb (mg/l)	0.1	0.01-0.15
Manganese (mg/l)	0.4	Below detection
Mercury as Hg (mg/l)	0.006	Tr-0.009
Molybdenum (mg/l)	0.07	Nil
Selenium (mg/l)	0.01	Nil
Sulfates (mg/l)	150	310-730
Uranium (mg/l)	0.015	Nil
Zinc as Zn (mg/l)	5.0	0.003-0.42
Remarks	Desirable-permissible limit	Needs treatment for industrial reuse only
Source	WHO report 2008 and others	Compiles from various sources

for respective uses. The following water management program can be envisaged:

(a) Identify all surface and subsurface sources of all types of water for adequate availability.

(b) Introduce oil and grease trap and separator.

(c) Construction of check dams, garland drains all around mine pit and waste dumps, soak pits, septic tanks, domestic sewage water and other water harvesting practices to arrest seasonal rainwaters and any discharge of industrial effluent water for reuse in industry and plantation.

FIGURE 14.8 Water management by sharing between industry, domestic and agricultural purposes from Jhamri dam at Jhamarkotra rock-phosphate mine, India.

(d) "Zero" discharge water management for mine pumps and recoup from tailing dam followed by sand bed filtering, treatment for pH and recycle mainly for industrial uses.
(e) Low-density polyethylene (LPDE) lined for seepage control in and around the tailing dam and other mine water storage.
(f) Minimize applications of fertilizers, herbicides, pesticides and other chemicals.
(g) The dam and reservoir water (Fig. 14.8) is partially used for industry and domestic purpose of the township. A major portion is diverted to the surrounding villages for agriculture and drinking through a long-term water management master plan.

14.2.2.10. Hazardous Process Chemicals and Management

Various process chemicals are used in the froth-flotation of metallic ore of uranium, copper, zinc, lead, and iron. The flotation chemicals are mainly Isooctyl Acid Phosphate, Sodium Isopropyl Xanthate and Potassium Amyl Xanthate as conditioner and collector, MIBC as frothers, and sodium cyanide and copper sulfate as depressors. Cyanide is a useful industrial chemical and its key role in the mining industry is to extract gold. Acid leaching of low-grade Cu, Au, Ag, Pt ore and tailing pad is a common practice. The main leaching reagents are diluted hydrochloric, sulfuric and nitric acids. These hazard process chemicals, disposed to the tailing dam and downstream are fast-acting poisons. The intake of these diluted chemicals over long time by gas inhalation, skin contact and through water, milk, vegetables and food pose toxic effect on human, animals, birds and insects. The chronic sublethal exposure, above the toxic threshold or repeated low doses, may cause significant irreversible adverse effects on the central nervous system.

Acid mine drainage is similar naturally formed chemical hazard usually associated with weathering of hard rock metalliferous mines. It can also manifest itself in pyritic black shale-hosted sulfide ore, coal and in mineral sand mines.

Managing these chemical hazards of Acid Mine Drainage:

(a) Geological mapping, modeling and control to separate acid-generating rocks. Risk of miners can be avoided by use of proper safety shoes and mine dress code.
(b) Many gold mines practice cyanide destruction method to reduce the risk of environmental impact in tailing storage facilities (TSF) or mined out pit voids.
(c) The bottom and sidewalls of the tailing ponds are sealed by concrete to arrest any percolation of water to the surrounding water channels.
(d) Entire seepage water passing through sand-gravel bed below the dam is collected, treated and then to recycle and reuse for industrial purposes.
(e) The tailing production should be reduced and reused wherever possible.
(f) Routine sampling of water bodies 10 km around the mines/tailing ponds/smelters/refinery plants to assess any contamination and remedial measures.
(g) Adaptation of hazardous chemical management code.

14.2.2.11. Biodiversity Management

Ecology is concerned with relations between living organisms and their changes with respect to change in the environment. The direct ecological impact of mining to the surface of the land is usually severe due to the removal of natural topsoil that effects humans, animals, and plants. The net consequence is the likelihood of destruction in biodiversity within the natural ecosystems. The natural plantation and forest are likely to demolish due to removal of topsoil and indiscriminate cutting of forest for infrastructure development. The composite effect of mining changes the wild life ecology, causes species to become endangered, and changes the travel route of wild animals and migratory birds. The social and legislative context of mining in many parts of the world enforces and sets some form of land rehabilitation goals at mine closure situation. It is often anticipated prior to the granting of ML. Rehabilitation considerations are incorporated into mine planning such that it becomes a major governing factor during routine mining operations and waste disposal.

The management of ecosystem restoration are as follows:

(a) Encourage, incentives and data sharing among Government, mining organizations and Universities for research in Biodiversity,

(b) Community liaison and involvement to address the issue,
(c) Topsoil restoration by removal, storage and replacement with technical and scientific skill,
(d) Planned waste disposal with stabilizing and binding the dump slopes by low-cost eco-friendly biodegradable jute or coir mat cover and plantation to protect the movement of fines to nearby land and drainage,
(e) Minimize plant cutting and grow more plantation all around for beautification including open-pit benches, waste dumps and overdried tailing ponds,
(f) Routine restoration and development of forest as practiced at bauxite mining of Western Australia and South African coastal dune mining of heavy minerals,
(g) Ensure availability of sufficient safe potable water.
(h) Finally "Ecological awareness is truly spiritual."

14.2.2.12. Social Impact Assessment and Management

It is by chance that most of the mineral deposits exist in remote inaccessible location inhabited by tribal communities who lead their unique life style in close society. The people who fall within the lease area are directly affected. Those who live outside the mining limit have indirect impact of the activities. The livelihood of these groups of people depends on farming their own land, tenants, and sharecropping or as landless firm laborers. Most of the tribal people depend on forest products. Many of them stand to loose their home. However, mining activities make marked differences across all levels of society. The government and mining companies understand that "Globalization is economic welfare for all".

The socioeconomic profile of the area is assessed before the mine starts. The compensation packages that are framed include the following:

(a) Assessment of possible general social impact with respect to particular effected community and region,
(b) Assessment of social impact due to mining operations, expansion and closure to the associated community and region,
(c) Compensation and rehabilitation of effected families by reaching the benefits to the poorest amongst the poor,
(d) Impart formal education and vocational training with particular reference to younger generation with motto that "Education for all and all for Education",
(e) Direct and indirect employment with first preference to displaced persons and next choice to engineering, skilled, semiskilled and unskilled workers from nearby locality,
(f) Community development, better living condition, housing, roads and transport facilities,
(g) Improved health care and hygiene, and safe potable water,
(h) Community participation in sports and recreation,
(i) Supporting women's literacy, welfare, community participation and child care,
(j) Improve quality of life in many respects, except affluence industrial environment causes addiction like smoking, alcohol, drugs and increase crime rates.

14.2.2.13. Economic Environment

The impacts of mining are believed to be harmful in general. But it is partially true in practical wisdom. The damages can be prevented with due care and timely manner. The economic issues can be addressed by covering the broad aspects of land, water, ecosystem and society. In most of the cases mining is beneficial to the society with measurable economic parameters. The project cost invariably includes large amount of resources for environmental management and mitigation measures. The affected and related people of mining area are benefited by rehabilitation and resettlement programs in self-sustained pollution-free township and community development with essential amenities like potable water, health care, educational, banking, sports, recreational and other high-quality infrastructural facilities. This improves the standard of living compared to others in rural areas. The on job vocational training and managerial skill development programs improve the efficiency of the workforces resulting in higher effectiveness and productivity at lesser costs. Environmental economics provide the road map to achieve sustainable uses of the mineral resources.

14.2.2.14. Environmental Impact Assessment

EIA is the systematic evaluation and identification of potential environmental changes on account of establishing a new project say mine, smelter, etc. EIA is conducted during investment decision, planning and design stage. In case of mineral industry the broad impact areas can be envisaged for land use, landscape, alternative mining technology, waste assimilation, ground subsidence, mine fire, air quality, dust and noise pollution, vibration, water resource and quality, ecology (flora and fauna), public health, safety, activity-related risk and hazards, and socio-economic setting. The individual area has been discussed in the previous paragraphs along with remedial measures to mitigate environmental impacts.

It is desirable to identify, well in advance, the public psychology, their interest or fear about the proposed development and local needs to safeguard the long-term interest of the project. It is always better to timely educate and counsel them about the benefits of the project outcome to the general improvement and quality of life. The management must integrate the affected people and the local administrative body into the project to create an inherent feeling about their profound responsibility in program formulation and successful implementation. The assistance can be opted from experts in the trade and Non-Government Organizations (NGOs). Public hearing can be arranged to ensure participative involvement of local elected representatives, local people and debated on major issues. Participation of locals will defuse misunderstanding, confusion and conflicts.

Monitoring mechanisms are evolved to effectively implement and introduce corrective measures for an ongoing mining project. There are various ways of monitoring measures such as:

(a) Ad hoc method
Ad hoc method is simple, preliminary and general type in nature. It identifies broad areas of possible impacts and states in subjective and qualitative terms such as "low", "moderate", "high", "significant", "insignificant", "no effect", "beneficial" etc. It does not quantify the impacts.

(b) Check list method
Check method primarily lists all possible broad potential impact areas that can exist in a given situation. The list extends with identifying subareas under each major heading. A major area can be water resource with subareas as source (surface, lakes, rivers and groundwater), quantity and quality. The measure of impact is subjective and qualitative like ad hoc method. Each item will be identified as "adverse", "beneficial" or "none".

(c) Overlay method
Overlay method prepares a series of base maps of the project area. The maps display the georeferenced distribution of physical, demographic, social, ecological and economic aspects of the area. The individual impact layers are overlaid on transparent sheets or digital layers (refer Fig. 6.12) to produce composite scenario of the regional environment.

(d) Matrix method
This is open-cell matrix approach with identified project activities and possible magnitude and significance of environmental impact. The magnitude can be subjective expressed by numbers between 1 and 10. It can be expressed as judgment like "+" or "−" sign. It is relatively a complex algorithm of impact assessment based on facts and judgments.

14.2.2.15. Environmental Management Plan

The report ensures control measures for all identified environmental related problems. This part of the report is EMP by better technology alternatives. The total report is prepared by interdisciplinary groups involved in the entire activities. The group includes persons from planning, execution, Research and Development, finance and Board of Management. The clearance from respective authorities like Ministry of Environment and Forests is obtained and attached with the ML application.

Industrial development or development of the Nation and protection of environment are the two sides of a coin. A judicial balance is to be made between environmental damage and opportunity loss for the progress of human society—every living and nonliving entity. Each member of the society is to share the responsibility.

14.2.2.16. Mine Closure Plan and Management

The mine closure plan is a long-term process. A preliminary closure plan is required by the regulatory authorities as part of the mine lease approval process. The mine closure plan may have two components namely, (a) Progressive or Concurrent and (b) Final. The progressive mine closure plan is integrated into the total program of operating life cycle and activated at an appropriate stage rather than being attended at the end. It includes various land-use activities to be performed continuously and sequentially during the entire period of mining operations.

The final mine closure activities would start toward the end of mine life. The process will continue even after the last tonne of ore is produced and carry on till the mining area is restored to an acceptable level a self-sustained ecosystem. All mine owners, even if he has accorded approval along with ML, are required to obtain an approval of the final mine closure plan as per the guidelines and format by the competent authority within a period of 1 or 2 years in advance. The continuity of orebody in strike, depth and nearby area must fully be explored before final closure decision. Alternative opportunities must be examined to rehabilitate the community. The mine closure plan and procedures should be in the knowledge of the community and stakeholders well in advance.

14.2.2.17. Mining Rehabilitation and Measurement

Rehabilitation is the process used to mitigate the impacts of mining on the environment at the time of closure. The rehabilitation process can vary between simply converting an area to a safe and stable condition and restoring the premining conditions as closely as possible to support the future sustainability of the site. The key rehabilitation processes are as follows:

(a) Land rehabilitation is the process of returning the land in a given area close to its former state to bring some degree of restoration. Modern methods attempt to restore the land in improved condition after treatment.

(b) Characterization of topsoils and overburden at the early exploration phase and continue through the prefeasibility and feasibility phases as a basis for mine planning. The topsoil and overburden are preserved at a suitable place, protected with adding organic

FIGURE 14.9 An abandoned open-pit magnesite mine stored rainwater for agriculture and fisheries.

FIGURE 14.11 Six- to eight-month-old plantation of Vetiver grass and shady trees to cover surface at Lennard Shelf zinc-lead deposit, Australia.

fertilizer and revert back to the original place and shape at mine closure.

(c) Waste dumps are flattened to stabilize condition. The bulky broken dumps are protected against erosion by spreading biodegradable nets (jute/coconut coir) and planting vegetation all around the slope.

(d) Waste dumps, open pits and underground entries are fenced off to prevent livestock inviting dangers.

(e) Open-pit mines are filled either with mine waste rocks for stability or with water (Fig. 14.9) for agriculture, fishery and water sports. Similarly, the underground mines are filled either with cement-mixed tailings for stability or store water, particularly in dolomite country, for agriculture and drinking purposes.

(f) Leftover sulfide ore is usually covered with a layer of clay to prevent access of oxygen from air and rainwater which oxidizes the sulfides to produce sulfuric acid.

(g) Mine spoil area is vegetated by fast growing pasture, Vetiver grass (popularly known as "Khus" in India, Fig. 14.10, a perennial grass of the Poaceae or Gramineae family), and shelter trees with tolerance to extremely high levels of nutrients as rehabilitation program.

The Vetiver grass family grows under hostile dry and hot climate with low rainfall. The plants develop fast, stabilize the soil and protect it against erosion, pests and weeds. The grass is favored for animal feed. This species with all qualities and less care is preferred along with shady shelter trees to cover remote averse surface in remote areas (Fig. 14.11).

(h) Tailing dams are mainly pumped out for water recycle and reuse and left to partial evaporation. The top surface is covered with waste rock and thin soil layer which is planted to stabilize (Fig. 14.12).

(i) The removal of township, hospital, school, college and other educational institutions, recreation centers, markets, banks, workshops, mine infrastructures and beneficiation plant are not always part of the rehabilitation program. They possess heritage and cultural

FIGURE 14.10 Vetiver grass family ("Khus" in India) grows fast in hostile climate, stabilizes soil and protects from erosion, pests and weeds at no care.

FIGURE 14.12 Plantation of Vetiver grass and tall shady Eucalyptus trees (Myrtaceae family) over abandoned old tailing dam at Zawar Group of mines, India.

values and make use by the local municipality for tourism and alternative purposes.

(j) And finally, the rehabilitation of community is provided by alternative employment nearby or elsewhere.

14.2.3. Smelting and Refining

(A) Smelting waste

The smelting waste is generally in the solid form as slag that sometimes contain precious metals. In such case the slag is stored carefully for future recovery of value-added elements. Otherwise it can be used as waste filling or as road material. A small amount of fine dust passes though the gas chimney. The dust can be arrested by installing Electrostatic Precipitators, Waste Heat Boilers and cyclones. The other smelting waste is in gaseous form. Sulfur dioxide (SO_2) is a major air pollutant emitted in the roasting, smelting and converting of zinc, lead, copper and nickel sulfide ores. The sulfur dioxide emission is controlled by conversion to sulfuric acid, recovery as liquid sulfur dioxide or elemental sulfur. The remaining part of gas is dispersed and defused to atmosphere through extra tall chimney. A large amount of solid waste is created by leaching process of low-grade copper and gold ore (Fig. 12.25) and tailing from gold recovery. The final waste is disposed carefully and separately.

(B) Refining waste

The refinery waste is generally in the fluid form containing precious trace elements like Ag, Au, Co, Pt, Pd, etc. The value-added metals are recovered by electrolytic metal/acid refinery process. The refinery discharge water contains large quantities of arsenic, antimony, bismuth, mercury and other hazardous elements. It must be neutralized and treated for effluent removal. The effluent water treatment plant (Fig. 14.13) is designed to remove heavy metals and other toxic components. The water discharged from various plants is collected in ponds, tanks and chambers. The water is often recycled for industrial purposes after neutralizing with lime following environmental compliance.

FIGURE 14.13 Effluent treatment plants to recover metals and recycling of water for industrial uses, Rajasthan, India.

14.2.4. International Organization for Standardization

International Organization for Standardization (ISO) is the World's largest and most accepted developer and publisher of international standards. ISO is a voluntary NGO with network of the National Standards Institutes of 164 member countries with Central Secretariat operating from Geneva. ISO is developed over years to focus "Excellency" in industries and forms a bridge between public and private sectors. The standard framework for certification of total quality achievement and long-term sustained maintenance is assured. ISO facilitates an agreement to be reached on solutions that satisfy both the business houses and consumers or society as a whole.

The model certifications that are commonly adopted by the exploration, mining and smelting companies are as follows:

(1) ISO 9000 Series of Quality Manual and Certifications ensure quality management systems and designed to help organizations that they satisfy the needs of customers. ISO 9001:2000 makes certain for a quality management system to demonstrate its ability to consistently provide product that fulfills the regulatory requirements and customer's need. It aims to enhance customer satisfaction throughout by effective application of the system.

(2) ISO 14000 families of Certifications ensure framing the "Excellency" in ESM of the organization to minimize the negative effect of operations on environmental aspects and comply with applicable laws, regulations, and other environmentally oriented mitigations. The model entrusts on continual improvement in the system. ISO 14001 addresses exclusively the environmental issues and assists organizations to achieve environmental and financial gains through the implementation of effective environmental management. The standards provide both a model for streamlining environmental management and guidelines to ensure that environmental issues are considered within decision-making practices. Many large business houses have obtained certification under ISO 14001 for their Environment Management Systems IS 10500:1991 deals with Indian standards of drinking water specifications in line with WHO.

The benefits of ISO 14001 or IS 10500 Certification are mainly realized by large organizations. Small to medium enterprises (SMEs) have a smaller turnover and thus a correspondingly small return on the costs of certification. A fully certified ISO EMS may not be suitable for smaller organizations. But the system provides guidelines that assist them (each of the entrepreneur) to consider all relevant issues and thus gain the most benefit from their ESM. SMEs can use ISO 14001 as a model for designing their own ESM.

(3) ISO 18000 family covers Occupational Health and Safety Management systems in order to assist business houses to focus their sincere concern toward protection of the employees. It also ensures that they are operating according to their stated health and safety policies.

The larger organizations find certification more valuable when considering the potential trade and market advantages of an internationally recognized and certified ESM. This is a significant factor for companies seeking certification under the ISO 9000, 14000 and 18000 families for quality assurance standards and is likely to be a factor in decisions regarding ISO 14001.

14.2.5. Benefits of ESM

ESM requires the spirit of an organization to take an active participation in reviewing the existing mitigation practices, compensation packages and the results. The implementation of an ESM is essentially a voluntary initiative jointly for management, workers and community. It can act as an effective tool for governments to protect the environment through regulatory systems. The benefits of ESM can be summarized as below:

(a) ESM minimizes the environmental liabilities through well-designed mitigation procedures.
(b) It maximizes the efficient use of resources.
(c) It reduces waste generation by proper planning.
(d) It demonstrates a well-accepted corporate image.
(e) It motivates awareness of environmental concern among all level employees.
(f) It increases better understanding of environmental impacts of business activities.
(g) Finally, a good work environment and quality people increase the skill and efficiency resulting higher productivity at lesser costs and higher profits.

14.3. SUSTAINABLE DEVELOPMENT IN MINING

The growth of primitive people, for its very existence, was development through inventions for food, shelter and movement. Mining of minerals and other natural resources became significant to human survival and progress. Mining industry is fundamental and vital for economic growth. It creates wealth and adds value to the economic well-being of the country as a whole and to our daily life in particular. Yet, the notion persists that the mining industry is environmentally destructive, socially irresponsible and many a times illegal rampant in developing and underdeveloped countries. The large and multifaceted industry can be made sustainable, minimizing its harmful impacts and maximizing its social and economic contributions. Sustainability is a concept of optimum conservation of resources and a balance between prosperity level of well-being for the present and the future. The concept of sustainable development is a dynamic fine-tuning in institutional, economic, scientific and technological factors satisfying needs and aspirations. The most widely accepted definition of sustainable development by World Commission of Environment and Development is "development that meets the needs of the present without compromising the ability of future generations to meet their own needs". One must understand in clear term that "there is sufficient mineral resources in the world for man's need, but not enough for man's greed".

The principle of sustainable development promotes the thought of optimal resource utilization and leaving behind adequate resources for the future generations. The concept works on six key words "Resource, Regenerate, Reduce, Reuse, Recycle and Replace", as articulated by a little girl.

We want to take care of our 'Earth'
We want to 'Recycle' our things
We don't want to 'Waste Water'
So let's take care of our 'Earth' and make it clean
Everyday and now.
All the people that are here on this planet called 'Earth'
The 'Earth' is the best planet
You 'Recycle' plastic and paper and other stuff too -
To make some-thing new
We want to help our pets and animals too
We want to help other people
and poor people on 'Earth' as well.

—Srishti (6 years), Los Angeles, September 2008.

(a) Resource
Mineral "resources" and "reserves" are natural concentration of inorganic and organic substances including major and minor minerals and their by-products, fuels and underground water on Earth. These resources cannot be renewed in the laboratory and ever lost once taken out of the ground. The quantity of mineral resource is finite as outlined by exploration. Some of the mineral commodities are rare and scare in many occasions. The resources, in raw form or transformed, are the source of supply for consumptions and benefits of the society.

(b) Regenerate
In biological science the "regeneration" means the continuous process of renewal, restoration, reproduction

and growth of organisms. In case of mineral science regeneration is not likely in true sense. The mineral cannot be regenerated. But the resource can be augmented by identification of new areas and extension of existing mineral bodies formed by geological process over millions of years earlier in the Earth's crust.

(c) Reduce

As the mineral resources are nonrenewable, rare and scare natural commodity with finite quantity the consumption must be reduced to minimum to satisfy the most select necessity. One has no option left to waste this valuable asset and makes a balance between demand and supply. Little reduction in use can make everything different as echoed by our childhood little rhymes:

Little drops of water
Little grains of sand
Make the mighty ocean
And the pleasant (beauteous) land ...

—Mrs Julia A. Carney (1845).

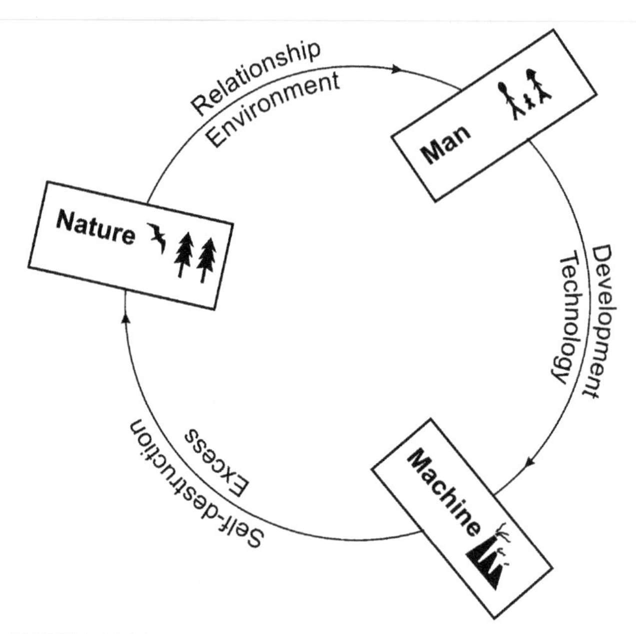

FIGURE 14.14 Nature, man and machine—a continuous circle for survival and/or destruction.

In this context the mines must reduce the amount of waste generation by innovative mining technology to reduce the waste handling and improve the environmental damage control. Annual waste audit is a common practice in mining industry.

(d) Reuse

In this pursuit of sustainable development adaptation of "reuse" by the community must be encouraged to move toward lowest consumption of material.

(e) Recycle

Recycling of process water by chemical treatment and conditioning is common practice and used for industrial purposes. The wastewater from mines, workshop, process plants, tailing dam is reclaimed as "0" discharge. Similarly, all metallic scraps of copper, zinc, lead, aluminum, iron etc. are reclaimed and metals recovered.

(f) Replace

The use of metals can be substituted by alternatives. Many of the home appliances, machine parts and automobile bodies are replaced by plastics and other by-products of petroleum refinery. This will reduce the consumption of primary metals.

Man was born as part of the nature. However, with more and more development and luxury in living style he is unable to appreciate the role of mineral resources in day-to-day life. In the process he moves away from the nature i.e. the three fundamental ingredients comprising of air, water and minerals. Excessive development, production and consumption ultimately became self-destructive and gradually depriving the future generations from those three fundamental gifts of nature. The gravity of the situation should be realized, respected and resolved by concept of sustainable development. Mining is like any other industry and with the right technical approach and feeling for the community it can be run in a modern, responsible fashion for the benefit of a wide range of society.

Technology and economic development is never to destroy the environment, but to sustain the continuity of human race as a whole. The binding relation between nature and man is the environment. Man develops technology using machines for his growth. "Nature-man-machine" relation is depicted in Fig. 14.14. The nature of forward movement should be at minimal and optimal level for the development of human society and become sustainable.

Sustainable development is a pattern of social and structured economic transformations (development). It optimizes economic and social benefits for the present. It does not impair future ability of the environment to provide sustenance and life support for every one. It implies equitable distribution level of economic well-being that can be perpetuated continually for many generations.

There are some standard indicators which can sensibly indicate the status of sustenance of a country as a whole and society in particular.

14.3.1. Indicators

There are few basic areas that can visualize some measures of indicator:

- "Gross National Product" (GNP) or "Gross Domestic Product" (GDP) is universally adopted yardstick or indicator for quantifying economic development of

the nation. Other alternative indexes are "per capita income" and "per capita consumption".
- Man-made or natural calamities e.g. Tsunami—War add spending to GDP.
- Loss to society or emotional trauma of individual is never accounted in this system.
- Indicators provide information inherent in long-term sustainability.
- Role of minerals and mining endorse major contribution in achieving sustainable development.
- Nonrenewable natural resources play significant place in the following five indicators.

(A) Environmental indicator

Status of reserves (million tonnes/billion gallons) of all minerals including groundwater, annual withdrawals, depletion of resources, rate of erosion/removal of topsoil, reckless mining, generation of industrial and hazardous waste, waste disposal, wastewater treatment (total and treatment type), industrial discharges into freshwater, land-use changes, protected area as percent total land area, use of chemical fertilizer, deforestation rate and emission of CO_2, Sulfur Oxide (SOX), and Nitrogen Oxide (NOX) will depict a clear picture of the environmental standard in the region.

(B) Economic indicator

The economic indicators are GDP growth rate, Gross export, Gross import, reserve of natural resources (oil, natural gas, coal and lignite, and other all major and minor minerals), minerals/concentrate/metals produced, transacted in market, contribute to GDP, energy consumption per capita, energy used from renewable resources, distribution of jobs and income in mineral sectors and annual growth rate of mining sectors. Energy minerals are vital to economic growth and environment.

(C) Social indicator

The social indicators related to mining sectors are employment/unemployment rate, poverty ratio, population growth rate and density, migration rate, adult literacy rate, percent GDP spent on education/health care, females per 100 males in population, primary/secondary school, college, University, percent people access to safe drinking water, infant mortality rate, life expectancy, child abuse/neglect/abandonment, crime rate/damage/money spent, incidence of environmental-related diseases, motor vehicles in use, loss of leisure due to long extra hours of work and travel time to workplace.

Mining industry improves the standard of living condition by availability of potable water, food, shelter, health care, education, sports and recreation to the ethnic inhabitants. However, exposure to mining and processing of minerals like galena, uranium, asbestos, fluorite, silica, and mica can cause diseases like TB, pulmonary and kidney disorders, cancer, restlessness and insomnia. High income and incentives invite addiction to alcohol and drugs with increase of crime rates and abuses.

(D) Institutional indicator

The Federal and State Government institutions and the mine owners formulate the strategies for sustainable development and programs for information on national environmental statistics. General awareness is growing with research programs toward treatment of solid, liquid and gaseous effluents and prevention of degradation of forest and wild life. National and International seminars and research publications are often organized for the growth and harmony on sustainability.

(E) Human happiness—a new concept

It is important to understand the relation between human happiness and well-being in the mission for environmental sustainability. The assessment of human happiness is designed in an attempt to define an indicator that measures quality of social progress in more holistic and psychosocial terms than only the economic indicator of GDP. The relationship between economic growth measured by GDP and personal levels of happiness indicates that happiness increases with GDP up to a certain level. The increase of GDP beyond this level does not reflect to more personal happiness. This was well realized and sermonized by Ancient Philosophers cutting across the countries and religions. Happiness is a function of nonmaterial factors and "very happy people" belong to a certain annual per capita income in Purchasing Power Parity (PPP) exchange rate (Fig. 14.15). His Majesty Jigme Singye Wangchuk,

FIGURE 14.15 Human happiness and well-being—a new thinking in indicator. (Source: Times of India.)

king of Bhutan, quotes "Gross National Happiness (GNH) is more important than GNP". Research in this aspect is continuing all over the World that includes project survey of United States, Japan and remote rural villagers in Andhra Pradesh, India.

14.3.2. Minerals and Mining as Means of Achieving Sustainable Development

(1) Science and technology
  (a) Focus on pollution prevention, energy saving and health care.
  (b) Clean technology that minimizes undesirable effluents, emissions of noxious gases and waste from products and processes.
  (c) eficiency/excess of calcium, magnesium, potassium, iodine, zinc, selenium has to be optimized through grains, vegetables and fruits by using less chemical fertilizers and insecticides.
(2) Fiscal measures
  (a) Tax formula aims at minimizing damage to environment and ecological balance.
  (b) Incentives to encourage reinvestment of income generated from mining in other mineral enterprises for sustainability.
(3) Legislations

Legislation is universal means to enforce any policy.

(4) Preservation of environment and forest
  (a) Clean Water (Prevention and control of Pollution) Act, 1972.
  (b) Clean Air (Prevention and control of Pollution) Act, 1970.
  (c) The Environment (Protection) Act, 1986.
  (d) The Forest Act, 1927.
  (e) Wild Life Protection Act, 1972.
  (f) Tribal's in Mining Projects.
  (g) The Environment and Sustainable Development Act.
(5) Regulated exploitation of mineral resources—sustainability and longer life
  (a) National Mineral Policy (NMP)
  (b) Mineral Concession Rules (MCR)
  (c) Mines Act
  (d) Mines Rules (Health and Safety)
  (e) Mines and Minerals (Development and Regulation) Act.
  (f) Mineral Conservation Act.
  (g) Oil Fields (Regulation and Development) Act, 1948.
  (h) Coal Mines (Conservation and Safety) Act, 1952.

Sustainable mining is not merely about complying with the applicable regulations. Compliance is just the basic foundation of sustainability and more often it remains hidden from the eyes of most of the community and stakeholders. The visible issues are superstructure, track record of environmental care, biodiversity conservation, the socio-community development efforts, transparency and delivery of good governance. All these dimensions are relevant and integral to sustainable mining. The key management tasks in achieving sustainability in mining industry are as follows:

(1) Mining sustainability focuses around two themes: (a) concern about well-being of future generations and (b) community development with humility. Our mission would be eradication of illiteracy, hunger, poverty and diseases reaching to the poorest amongst the poor.
(2) Let us live with happiness for the present and leave enough for the future generations.
(3) Mineral resources are limited, finite and nonrenewable. Once out of ground—lost forever.
(4) Mineral exploration is continuous process to augment the resource within certain limit.
(5) Promoting environmental awareness within exploration and mining companies. Spread the message to the community people through programs. Share the concerns and commitments with them.
(6) Educate and train employees and contractors. Adopt the method in practice.
(7) Educate the local community people for economically sustainable program to achieve self-support in short period. Education is the seeds and economy is the fruits.
(8) Early dialogue for community development to establish trust and confidence. Encourage to work together. Build partnerships between different groups and organizations so that there is a sense of integrity, cooperation and transparency for shared focus to achieve mutually agreed common goal.
(9) Developing community engagement plan involving employment with flexible work rosters, collaborative participation in decision making, services to the society, health care and medical advice, women education and child care, participation in community and spiritual festivals and handle with deep sense of humility.
(10) Ensure sustainable post-mine closure uses of land and all infrastructure toward creation of alternative employment.
(11) Full adaptation of compliance of national and international Impact Management Codes supported by independent audit.
(12) Transparency and good governance much reflect in every plan and action.
(13) Research, publication, knowledge sharing seminars and participants at workshop.

FIGURE 14.16 "Little deeds of kindness, little words of love, make our earth an Eden, like the heaven above", Julia A. Carney (1845).

(14) Leave the area in much more environmentally beautiful, progressive and sustainable. Much has been done, a lot more remains to be completed.

(15) Let the future generations grow in an environment of love, affection, compassion, happiness, trust, genuineness and transparency …. (Fig. 14.16).

(16) And finally, sustainability leads us to long term prosperity and eternal peace. Let us repose our faith in the invocation and verses from Ancient Indian Philosophy:

"May God protect us together
May God nourish us together
May we work conjointly with great energy
May our study be brilliant and effective
May we not mutually dispute
(or may we not hate any)

Let there be Peace in me
Let there be Peace in my environment
Let there be Peace in the forces that act on me
Peace must be our ideology, progress our horizon.

Lead us from the unreal to the real
Lead us from the darkness to the light
Lead us from the death to the immortality
Let there be 'Peace, Peace, and Peace'."

— Upanishad

FURTHER READING

Books by Evans (1997) [24] and Chamley (2003) [9] are suggested for general reading of the subject. Saxena et al. (2002) [64] provides a clear description related to environmental management in mining areas. Sen (2009) [67] highlighted the managerial and decision-making aspects of technology and economics. "*A Guide to Leading Practice Sustainable Development in Mining*" by David Laurence (2011) [15] as Principal Author, Government of Australia, is an outstanding report with enormous case studies covering zinc, lead, copper, nickel, aluminum, diamond, uranium, coal and silica sand engaged in open pit and underground all over the World.

Chapter 15

Mineral Exploration—Case Histories

Chapter Outline

Mineral Exploration. http://dx.doi.org/10.1016/B978-0-12-416005-7.00015-5

Exploration case histories enrich our knowledge-base for future mineral search.

—Author.

15.1. DEFINITION

The case history of mineral exploration, mining, processing, extraction, geostatistical evaluation, environmental sustainability and past discovery trend helps conceptualizing the possibility of mineral occurrences and its commercial management in new matching environments. The following sections present the case histories of several zinc-lead-copper-platinum-nickel-chromium and rock-phosphate deposits along with discovery trend of base metal deposits. Each of the deposits is unique in its own characteristic and symbolizes a type.

15.2. ZAWAR GROUP, INDIA: AN AGE-OLD ANCIENT ZINC-LEAD-SILVER MINING-SMELTING TRADITION

15.2.1. An Ancient Mining Tradition

A rich, impressive and traditional ancient mining and smelting history was well developed throughout the country in the vicinity of major mining center of modern India viz., Zawar (180 ± 35 BC), Rajpura-Dariba (1300-350 ± 120 BC), Rampura-Agucha (860 ± 100 BC), Ambaji-Deri, Khetri (3000-1500 BC), Singhbhum (~3000 BC) and Kolar (second century AD). The presence of extensive mine workings and debris, large heaps of slag and retorts (Fig. 3.7), furnaces (Fig. 3.5), ruins of temples and townships (Fig. 3.8) bear mute testimony to the art of exploitative and extractive metallurgy during the early period. Ancient mine openings are represented in the form of trenches, open pits, inclines, shafts, open stopes, chambers and galleries. The technology was probably developed around 3200-2500 BC. The importance of mining was well established as a source of revenue for the States during 1000 BC. A distinct urban economy was developed from 600 BC.

Once the ore was located on the surface based on the presence of gossans and/or mineralized veins and striking of lightning during thunderstorms the miners followed the downward extension of the ore shoot (Figs 3.1 and 3.2) and developed huge stopes (Fig. 3.3) and chambers. Mining was carried out by fire setting and supported by arcuate rock pillars (Fig. 3.3) and timbers. Timbers were also used as ladder, basket for ore carrier, and launders for underground drainage. Conical clay pots are used as lamps to illuminate underground working area. Mineral processing was done by manual crushing in hard rock mortars (Fig. 3.4) and hand sorting. Metal extraction was performed by traditional simple shaft furnace (Figs 3.5 and 3.6) in which the smelted metal settles at the bottom. The mining tradition was discontinued from time to time due to feudal war, draught, epidemic and other various reasons. These remnants of mining-smelting activities play significant guide to future mineral exploration. American Society of Metals International (ASM) recognized the process and designated zinc distillation furnace a historical landmark through a plaque reading. "This operation first supplied the brass for the instrument making in Europe, a forerunner of the industrial revolution".

15.2.2. Location and Discovery

Zawar group of deposits (24°20′56″N, 73°42′49″E) are one of the oldest mines in the World and are located ~40 km south of Udaipur City, Rajasthan, India. Zawar is very well connected by road, railways (0 km) and airport (90 km). The belt extends over 16 km and comprised four major mining blocks (Fig. 3.10) viz., Balaria in the east (Fig. 15.1), Mochia in the west, and Baroi and Zawarmala in the south. The mineralization is manifested on the surface by long abandoned workings that constitute regional guide for exploration.

In the present phase of mining episode, GSI started exploration in 1942. Mewar Mineral Corporation India Limited (MCIL) revived and modernized India's ancient zinc-lead technology by reopening the Mochia mine for

FIGURE 15.1 Topography of Balaria mine block in the center, the Aravalli Mountain range in the far background and undulating landscape in the foreground with scattered residential houses (image in 1978).

175 tpd after a long dormant period of more than a century. In 1945, MCIL increased the capacity to 500 tpd mine and beneficiation plant. The capacity remained so till the formation of Hindustan Zinc Limited (HZL, A Government of India Enterprise) in 1966. HZL obtained ML for 52 km^2 and continued underground exploration and ore production at ~1 Mt/a. The capacity increase in successive years is due to enhancement and commissioning of 2000 tpd from Mochia (1972), 2000 tpd from Balaria (1979), 600 tpd from Zawarmala (1983) and 400 tpd from Baroi (1988). Disinvestment process completed in favor of M/s Sterlite Opportunities and Ventures Limited in 2002. The mine production capacity growth planned to 1.5 Mt/a in 2009. Environment clearance was obtained from Ministry of Environment and Forest in 2009.

15.2.3. Regional Setting

The geology is represented by metasedimentary rocks of Middle Aravalli Group, resting over Banded Gneissic Complex (BGC), in a sequence of phyllite (carbonaceous and dolomitic), graywacke, dolomite, quartzite and phyllite. The sediments, deposited in a basin marginal environment, were subjected to several phases of deformation. The regional fold of Balaria-Mochia-Sonaria-Ruparia along with the north plunging major fold at Zawarmala was formed in the second stage of deformation (F2) on the bedding and axial planar schistosity of the first fold (Fig. 3.10).

15.2.4. Host Rock

The principal host rock throughout the belt is pure dolomite and partially black phyllite. The footwall and hanging wall rocks are represented by graywacke and phyllite/quartzite respectively.

15.2.5. Mineralization

There are numerous ore lenses extending for 100-200 m and disposed in en-echelon pattern. Variation in grade and metal zoning exist in various blocks. The principal OFM are sphalerite, galena, and pyrite in different proportions at different mining blocks. Arsenopyrite is a minor but ubiquitous mineral. Presence of chalcopyrite and pyrrhotite are in traces. Silver and cadmium are two value-added by-products obtained from Zawar ore. Cadmium occurs exclusively in the sphalerite structure. Silver occurs with galena and as minute native flakes (Fig. 1.3B).

15.2.6. Genetic Model

The orebodies are exclusively "strata-bound" in pure dolomite and hanging wall carbonaceous dolomitic pyrite-rich phyllite. The mineralization is locally stratiform. The grade of metamorphism is low. The mineralization is placed in the SEDEX type of 1710 Ma. The host rock, depositional features, rift-related intracontinental basin structure, strata-bound and occasionally stratiform nature, size and grade, high process recovery and other features are comparable with Mississippi Valley-Type (MVT) mineralization of the type area in USA.

15.2.7. Size and Grade

Total Reserves and Resources stand at 61.2 Mt at 4.4% Zn, 2.3% Pb and 35-40 g/t Ag as on 31 March 2012. A typical surface geological plan and cross section showing the prototype surface diamond drilling at Balaria mine block are given in Figs 3.11 and 15.2 respectively.

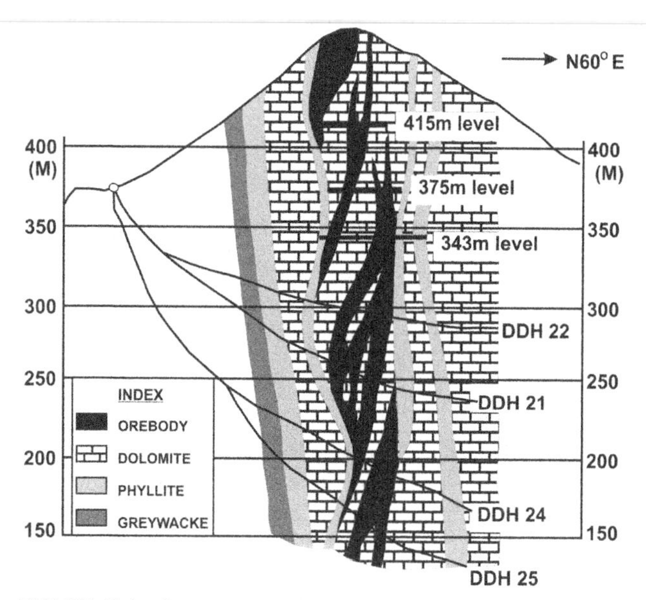

FIGURE 15.2 Cross section of Balaria mine block showing the prototype orientation surface diamond drilling during the initial phase of exploration.

15.2.8. Salient Features of the Mine Blocks

The summary information of individual mine block comprising exploration type, host rock, mineralization, metamorphism, surface oxidation/gossans, metal zoning, halos, genesis, age, deposit size and grade, contained metal, minable reserve, capacity and life of mining are given Table 15.1.

15.3. BROKEN HILL, AUSTRALIA: THE LARGEST AND RICHEST ZINC-LEAD-SILVER DEPOSIT IN THE WORLD

Australia hosts rich zinc-lead-silver deposits in the Middle Proterozoic metasediments with exhalative association (SEDEX). The deposits are mainly located in McArthur-Mount Isa basin in the Northern Territory, Broken Hill in New South Wales, medium size mines in Tasmania and small deposits in Western Region. The former intra-continental sedimentary basin system exposed over 1200 km in NW-SE trend with Mount Isa in the south and McArthur in the north. The 5- to 10-km-thick basin formation was active through a series of rift-sag cycles between 1800 and 1550 Ma. The basin hosts five world-class stratiform deposits [from south to north: Mount Isa, Hilton, George Fisher (1653 Ma), Century (~1575 Ma) and HYC (Here is Your Chance) (1640 Ma)] each with over 100 Mt of ore reserves at +10%, Zn + Pb. There are few other deposits namely Lady Loretta, Dugald River, Cannington, Grevillea, Mt Novit, Kamarga and Walford Creek either with low tonnage and high-grade or no published reserve available (Large et al., 2004) [50]. The Broken Hill Zn-Pb deposit consists of a cluster of orebodies within Willyama Supergroup having large global reserves of +300 Mt. Rosebery silver-lead-zinc deposit was discovered in 1893 at Tasmania's west coast on the slopes of Mt Black.

15.3.1. Location and Discovery History

Broken Hill deposit (31°56′S:141°25′E) is located 511 km northeast of Adelaide (NH-A32) and 1160 km west of Sydney (NH-32). Broken Hill (The Silver City) is an isolated mining city, far west of New South Wales Province, Australia.

The deposit was discovered by Charles Rasp, a boundary rider, while mustering sheep around Broken Hill area in 1883. Being trained in chemistry he was fascinated by the mineral appearance and formation (gossans). Rasp, joined by others, submitted ML and initiated prospecting for tin in gossans. The initial assaying from small shaft indicated low-grade lead and silver and finally reported finding of massive galena, sphalerite, cerussite and rich silver. Small scale mining was gradually accelerated to increase in tenure, mine size and efficiencies by consolidation of claims during last part of twentieth century. The Broken Hill Proprietary Company Limited (BHP Co. Ltd) was incorporated in 1985 for operating zinc, silver and lead mines. BHP Billiton merged in 2001 and became the largest global mining company measured by revenue in 2011.

15.3.2. Regional Setting

The Willyama Supergroup (7-9 km thick) has been divided into six principal packages defined by litho-stratigraphy (Table 15.2) probably deposited on existing continental crust. The mineralized packages from an arcuate belt of deformed, high-grade amphibole granulite rocks of Paleoproterozoic age represent the regional setting of Broken Hill deposits.

The Broken Hill mineralization occurs within the Broken Hill Group. This is marked by a widespread development of metasediments, interpreted as sudden deepening of the rift and the onset of more significant hydrothermal activity giving an interpretative magnetic age of 1680-1690 Ma (Page and Laing, 1996) [56] which is widely quoted as inferred age of mineralization.

15.3.3. Host Rocks

The orebodies are confined within a single unit of the mine sequence i.e. the Lode Horizon which is subdivided into four units such as the clastic and calc-silicate, garnet quartzite, C-lode and mineralization horizon. Orebodies are rich in calcite, fluorite, lead and rhodonite and to a lesser extent within clastic and calc-silicate horizon, a unit dominated by clastic psammopelitic to pelitic rocks with some well-developed calc-silicate layers, weak amphibolites and Potosi gneiss.

TABLE 15.1 Salient Features of Four Mining Block at Zawar Group

Features	Balaria	Mochia	Baroi	Zawarmala
Strike length	1000 m			
Exploration	21 km surface drilling, UG drilling at 15 m	Development and UG drilling	Surface and UG drilling	Surface and UG drilling
Host rock	Dolomite, carbon phyllite	Pure dolomite	Pure dolomite	Pure dolomite
Mineralization	En-echelon, strata-bound and often stratiform, Sp, Ga, Py	Strata-bound and often stratiform, Sp, Ga, Py	Strata-bound, fine-medium-grained Ga, minor Sp, Py	Strata-bound and often stratiform, Sp, Ga, Py
Metamorphism/deformation	Penetrative deformation under green schist facies	Penetrative deformation under green schist facies	Penetrative deformation under green schist facies	Penetrative deformation under green schist facies
Gypsum/Baryte	Gypsum, Baryte	Gypsum, Baryte	Gypsum, Baryte	Gypsum, Baryte
Surface oxidation	Fresh sulfide exposed to surface	Fresh sulfide exposed to surface	—	—
Zoning	Metal zoning	Metal zoning	Metal zoning	Metal zoning
Halos	Rich pyrite	Pyrite	—	—
Genesis	SEDEX/MVT	SEDEX/MVT	SEDEX/MVT	SEDEX/MVT
Age (Ma)	1700	1700	1700	1700
Size and grade (March 2012)	16 Mt at 5.8% Zn, 1.2% Pb, 40 g/t Ag	17 Mt at 4.26% Zn, 1.7% Pb, 40 g/t Ag	6 Mt at 1.7% Zn, 4.6% Pb, 40 g/t Ag	7 Mt at 5.0% Zn, 2.2% Pb, 40 g/t Ag
Contained metal	1.12 Mt	1.01 Mt	0.38 Mt	0.50 Mt
Minable reserves	Estimated at 30.866 Mt at 3.67% Zn and 2.05% Pb			
Infrastructure	Water requirement is met from Captive Tidi Dam. Power is adequately sourced from 80 MW coal-based captive thermal power plant, 6 MW diesel generator set and shortfall, if any, is met by State grid.			
Mining methods	Sublevel open stoping ± paste, rock and cemented filling			
Modernizations	Drill jumbo with Low Profile Dump Truck (LPDT)-LHD combination for mechanized faster mine development, operational efficiencies, cost reduction and improve safety.			
Depth of mine (m)	418	422	86	260
Mine capacity (Mt/a)	0.350	0.450	0.400	0.300
Life of mine	+20 Years			

TABLE 15.2 Stratigraphy of the Willyama Supergroup Showing Depositional Trend and Mineralization and Exalites

	Depositional trend	Mineralization and exalites
Paragon Group	Platform deposition	
Sundown Group	Rift fill	
Broken Hill Group	Rift stage	Broken Hill-type Pb-Zn-Ag
Thackaringa Group		Banded iron formation
Thorndale composite gneiss	Early rift stage	
Clevedale migmatite		

15.3.4. Mineralization

The mineralization is essentially strata-bound and stratiform and has been traced for 25 km along strike and up to a depth of 2000 m. The Broken Hill mineralization is interacted with exhalites like Qtz-mgt ± Fe, Cu sulfides, Qtz-Fe oxide/sulfide ± Cu and stratiform and strata-bound scheelite. The orebodies have been divided in two categories i.e. lead lode and zinc lode, based on Pb:Zn ratio. The former type is the calcitic orebodies that contain calcite, rhodonite-bustamite, apatite, garnet and fluorite, abundant lead and largely hosted within clastic metasediments. The second type is the primary quartz orebodies rich in primary quartz and garnet, gahnite and cummingtonite, little or no calcite or barium-rich calc-silicate component and rich zinc metal. The primary quartz orebodies mostly lie in the upper part of the sequence while the calcitic orebodies lie in the lower part (Fig. 15.3). The Thackaringa Group primarily hosts the Banded Iron Tourmaline ore.

15.3.5. Genetic Model

The regional field relationships have established an essentially strata-bound and stratiform syn-SEDEX model of Broken Hill type. The metal deposition has been conceptualized as a result of high heat flow within the sedimentary basin in which the Willyama Supergroup was being deposited. This high heat flow eventually led to the high-grade regional metamorphism of the enclosing sediments.

15.3.6. Size and Grade

There are nine separate but closely related orebodies stacked within a single package of stratigraphy. The preproduction reserve has been estimated as 300 Mt at 12.0% Zn, 13.0% Pb, and 175 g/t Ag. There are seven mine blocks from SW to NE as Southern Operations, North Mine, Potosi North, Silver Peak, Central block, Flying Doctor and Henry George.

The combined Mineral Resource (Measured, Indicated and Inferred) of Broken Hill Operation as on June 2011 stands at ~22.7 Mt at 9.0% Zn, 7.0% Pb and 86 g/t Ag. The

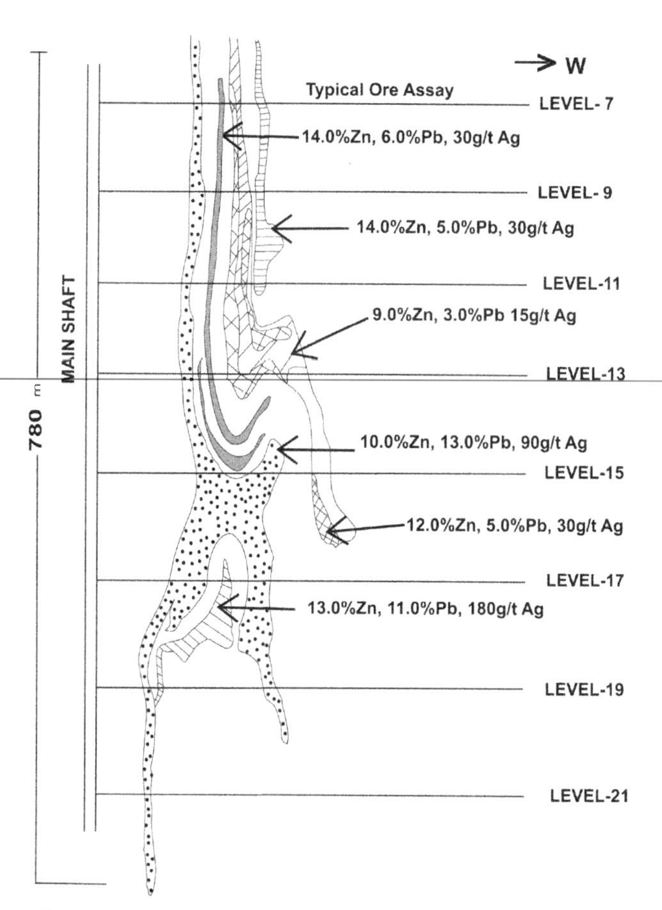

FIGURE 15.3 Schematic geological cross section no. 3C, through main shaft, S150 E at Broken Hill deposit (Source: Haldar, 2007) [33].

Ore Reserve, which only applies to the Southern Operations, stands at 14.7 Mt at 5.3% Zn, 4.0% Pb and 43 g/t Ag. The statement is based on JORC Code and expected mine life is more than 10 years. Resource and Reserve Drilling is now recommenced to increase confidence around existing resources and reserves and potentially enhancement. (Source: ASX and Media Release 28 December 2011.)

15.4. MALANJKHAND: THE SINGLE LARGEST PORPHYRY COPPER-MOLYBDENUM OREBODY IN INDIA

Copper is an important nonferrous metal having tremendous industrial applications. India is not self-sufficient in the resources of copper ore in terms of tonnes as well as grade. The domestic demand of copper metal is met through import of metal and concentrate. Hindustan Copper Limited, A Public Sector undertaking incorporated in 1967, is the only integrated producer of primary refined copper. The two other producers in the Private Sector viz., Hindalco Industries Limited and Sterlite Industries Limited import copper concentrate for their smelters.

Major commercial deposits are located at Shingbhum Copper Belt (Jharkhand), Malanjkhand (Madhya Pradesh), Khetri Copper Belt (Rajasthan) and Rongpo poly-metallic deposit (Sikkim). The Khetri and Singhbhum copper deposits are medium to low grade and deep seated. The orebodies are no longer supporting large-scale mechanization in underground mines due to the nature of the narrow width and flatter inclination. Manufacture of primary copper based on indigenous ore is characterized by high energy consumption due to low scale of operation. The by-product contribution from Indian copper ore is 4-5% as compared to 20-40% in other parts of the World. Malanjkhand copper ore is exposed to the surface. The deposit is initially developed by low-cost high-tonnage open pit with future adaptation to underground mining in combination.

The total in situ copper resources have been estimated at 712 Mt (9.4 Mt Cu metal). The Malanjkhand alone contributes 295 Mt (3.9 Mt Cu metal, 45% share of resource and 80% share in production). This is followed by 195 Mt (2.5 Mt Cu metal) from Rajasthan and 179 Mt (2.3 Mt Cu metal) from Jharkhand.

15.4.1. Location and Discovery

Malanjkhand is a classic porphyry copper deposit and is the single largest known copper-molybdenum orebody in India. The deposit is exposed to the surface and that enables low-cost open-pit mining at the shortest time. The deposit (22°12′N:80°42′E) is located at a distance of 90 km NE of Balaghat, 182 km from Gondia, Madhya Pradesh, and

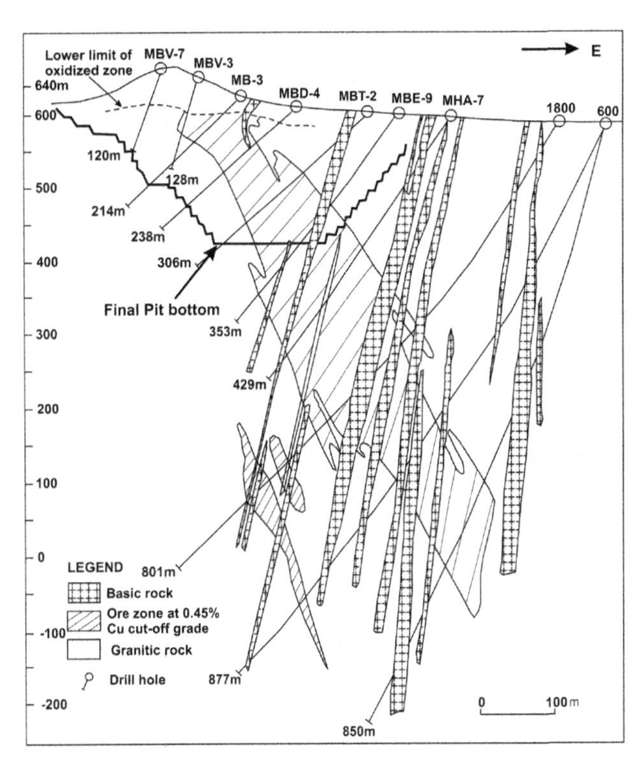

FIGURE 15.4 Geological cross section of Malanjkhand copper orebody showing the surface drilling pattern and ultimate open-pit bottom.

259 km from Nagpur, Maharashtra, at an altitude of 576 m above mean sea level.

The deposit was discovered by Colonel Bloomfield in 1889. GSI carried out systematic exploration during 1969. In 1970, Hindustan Copper Limited became interested in the economics of the property after studying the drill cores of seven boreholes totaling 613 m and submitted ML in 1973. Subsequently detailed exploration was conducted by 101 km of surface drilling up to a depth of 700 m from surface and associated geological work. The orebody dips 65°-70° → east (Fig. 15.4).

This resulted in delineation of the orebody with high confidence. The orebody extends over 2.60 km (Fig. 15.5). The deepest intersection is reported at 900 m vertical depth from surface.

15.4.2. Regional Setting

The rocks associated with the porphyry Cu-Mo deposit are represented by an elongated oval-shaped dome. The rocks are granodiorite, quartz monzonite pluton comprising biotite granite and tonalite. The Malanjkhand granite along with the mineralized quartz veins and basic dykes directly underlie the basal conglomerate of Chilpi Ghat Group. The deposit is located at the intersection of two major sets of lineaments trending N-S and E-W. The N-S lineament within the pluton represents the major mylonitized, sheared

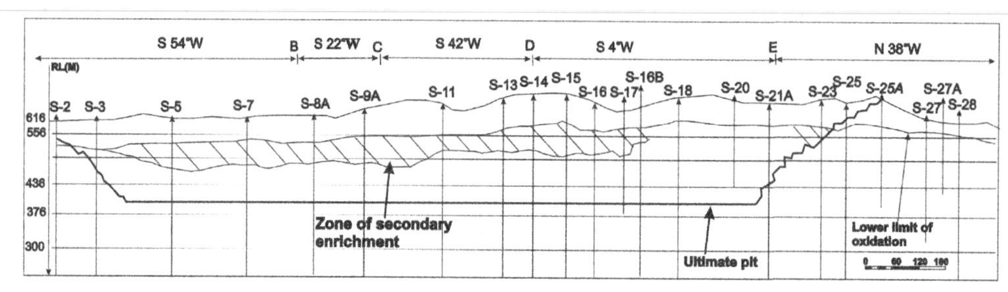

FIGURE 15.5 Composite longitudinal section along Malanjkhand copper orebody showing the surface drilling interval and ultimate open-pit bottom.

and crushed zone and acts as a favorable structural loci for the emplacement of ore-forming hydrothermal fluids to ascend, some of which are mineralized for varying extent.

15.4.3. Mineralization

Mineralization occurs as stockwork of thin quartz veins in granitoids and thick veins emplaced in sheared and mylonitized zones within the pluton. The reef varies with attitude from a single thin tabular body at the northern end to a composite of parallel-sheeted interconnected veins at the central and southern part attaining a maximum width of 200 m. The orebody has been oxidized near the surface at the south with complete removal of copper and reconcentrated as supergene sulfide enrichment at deeper levels. In the northern side copper is retained in the form of cuprite, chalcocite, covellite and native copper within the oxidation zone without any supergene enrichment. The Cu-Mo mineralization occurs extensively in the entire Malanjkhand pluton in the form of reef-quartz, stringer, pegmatitic and disseminated ore.

Major OFM in the primary zone are chalcopyrite, pyrite, magnetite, molybdenite, sphalerite and gold in order of abundance. Limonite, hematite and goethite commonly occur along fractures in quartz vein in the oxidation zone. Chloritization, sericitization, epidotization, potassic alteration and silicification are frequently present in order of occurrences.

15.4.4. Genetic Models

The entire magmatic episode has been emplaced as a single large pluton with progressive differentiation and repeated tectonic activity which is responsible for variation in composition and generation of sheared lineaments. The last but one phase of differentiation resulted in Cu-Mo-bearing quartz veins of economic significance located at the structure-controlled fracture zones. This is evidenced by field relation and geochemical analysis of granitoid. The likely age from today ranges between 2270 ± 90 and 2458 ± 26 by Rb-Sr method.

15.4.5. Size and Grade

The ore reserve is estimated at 145 Mt at 1.35% Cu that includes 20 Mt at 1.13% Cu in the open-pit limit and 125 Mt at 1.39% Cu between ultimate open-pit bottom and 0 m level. The ore contains 100 g/t Mo, <1 g/t Au and <10 g/t Ag. The orebody is open at depth.

M/s Hindustan Copper Limited is mining the deposit since 1982 by mechanized open pit up to a depth of 200 m from surface. The total ore production till 2006 has been 44.67 Mt at 1.15% Cu at 2 Mt/year with ore: waste ratio of 1:5.75. Heap leaching of oxidized as well as disseminated ore from the hanging wall zone is planned to augment the production to 3 Mt per annum. The future plan is to enhance the production to 5 Mt per annum from the combined open-pit and deep underground mining operations up to a depth of 1000 m. The underground development is in progress.

15.5. SINDESAR-KHURD: ROUTINE DRILLING DISCOVERED CONCEALED ZN-PB-AG DEPOSIT IN INDIA

The State of Rajasthan possesses more than 95% of zinc-lead-silver Reserves and Resources in the country. The three major groups of deposits are clustered along a linear trend of NE-SW. The major groups, from southwest to northeast, are represented by Zawar (Balaria, Mochia, Baroi and Zawarmala), Rajpura-Dariba (Dariba, Mokanpura, Sindesar-Kalan (E), Sindesar-Khurd, Bamnia and Bethumni) and Rampura-Agucha. The minor deposits are Sawar, Ghugra and Kayar (development in 2012), all in Rajasthan.

15.5.1. Location and Discovery

Rajpura-Dariba-Sindesar-Khurd-Bamnia belt is located at a distance of 75 km northeast of Udaipur City in the State of Rajasthan. It is well connected by roads. The nearest rail station "Fateh-Nagar" and airport "Dabok" are at a distance of 22 and 50 km from Rajpura-Dariba mine respectively.

Since 1000 BC ancient miners carried out extensive mining periodically for many centuries employing both open-pit and underground methods. The ancient mining, beneficiation and smelting manifestations were rediscovered by GSI as early as 1934. The presence of unique gossans (Figs 4.4, 4.5, 4.6, 4.8 and 4.9) coupled with ancient open-pit and underground (Figs 3.2 and 3.3) workings, grinding potholes (Fig. 3.4), heaps of mine debris and slag all served as guides to sulfide mineralization in the belt. GSI initiated systematic exploration by geochemical sampling in 1962 at the southern closure of the belt. Surface drilling at Rajpura-Dariba, Bethumni and few other locations were completed by 1970. HZL acquired ML of 11.42 km^2, conducted detailed delineation, Feasibility study and commencement of underground mining at Rajpura-Dariba. Surface drilling by GSI (1962-1970) and subsequently by HZL (1970-1980) is of the order of 34,100 m at 50 m spacing. Underground definition drilling by HZL till March 2003 is 68,700 m at 25-30 m spacing up to "0" m level. GSI continued exploration in other blocks and identified large volume of low-grade graphitic schist-hosted deposits at Sindesar-Kalan (E) and Mokanpura. Later drilling discovered concealed dolomite-hosted deposits at Bamnia-Kalan in 1982. The geological map of Dariba belt is given in Fig. 15.6.

The investigation so far has been planned by tracing the signature of surface oxidation (gossans), host rock assemblages and ancient mining-smelting remnants between Rajpura-Dariba in the south and Bethumni in the north. The orebodies are expected to trend NE-SW and dip toward E to SW as experienced for all zinc-lead-copper deposits of Aravalli mountain range hosted in Aravalli and overlying Delhi Supergroup. The exploration drilling was designed accordingly and identified deposits like Mokanpura, Sindesar-Kalan (E) and Bamnia-Kalan. Most of the drill holes were terminated as soon as quartzite/quartz mica-schist was encountered at the footwall beyond east dipping mineralization. The existence of Sindesar-Khurd was unknown until 1987-1988.

As it happened, drill holes in Mokanpura, situated east of Sindesar-Khurd village, established calc-silicate-bearing dolomite horizon underneath the quartzite ridge. One of the terminated hole intersected attractive mineralization in this host rock similar to Bamnia-Kalan. The subsequent routine drilling intersected huge zinc-lead-silver-rich mineralization beneath the stony barren massive quartzite ridge. The orebody represents the single western dip in the Aravalli range. The massive (61 Mt, March 2012) silver-rich (215 g/t Ag) zinc (5.8% Zn) and lead (3.8% Pb) deep-seated hidden "Sindesar-Khurd" was discovered in 1987. The orebody occupies the local western anticlinal limb of the regional synclinal fold system. HZL explored the deposit, obtained the ML for 2 km^2, Feasibility studies completed, infrastructure and

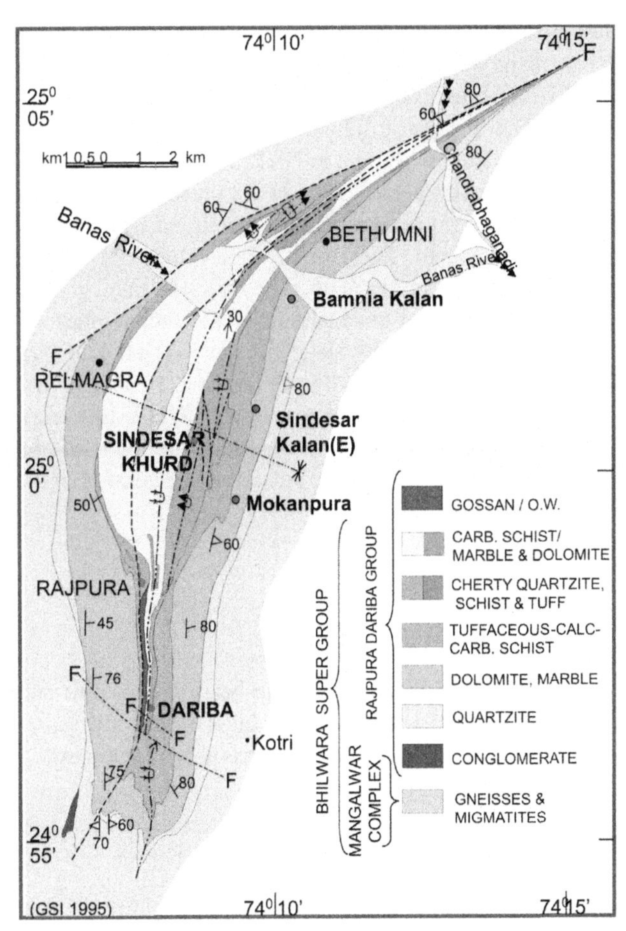

FIGURE 15.6 Geological map of Dariba-Sindesar-Khurd-Bamina-Bethumni Zn-Pb-Ag metallogenic belt (Source: Haldar, 2007) [33].

mine development made and finally commissioned in 2006 by trackless underground mining with main pathway for men, materials, machineries and ore. Sindesar-Khurd mine will be the richest silver source and produce the maximum silver in the country.

15.5.2. Regional Setting

The metallogenic belt comprised medium- to high-grade meta-volcano-sedimentary rocks equivalent to ortho-quartzite, carbonates and carbonaceous facies metamorphosed to medium-grade amphibolite facies of Proterozoic Middle Aravalli Group. Banded Gneissic Complex rocks of Archean age unconformably underlie the cover sequence. The belt extends over 19 km in north-south direction as crescent-shaped regional synform with closure at south of Dariba and the fold limbs continue to Surawas-Bharak in the north. The sequence continues in a north-south trend to the south of Dariba as the Wari antiform and further south as Bhinder synform. Sulfides bearing predominantly chemogenic rocks, admixed with minor clastics, are deposited under reducing euxinic conditions in linear inland basin/trough.

The rocks have suffered at least three phases of deformation. The first phase fold (F1) is tight symmetrical and overturned. The axial trend is NNE-SSW with sharp easterly dipping axial plane. Folds of F2 phase having NNE-SSW axis are tight to open and asymmetrical with easterly dipping axial plane. These are Z-type folds with steeply dipping limb of shorter length. The major Dariba syncline with NNE-SSW axial trace belongs to the second phase of deformation and exhibits steep plunge (55°-60°) to NE. The eastern limb of the fold is overturned. F1 and F2 are coaxial. The F3 folds are large open with WNW-ESE axis. The most conspicuous fault transects the belt in the southern end and is likely to be an eastward extension of the Banas dislocation zone.

15.5.3. Host Rocks

Zinc-lead deposits of various sizes and grades occur throughout the belt in calc-silicate-bearing dolomite and graphite mica-schist horizons, the latter in general represents low-grade disseminated sulfides of large volumes.

15.5.4. Mineralization

The ores from the different deposits in the belt have more or less similar mineral assemblage, differing mainly in their relative proportions from deposit to deposit. The strata-bound and stratiform orebodies are mainly comprised sphalerite, galena, chalcopyrite, pyrite-pyrrhotite, arsenopyrite and some fahlore (tetrahedrite-tennantite). Preferred concentration of silver has taken place in fahlore. Sindesar-Khurd and Bamnia-Kalan have an excessive predominance of pyrrhotite. The other deposits in the belt are dominated by pyrite. The stratiform ores of Rajpura-Dariba are characterized by the presence of different verities of laminated (Fig. 2.12) sphalerite such as lemon yellow (Fig. 1.3G), light brown and dark brown (Fig. 1.3F).

15.5.5. Genetic Model

The base metal sulfides of the belt deposited in shallow marine, volcano-sedimentary basins, formed part of the larger intracontinental incipient rift. The stratiform sulfides are sedimentary-diagenetic and hydrothermal emissions of metal-rich brines in which the sulfur was probably derived from thermochemical reduction of circulating seawater and partly leached from mafic source. The deposits of this metasedimentary basin are metamorphosed to medium-grade amphibolite facies and exhibit characteristic features of SEDEX class mineralization and dated as 1804 Ma.

15.5.6. Size and Grade

The resource of the belt is around 290 Mt estimated at 3-4% Zn + Pb or +100 Mt of 6.5% Zn and 2.4% Pb. The probable reserves of the in situ gossans, up to a depth of 40-50 m, developed over Dariba orebodies was estimated at 31.82 Mt, grading up to 1.2% Zn, 0.4% Pb, 0.27% Cu, 200 g/t Ag, 500 g/t Hg, 1000 g/t Sb and As, and less than 0.5 g/t Au.

There are four blocks under ML of M/s HZL. The blocks are at various stages of production, development and decision making.

15.5.7. Rajpura-Dariba Mine Block

Rajpura-Dariba (24°56′51″N, 74°7′53″E) is an underground mine commissioned in 1983. It is located at the southern extremity of the belt. Surface manifestation of mineralization is by way of a well-developed gossans (National Gossan Monument, Fig. 4.8). The zone of oxidation extends to a depth of 40-50 m from the surface. This mine perhaps represents the oldest mining and smelting operations in the World for the extraction of zinc metal. All along the strike of the lodes, there are numerous open pits of varying dimensions and depths. A large number of narrow openings provided access and ventilation to the prolific underground ancient mining operations. A deep ancient open pit exists at the top of the East lode where massive wooden revetments are holding the collapsing hanging wall. This speaks of the marvelous ongoing skill of the ancients. Timber support, baskets and bamboo are recovered from ancient workings and at drill hole intersection at 265 m depth from surface. A large slag spread is also found around the mine.

The mine area is constituted mainly by a sequence of metamorphic equivalent of volcano-sedimentary rocks with minor tuffaceous components and mafic sills, now represented by amphibolites. The rocks, from footwall to hanging walls, are comprised mica-schist, calcareous biotite schist, calc-silicate dolomite and graphite mica-schist. The calc-silicate rocks and the graphite mica-schist at the contacts are the main host for mineralization. Barren portions of this dolomite at places constitute the footwall or hanging wall of the lodes. Cross-bed lamination is frequent in mineralized schist and dolomite bands. The presence of bands of water-lain tuffs within the graphite mica-schist is reported. The formations strike N-S with moderate to steep easterly dips. Joints are grouped into four sets. Narrow zones of crushing, brecciation and gouging, mostly 0.1-2.0 m wide, represent shears. These are persistent along strike and dip, along the hanging wall contacts and within the ore zones.

Zn-Pb-Cu-Ag sulfide bands are strata-bound and occasionally stratiform (Fig. 2.3) exhibiting bedded geometry

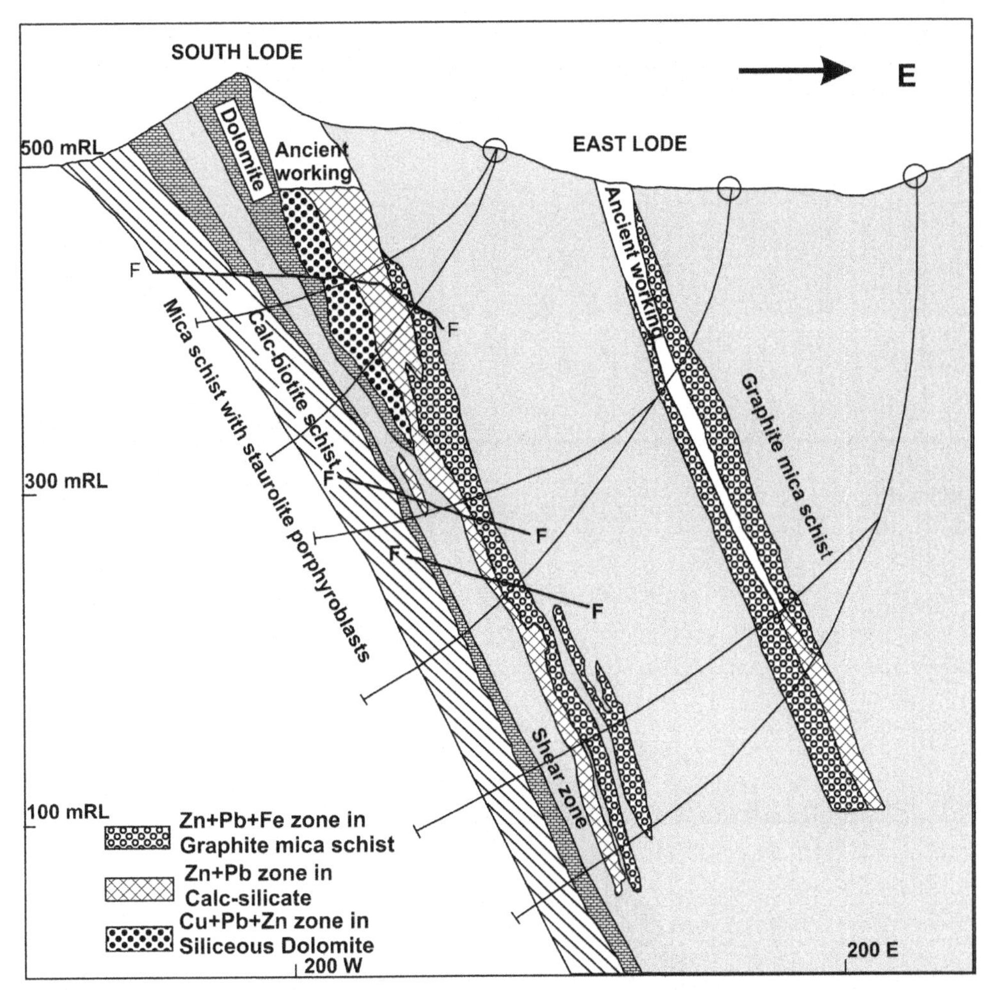

FIGURE 15.7 Cross section of Rajpura-Dariba South and East lodes showing the prototype of orientation surface diamond drilling (Source: Haldar, 2007) [33].

and show relics of primary structure. Later polyphase deformation, metamorphism, recrystallization and remobilization obliterated the original sedimentary geometry, primary synsedimentary and diagenetic fabrics. This generates secondary structure, and locally, discordant strata-bound sulfide bodies along fractures, shears, fold closures and flexures.

The main lode extends over 1700 m and separated into two orebodies, South and North, by a barren stretch of 300 m. The South lode strikes N-S and dips 60°-70° toward east (Fig. 15.7). The South lode extends over 500 m and continues below 0 m level at reduced width. The North lode has a strike length of 900 m. It strikes N-S and dips 70°-75° toward east. The East lode, with a length of 600 m, strikes N-S and dips easterly at 60°-70°. It is located about 150-200 m away on the hanging wall side of the South lode. The lodes occur in lensoidal shape with pinching and swelling style. Average width of South, North and East lodes are 24, 18, and 18 m respectively.

The South lode has sharp ore contacts in calc-silicate dolomite and graphite mica-schist. The dolomite unit is more siliceous at the footwall with dominance of chalcopyrite. This unit is followed by enrichment of sphalerite and galena in the middle and concentration of pyrite and pyrrhotite toward hanging wall graphite mica-schist. Thus, a metal zoning of Cu → Zn-Pb → Zn-Fe is observed from footwall to hanging wall in poly-metallic sedimentary-diagenetic sulfide deposits. The North lode is hosted by calc-silicate dolomite only.

Total Reserves and Resources stand at 42.2 Mt at 6.6% Zn, 1.7% Pb and 82 g/t Ag as on 31 March 2012. The mine has produced about 8.50 Mt of ore till March 2004 mainly between 0 and 400 m levels.

15.5.8. Sindesar-Khurd Mine Block

The silver-rich Sindesar-Khurd deposit (25°01′:74°08′, 45 K/4) is 6 km north of the mining town of Rajpura-Dariba Mine. The block constitutes a ridge of stony barren land with quartzite ridges rising to an elevation of 570 m showing no indication of mineralization.

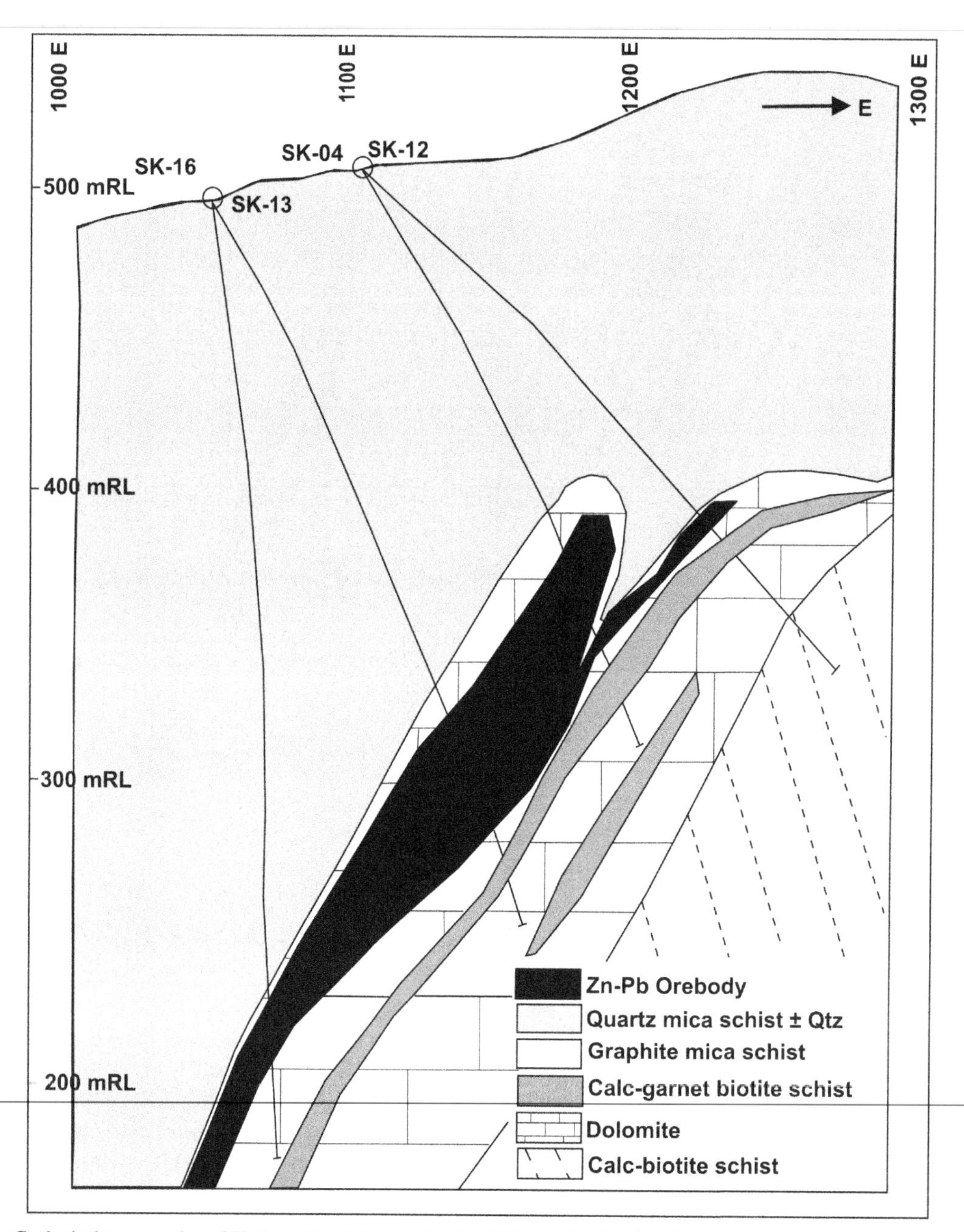

FIGURE 15.8 Geological cross section of Sindesar-Khurd deposit showing hidden orebody below 120 m from surface and steep westerly dip in sharp contrast to the usual easterly dip (Source: Haldar, 2007) [33].

The strata-bound single massive orebody is located in the central part of western limb of Dariba-Bethumni regional fold. The best exposed rocks are the inter-banded mica-schist/chert/quartzite and form a prominent NNE-SSW trending ridge. Calc-silicate-bearing dolomite horizon, principal host rock, is completely concealed under the quartzite/chert and inter-banded mica-schist at a depth of about 120 m (Fig. 15.8). The hanging wall rocks, graphite mica-schist and calcareous quartz biotite schist are intersected only in drill holes.

The average strike is N10°W with dip swings from low to moderate angles toward west to nearly vertical. A conspicuous structural feature of the deposit is the presence of a concealed NNE-SSW trending broad, open and asymmetric antiformal fold with sub-horizontal to gently plunging fold axis. The orebody constitutes the steeper western limb of the antiformal structure. The westerly disposition of orebody is in sharp contrast to the usual easterly dips exhibited at Rajpura-Dariba, Bamnia-Kalan and entire Aravalli-Delhi zinc-lead-copper deposits.

GSI has drilled 9267 m in 29 boreholes at 200 m spaced sections with one or two level intersection. HZL commenced surface drilling in December 1992 and completed 6786 m in 22 holes over 400 m strike extension up to a depth of 350 m from surface.

The strike extension of the main body has been traced over 1000 m. However, bulk of the ore reserves is confined within 400 m with an average width of 34 m (ranging between 6 and 59 m) up to a known depth of 400 m from surface. Additional 64,884 m of surface drilling completed between 2005 and 2010 that has established continuity of mineralization at a depth of +800 m from surface. Principal OFM are sphalerite and galena. Pyrrhotite is the most abundant and ubiquitous sulfide gangue, while pyrite and arsenopyrite are rare. RFM are calcite, dolomite, quartz, mica, garnet, actinolite, tremolite, argillaceous and carbonaceous schist materials.

About 60.8 Mt of ore reserves and resources at 5.8% Zn, 3.8% Pb, and 215 g/t silver have been outlined at Sindesar-Khurd as on March 2012. The deposit is under mining by M/s HZL for a production target of 0.3 Mt/annum.

The other two leasehold blocks await either decision (Bamnia-Kalan, 25°03′:74°09′, 45 K/4) or kept in abeyance (Sindesar-Kalan-E, 25°01′:74°09′, 45 K/4) due to either low tonnage or low ore grade.

15.5.9. Salient Features of the Leasehold Blocks

The summary information comprising exploration, host rock, mineralization, metamorphism, surface oxidation/gossans, metal zoning, halos, genesis, age, deposit size and grade, contained metal, minable reserve, capacity and life of mining are given Table 15.3.

15.6. NEVES CORVO, PORTUGAL: DISCOVERY OF DEEP-SEATED ZN-CU-SN DEPOSIT: A GEOPHYSICAL SUCCESS

Portugal or Portuguese Republic is the southwesternmost country in Europe with its capital at Lisbon. It is situated on the Iberian Peninsula bordering Atlantic Ocean to the west and south and Spain to the east and north. Portugal is essentially an agricultural country. The mineral industry is modest by World standards in terms of production. The mineral industry growth became significant mainly because of the discovery and development of the rich copper-zinc-tin deposits at Neves Corvo by Sociedade Mineira de Neves Corvo S.A. (Somincor). The mineral commodities are copper (São Domingos open-pit Mine), zinc-lead (Aljustrel mine), copper-zinc-tin (Neves Corvo), tungsten (Panasgueira), gold (Góis region) and coal (Guimrota). There are two metallic mines in operation: (a) Neves Corvo in Castro Verde, the largest producer of mined copper in the European Union, and (b) Panasgueira in Covilhã, Castelo Brancho, one of the largest tungsten mines in the World. The discovery and successful implementation of the Neves Corvo project in Iberian Pyrite Belt (IPB) has brought significant benefits to the local and regional communities in the southwestern Portugal.

15.6.1. Location and Discovery

The Neves Corvo copper-zinc-tin deposit, Castro Verde, Beja District, Portugal, is located 220 km southeast of Lisbon and 100 km north of Faro. Neves Corvo has good connections to the national road network which links with Faro and Lisbon. The mine has a dedicated rail link into the Portuguese rail network. The deposit (37°34′23″N:7°58′15″W) forms part of the 250-km-long and 30- to 50-km-wide IPB that trend NW-SE from Alcácer do Sal (Portugal) to Sevilla (Spain). The Iberian pyrite belt, in the Spain part, is known for more than 250 pyrite deposits/active mining over 1000 years. Neves Corvo, in the Portugal side, is the highest grade copper-zinc-tin deposit ever found on the IPB (Fig. 15.9).

The discovery of deep-seated concealed richest poly-metallic deposit at Neves Corvo can be credited to the success of modern sophisticated geophysical techniques. A Joint Venture Consortium between 'Bureau de Recherches Géologiques et Miniéres' (BRGM), France, Penarroya and Portuguese states mineral company "Empresa de Desenvolvimento Mineira (EDM)" carried out the first phase of prospecting between 1969 and 1973 in southern half of IPB. The investigation was mainly by geological mapping and gravimetry survey leading to identification and ranking of several anomalies. The first exploratory drill hole was sunk in 1973 achieving nothing of any significance. The negative drilling result held the project in abeyance. The exploration team members were not in agreement with the setback and renewed the second phase including diamond drilling between 1973 and 1977. They accumulated data from the rest of the Baixo Alentejo province, compared with drill information and reinterpreted geophysical anomalies. They were seriously convinced that the holes so far drilled had not gone deep enough. The fifth drill hole of the third phase intersected 50 m of massive sulfide from Neves orebody. The discovery of Neves orebody at a depth of 330 m from surface was a gift to the country in April 1977. The drilling continued and identified three other orebodies starting at a depth between 250 and 630 m, viz., Corvo, Zambujal and Graca during 1977 and 1978. Tin orebody was discovered during the development of the mine, which led to the construction of a tin plant. Routine surface exploration drilling close to the mine discovered a new high-grade copper-rich massive sulfide deposit in October 2010 and named as "Semblana".

TABLE 15.3 Salient Features of Four Mining Block at Rajpura-Dariba Belt

Features	Rajpura-Dariba	Sindesar-Khurd	Bamnia	Sindesar-Kalan
Location	75 km northeast of Udaipur City	6 km NE of Rajpura-Dariba	4 km NE of Sindesar-Khurd	1 km East of Sindesar-Khurd
Exploration	34 km surface and 75 km under-ground drilling	81 km surface drilling by GSI and HZL	17 km surface drilling by GSI and HZL	Surface drilling at 100 m by GSI
Host rock	Calc-silicate, dolomite, GMS	Calc-silicate, dolomite, GMS	Calc-silicate and dolomite	Graphite mica-schist
Mineralization	Strata-bound and stratiform, Sp, Ga, Ag, Cu, Py	Strata-bound, Sp, Po. Ga, Ag, Py, Cp	Strata-bound, Sp, Ga, Po and Py	Strata-bound and lensoidal, Sp, Ga, Po, Py, As.
Metamorphism/deformation	Strong deformation, amphibole facies	Strong deformation, amphibole facies	Strong deformation, amphibole facies	Strong deformation, amphibole facies
Gypsum present	Gypsum present	Gypsum	Gypsum	Gypsum
Surface oxidation	Excellent oxidation, multicolor gossans	Ore at 120 m below massive barren quartzite	Ore at 120 m below 20 m thick soil	Near surface, oxidation present
Zoning	→ Cu-Zn-Pb-Fe	Metal zoning	—	—
Halos	Pyrite at H/W	Pyrrhotite	Pyrite present	Pyrite present
Genesis	SEDEX	SEDEX	SEDEX	SEDEX
Age (Ma)	1800	1800	1800	1800
Size and grade, 31 March 2012	42.2 Mt at 6.6% Zn, 1.7% Pb, 82 g/t Ag	60.8 Mt at 5.8% Zn, 3.8% Pb, 215 g/t Ag	Three lenses, 4 Mt at 5.7% Zn, 2.5% Pb, 100 g/t Ag	Near surface, 94 Mt at 0.6% Pb and 2.1% Zn
Contained metal	2.05 Mt	3.55 Mt	0.33	2.54
Minable reserves	~20 Mt	~30 Mt	Economic-uneconomic border line	Abeyance due to uneconomic metal grade
Infrastructure	Water requirement is met from Matrikundia dam on Banas River. Power requirement met by captive power plants and shortfall is met by State grid.			
Mining methods	Vertical crater retreat and blast hole Stoping mined out stopes backfilled with cemented tailings.			Open pit
Modernization	Equipped with world-class infrastructural facilities including the latest and the best machineries			Open pit
Depth of mine (m)	500	500	300	40
Mine capacity (Mt/a)	0.900	1.500	0.300	1.500
Life of mine	+20 Years	+20 Years	+10 Years	+40 Years

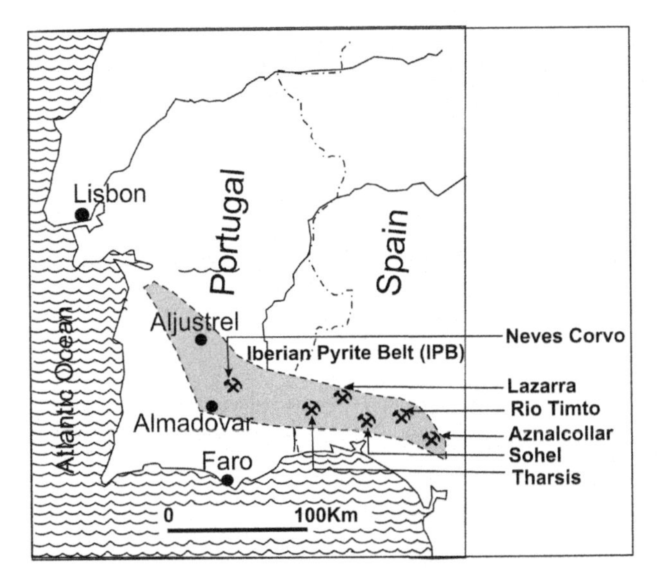

FIGURE 15.9 Location map of IPB in Portugal and Spain showing Neves Corvo poly-metallic deposit (Source: Haldar, 1990) [32].

The company, "SOMINCOR" (Sociedade Mineira de Neves Corvo S.A) was formed in 1980 as operator. One hundred fifty kilometer of surface and 50 km of underground drilling, delineation and evaluation identified copper-rich Corvo and Graca to be viable. RTZC (Rio Tinto Zinc Corporation) acquired the share of Penaroy and BRGM in 1985. In March 2004 EuroZinc acquired 100% of Somincor, giving it full ownership of the Neves Corvo mine. EuroZinc merged with Lundin Mining in October 2006.

A 3.50-km decline ramp for men and services and a 500-m vertical shaft for ore hoisting started in 1981 and 1982 respectively. The mine ventilation is supported by two vertical shafts. The mine design was completed between 1983 and 1984. The underground mine was commissioned on 10 December 1988. The commercial production of metal commenced in January 1989. The company made small copper biscuits from the first batch of metal to commemorate the occasion. The author is privileged to receive one such memento during his visit to the mine (Fig. 15.10A and B) in early 1990. The tin plant was commissioned in 1990. The copper and tin concentrate was processed at site and shipped to smelter overseas till 2006 with the commencement of treating zinc ore. Zinc production was restarted at a limited rate in 2010 and a new zinc expansion project was completed in July 2011. The expanded plant has the flexibility to process zinc or copper ores.

15.6.2. Regional Setting

The Neves Corvo deposits occur within the volcanic sedimentary complex (IPB). The host rock sequence consists of acid volcanic rocks separated by shale units. A discontinuous black shale horizon lies immediately below the ore horizon. There is a thrust-faulted repetition of volcano-sedimentary units above the mineralization. The whole assemblage has been folded into a gentle anticline oriented NW-SE and plunges → SE. This has resulted in orebody distributed on both limbs of the fold (Fig. 15.11). Entire sequence has extensively been affected by both sub-vertical and low-angle thrust faults causing repetition.

15.6.3. Exploration

Surface and underground exploration drilling is an ongoing operation for sustainable reserve growth. The drilling is carried out both by in house and contractual. The surface drilling of NQ series is spaced at either 100 or 75 m. The underground drilling is in fan pattern with intersections at

FIGURE 15.10 Small slab of first copper metal produced from Graca orebody, Neves Corvo mine in 1988 commemorating the occasion: (A) emblem of the owner Somincor and (B) the year of production.

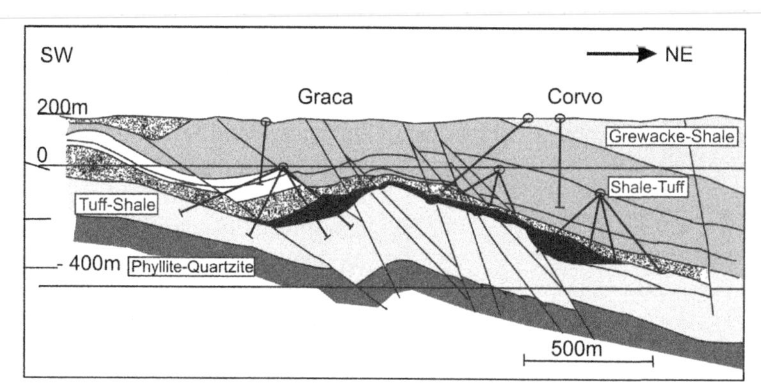

FIGURE 15.11 Surface and underground exploration by fan drilling design at Neves Corvo mine (Source: Haldar, 1990) [32].

either 17.5 or 35 m spacing (Fig. 15.11). All drill holes are surveyed by multi-shot reflex camera at 30 m intervals for an accurate location of the drill intersections. The average annual surface and underground drilling are in the order of ~50,000 and ~25,000 m respectively.

15.6.4. Mineralization

Six laterally linked massive stratiform sulfide orebodies have been defined viz. Neves (North and South), Corvo, Graça, Zambujal, Lombador (North, South and East) and Semblana. The orebodies are deeply rooted stockwork and stringer system. The metal grades are segregated by the strong zoning into copper, tin and zinc zones, as well as barren massive pyrite (Fig. 15.12). The sulfide deposits are typically underlain by stockwork sulfide zones which form an important part of the copper orebodies. The OFM are pyrite, chalcopyrite, sphalerite, galena, cassiterite and silver minerals. The primary and secondary commodities from the mine production are copper and zinc with subsidiary product of tin and silver.

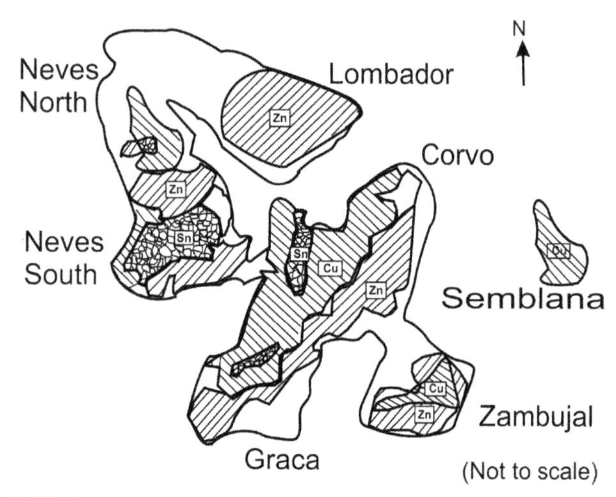

FIGURE 15.12 Distribution of orebodies and metal concentration at Neves Corvo group of orebodies.

Corvo orebody has a strike extension of ~1100 m and dips at 10°-40° → NE. The orebody appears at a depth of 280-630 m below surface with maximum true thickness of 95 m. The massive sulfide lens consists of a basal layer of copper ore up to 30 m thick and overlain by barren pyrite. The orebody is overlain by a complex mineralized sequence called "Rubane". This consists of an assemblage of chloritic shale, siltstone and chert-carbonate breccias which are all mineralized with crosscutting and bedding parallel veins and occasional thin lenses. The sulfides are cupriferous over much of their occurrences and Rubane copper ore constitutes over 15% of the total cupriferous content of Corvo.

Tin occurs closely associated with copper ore, both in the massive sulfide lens and Rubane. Massive sulfide tin ore containing high copper grades is distributed through the copper ores of Corvo defining a trend from north to south. Rubane and the underlying stockwork are mineralized with high grades of tin ores at the north end of the massive sulfides.

Graca orebody lies on the southern flank of the anticline and dips at 10°-70° → S. The orebody extends over 600 m along strike and 450 m in the dip direction. The orebody appears between 250 and 480 m from surface. The lens is up to 80 m thick and is linked to Corvo by a bridge of continuous thin sulfide mineralization. Much of the copper ore occurs as basal layer overlain by barren pyrite.

Neves orebody consists of two lenses, North and South, joined by a thin bridge and dips at 5° and 25° → N. The orebody extends over 1500 m in strike and 700 m in dip direction with maximum thickness of 50 m. The southern lens contains mostly zinc with significant lead, silver and copper and underlain by tin-bearing copper ore. The northern lens contains a significant resource of massive sulfide copper with minor complex zinc ore. There is a thick zone of cupriferous stockwork underlying the northern lens.

Zambujal is a small thin flat dipping massive sulfide orebody containing some low-grade zinc ore, minor copper ore and barren pyrite.

Semblana deposit is located only 1.3 km northeast of the Zambujal orebody at a depth of 850 m. Exploration drilling has outlined an area of 600 by 250 m of massive sulfide and stockwork mineralization in seven drill holes. This new deposit remains open in all directions and appears to be flat-lying. The mineralized thickness varies between 4 and 36 m. It is copper rich with grades varying between 1.5 and 4.5% Cu, 0.2 and 2.6% Zn, 0.1 and 1.0% Pb, and 0.1 and 0.3% Sn.

15.6.5. Genetic Model

The IPB was formed 350 million years ago on account of active hydrothermal volcanism forming volcano-sedimentary complex. Volcanic activity in the region led to several giant VMS ore deposits associated with poly-metallic massive flanks of volcanic cones. More than 250 deposits of the IPB are placed at volcanic-sediment-hosted massive sulfide (VSHMS) deposits, a complementary between the VMS and the SEDEX types.

15.6.6. Size and Grade

The proved and probable ore reserves as on March 2011 stand at:

Copper-rich ore 27.7 Mt at 3.0% Cu, 0.9% Zn, 0.3% Pb and 44 g/t Ag

Zinc-rich ore 23.1 Mt at 7.3% Zn, 0.4% Cu, 1.7% Pb and 66 g/t Ag

Tin-rich ore 2.68 Mt at 13.62 % Cu, 1.27% Zn and 2.42% Sn *

Neves Corvo mines copper, zinc and tin ores. The tin resource* is nearing exhaustion and forms no part in future mining plans.

Underground mining is based on mechanized stoping using primarily "bench and fill" and "drift and fill" methods with sand and paste backfill. The processing facilities are (a) a copper plant with 2.5 Mt/a capacity at 25% Cu in concentrate, (b) a zinc plant with 1 Mt/a capacity at 50% Zn in concentrate, (c) tailings impoundment and (d) backfill plants. The mine life is +10 years at current minable reserves.

Neves Corvo is certainly the significant richer copper mine in the World and will remain Europe's "El Dorado" ("the golden one") for many years.

15.7. BUSHVELD, SOUTH AFRICA: THE LARGEST PLATINUM-CHROMIUM DEPOSITS OF THE WORLD

The resource base and production of platinum group of elements (PGE) and chromium metals of South Africa rank number one in the World. The country produced 163 tonnes of platinum metal and 7.490 Mt of chromite ore in 2005. The World's share of platinum, palladium and chromium production was 78, 39 and 39% respectively during 2005. Three of the five largest PGE-chromite deposits of the World occur near the stratigraphic middle of large layered intrusions, namely the Merensky and UG2 Chromitite Reefs of the Bushveld Complex, South Africa, and the Main Sulfide Zone of The Great Dyke, Zimbabwe.

15.7.1. Location and Discovery

The Bushveld Igneous Complex (BIC) is the largest layered igneous intrusive in the World. The complex (25°S:29°E) is located in South Africa and hosts some of the richest and largest chromium-PGE deposits on Earth.

The igneous complex was discovered in 1897 by Gustaff Molengraff, a Dutch geologist, biologist and explorer. He discovered the Bushveld Complex while mapping the Transvaal region as a working state geologist. Chromite mining started thereafter. The igneous complex had been tilted, eroded and outcrops around the edge of a great geological basin. Merensky Reef was an outstanding discovery of platiniferous orebody in the Eastern Bushveld Complex in 1924 by Dr Hans Merensky and Andries Lombaard. In 1925 Dr Hans discovered another promising orebody in the Northern limb and named it Platreef. Under Ground-2 (UG2) chromite layer, discovered in 1970s, surpassed the Merensky Reef in platinum reserve. Mining of these lodes for platinum-palladium started from 1970 by Lonmin, Anglo American Platinum Ltd and others by underground method.

15.7.2. Exploration

The chromium-platinum group of deposits has extensively been drilled over the years from surface and underground. The continuity and delineation have been established by drill holes up to a depth of 1500 m for Platreef and 3300 m for Merensky and UG2.

15.7.3. Regional Setting

The structure represents a huge saucer shape layered mafic to ultramafic igneous intrusion (2060 Ma age) forming

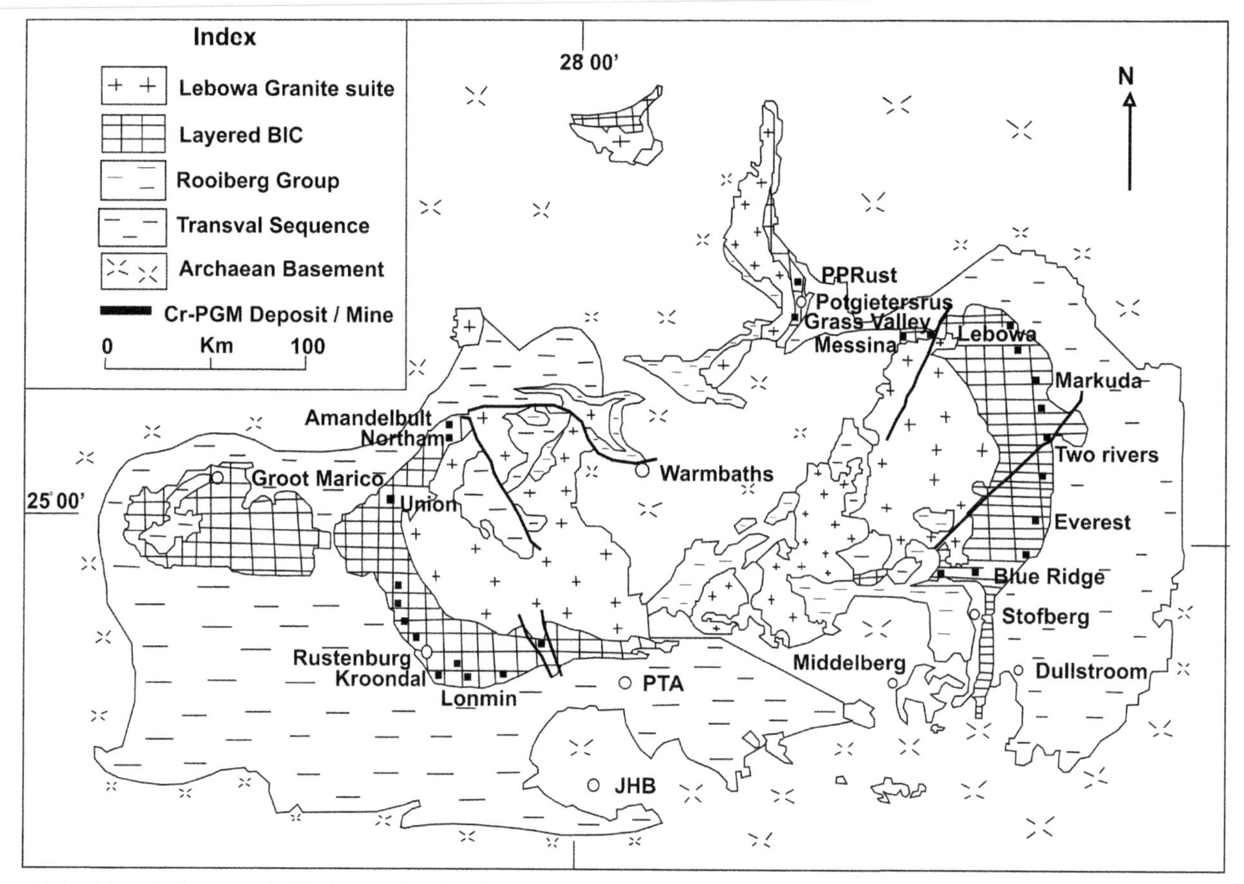

FIGURE 15.13 Surface map of BIC showing the chromium-PGE deposits/mines (Source: modified after many from Internet and Haldar 2011) [34].

a great geological basin. The BIC is divided into two prominent lobes—an eastern and western, with further northern extension (Fig. 15.13).

All three sections of the system are remarkably similar. The rock type is comprised ultramafic (peridotite, chromitite and harzburgite) in the lower to mafic (gabbro, norite and anorthosite) in the middle and felsic phase (granite) in the top section (Table 15.4).

15.7.4. Genetic Model

The Complex was formed by repeated injection of magma over large time span into a huge sub-volcanic chamber. Cooling and differential crystallization was a slow process forming sub-horizontal shallow level layers from the base of the chamber. The chamber covers an areal extent of +65,000 km^2 as preserved today. These

TABLE 15.4 Generalized Geological Succession at BIC

Formation	Period	Rock type
Lebowa Suite	1790 ± 114 to 1604 ± 70 Ma	Younger sedimentary cover and granite intrusion
Rustenburg Layered Intrusive Suite, host rock	Proterozoic	Mafic and felsic intrusive rocks: norite-gabbro-pyroxenite-harzburgite-chromitite, granite.
Rooiberg Group	Early Proterozoic	Volcanic flows and pyroclastics
Transvaal Supergroup	Late Archean to Early Proterozoic	Metasediments: shale and quartzite.
Basement	Archean	Granite and gneiss.

processes were repeated reaching ~9 km thickness by the intermittent replenishment and addition of existing and new magma producing a repetition of the mineral layering. Some individual layers or groups of layers can be traced for 100s of kilometers. This layered sequence, the Rustenburg Layered Suite, comprises five principal zones: Marginal, Lower, Critical, Main and Upper Zones. The intrusive in general dips to the center of the complex. BIC consists of four compartments, the western, eastern, northern and southern limbs, in order of economic importance.

15.7.5. Mineralization

There are more than 20 chromitite horizons of varying thickness enriched in PGE and Fe-Ni-Cu sulfides. The three major, distinct and rich orebodies in Bushveld Complex are Merensky Reef, UG2 and Platreef. The former two reefs can be traced over 300 km in two separate arcs and the later extends over 30 km. The grade and thickness variation in strike for kilometers and down dip are negligible. Some layers consist of 90% chromite making the rare rock type, chromitite.

The chromite reefs are named from stratigraphic bottom to top as Lower Group (LG1 to LG6, LG6A and LG7), Middle Group (MG1 to MG3, MG4A and MG4B) and Upper Group (UG1, UG2, Merensky Reef) located in East and West lobe. The LG contains 0.2-1.2 g/t Pt. The MG contains relatively higher grade varying between 0.9 and 5.5 g/t Pt. The Pt content of UG1 varies between 1.2 and 3.4 with up to 43.5% Cr_2O_3.

UG2 occurs both in west and East lobe. The layer has sharp lower and top contacts with thickness ranging between 40 and 120 cm. The UG2 is usually underlain by a coarse-grained feldspathic pyroxenite and rarely by anorthosite. It contains massive chromite grading between 60 and 90% Cr_2O_3 and 4.5-8.0 g/t Pt.

Merensky Reef occurs in East and West lobe with thickness varying between 30 and 90 cm. The reef rests on anorthosite and overlain by pegmatitic feldspathic pyroxenite. The host rock is norite with extensive chromitite and sulfide layers or zones containing the ore. The Merensky Reef contains 3.5 and 9.5 g/t Pt in pyrrhotite, pentlandite and pyrite as well as in rare platinum group minerals.

Platreef is located close to the Thabazimbi-Murchison Lineament in the northern lobe with thickness ranging between 50 and 400 m. It is formed as a result of interaction between a new gabbroic parental magma of Main Zone and a suit of sulfur-bearing sediments and preexisting Lower Zone cumulates. The chromiferous ore is rich in Fe-Ni-Cu sulfide and PGE.

15.7.6. Size and Grade

The Bushveld Complex constitutes 75% of the World's resources of Pt; 58% of it occurs in one UG2 chromitite unit. The area contains some of the richest chromite and PGE ore deposits. The reserves of platinum group of metals (platinum, palladium, osmium, iridium, rhodium and ruthenium) are the World's largest. In addition there are vast reserves of chromium, iron, tin, titanium and vanadium.

The PGE reserve and resource base of Bushveld Complex (Merensky, Platreef and UG2) stand at 13,510 million tonnes with Pt ranging between 1.3 and 3.2 g/t and Pd ranging between 1.4 and 2.0 g/t. The chromite reserve and resource base estimated at ~3100 and ~5500 million tonnes respectively with Cr_2O_3 grade vary between 40 and 47%. The country rocks i.e. gabbro and norite are used as dimension stones.

15.8. SUDBURY, CANADA: THE LARGEST NICKEL-PLATINUM-COPPER DEPOSITS OF THE WORLD

Canada ranks first in exploration expenditure in the World. It stands within top five in production of Ni, Zn, PGE, Au, Mo, Cu, Co, Pb and Cd. Canadian mining industry is equipped with knowledge-base high technology and supported by more than 2200 mining-related companies. The various metallogenic provinces include Proterozoic magmatic Ni-Cu deposits in Sudbury region (1646 Mt), Neoarchean Intrusive Pd deposits in Lac des Iles (64 Mt) near Thompson, massive volcanogenic-type sulfide and quartz-carbonate gold deposits from Archean greenstone belts of Quebec and Ontario, Proterozoic volcanic belts of Manitoba and Paleozoic volcanic rocks of New Brunswick. There are two world-class large and rich zinc deposits viz. Red Dog (150 Mt at +16% Zn + Pb) located in British Columbia and Sullivan (160 Mt at +12% Zn + Pb) located near the town of Kimberly in southeastern British Columbia. Uranium and diamond are relatively new, but growing mineral commodity in Canada.

15.8.1. Location and Discovery

The "Sudbury Basin", "Sudbury Structure", the "Sudbury Nickel Irruptive" or "Sudbury Igneous Complex (SIC)" (46°27′50″N, 81°10′29″W) is a major geological structure in the city of Greater Sudbury, Ontario, Canada. Sudbury Basin (200 km diameter, 1840 ± 21 impact age) is the second largest known impact crater on Earth after Vredafort (250-300 km diameter, 2023 ± 4 Ma impact age) of South Africa. The former is known for the largest resources of nickel-copper-cobalt + PGE over centuries. The later is at the preliminary search for gold and uranium occurrences. The Sudbury Structure is the remnant of a deformed multi-ring impact basin and hosts to a vast amount of Ni-Cu-PGE sulfide mineralization (Doreen, 2008) [23]. The melting effect of impact at Sudbury Basin could be traced up to 15 km depth from surface.

Alexander Murray of Geological Survey of Canada (GSC) first reported the presence of sulfide minerals in 1856 at the present-day site of the Murray mine. Yet, the discovery of this significant Ni-Cu-PGE resource could be documented only by a blacksmith in 1883 during construction of the first transcontinental Canadian Pacific Railway. The identification of Ni-Cu metals resulted in a prospecting and staking rush. The PGE realization came much later. Falconbridge mined nickel-copper ores in the Sudbury area since 1929. Xstrata acquired the Sudbury mine through its acquisition of Falconbridge in 2006.

15.8.2. Regional Setting

The Sudbury mining district is represented by five different geological environments ranging between Archean and Paleozoic. SIC is structurally placed between Archean Levack granite gneiss of Superior Province in the north and east, Paleoproterozoic metasedimentary and meta-volcanic Huronian Supergroup rocks of the Southern Province and White Water Group of sedimentary package in the central part (Table 15.5). The Huronian Supergroup of rocks deposited in an intercontinental rift basin environment.

The main rock units of the SIC range from an outer ring of norite through a transition zone of gabbro to an inner zone of granophyre. SIC represents a prominent elliptical multilayer ring structure (Fig. 15.14). The outer rims are popularly known as North and South Range. The North, East and South Ranges are dotted with more than 100 deposits and +150 occurrences. The producing and abandoned Ni-Cu-PGE mines are over 80. More than 11 Cu-Ni-PGE projects are in advance stages of exploration.

TABLE 15.5 **Stratigraphic Succession of Sudbury Mining District**

Stratigraphy	Formation	Rock type
Paleozoic	—	Limestone and dolomite
Mesoproterozoic	—	High-grade metamorphic rocks
Paleoproterozoic (1840 ± 21 Ma)	SIC/Structure	Granophyres
		Quartz-rich gabbro
		Norite Ni-Cu
		Sub-layer norite and gabbro Cu-PGE
Paleoproterozoic (2500-1600 Ma)	White Water Group	Chelmsford formation
		Onwatin formation Zn-Cu-V-Cd-Ag-Ni
		Onaping formation Zn-Cu-Pb-Ag
	Huronian Supergroup	Murray granite, gabbro-anorthosite
		Quartzite
		Graywacke and volcanic rocks
Archean (+2500 Ma)	Levack Basement	Granite gneiss, felsic plutons, meta-volcanic, metasediments

15.8.3. Exploration

The economic recognition of the Sudbury mining district started with Cu-Ni resources. Exploitation and exploration for new deposits continued in the complex and promoted to a poly-metallic Ni-Cu-PGE program. The current exploration expenditures are mainly focused to the footwall-hosted Cu-Ni-PGE systems containing high Cu and PGE. This has been rewarded by the discovery of many new deposits like Victor, Levack Footwall, etc. New exploration models for low-sulfide, high-PGE poly-metallic deposit style of mineralization are refined for both the North and South ranges. The new discovery of base metal deposits are frequent in Sudbury camp by evolving deposit model, advance exploration techniques, introduction of deep directional drilling (Fig. 15.15), and genetic model.

Knowledge-based 3D exploration modeling coupled with application of electromagnetic borehole geophysics and interpretative skill resulted identification of deep-seated target areas for Ni-Cu-PGE systems at Sudbury Structure. The follow-up single collar multiple wedge orientation deep drilling over 2000 m depth paid dividend to many success stories in past decades and in the recent century. The geophysical program employing "University of Toronto Electro-Magnetometer" System (UTEM) in old and new drill-holes credited to the discovery of Victor contact Ni-Cu-PGE (1970s and 1980s), Fraser-Morgan contact and McCreedy East 153 zone (1990s), Kelly Lake Ni-Cu-PGE in the Copper Cliff Offset (1995), Totten Depth Ni-Cu-PGE (1997), Ni Rim South (2001) and Podolsky (2002). The discovery and exploration of high-grade copper-nickel-precious metals at Morrison (Levack Footwall) is a significant addition of 2005. Probable reserve as

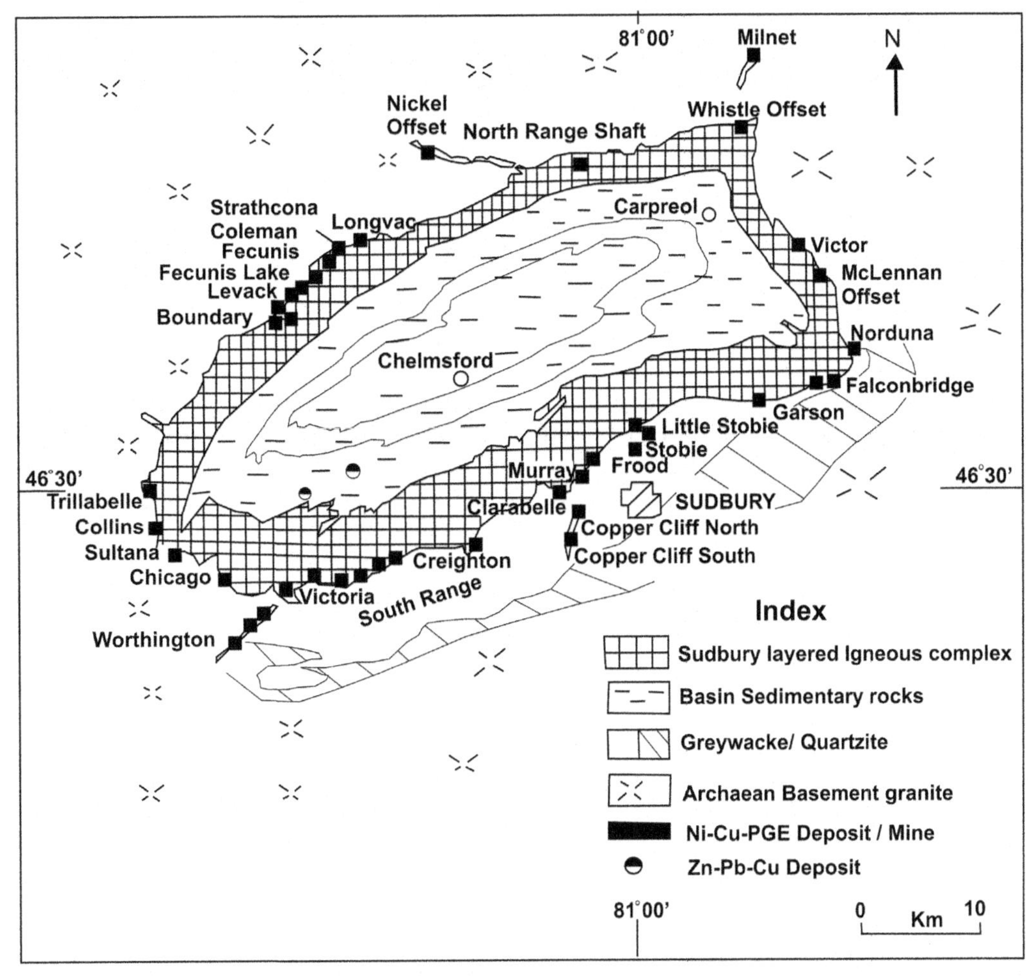

FIGURE 15.14 Geological map of SIC showing Ni-Cu-PGE and Zn-Pb-copper deposits/mines (Source: modified after many from Internet and Haldar 2011) [34].

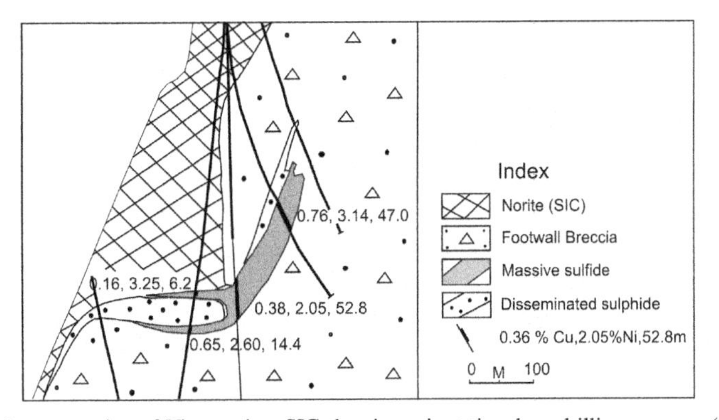

FIGURE 15.15 Geological cross section of Victor mine, SIC showing orientation deep drilling program (Credit: INCO Exploration).

of December 2010 stands at 1.6 Mt at 10.45% Cu, 2.1% Ni and 8.4 g/t Pt + Pd + Au. The Morrison deposit is open in all directions and is being expanded by FNX Mining with ongoing drilling program centered at the 4000-foot level exploration drift. The mine is in production from 2010.

15.8.4. Mineralization

The mineralization is classified into three types viz. (a) contact, (b) offset, and (c) breccias. All contain Ni-Cu-PGE.

Contact deposits are typically located at the interface between the SIC and Archean or Paleoproterozoic basement rocks (e.g. Murray Mine and Creigton Mine, South Range). The massive pyrrhotite-rich Ni-Cu-PGE sulfide ores are placed within the contact megabreccia zone. The basal igneous sub-layer norite ± mafic-ultramafic inclusion hosts the disseminated sulfides.

Offset-type deposits are hosted within radial or concentric quartz diorite offset dykes that may extend more than 30 km from the parent source of SIC. Economic mineralization are found at Kelly Lake deposit on the Copper Cliff offset dyke, South Range, Nickel offset dyke at North Range, Totten deposit on the Worthington offset dyke, Ni-PGE mineralization on the Trill offset dyke and Milnet deposit at Whistle offset dyke (Fig. 15.14). The PGE-Cu-Ni content increases at distance from the SIC. Massive and vein-type sulfides typically occur in steeply plunging orebodies along the length of the offsets. The deposits are dominated by pyrrhotite with less abundance of pentlandite and chalcopyrite.

Breccia-type deposits are primarily Cu-Ni mineralization within fractured country rocks at the base of the SIC. The Sudbury breccia consists of fragments of ultramafic inclusions and norite in a quartz-feldspar matrix. The sulfides are disseminated and massive stringers within footwall breccias and disseminations within overlying sub-layer norite. Strathcona Cu-rich ore at North range is located in fractures up to 100 m away from the basal contact.

15.8.5. Genetic Model

The structure is postulated to be formed as the result of a meteorite impact (1840 ± 21 Ma) that produced a 150- to 280-km multi-ring crater, containing a 2- to 5-km-thick sheet of andesitic melt. The immiscible sulfide liquid collected into topographic lows, where it was differentiated into Ni-PGE-dominated contact deposits by crystallization of a mono-sulfide solid solution. Residual sulfide liquid migrated into the footwall to form a variety of vein and disseminated type of deposits that underwent remobilization of metals. The footwall of the SIC and the older basement rocks are extensively brecciated, deformed and metamorphosed due to the impact. These breccias host the remobilized high-grade Ni-Cu-PGE mineralization. The period of mineralization and ore-forming process related to cooling of the SIC can be over 100 to 1 million years. Many deposits in the South Range have been modified by deformation. The whole episode prompted Sudbury as the principal base and PGE metal mining district of Canada.

15.8.6. Size and Grade

The total resources including past production are estimated at 1648 Mt at 1% Ni, 1% Cu, and 1 g/t Pd + Pt. The figure includes past production, reserve, and resources (Doreen, 2008 [23]). The district also hosts significant poly-metallic resources comprising U-Ni-Cu-PGE-Au-Cu and Zn-Pb-Cu deposits.

McCreed east is producing +400,000 tpa of HGO. Some other deposits, namely, Kelly Lake, Totten Depth, Ni Rim South (2.17% Ni, 6.45% Cu, and 15.40 g/t Pt + Pd + Au), Podolsky (13.8% Cu, 1.0% Ni, and 8.2 g/t Pt + Pd + Au) and Morrison are either in production or being evaluated for production planning purposes.

The twenty-first century witnessed banner years for exploration, active mining claims and new discoveries with a marked shift in exploration focus to the high-grade Cu-Ni-PGE deposits. The potential for Ni-Cu-PGE discoveries in the Sudbury Structure is considered very high, both at surface and at depth. There have been unprecedented exploration expenditures and activities over the past 2 years by both junior and established mining companies. Recent exploration success by major (INCO, Falconbridge), mid-sized (FNX), and junior (Wallbridge) exploration companies enhanced outstanding resource potential with considerable smelting and refining infrastructure.

15.9. RAMPURA-AGUCHA: THE SINGLE LARGEST AND RICHEST ZN-PB-AG DEPOSIT IN INDIA: GEOSTATISTICAL APPLICATIONS IN MINERAL EXPLORATION

Rampura-Agucha Mine is the World's largest open-pit zinc mine today with an annual ore production capacity of 6.0 Mt and matching beneficiation facility. The mine was commissioned in 1991. Rampura-Agucha is the single largest and richest known zinc-lead-silver orebody in India and comparable to world-class deposit. Rampura-Agucha is also one of the lowest cost zinc producer globally.

15.9.1. Location and Discovery

Rampura-Agucha deposit (25°50′20″N, 74°44′47″E) is located 210 km SW of Jaipur and 230 km NE of Udaipur in the State of Rajasthan. It extends over a strike length of 1.6 km along NE-SW with steep dips (60°-70°) toward SE. The vertical extension continues +1200 m from surface.

Mr T. C. Rampuria, Geologist of the State Department of Mines and Geology, Rajasthan, picked up few pieces of colored weather rock samples (gossans) from a shallow depression, out of curiosity, during a routine inspection visit of garnet mine in August 1977. The samples indicated the presence of significant zinc and lead values and the deposit rediscovered by chance. The same year Neves Corvo deposit was discovered as unique geophysical success. Archaeo-metallurgical investigations in and around Rampura-Agucha deposit unraveled the mysteries of exploitation and extraction practices for silver and lead metals by the ancients, circa fourth century BC.

15.9.2. Regional Setting

Rampura-Agucha is strata-bound and stratiform sediment-hosted high-grade zinc-lead-silver deposit. The geological sequence of rocks from east to west can be grouped under:

4. Garnet-biotite-sillimanite gneiss (GBSG) (hanging wall)
3. Graphite mica sillimanite gneiss/schist (Zn-Pb-Ag ore bearing)
2. GBSG (Footwall)
1. Granite gneiss and mylonitic rocks (Basement).

The GBSG forms the predominant hanging wall unit in and around the deposit. The footwall section is composed of garnet-biotite gneisses with lenses of aplite and pegmatite. The host rock belongs to the Middle Aravalli Group of Paleoproterozoic age.

The main structural feature of the deposit is interpreted as northerly plunging isoclinal synform with core occupied by the host rock. The rocks underwent three successive phases of deformation. Rampura-Agucha deposit occurs in highly metamorphosed rocks conformable with the surrounding gneisses by the presence of garnet, sillimanite and microcline and lack of muscovite. Garnet has been altered to biotite indicating retrograde metamorphism. All mineral assemblages represent high-grade metamorphic condition.

15.9.3. Mineralization

The orebody is massive and single lens shape. The major opaque phases are sphalerite, pyrrhotite, pyrite, galena and graphite with minor chalcopyrite, arsenopyrite and tetrahedrite-tennantite.

15.9.4. Genetic Models

High-grade metamorphism of upper amphibolite facies and strong deformation has obliterated all primary features. Carbon isotope study indicates the origin of abundant graphite content in the ore at Rampura-Agucha as biogenic derivation. Sulfur isotope study of sphalerite suggests that the orebody possibly recorded a profuse influx of metalliferous fluids in a proximal paleo-trough from which sulfides rapidly precipitated. It also incorporated relatively undiluted hydrothermal sulfur. This sulfur probably originated from higher degree of inorganic reduction of seawater sulfate or basaltic sulfur from the abundant tholeitic sills in the footwall of the mineralization. Large carbonaceous constituents of the graphite mica sillimanite schist host along with strata-bound and stratiform nature of orebody suggest a sediment-hosted Zn-Pb-Ag (SEDEX) type of ore deposit. The model Pb age of the deposit is 1804 Ma.

15.9.5. Size and Grade

Reserve and Resources of Rampura-Agucha Mine as on 31 March 2012 stand at +120 Mt at 13.90% Zn, 2.00% Pb, 63 g/t Ag and 9.50% Fe. Global metal grade is consistent and is not declining as mine operation moves down.

15.9.6. Mine Operation

The mine is highly mechanized with 34 m^3 excavator and 240 tonnes dumpers, for excavation of ore and waste. Processing facilities use both rod mill-ball mill and semi-autogenous grinding mill in combination and froth flotation to produce zinc and lead concentrates. The concentrator is equipped with process control system and multi-stream analyzer to facilitate effective QC. Metal extraction is by in-house smelters. The tailing is disposed to specially constructed on-site tailing dam with layer of impervious soil at the bottom. The tailing water is settled, recovered and recycled for industrial uses as zero discharge principle. Underground development work beyond the ultimate open-pit depth of 372 m from the surface is in progress.

15.9.7. Geostatistical Applications in Mineral Exploration

Mineral exploration requires high investment, sustained cash inflow and large time with inherent high element of risk. Exploration drilling is usually planned in grid pattern to be conducted in sequential and dynamic manner. The mechanism in such campaign is midterm assessment of quality, quantity and reliability of estimates. The traditional procedures are unable to provide the degree of reliability. Application of mathematical models can quantify the global precision and bring forth decision-making criteria at the end of each stage. Various statistical and geostatistical procedures are discussed in Chapter 9. These techniques can be used to evaluate sequential exploration data with an aim to optimize sampling for specific objectives. It helps in decision making to continue or to keep the project in abeyance (Haldar, 2007 [33]). The standard procedures are the following: Theory of Probability Distribution, Frequency, Mean, Variance, Standard Deviation, Trend surface, Semi-variogram and Kriging. Drill samples share major part of the investment during exploration and justify critical analysis at every phase. The ongoing drilling program can be modified accordingly. Statistical-geostatistical methods are equally appropriate during mine production and at the time of mine closure.

TABLE 15.6 Phased Exploration Scheme to be Conducted in Sequence

Phase	Drill interval (m × m)	Meter drilled	No. of boreholes	Objectives
1	200 × 50	2948	18	Establish broad potential (25-30 Mt) over the strike length and laboratory scale metallurgical tests.
2	100 × 50	2939	18	Firmly establish reserves, grade and bench scale metallurgical tests, preparation of DPR, conceptual mine planning and investment decision.
3	50 × 50	6011	36	Precision in estimates, database for detail production planning, grade control and pilot plant tests.
4	Close space near surface, wide space at lower levels	6197	61	Delineation of gossans, old workings, extension of orebody up to 300 m vertical depth from surface.
	Total	18,095	133	2 Years time at (Rs) ₹ 12.5 million

15.9.8. Exploration Scheme

The surface manifestations of economic mineralization at Rampura-Agucha are deciphered by weathered and gossanized cover rocks, malachite encrustation, mine debris on either side of the shallow depression, slag heaps, shaft and drive like excavations reflecting ancient mining and smelting activities. HZL formulated a systematic exploration scheme (Fig. 1.5 and Table 15.6) guided by the surface evidences and results of the first borehole of DMG. The program planned conceptually with sequential evaluation model approach to achieve specified objectives.

The first three phases cover the upper 100 m depth followed by Phase-4 for probing depth extension of gossans and orebody (Fig. 15.16). The exploration program comprising 18,095 m of surface drilling in 133 boreholes was completed in 2 years (1980-1982).

15.9.9. Database

The final exploration input, up to 400 m vertical depth, includes 24,900 m of surface drilling in 229 boreholes at 50 × 50 and 25 × 25 m grid, 256 m drilling in six underground boreholes and 567 m exploratory development (1980-1982). The raw sample data comprising survey, rock type, assay and specific gravity are codified. The sample data are transformed to equal length of 1 m for general statistics and 5 m composite for geostatistical studies related to subblock grade estimation using in-house and commercial software facilities. The data files are continuously updated with ongoing borehole assay input. Additional 24,000 m drilling completed during 2007-2008 to explore and establish the continuity of ore up to 1200 m depth from surface.

15.9.10. QC and QA

The QC and QA analyses are maintained following international protocol by insertion of blanks and standards at regular intervals. Three statistical tests are conducted before accepting the sample data (Table 15.7).

- Sample bias (Original half against duplicate half of core)
- Assaying bias including blanks and standards
- Inter-laboratory bias (India and Canada).

The results, statistical analysis and scatter plots (Fig. 15.17) obtained indicate statistically significant with high accuracy. The exploration database is intrinsically reliable and accepted for estimation of reserves and resources.

15.9.11. General Statistical Applications

Histograms of 8570 numbers of drill core samples at 1 m length for zinc, lead and iron have been plotted to understand the distribution pattern. The plots have revealed distinct bimodal, positively skewed and normal distribution for zinc, lead and iron respectively at 1 m sample length. The samples are composited to 5 m (1181 numbers) for mine subblock estimation and represented similar distribution pattern. The 1 m samples are also utilized for estimation of deposit grade and CLs applying normal statistical

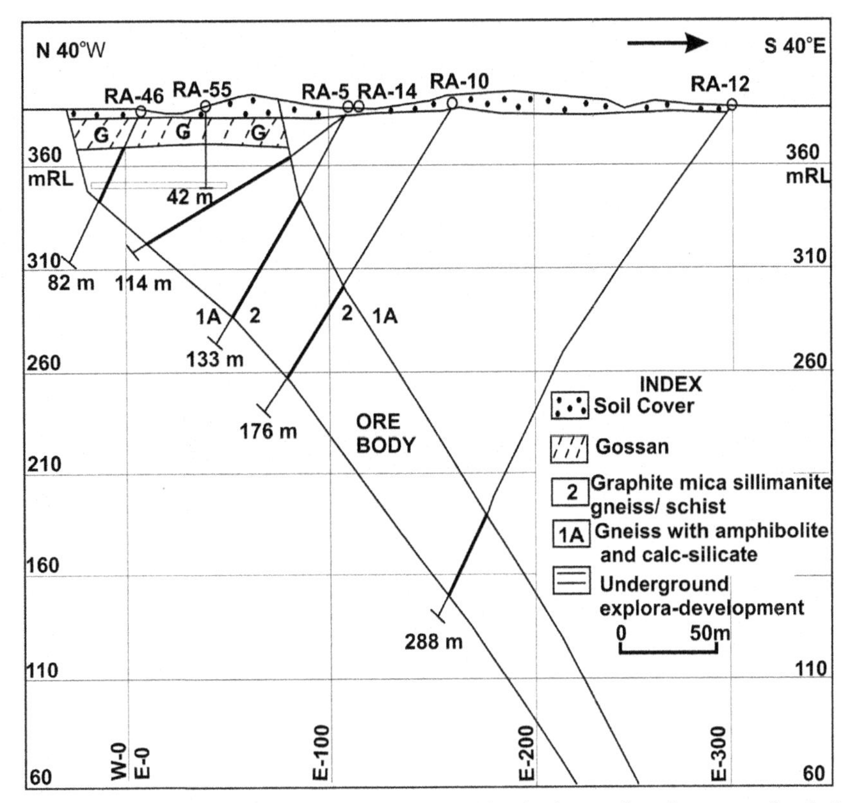

FIGURE 15.16 Drill section of Rampura-Agucha deposit probing depth extension of gossans and orebody (1982).

TABLE 15.7 Results of QC/QA Check Studies between Two Halves of Drill Core, Repetitive Analysis in the Same Laboratory and Other Standard Laboratories

Parameters	Original versus duplicate half core	Random repeat including blanks and standards	Inter-laboratory (India/ Canada)
No. of samples	123	990	46
Average of first set (% Zn)	10.99	11.04	10.27
Average of second set (% Zn)	11.11	11.04	10.15
F value	1.02	1.00	1.13
Paired *t* value	0.66	0.09	0.35
Correlation coefficient (*r*)	0.97	0.97	0.97

mode. Probability plot of cumulative distribution of zinc (Fig. 9.4) and logarithmic probability of lead have been attempted for comparison. Summary statistics by various methods are shown in Table 15.8. The global average grades and CLs have been accepted as estimated by normal statistical procedures.

15.9.12. Isograde Maps

Isograde maps along composite longitudinal section indicate identical disposition of primary metals, broad geological domains and correlation imprint within each. Zn (Fig. 15.18), Pb and Ag grades increase toward north, while

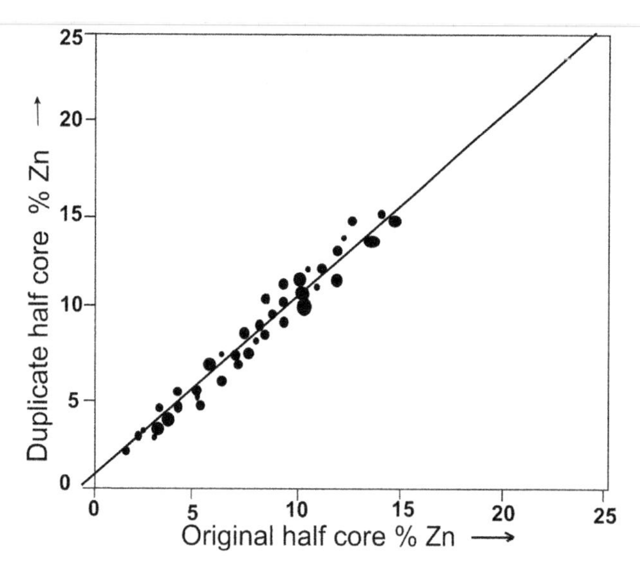

FIGURE 15.17 Scatter plot of original and duplicate half core sample values confirms significant statistical reproduction.

iron and mineralized width decrease. The periodicity, repetitive north plunging high and low or peaks and troughs, of grade at 200-300 m intervals along strike is predominant. Zn and Pb have a fairly good affinity ($r = 0.75$) in the north half of the orebody. Silver is having close association with either lead or occur independent.

15.9.13. Semi-variogram

Three dimensional semi-variography of 5 m composite assay value for zinc, lead and iron are computed along and across strike, plunge and vertical depth of orebody on different options of step tolerance and angular regularization. The experimental semi-variograms and the fitted spherical model for zinc and spatial variability parameters along four directions are given in Table 15.9.

15.9.14. Sequential Evaluation Model

The 1-m drill core samples from upper 400 m vertical height have been analyzed by sequential evaluation models at the end of Phase-1 (200 × 50 m), Phase-2 (100 × 50 m) and Phase-3 (50 × 50 m) by (A) frequency, (B) probability and (C) semi-variogram.

(A) Frequency model

The 1-m samples at optimum cutoff have been analyzed employing statistical techniques at the end of each exploration phase. The zinc frequency plot represents a bimodal distribution with two modes around 5 and 20% Zn. It is evident from the frequency plot (Fig. 15.19A-C) that the pattern of population distribution is identical in three phases of drilling.

The variation over mean grade has changed from ±3.43% at 200 × 50 m space to ±1.71% at 50 × 50 m space of drilling at 95% probability level. The percentage variation in tonnage confidence has improved from 18% to 9% respectively. The comparative statistics is summarized in Table 15.10.

The grade precision is marginal with sequential drilling and has supported confidence on global estimation. The tonnage confidence has improved in progressive drilling and is adequate at 100 × 50 m space operation for investment decision.

TABLE 15.8 Summary Statistics of Estimates for Zinc-Lead and Iron by Normal, Probability and Lognormal Procedure

Estimation method	Zinc	Lead	Iron
No. of 1 m drill core sample	8570	8570	8570
Arithmetic			
Mean grade %	13.48 ± 0.17	1.93 ± 0.03	8.20 ± 0.07
Variance	63.20	3.26	10.42
Normal probability			
Mean grade %	13.20 ± 0.15	—	8.20 ± 0.08
Variance	51.84	—	14.44
Logarithmic probability			
Mean grade %	—	1.65 ± 0.02	—
Variance	—	0.62	—

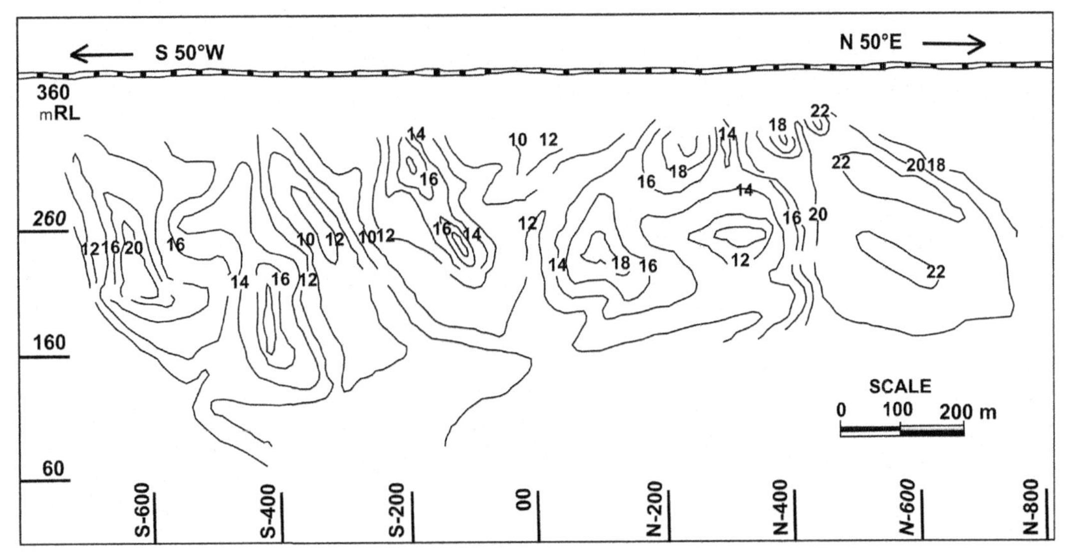

FIGURE 15.18 Isograde map of zinc indicates repetitive high and low metal concentration and periodicity along longitudinal direction of the orebody.

(B) Probability model

The mean assay and mineralized width, corresponding standard deviation, and their CLs are computed (Table 15.11).

These limits, around their respective mean, when plotted on an *x-y* coordinate system (assay, width and tonnage on *x*-axis and their respective standard deviation on *y*-axis) would delimit an area statistically defined as the **"grade-tonnage space"** in which the true population mean and the standard deviation are expected to exist (Fig. 15.20A-C). The probability area approximates an elliptical configuration defining a region of 90% probability indicating that the point representing the true population mean and standard deviation for the deposit would plot within this region. The size of such a probability region reduces with increase of samples. Such progressive probability regions for cumulative CLs of the mean and standard deviation would indicate the improvement in the estimate of the average grade, width and tonnage of the deposit. The reduction in the probability area with an increase in drilling is indicative of improvement in precision. However, it is evident that after Phase-2 drilling, the project does not necessiate any further improvement for investment decision.

TABLE 15.9 Global Semi-variogram Parameters for Zinc

Parameters	Strike	DTH	Plunge	Down dip
Step interval (m)	25	5	25	25
Angular regularization	45°	45°	45°	45°
Population variance	57.92	57.92	57.92	57.92
Random variability	14.00	7.10	38.00	7.10
Structured variability	43.92	50.82	19.92	50.82
Range (m)	115	35	100	90

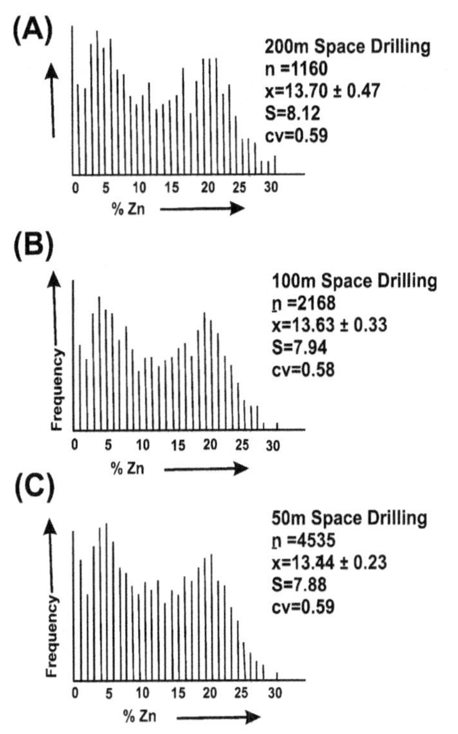

FIGURE 15.19 Zn frequency distribution plot of 1 m and statistical parameters of upper 400 m vertical depth: (A) 200 × 50 m, (B) 100 × 50 m and (C) 50 × 50 m grid (Source: Haldar, 2007) [33].

TABLE 15.10 Grade-Tonnage Parameters at the End of Each Phase of Exploration

	Borehole spacing		
	200 × 50 m	100 × 50 m	50 × 50 m
No. of boreholes	18	36	72
Meter drilled	2948	5887	11898
No. of 1 m samples	1160	2168	4535
Mean grade % Zn	13.70	13.63	13.44
Standard deviation	8.12	7.94	7.88
CL at 95% level	±0.47	±0.33	±0.23
% Variation	±3.43	±2.42	±1.71
Tonnage in million tonnes	24.78	24.42	25.22
CL at 95% level	±4.57	±3.53	±2.29
% Variation	±18.4	±14.5	±9.0

(C) Semi-variogram model

The statistical models enumerated earlier are based on the firm assumption that continuity of ore between holes is certain. A geostatistical model, such as semi-variogram of grade and width on mineralized intersections along different directions, can confirm this ore continuity or otherwise. The global semi-variogram along the strike for zinc composite grades (Fig. 15.21) is fitted to the spherical model. The semi-variogram shows moderate unexplained variance ($C_0 = 1.89$) and range of sample influence ($a = 172$ m) for estimation.

Similarly, the global semi-variogram along the strike for mineralized widths (Fig. 15.22) is fitted to the spherical model. The semi-variogram shows no nugget effect ($C_0 = 0$) and high range of sample influence ($a = 555$ m) for estimation.

15.9.15. Estimation Variance and Optimization of Drill Hole Spacing

Grade and tonnage precision with increasing number of drill hole samples using semi-variogram model have been depicted in Table 15.12 (refer Chapter 9 for estimation procedure).

The number of boreholes in the x-axis and corresponding precisions in the y-axis at each drill intervals from Table 15.12 are plotted in Fig. 15.23. The curve becomes flat and stable after drilling at 100 × 50 m space drill grid. It clearly suggests that further drilling improves the confidence marginally. The additional drilling beyond this stage is necessary for precise mine planning and dilution control.

TABLE 15.11 Mean and Standard Space (Probability Model) for Zinc, Mineralized Width and Tonnage at the Completion of Three Phases of Exploration Program

	Mean	$\overline{X}$ Limits at 95% level		Standard Deviation	Standard limits at 95% level	
No. of boreholes	$\overline{X}$	Lower	Upper	S or σ	Lower	Upper
Zinc grade (%)						
18	13.70	13.22	14.16	8.12	7.79	8.45
36	13.63	13.30	13.96	7.94	7.70	8.18
72	13.44	13.21	13.67	7.88	7.72	8.04
Mineralized width (meter)						
18	61.22	50.52	71.92	23.16	15.59	30.73
36	59.76	51.65	67.87	24.82	19.09	30.55
72	61.51	55.98	67.04	23.96	20.05	27.87
Tonnage in upper 100 m (million tonnes)						
18	24.78	20.27	29.29	23.16	15.59	30.73
36	24.42	20.89	27.95	24.82	19.09	30.55
72	25.22	22.93	27.51	23.96	20.05	27.87

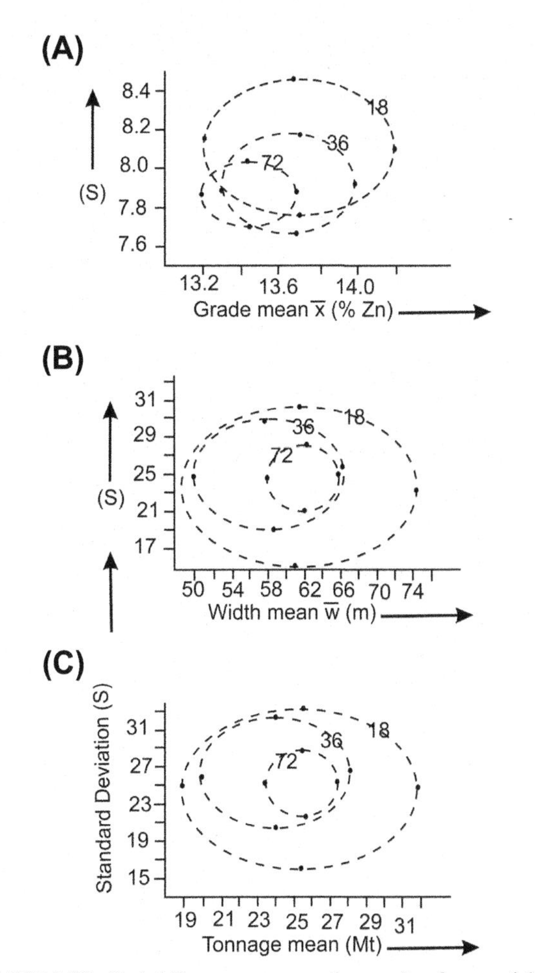

FIGURE 15.20 Probability areas representing results of sequential drill analysis: (A) zinc, (B) width and (C) tonnage with increased sample (Source: Haldar, 2007) [33].

The exploration data at the end of each phase are processed using statistical and geostatistical models for precision and adequacy to achieve specific objectives. The detailed analysis of Rampura-Agucha exploration data employing frequency, probability and semi-variogram

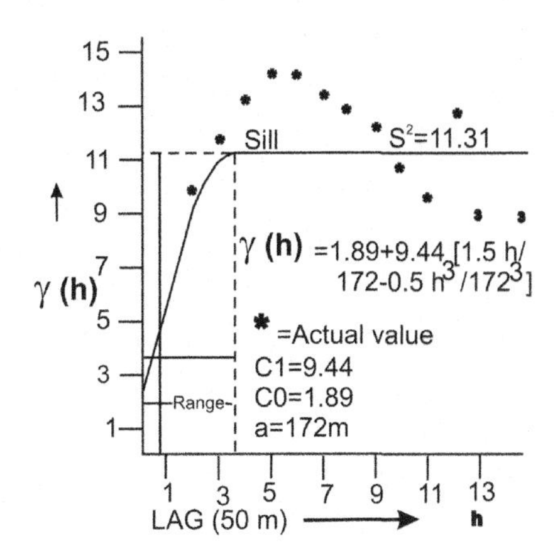

FIGURE 15.21 Global semi-variogram of zinc along strike showing unexplained variance and influence of samples.

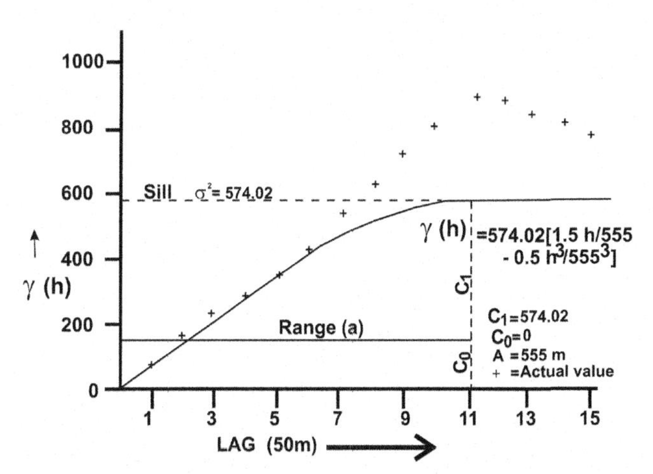

FIGURE 15.22 Global semi-variogram of mineralized width along strike showing no nugget effect and long range of sample influence.

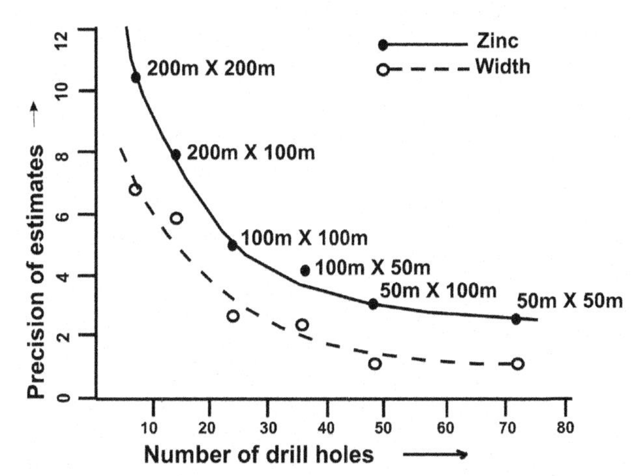

FIGURE 15.23 Precision of estimation vs. drill hole spacing that indicates the adequacy of sampling leading to investment decision (Source: Haldar, 2007) [33].

models indicates that drilling at 100 × 50 m space provides adequate confidence for definitive feasibility report preparation leading to an investment decision. The additional drilling at 50 × 50 m was carried out with a view to mine design and subblock grade estimation/control.

15.10. JHAMARKOTRA, INDIA: DISCOVERY OF THE LARGEST STROMATOLITIC ROCK-PHOSPHATE DEPOSIT BY GEOLOGICAL MODELING AND EXPLORATION TO ENVIRONMENT MANAGEMENT PRACTICES: A HOLISTIC APPROACH

Phosphorus (P) or Phosphate (P_2O_5) is essential as important plant nutrients i.e. growth of agricultural productivity Single Super Phosphate (SSP) and phosphoric acid

TABLE 15.12 Errors of Estimation at Different Drill Hole Spacing

Drilling grid			Estimation variance	Standard deviation	CL	% Variation around mean
Strike "*h*"	Depth "*l*"	No. of boreholes	$\sigma^2 e$	σe	(±)	
Zn grade						
200 × 200 m		7	3.54	1.88	1.39	10.37
200 × 100 m		14	4.15	2.03	1.07	7.95
100 × 100 m		24	2.83	1.68	0.67	5.00
100 × 50 m		36	3.09	1.76	0.57	4.20
50 × 100 m		48	2.15	1.47	0.42	3.10
50 × 50 m		72	2.30	1.51	0.35	2.60
Mineralized width						
200 × 200 m		7	30.42	5.51	4.08	6.67
200 × 100 m		14	45.92	6.78	3.54	5.79
100 × 100 m		24	16.07	4.01	1.60	2.68
100 × 50 m		36	20.09	4.48	1.46	2.45
50 × 100 m		48	6.31	2.51	0.71	1.16
50 × 50 m		72	9.18	3.03	0.70	1.13

as fertilizer, chemicals and medicinal uses. There is neither any effective substitute for phosphate rock nor any scope of their recycling. It occurs as igneous deposits (inorganic) of fluor-apatite (CaF) $Ca_4 (PO_4)_3$ with 42% P_2O_5 and chlor-apatite (CaCl) $Ca_4 (PO_4)_3$ with 41% P_2O_5. The other type is of sedimentary (organic) rock-phosphate or phosphorite with 14-34% P_2O_5. The fertilizer resources and production play a pivotal role for the economic growth of agriculture-based countries like India. About 35-40% of the requirement of raw material for phosphatic fertilizer production is met through indigenous sources. The rest is imported in the form of rock-phosphate, phosphoric acid and direct fertilizers. The deposits are situated at Udaipur, Mussoorie, Latitpur, Hirapur and Jhabua. Jhamarkotra rock-phosphate mine from Udaipur group of deposits contributes 98% of production in the country.

15.10.1. Location and Discovery

The Jhamarkotra deposit (24°28′N, 73°52′E) is the largest rock-phosphate-bearing stromatolite colony in the World. It is one of the largest and fully mechanized open-pit mines in India. It is located at a distance of 26 km southeast of Udaipur city.

GSI mapped the so-called phosphate belt during 1950s as limestone bed showing typical elephant skin color and crocodile skin texture. GSI missed the phosphate belt unsuspected in Arcean/Proterozoic host rocks of +2000 million years old. Prof. R. P. Sheldon, noted phosphate authority of USGS, conceived and postulated a knowledge-based model that was unconceivable at that point of time. The proposal of existence of stromatolitic phosphates of 1300-2000 Ma around Udaipur came as a complete surprise. Thereafter, GSI discovered Kanpur and Maton deposits during 1968 (adjacent to Jhamarkotra village) on a routine geochemical survey for phosphate during mid-1960s. Hunting for crocodile skin looking stromatolitic phosphorite rewarded State Department of Mines and Geology, Rajasthan, with discovery (30 June 1968) of Jhamarkotra, the oldest (2000 Ma) and largest deposit of real commercial significance (richest) having more than 80% proven reserves in India.

Eighty-two thousand tonnes of High Grade Ore (HGO) production commenced within a year of discovery (1969-1970). Surface outcropping and compulsive acute necessity of phosphate prevailed at that point of time resulted in unconventional primitive manual pick axe mining operation of HGO concurrently with exploration following "earning while learning" from 1968.

15.10.2. Regional Settings

The regional settings of Jhamarkotra deposit belong to Lower Aravalli age (Early Proterozoic) and stand as:

Post Aravalli granite

Graywacke

Dolomitic limestone with Phosphate beds

Ortho quartzite

~ ~ ~ Arkose and conglomerate ~ ~ ~

Basal quartzite and volcanics

Banded Gneissic Complex

15.10.3. Host Rocks

The host rock is stromatolitic blue green algae-rich dolomite limestone in closed basin. Polyphase folding, faulting, deformation, weathering and leaching have resulted enriched HGO.

15.10.4. Mineralization

The stratified phosphate horizon is sandwiched between dolomite-carbonate sediments and represents columnar biogenic sedimentary structure called "stromatolite" (Fig. 1.2V). The columnar interspace is made of dolomite.

15.10.5. Genetic Model

The stromatolitic phosphate-bearing bed in Jhamarkotra represents a thick pile of sedimentary package formed under a shallow epicontinental sea of lower Aravalli age. The origin is linked to the metabolic activity of a certain primitive life form.

15.10.6. Exploration

Progressive exploration between 1970 and 1980 by State Department of Mines and Geology concurrently with mine development and production by Rajasthan State Mines and Minerals Ltd include regional mapping of Jhamarkotra basin, plain table précised mapping of 16-km-long phosphate bed, 500 pits and trenches, 18,000 channel samples and 60,000 m of diamond drilling in 600 boreholes at an interval of 100 m strike and 50 m inclined depth (Fig. 15.24). The exploration established 16 km linear stretch of phosphate bed and outlining major reserves at "A" to "D" blocks. The highest elevation varies between 600 and 200 mRL. The thickness of phosphate bed varies between few centimeters to 35 m with an average of 15 m.

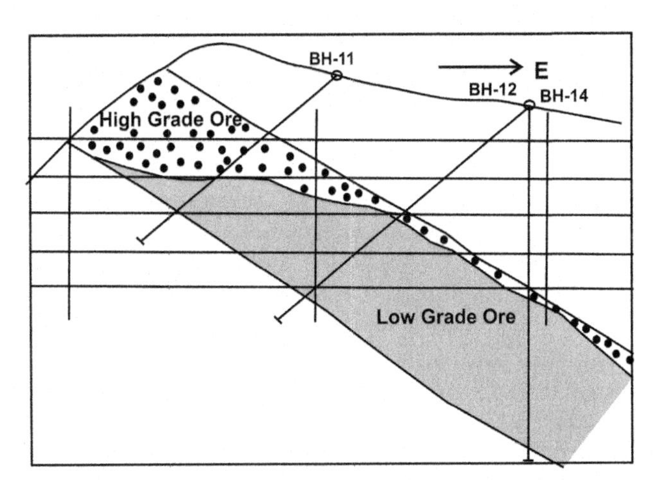

FIGURE 15.24 Standard cross section showing the drilling pattern at Jhamarkotra rock-phosphate deposit.

15.10.7. Size and Grade

HGO is represented by 16 Mt at 33% P_2O_5, 12% SiO_2, 1.6% MgO and 2.6% R_2O_3. LGO is comprised 36 Mt at 18% P_2O_5, 8% SiO_2, 10% MgO and 2% R_2O_3 making a total of 52 Mt at 22% P_2O_5, 9% SiO_2, 7.5% MgO and 2% R_2O_3 as on April 2008.

15.10.8. Mining

Conventional open-pit mining (Fig. 11.6), blending and processing of HGO and LGO started from 1985. Pit slope at footwall is between 32° and 50° and the same at hanging wall is between 42° and 50°. Overall ore to overburden ratio is 1:8.7. The mining machineries deployed are Jackhammer, Wagon (Fig. 11.29), IR and RC pit drilling units, Hydraulic front-end loader, and 50-90 t dumpers (Fig. 11.39).

Cumulative rock handling between 1970 and 2008 is in the order of 26 Mt (HGO + LGO) ore, 272 Mt of overburden making a total rock handling of 298 Mt at ore to overburden ratio of 1:10.50.

15.10.9. Beneficiation

The beneficiation of low-grade rock-phosphate includes crushing, grinding, and milling followed by conventional froth flotation route. About 0.60 Mt fine tailing fraction in slurry form is generated annually.

15.10.10. Environment Management

The environmental matters have been conceived by the management from the inception of the mine with a view to address it applying human face. The issues, as and when identified, are resolved by appropriate matching actions. The salient features are:

- Geomorphic and demographic aspects—land use

Sixty plus years of ultimate mine life will result formation of huge serpentine gorge, 7 km long, 700 m across, elevation height between 600 and 320 mRL and formation of few artificial massive waste mounds flanked by two micro watershed regimes. The mine lease area of 1379 hectare includes 270 hectare of de-reserved forestland. There are 12 surrounding villages occupying 30% of 6946 hectare land. The villages are populated by 7394 tribal inhabitants at subsistence economy. Migration and rehabilitation were mandatory obligation and social responsibility. This was taken care by least displacement and disturbances of villages and villagers, even falling within the leasehold.

- Waste rocks disposal—overburden

The classified forest area within the leasehold is de-reserved to accommodate 450 Mt waste (173 million m^3) right on the edge of D-Block considering minimum haulage, maximum waste accommodation capacity with least horizontal spread and withstanding the load. The boundary limits of progressive dumping ground in de-reserved forest area are demarcated by stone walls. Transportation made by large dumpers and bulldozers. The stages of dumping are maintained by horizontal movement and vertical lift. The west dump areas do not affect the agricultural land or village premises. Reclaimed land has been used for infrastructure development, beautification, parks and playground.

- Fine waste disposal—mine and beneficiation plant

Handling of 0.60 Mt fine wastes from mine and beneficiation plant was a matter of concern. This could be handled by fixed sprinkler ensuring effective dust suppression in the dumping and crushing areas. Mobile sprinklers are deployed along the haulage road to suppress dust emission. The rejects of the beneficiation plant is filled to the tailing pond in slurry form. The tailing pond is designed with "0" water discharge. The recovered water is recycled for industrial usage. Extensive plantation is made on the old tailing dam to arrest flying dust, particularly in the summer months.

- Air

The effect of relay blasting gas is minimized by explosives selection, evacuation of people from blast areas and dust settling.

- Noise and vibration

The effect of noise and vibration is minimized by proper selection of machineries.

- Effluent treatment from tailing ponds

Acid-water treatment plant has been installed to recover from tailing pond and used for industrial purposes.

- Water management plan

A captive dam (Fig. 14.8) has been constructed to arrest following monsoon water of Jhamri river catchment area and storage in reservoir. The water is drawn from the intake wells and filtered. The safe potable water is supplied to mine township, Jhamarkotra and other surrounding villages and to Udaipur city. A garden and guest house are developed around the dam site for picnic and other social amusement.

- Afforestation

The Company made an eco-friendly face lift by extensive plantation of Neem (*Azadirachta indica*) saplings on level dumping ground at backdrop raised waste dump, Jatropha (Euphorbiaceae family) plantation for fuel oil, and other types of plant species that can grow well around the mine and road sites.

- Health care

A fully equipped medical dispensary with senior doctors exists at the mine site to attend the routine health care issues of the employee's family members and the local tribal villagers at no cost. The management has established a cardio thoracic and a spiral CT scan units at Udaipur city Government Medical College and Hospital to treat on priority employee's family and tribal villagers as and when referred by mine medical officers.

- Education

The Company has set up a Higher Secondary school at the mine colony to spread education around the villages and organizes computer awareness. One Industrial Training Institute and vocational training center have been established with hostel facilities for tribal students.

- Economy

Direct and indirect employment to all class of people raised significant living standard of common mass that could be observed commencing with the discovery, exploration, mine development and reaching to full capacity production. The other benefits to the Nation is by payment of (Rs) ₹ 50 million to State coffer in 37 years for state developmental works and (Rs) ₹ 12 million per annum as royalty to the State Government for infrastructure development, and many more ....

The total management philosophy of care with a human face is reflected by Fig. 15.25.

15.11. BASE METAL DISCOVERY TREND: THE LAST 50 YEARS IN INDIA

The last half of the twentieth century has been regarded as the golden age for geosciences—a time of major geoscientific breakthrough in mineral discovery. The global development of mineral-based industries during the

FIGURE 15.25 Mother protects the child with warm comfort and utmost care. Similarly the mining companies must treat the affected community with human face.

preceding 50-60 years experienced phenomenal growth and India is no way different. Indian mineral explorations, for centuries, had been historically focused mainly on coal, iron ore, manganese, mica and such. A number of interesting events occurred in the 1940s—World War II ended and at last free new India was born. The emerging Republic of India visualized a new perspective and formulated 5-Year Plans in tune with the infrastructure development of the country as a whole. Over the years, the Government incorporated various Public Sector Undertakings such as ONGC (1956), NMDC (1958), BALCO (1965), HZL (1966), Hindustan Copper Limited (HCL) (1967), Uranium Corporation of India Limited (UCIL) (1967), BGML (1972) and Steel Authority of India Limited (SAIL) (1973). These Public Sector Undertakings were entrusted with the mission of exploration, development and production of public consumable and marketable mineral base end product. This triggered extensive new searches for petroleum, coal, bauxite, gold, iron, zinc-lead, copper, uranium and many others. Perhaps for the first time in Indian history the earliest recordings of Kautilya's Arthashastra realized. Kautilya (Chanakya) was a scholar, teacher and guardian of Emperor Chandragupta Maurya, the founder of Mauryan Empire. This treatise on Economics (327 BC) described an extensive search and research undertaken in mineral industries in ancient India. This topic focuses on the base metal (zinc-lead-copper) discovery trend rather than sampling a few significant singular giant-type deposits.

15.11.1. Significance of Investment Framework

The process of mineral discovery passes through high risk compared to any other investment opportunity. The minerals are indispensable and their discovery is essential for the growth of any nation. The fundamental shift in the postindependence national character empowered the new generations looking for adventure in mineral sector. The amount of investment and associated risks walking through discovery type of Greenfield-Brownfield and Exploration Stages such as Reconnaissance, Prospecting and Detailed Exploration has been discussed in Chapter 1.

15.11.2. Reserve/Resources

After independence, the continued efforts by various exploration agencies have resulted in significant augmentation of Indian zinc-lead-silver-copper resources. In 1947 the known zinc-lead resource was a meager 0.5 Mt from Mochia Mines of Zawar Group, Rajasthan, with 175 tpd production capacity. Since then the resource had risen to 418 Mt by 2000 and further to 510 Mt in 2010 (Fig. 15.26). The ongoing exploration during 2010 and 2012 outlined additional ~100 Mt reserves and resources from the working deposits. Indian subcontinent is blessed with +95% zinc-lead-silver deposits in the northwestern Proterozoic sediments of Rajasthan. The total reserves and resources base in producing/developing mines and deposits under detailed exploration as on 31 March 2012 stand at 313 Mt containing 34.7 Mt of zinc-lead metal and 885 million ounces of silver.

The total in situ copper reserves and resources was estimated at ~10 Mt in 1947 with production of 310,000 tonnes of ore at 2% Cu per annum from a single mine at Mosaboni, Singhbhum Group of deposits in Jharkhand. The reserves and resources are increased to 1390 Mt in March 2005 by continued exploration. A 369 Mt (26.5%) fall under "Reserves" (proved and probable) and the balance of 1020 Mt (73.5%) are "Resources" (Feasibility, Pre-feasibility, Measured, Indicated and Inferred). Largest resources of copper ore to a tune of 668 Mt (47.9%) are in the State of Rajasthan followed by Madhya Pradesh with 404 Mt (29%) and Jharkhand with 226 Mt (16.2%).

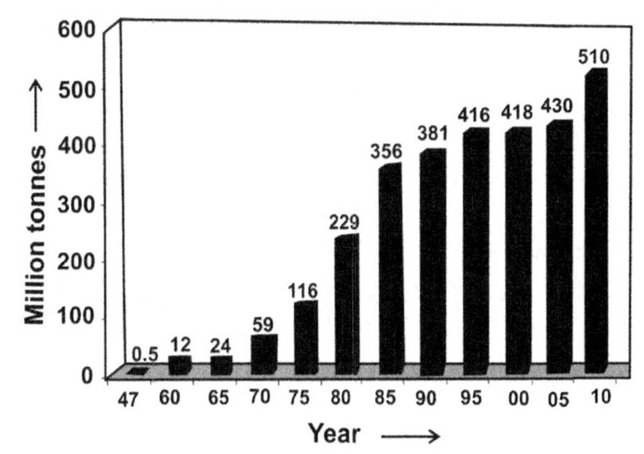

FIGURE 15.26 Reserves and resources buildup of Indian zinc-lead deposits over last six decades.

The status of all mineral deposits changes with exploration in and around the existing deposits, mine depletion and new discoveries. The questions frequently arise are the following:

(a) What is the reserve and resource which should be reported for a mineral body?
(b) Should it continue to reflect the initial discovery?
(c) Should we assess the current state of the reserve after depletion?
(d) Report the status quo after accounting for both depletion and new finds?

The author strongly advocates the last option. Other than the obvious benefits to local explorers and business alike from accurate reporting, this update will qualify the deposit for global comparison and merit-based ranking.

15.11.3. Reserve Adequacy

By and large the Indian zinc-lead deposits are comparatively richer in zinc with low lead content. The exclusive lead deposits are few like Sargipalli in Orissa and Agnigundala in Andhra Pradesh. The reserve contribution was too low tonnage and both have already been phased out.

HZL is one of the World's top producers of zinc (0.879 Mt pa), silver (500 t) and lead (0.185 Mt pa) metals. Binani Zinc Limited produces another 30,000 tonnes of zinc metal based on imported concentrate. These Companies ensure India's significant self-sufficiency in zinc (Fig. 1.1) and become a leading net exporter of zinc metal.

The Indian copper deposits are either narrow with low-grade orebody (Khetri belt) or low-angle deep-seated body and do not support large-scale mechanization in underground mining (Singhbhum belt), or changing from surface to underground method as at Malanjkhand, Madhya Pradesh. India depends primarily on import of copper concentrate and metal.

15.11.4. Resource Requirement

Zinc is the fourth most widely used metal after iron, aluminum and copper in the World. Due to its resistance to nonacidic atmospheric corrosion zinc is instrumental in preventing the corrosion of steel and extending the life of building, vehicle, ship, steel goods and structures, ranging from telecom towers to hi-mast lighting. In a booming economy which encourages construction, the demand for zinc literally galvanizes a nation. Most of the zinc demand in the country (+70%) is in the construction sector. The current output of zinc-lead ore being mined is about 9 Mt per annum at 7-9% Zn + Pb. This satisfies +100% of the national demand for zinc metal.

Copper metal production is yet a constraint as it is heavily dependent on imported concentrate and metal. This is a good time to pause and remember that mineral resource is a nonrenewable asset and that the gestation period between discovery and production takes 10-15 years. The question arises—What discovery rates would be required, at the current rates of production, if the industry had to replace exhausted reserves with new grassroot discoveries? Regional exploration plans would need to maintain a rolling in situ reserve inventory of 100-190 Mt to maintain self-reliance on long-term basis. If we fail to meet these numbers, we would be left with the alternatives of import of concentrate, metal, secondary generation etc. The country is fast approaching crisis state, which can only be overcome by moving toward new opportunities beyond state/national boundaries.

15.11.5. Discovery Trend

Zinc was probably extracted in India as early as in the tenth century Before the Cristian Era (BCE). The reduction of zinc oxide using a condenser and oxygen-devoid container on an industrial scale took place at Zawar mine area, India, around the thirteenth century AD, after which the technology spread to China and other countries.

India has always been one of the primary copper metal producers in the World. There are ample evidences of ancient mining and smelting remains around all the present zinc-lead-copper mining belts. Carbon dating of wooden implements recovered from the ancient workings indicates that these operations are +2100 years BP. Therefore, all the "finds" of Zn-Pb-Ag-Cu deposits are, in a strict sense, "rediscoveries" of abandoned mining belt and may be termed as "Brownfield discovery". Barring a few hidden deposits at a depth of ~130 m from the surface at Rajpura-Dariba-Bethumni belt, most of these "rediscoveries" are by chance (Fig. 15.27). New investigation over the years has enhanced significant economic reserves in and around the known orebody and cannot thus be termed as "New" discoveries. The availability of existing infrastructures enhances the value of these reserves.

A quick scan of historical discovery trend reveals some interesting results. The overall records of mineral discoveries by number of deposits and associated reserve base created over last 60 years have been depicted in Fig. 15.28 respectively. The point to note is that the base metal discovery rates picked during 1960s and 1970s and declined thereafter. Overall, this major reserve enhancement should provide for immediate (short/medium-run) mine development and production. Approximately 100 Mt of Zn + Pb reserves have been added between 2008 and 2012 from the existing mining projects.

Most of the Indian deposits are medium size, the exceptions being the three world-class "BIG" deposits: Malanjkhand Copper (145 Mt at 1.35% Cu), Madhya Pradesh, discovered in 1970, the often-mentioned Rampura-Agucha Zinc-Lead-Silver (+120 Mt at 13.90% Zn,

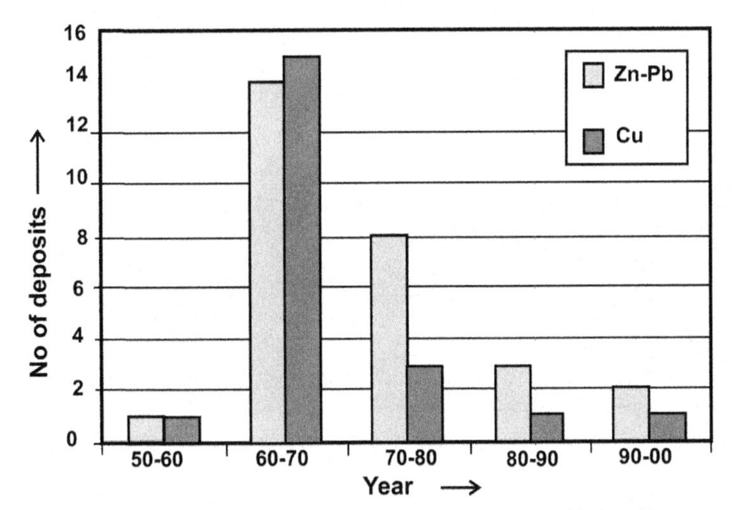

FIGURE 15.27 Number of Zn-Pb and Cu deposits discovered in last 50 years.

2.00% Pb, 9.50% Fe and 63 g/t Ag), Rajasthan, discovered in 1977 and Sindesar-Khurd Zinc-Lead-Silver (61 Mt at 5.8% Zn, 3.8% Pb, 215 g/t Ag), adjacent to Rajpura-Dariba Mine, discovered in 1987. The first two deposits are large reserve high-grade open-pit mine with low cost of production. Thus these are significant for lowest payback period and high return in investment over long mine life in the Indian mining sector. Sindesar-Khurd deposit is deep-seated underground mining opportunity and will payback fast due to rich silver content. One other deposits worthy of mention is the Banawas Copper Block (18 Mt at 1.74% Cu), an extension of Khetri Copper Complex, discovered in the 1990s. The deposit is rich in metal content and amenable to underground mining. As we can see in these cases, "the real voyage of discovery consists not in seeking new landscapes but in having new eyes".

15.11.6. A New Beginning

Even as the technology development enables better and faster method of mineral discovery, the World's nonrenewable mineral reserves are steadily depleting with progressively increasing rates of production and consumption. The economic growth, in the process, has assumed the softer name of revenue! Greenfield discoveries and expansion of reserves in and around the existing operating blocks must need to be intensified for "a tonne-to-two tonnes" substitution of the depleted ore. In India, surface orebodies have become particularly hard to find any more. Therefore, traditional exploration techniques may no longer yield the required results. We need to take the plunge, and target deep-seated, and hitherto hidden, deposits.

Future exploration will integrate several approaches and techniques in all the phases of discovery. Multidisciplinary approaches will utilize "in-depth" knowledge of regional and deposit geology, studies of structures like lineaments, faults and fractures which control the mineralization in an area, application of rock and soil geochemistry, and interpretation of high-resolution aeromagnetic (magnetic and electromagnetic) and gravity surveys with deep penetration capacity. These are the new divining roads which will lead us to the hidden metallic bodies.

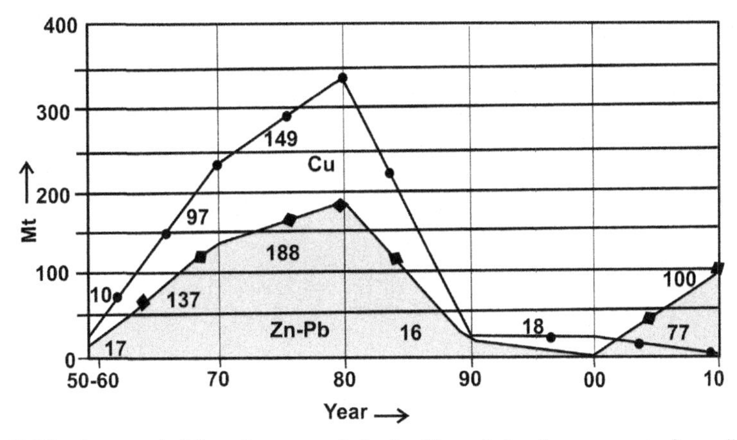

FIGURE 15.28 Resource buildup of copper and zinc-lead in each decade on account of new discoveries.

FIGURE 15.29 "Rising Sun brings the hope": one early morning in the exploration camp at Rampura-Agucha in 1978, just after rediscovery of the giant deposit.

The exploration projects would still need to be sequential and dynamic, with controlled objectives, schedules and budgets for the intermediate phases of activity. The data collected and computed must be interpreted and evaluated at the end of each intermediate phase, with the subsequent phases being rescheduled as and when necessary. In the event that intermediate results do not meet the "Go-No Go" criteria, the project can be shelved till adequate resources are available for the next stage or the market price increased in between. Thus, today's explorer needs to wear the hard-hat of the engineer, the stopwatch of the project manager and the pocket protector of the statistical analyst. Many deposits such as the Neves Corvo copper-zinc-lead-tin orebodies in Portugal, the Cannington zinc-lead-silver orebodies in Australia and Sindesar-Khurd zinc-lead-silver orebodies in India are the results of reinterpretation, and application of new approaches to existing information.

15.11.7. Conclusions

The discovery rate of base metal deposits throughout the World rose during the 1960s and 1970s. It declined during 1990s and no significant deposits have yet been reported in the twenty-first century. Indian base metal deposits are rediscoveries of ancient abandoned mining sites and extensions in the known belts. Most of the orebodies have cropped out at the surface, either unaltered or exposed as oxidized gossans. Traditional prospecting combined with geology, some limited geophysics and geochemistry gave rise to the discoveries in the recent past.

Chances of finding new deposits at shallow depths are remote in India. Forthcoming discoveries will depend on modern exploration technologies applied to an integrated geological model. The international race to the moon is already begun again—this time with the stated objectives of mining the moon for resources as varied as rocket propellant, helium-3 and lunar metals! But that is another story ....

To emerge victorious in this new environment of increasing discovery risk requires extraordinary creativity, innovation, technical skill and perseverance coupled with commercial discipline. The incentives are global economic demand, driving continuously high trends in metal prices, technological necessity, and the primeval joy of unveiling the opening chapters of creation. These are the factors that motivate corporations to pioneer and unearth new deposits and to upgrade the marginal low-grade deposits for production purposes. To accept the challenges ahead, the mineral exploration and mining companies need fortitude, foresight, fortune, fearlessness—and a fair share of humor (Fig. 15.29).

FURTHER READING

Exploration Modeling of Base Metal Deposits by Haldar (2007) [33] will be adequate. Additional reading suggested such as Haldar et al. (1985) [31], Haldar (1990 [32], 2011 [34]), Page et al. (1996) [56], Large, Ross (2004) [50], Doreen (2008) [23] and Lundin (2011) [53].

References

[1] J.C.D. Arnaud Gerkens, Foundation of Exploration Geophysics, Elsevier, 1989, p. 668.

[2] P.K. Banerjee, S. Ghosh, Elements of Prospecting for Non-fuel Mineral Deposits, Allied Publishers Ltd, 1997, p. 320.

[3] A.A. Beus, S.V. Grigorayan, Geochemical Exploration Methods for Mineral Deposits, Applied Publishing Limited, 1975, p. 287.

[4] A.M. Bateman, Economic Mineral Deposits, John Wiley and Sons, Inc., 1950, p. 916.

[5] G.F. Bonham Carter, R.K.T. Reddy, A.G. Galley, Knowledge-driven modeling of volcanic massive sulfide potential with a geographic information system, in: Mineral Deposit Modeling, Geological Association of Canada, Special Paper 40, 1995, pp. 735–749.

[6] H.T.M. Bremner, Garland Wayne, J.R. Savage, Trends in deep drilling in the sudbury basin, Short course on technologies, and case histories for the modern explorationist, Toronto, 1996, pp. 53–75.

[7] J.B. Campbell, Introduction to Remote Sensing, The Guilford Press, New York, 2007, p. 626.

[8] E.J.M. Carranza, Geochemical Anomaly and Mineral Prospectivity Mapping in GIS, Handbook of Exploration and Environmental Geochemistry, vol. 11, Elsevier Publication, 2008, p. 351.

[9] H. Chamley, Geosciences, Environment and Man, Elsevier, 2003, p. 527.

[10] K.K. Chatterjee, Introduction to Mineral Economics, revised ed., New Age International, New Delhi, 2004, p. 379.

[11] K.K. Chatterjee, Uses of Metals and Metallic Minerals, New Age, New Delhi, 2007, p. 314.

[12] K.K. Chatterjee, Uses of Industrial Minerals, Rocks and Freshwater, Nova Science Publishers, New York, 2008, p. 584.

[13] I. Clark, Practical Geostatistics, Applied Science Publishers, London, 1982, p. 129.

[14] E.D. Colin, Biochemistry in Mineral Exploration, Handbook of Exploration and Environmental Geochemistry, in: M. Hale (Ed.), Elsevier Publication, 2007, p. 462.

[15] D. Laurence, A Guide to Leading Practice Sustainable Development in Mining, Australian Government, Department of Resources, Energy and Tourism. http://www.ret.gov.au, 2011. p. 198.

[16] M. David, Geostatistical Ore Reserve Estimation, Developments in Geo-Mathematics 2, Elsevier Scientific Publishing Company, Amsterdam, 1977, p. 364.

[17] M. David, Handbook of applied advanced geostatistical ore reserve estimation, in: Developments in Geo-Mathematics, vol. 6, Elsevier Scientific Publishing Company, New York, 1988, p. 232.

[18] J.B. David, An introduction to geographic information systems, The GIS Primer, www.innovativegis.com/basis/primer/The_GIS_Primer_Buckley.pdf, 1997, p. 215.

[19] J.C. Davis, Statistics and Data Analysis in Geology, John Willey & Sons, New York, 1973, p. 550.

[20] M. Deb, T. Pal, Geology and genesis of the base metal sulphide deposits in the Dariba-Rajpura-Bethumni Belt, Rajasthan, India in the light of basin evolution, in: M. Deb, W.D. Goodfellow (Eds.), Sediment Hosted Lead-Zinc Sulphide Deposits: Attributes and Models of Some Major Deposits in India, Australia and Canada, Narosa Publishing House, New Delhi, 2004, pp. 304–327.

[21] D.J. Deshmukh, Elements of Mining Technology, vol.1, Vidyaseva Prakashan, 1995, p. 436.

[22] D.J. Deshmukh, Elements of Mining Technology, vol.2, Vidyaseva Prakashan, 1995, p. 514.

[23] E.A. Doreen, Mineral Deposits of Canada, District Metallogeny Ni-Cu-PGE: Metallogeny of the Sudbury Mining Camp, Geological Survey of Canada, Ontario. www.gsc.nrcan.gc.ca. 2008.

[24] A.M. Evans, Introduction to Mineral exploration, Blackwell Science, 1999, p. 396.

[25] R.V. Gaines, H. Catherine, W. Skinner, E.E. Foord, B. Mason, A. Rosenzweig, V.T. King, Dana's New Mineralogy, The System of Mineralogy of James Dwight and Edward Salisbury Dana, John Wiley & Sons, 1997, p. 1819.

[26] T.M.H. Gaudin, Principles of Mineral Dressing, Tata-McGraw-Hill Publishing Company Ltd, 1939, p. 554.

[27] H. Gocht, R.G. Zantop, P.G. Eggernt, International Mineral Economics, Springer-Verlar, 1988, p. 271.

[28] G.J.S. Govett, Handbook of exploration geochemistry, in: Rock Geochemistry in Mineral Exploration, vol. 3, Elsevier Scientific Publishing Company, 1983, p. 461.

[29] Ish Grewal, Introduction to mineral processing. www.met-solvelabs.com/index.php. 2010, p. 23.

[30] R.P. Gupta, Remote Sensing Geology, Springer, 2008, p. 655.

[31] S.K. Haldar, J.C. Khare, N.N. Singh, Report on mineral exploration and evaluation course at University of New South Wales, Australia, 1985, p. 161.

[32] S.K. Haldar, Report on International Symposium APCOM-1990, Berlin and visit to Megen and Neves Corvo Mine, Portugal, 1990, p. 49.

[33] S.K. Haldar, Exploration Modeling of Base Metal Deposits, Elsevier Publication, 2007, p. 227.

[34] S.K. Haldar, Platinum-Nickel-chromium: Resource Evaluation and Future Potential Targets, 17th Convention of Indian Geological Congress and International Conference, 2011, pp. 67–82.

[35] H.L. Howard, Introduction to Mining Engineering, John Willey & Sons, 1987, p. 633.

[36] H.E. Hawkes, J.S. Webb, Geochemistry in Mineral Exploration, Harper and Row Publishing, 1962, p. 415.

[37] S. Henley, Nonparametric Geostatistics, Applied Science Publishers, London, 1981, p. 145.

[38] W.A. Hustrulid, R.L. Bullock (Eds.), Underground Mining Methods-Engineering Fundamentals and International Case studies, Society for Mining, Metallurgy, and Exploration, Inc., (SME), USA, 2001, p. 718.

[39] IBM, Mineral Concession rules, 1960, amended up to April, 2003, p. 143.
[40] H.E. Isaak, R.M. Srivastava, An Introduction to Applied Geostatistics, Oxford University Press, Inc., 1989, p. 559.
[41] C.J. Johnson, in: A.L. Clark, C.J. Jhonson (Eds.), Financial Evaluation Techniques and Their Applications to Mineral Projects, Mineral Resource Management for National Planning and Policy Formulation, Asia Productivity Organization and East-West Center, 1988, pp. 53−102.
[42] JORC, Mineral resources and Ore Reserves. www.jorc.org, 2004.
[43] A.G. Journel, Ch.J. Huijbregts, Mining Geostatistics, Academic Press, London, 1978, p. 600.
[44] P. Kearey, M. Brooks, I. Hill, An Introduction to Geophysical Exploration, Blackwell Science, 2002, p. 262.
[45] F.W. Kellaway, Penguin-Honeywell Book of Tables, Penguin Books Limited, 1968, p. 76.
[46] G.S. Koch, R.F. Link, Statistical Analysis of Geological Data, second ed. Wiley, New York, 1986, p. 375.
[47] D.G. Krige, A statistical approach to some basic mine valuation problems on the Witwatersrand, J. Chem. Metall. Min. Soc. S. Afr. 52, 1951, pp. 119−139.
[48] D.G. Krige, Statistical application in mine valuation, J. Inst. Mine Survey S. Afr. 12 (2 and 3), 1962.
[49] D.G. Krige, Lognormal-de Wijssian Geostatistics for Ore Evaluation, S. Afr. Inst. Min. Metall., Monograph Series, Johannesburg, 1978, p. 50.
[50] R. Large, P. McGoldrick, S. Bull, D. Cooke, Proterozoic Stratiform Sediment-hosted Zinc-Lead-Silver deposits of Northern Australia, in: M. Deb, W.D. Goodfellow (Eds.), Sediment −hosted Lead-Zinc Sulphide Deposits: Attributes and Models of some major deposits in India, Australia and Canada, Narosa Publishing House, Delhi, 2004, pp. 1−23.
[51] A.A. Levinson, Introduction to Exploration Geochemistry, Applied Publishing Limited, 1974, p. 614.
[52] T.M. Lillesand, R.W. Kiefer, Remote Sensing and Image Interpretation, John Wiley & Sons, Inc., 2003, p. 722.
[53] Lundin Mining, Neves-Corvo Mine, Portugal. www.lundinmining.com/i/pdf/Summary_Report_Neves-Corvo.pdf, 2011, p. 7.
[54] G. Matheron, The Theory of Regionalized Variables and its Applications, Les Cahiers du Centre de Morphologie Mathematique de Fontainebleau, no 5, Published by the Ecole Nationale Superieure des Mines de Paris, 1971, p. 211.
[55] Ministry of Mines, Government of India, National Mineral Policy-2008, (For non-fuel and non-coal minerals), 2008, p. 12.
[56] R.W. Page, W.F. Laing, Felsic metavolcanic rocks related to the Broken Hill Pb-Zn-Ag orebody of the high grade metamorphism, Econ. Geol. 87, 1996, pp. 2138−2169.
[57] C.C. Popoff, Computing reserves of mineral deposits; principles and conventional methods, US Bureau of Mines Information Circular 8283, 1966, p. 120.
[58] J.M. Rendu, An introduction to geostatistical methods of mineral evaluation, S. Afr. Inst. Min. Metall. Monograph Series: Geostatistics 2, 1978, p. 84.
[59] E.S. Robinson, C. Coruh, Basic Exploration Geophysics, John Wiley & Sons, 1988, p. 562.
[60] A. Rose, H.E. Hawkes, J.S. Webb, Geochemistry in Mineral Exploration, Academic Press, 1983, p. 657.
[61] A.B. Roy, S.R. Jakhar, Geology of Rajasthan, (Northwest India), Precambrian to Recent, Scientific Publishers, India, 2002, p. 412.
[62] A.B. Roy, Fundamentals of Geology, Narosa Publishing House, 2010, p. 291.
[63] B.K. Sahu, Statistical Models in Earth Sciences, B, S. Publication, Hyderabad, India, 2005, p. 211.
[64] N.C. Saxena, G. Singh, R. Ghosh, Environmental Management in Mining Areas, Scientific Publishers, India, 2002, p. 410.
[65] B.C. Sarkar, Geographical Information System and its role in geoenvironmental issues, ENVIS Monograph No. 10, 2003, p. 141.
[66] S.C. Sarkar, S. Banerjee, Carbonate-hosted Lead-Zinc deposits of Zawar, Rajasthan, in the context of the world scenario, in: M. Deb, W.D. Goodfellow (Eds.), Sediment Hosted Lead-Zinc Sulphide Deposits: Attributes and Models of Some Major Deposits in India, Australia and Canada, Narosa Publishing House, New Delhi, 2004, pp. 350−361.
[67] R. Sen, Environmental Management- Economics and Technology, Levant Books, 2009, p. 233.
[68] P.V. Sharma, Geophysical Methods in Geology, Elsevier Publishing Co, Inc., 1986, p. 442.
[69] R.K. Sinha, N.L. Sharma, Mineral Economics, Oxford & IBH Publishing Co. Pvt. Ltd, 1993, p. 394.
[70] A.K. Talapatra, Modeling and Geochemical Exploration of Mineral Deposits − A Treatise on Exploration of Concealed Land and Offshore Deposits, Capital Publishing Company, New Delhi, 2006, p. 170.
[71] L.J. Thomas, An Introduction to Mining, Hicks Smith and Sons, 1973, p. 436.
[72] UNFC, United Nations framework classification for energy and minerals. www.world-petroleum.org/publications/A-UNFC-FINAL.doc, 2004, p. 35.
[73] USGS Bulletin 1450-A, Principles of the mineral resource classification system of the U.S. Bureau of Mines and U.S. Geological Survey, 1976, p. A5.
[74] USGS Circular 831, Principles of a resource/reserve classification for minerals, 1980, p. 5.
[75] F.W. Wellmer, Economic Evaluations in Exploration, Springer-Verlag, 1989, p. 163.
[76] B.A. Wills, An Introduction to the Practical Aspects of Ore Treatment and Mineral Recovery, Pergamon Press, 2006, p. 444.
[77] G.P. Zipf, Human Behavior and the Principle of Least Efforts, Hafner Publishing Co., New York, 1949, p. 573.

Index

Note: Page numbers followed by "*f*" and "*t*" refer to figures and tables, respectively.

D

E

S

T

U

Printed in the United States
By Bookmasters